ARCHAEOLOGICAL CHEMISTRY

Second Edition

THE WILEY BICENTENNIAL–KNOWLEDGE FOR GENERATIONS

Each generation has its unique needs and aspirations. When Charles Wiley first opened his small printing shop in lower Manhattan in 1807, it was a generation of boundless potential searching for an identity. And we were there, helping to define a new American literary tradition. Over half a century later, in the midst of the Second Industrial Revolution, it was a generation focused on building the future. Once again, we were there, supplying the critical scientific, technical, and engineering knowledge that helped frame the world. Throughout the 20th Century, and into the new millennium, nations began to reach out beyond their own borders and a new international community was born. Wiley was there, expanding its operations around the world to enable a global exchange of ideas, opinions, and know-how.

For 200 years, Wiley has been an integral part of each generation's journey, enabling the flow of information and understanding necessary to meet their needs and fulfill their aspirations. Today, bold new technologies are changing the way we live and learn. Wiley will be there, providing you the must-have knowledge you need to imagine new worlds, new possibilities, and new opportunities.

Generations come and go, but you can always count on Wiley to provide you the knowledge you need, when and where you need it!

William J. Pesce
President and Chief Executive Officer

Peter Booth Wiley
Chairman of the Board

ARCHAEOLOGICAL CHEMISTRY

Second Edition

Zvi Goffer, Ph.D.
Retired Senior Scientist
Soreq Nuclear Research Center
Israel

WILEY-INTERSCIENCE
A JOHN WILEY & SONS, INC., PUBLICATION

Published by John Wiley & Sons, Inc., Hoboken, New Jersey
Published simultaneously in Canada

Wiley Bicentennial Logo: Richard J. Pacifico

Book cover illustration and line drawings by the author.

Library of Congress Cataloging-in-Publication Data:

Goffer, Zvi.
Archaeological chemistry / Zvi Goffer. – 2nd ed.
p. cm.
Includes bibliographical references and index.
ISBN 978-0-471-25288-7 (cloth)
1. Archaeological chemistry. I. Title.
CC79.C5G63 2007
930.1028 – dc22

2005028992

Printed in the United States of America

10 9 8 7 6 5 4 3 2 1

To the memory of my parents

Rivka and Moshe

CONTENTS

TEXTBOXES

PREFACE

Modern archaeological research involves collaboration of archaeologists and natural scientists. A virtual alphabet of scientific disciplines, ranging from aerology to zymurgy, today contribute to the elucidation of archaeological issues; some of these disciplines, however, among them chemistry and materials science, are more closely associated with archaeology than others. Although the foundations of modern chemistry as a science were laid only about two hundred years ago, and those of materials science even more recently, humans have always come into contact with chemicals and materials, experienced chemical activity, and developed chemical processes.

All materials, whether of natural origin or made by humans – the air we breathe, the food we eat, the cement we use to construct buildings and roads, the metals we shape into working tools, decorative objects or weapons, and the fibers we spin into yarn and then weave or knit into cloth – all are either chemical substances or mixtures of chemical substances. Cooking involves chemical processes at high temperatures that alter the physical state and chemical properties of the food. The production of wine and beer involves chemical processes, as do dyeing yarn and cloth or extracting metals from their ores; and pottery- and glassmaking involve particularly intricate and complex chemical processes.

Thus chemicals and the processing of chemical substances have always been part of the human experience. It seems obvious, then, that the practice of archaeology, which is the study of the human past through the analysis of its material remains, requires a thorough understanding of the chemical nature and properties of these remains.

The application of chemical methods to archaeology, which can be conveniently referred to under the collective term "archaeological chemistry" began just about one hundred years ago, has greatly broadened since the mid twentieth century and is shedding ever increasing light on the nature of materials and the processes used to make and change them in the past. Today, archaeological chemistry provides not only ways to identify and characterize ancient materials, it also makes it possible to establish how and when materials were worked, altered or manufactured, ascertain their provenance, and trace the routes along which they were transported and traded.

The first materials used by early humans for building and for the making of artifacts and clothing were naturally occurring minerals and stones, as well as wood, bone, and fur, derived from plants or animals, all of which could be gathered, more or less easily, from the surface of the earth or the surrounding environment. With the passing of time and the gaining of experience, humans discovered processes for extracting other materials from those they gathered, and learned how to make from them new materials more appropriate to their needs. Thus began a process of developing technologies for extracting metals from mineral ores, deriving fibers and dyes from plant or animal tissues, and making glass from sand and ceramics from clay, all of which still continue today.

It seems appropriate, therefore, to begin a survey of archaeological materials with a discussion of inorganic materials – from minerals and rocks, the most abundant materials on the planet, to those extracted, derived, or made from them, such as metals and alloys, glass and ceramics (Chapters 1–7). Organic and biological materials produced by, or derived from plants or animals are discussed next (Chapters 8–15). Finally, the atmosphere and the hydrosphere, which make up most of the environment that affects all materials and determines the way they decay, are surveyed (Chapter 17).

The subject matter included in the chapters was compiled during many years of investigating and teaching about archaeological materials, and I believe the book provides an overview of the present standing of archaeological chemistry studies. This second edition of *Archaeological Chemistry*, like its predecessor, has been written in such a way as to provide an introduction to the subject. I have tried to furnish extensive references for those who wish to pursue particular aspects in greater depth; I did not, however, attempt to present an exhaustive survey of the literature, since many of the publications on the subjects discussed tend to be repetitive and often convey little that is new.

The field covered by archaeological chemistry, has expanded greatly since the first edition of *Archaeological Chemistry* was published over twenty-five years ago. There was at that time, relatively little information on the chemistry of ancient materials or the application of chemical methods to archaeology, whereas archaeological chemistry studies today attract the attention of large numbers of scientists. Not only the range but also the complexity of archaeological subjects to which chemical methods are now applied have expanded vastly since the appearance of the first edition. This has been due mainly, to the development of new, sophisticated analytical instrumentation and of techniques for the study – important in the archaeological context – of ever-smaller samples of a continuously growing range of materials. These developments, together with increasing knowledge about the processing of analytical results, have made possible a coherent interpretation of the results of archaeological studies. The growth in interest, and in the amount of information available, is reflected in the scientific literature and in the numerous meetings on archaeological studies organized by prestigious scientific organizations such as the Royal Society in Britain, the Centre National de la Recherche Scientifique in France, the Max Planck Gesellschaft in Germany, the American Chemical Society, and others throughout the world.

In this second edition I have updated much of the material contained in the first one, and introduced, throughout the text, modifications to include the latest research contributions in the field. As a consequence of advances in knowledge and interpretation, however, the archaeological chemistry literature is ever burgeoning; this compelled me to be particularly selective of the topics and the bibliographic material to be included in the book.

Aware of the need to maintain scientific accuracy while at the same time keeping the book accessible to readers not familiar with the complexity of scientific writing, I have divided the book into two intertwining and complementary sections: text and textboxes. In the text the uses and applications of chemistry to archeology are discussed, avoiding the use of highly technical chemical language. In the textboxes are expanded discussions of chemical subjects only concisely presented in the text, and explanations of chemical concepts and nomenclature. Reading them is optional and not much will be lost on the topical consideration of the subjects by ignoring them altogether. All chemical subjects considered in the book should, therefore, be easily understood by those not familiar with the terminology of chemical literature.

Zvi Goffer

gofferc z@013.net
Rehovot, Israel
June 2007

ACKNOWLEDGMENTS

Few scientific books, nowadays, can be written by just one person. Even though I am credited by the publisher with being the sole author of this book, Archaeological Chemistry is no exception. It is with great pleasure and appreciation, therefore, that I acknowledge my indebtedness to the people who encouraged me and provided a helping hand during its writing. Members of the Kimmel Center for Archaeological Science at the Weizmann Institute of Science, Rehovot, Israel, helped with many suggestions and also served as a sounding board for the initial writing; special thanks are due to Francesco Berna, Elizabetta Boaretto, Eugenia Mintz, Ruth Shahak, Sariel Shalev, and Stephen Weiner of the Center. My recognition and gratitude are also due to John Elmsley of the University of Cambridge and Dafydd Griffiths of University College, London, England, Motti Findler of the Hadassah Medical Center and Archie Kauffmann of the Weizmann Institute of Science, Israel, Milton Taylor of Indiana University and Robert H. Tycott of the University of South Florida, USA, and to my friends Helen and Bruce Christie of Australia and Lionel Holland of Israel, for their counsel and advice. Their comments and critique led me to correct errors and clarify passages or illustrations that otherwise might have made parts of the book vague or obscure. I am much obliged to Danny Kijel and Vicky Shklarman, who assisted me with the artwork and to Evelyn Katrak who edited my language and made it more comprehensible.

I also thank the following for permission to reproduce their photographs: the late I. Perlman (Fig. 14); the Israel Antiquities Authority (Figs. 19, 32, 35, 36, 38, 43, 54, 55, 57, 68 and 80); M. Prausnitz (Fig. 21); I. Friedman (Fig. 23); R. L. Fleisher (Figs. 24, 25); Eretz Israel Museum, Tel Aviv, Israel (Figs. 27, 28, 56); R. H. Brill (Fig. 29); Y. Beit Arieh (Figs. 31, 37 and 79); J. M. Curran (Fig. 44); F. de Korosy (Fig. 52); E. Boaretto and J. Heinemeir (Fig. 63); E. Aspillaga and B. Ahrensberg (Fig. 81). Thanks are also due to Z. Herzog for permission to photograph the archaeological finds on the cover illustration.

The facilities provided by the Kimmel Center for Archaeological Science at the Weizmann Institute of Science, Rehovot, Israel, and the Institute of Archaeology, University College, London, England, served as heavenly places in which to gather data and write; the libraries and librarians of both these institutions provided unlimited help and inexhaustible sources of information.

Chava, my wife, and Ailit, my daughter, acquiesced with love and understanding to the long hours when my mind was in archaeological chemistry rather than with my family. They both stood, as always, by my side, helping me create a better book, improving my language, and restraining excesses in my writing.

I am, however, the only person answerable for any faults, errors, and other shortcomings that may still remain in the book.

Z. G.

MINERALS
ROCK AND STONE: PIGMENTS, ABRASIVES, AND GEMSTONES

The outer crust of earth has provided the solid foundation for the evolution of human beings, who are the prime focus of interest and concern to archaeology. The main components of this crust are *minerals* and *rocks*, some *consolidated* and others occurring as *sediments*, nonconsolidated deposits, created by weathering processes from the minerals and rocks. All these minerals, rocks, and sediments, as well as everything else in the universe, are made up from just over 100 *chemical elements* listed in Appendix I. Most of the elements in the crust of the earth occur in extremely low relative amounts, and only a few, listed in Table 1, make up almost 99% of its total bulk (Bloom 1969).

1.1. THE CHEMICAL ELEMENTS

There are two basic types of *elements*: *metals* and *nonmetals*. The *metals*, such as copper, gold, and iron (see Chapter 5), make up more than three-quarters of the total number of elements; *nonmetals*, such as, for example, chlorine, sulfur and carbon, make up much of the rest. Other elements, however, known as the *metalloids* or *semimetals*, have properties intermediary between the metals and the nonmetals (see Appendix I). Only a few elements, such as the metals gold and copper and the nonmetal sulfur, which are known as the *native elements*, occur in nature uncombined. Most elements occur naturally combined with others, forming *compounds*. It is from these compounds, which occur in the crust of the earth as minerals, rocks, or sediments, that humans extract most of the elements that they require (Klein 2000).

Archaeological Chemistry, Second Edition By Zvi Goffer

TABLE 1 Chemical Elements in the Earth's Crust

Element	Abundance (wt%)
Oxygen	47.0
Silicon	28.0
Aluminum	7.5
Iron	5.0
Calcium	3.5
Sodium	2.8
Magnesium	2.6
Potassium	2.1
84 others including, in descending order of abundance, titanium, hydrogen, phosphorus, barium, and strontium.	1.5

TEXTBOX 1

MATTER AND MATERIALS; ELEMENTS AND SUBSTANCES

Everything on earth, as in the whole of the universe, is made up of matter, that is, anything solid, liquid, or gaseous that has mass and occupies space. Depending on their composition and internal structure, *materials*, as specific types of matter are known, can be of one of two fundamentally different types: either *homogeneous* or *heterogeneous*. *Homogeneous materials*, for example, water, sand, salt, and glass, have uniform, unvarying composition and properties throughout, while the composition and properties of *heterogeneous materials*, for example, granite and pottery, vary from one part to another. Heterogeneous materials are mixtures of two or more homogeneous substances (Henderson 2000).

Homogeneous Materials; Substances and Solutions

The homogeneous materials can also be classified into two distinct groups: *substances* and *solutions*. The *substances*, for example, gold, graphite, common salt, and limestone, have a definite and characteristic chemical composition and physical properties. *Solutions*, for example, the water of the seas and oceans, are mixtures, in any weight proportion, of two or more substances dispersed within each other and forming a single blend of unvarying constitution, although of no definite composition. Also the substances may be of one of two types: either *elements* or *compounds*.

Elements The chemical *elements*, or simply elements, such as gold, oxygen, and the 110 others listed in Appendix I, cannot be separated into other, simpler substances by any means. There are 112 known chemical elements (see Appendix I), only 92 of which occur on the earth and are the constituents of all the planet's matter. The other 26, which are unstable and undergo radioactive decay (see Textbox 13), have long ago disappeared. Some of the 92 elements, such as copper, silver, and carbon, are familiar to everybody, while others, such as cerium and dysprosium, are so rare that few people are likely to hear of them, let alone encounter them. Many of the lesser-known elements are, however, important in archaeological studies. The relative amounts in which they occur in nature vary from one location to another, and this variation can be of use to characterize many natural materials and to determine their provenance, that is, their place of origin (see text below).

Compounds *Compounds* consist of two or more elements combined in singular and characteristic weight proportions; the properties of compounds differ from those of their component elements and from other elements or compounds (see Textbox 4). Their composition is independent of their origin or how they were formed, and their properties are unique and characteristic of that compound. All compounds can be decomposed, by chemical means, to yield their elemental constituents. This means that any particular compound, regardless of its origin or location, is always composed of the same elements combined in a unique weight proportion. For example, *common salt* (whose chemical name is *sodium chloride*) is a compound made up of two elements: (1) sodium, which makes up 39.39% of its total weight, and (2) chlorine, which makes up the remaining 60.61%. Additional examples of compounds are *quartz*, a constituent of common sand and limestone. *Quartz* consists of silicon dioxide, a compound formed by the combination of two elements, silicon and oxygen; silicon makes up 46.66% of the total weight of any sample of silicon dioxide, and oxygen makes up the remaining 53.33%. *Limestone* is composed of calcium carbonate, a compound made up of three elements – calcium, carbon, and oxygen; in the compound, calcium makes up 40% of the total weight; carbon, 12%; and oxygen, the remaining 48%.

Compounds are customarily classified as *inorganic* and *organic compounds*. The term *inorganic* refers (with only very few exceptions) to compounds whose composition does not include the element carbon; they are made up of combinations of two or more of any of the 92 stable elements that constitute all matter, but excluding carbon. The *inorganic compounds* are, by far, the main components of most of the matter in the universe. Compounds that include carbon in their composition are, with few exceptions, known as *organic compounds*. In the organic compounds, carbon is combined mainly with one or more of the elements hydrogen, oxygen, and nitrogen and sometimes also with a few others, such as sulfur and a few metals (see Textbox 51). The organic compounds are the main constituents of all living matter. The division between inorganic and organic compounds, which has become blurred in recent years, arose from an old misconception, held since the early days of chemical studies, when it was believed that a vital organic force was necessary for the creation of organic compounds.

Solutions *Solutions* are homogeneous forms of matter that may be composed of a solid dissolved in a liquid – such as common salt dissolved in water; a gas dissolved in a liquid – for example, oxygen dissolved in water; or a solid dissolved in another – for example, carbon dissolved in iron in some alloys of this metal. The composition and properties of each solution are determined by the nature of the components and the relative amount of each component in the solution (see Table 2).

Heterogeneous Materials

Heterogeneous materials are made up of more or less coarse mixtures of the grains of two or more homogeneous substances. Many rocks, for example, are mixtures of fragments of two or more different minerals (see discussion below). *Pottery*, another heterogeneous material, consists of a burnt mixture of clay and one or more type of particles of such varied materials as homogeneous sand and limestone or heterogeneous dung and

TABLE 2 Homogeneous and Heterogeneous Matter

Homogeneous matter	Heterogeneous matter
Elements	
92 chemical elements, including metals such as copper, gold, iron, lead, silver, and nonmetals such as, carbon, sulfur, and phosphorus	Most rocks (two or more mixed minerals)
	Bloomery iron (a mixture of iron, carbon and slag)
Compounds	Cast iron (a mixture of iron and carbon)
Inorganic compounds (water, salt, soda)	Lime cement (lime and sand)
Minerals (cassiterite (tin oxide), galena (lead sulfide	Pottery (fired clay and mineral or organic grains)
Monomineral rocks [(calcite (calcium carbonate), silica sand (silicon dioxide)]	Soil (clay, sand, and organic matter)
	Wood (cellulose, hemicellulose, and lignin)
Organic compounds (sugars, oils, fats.	
Solutions	Bone (organic and bioinorganic matter)
Seawater (salts in water)	
Gold–copper alloys (gold in copper)	Milk (fat droplets in a water solution)
Solder (alloy of lead and tin)	
Vinegar (acetic acid in water)	
Wine (ethyl alcohol in water)	

grass (see Chapter 7). Other heterogeneous materials are *wood* and *bone*, both examples of living forms of heterogeneous matter. *Wood* is a mixture of three organic substances – namely, cellulose, hemicellulose, and lignin (see Chapter 9); *bone* is an interspersed mixture of *collagen*, an organic substance, and *carbonated hydroxyapatite*, a bioinorganic substance (see Chapter 15). Heterogenous materials can be separated into their homogeneous components using physical techniques (see Fig. 1).

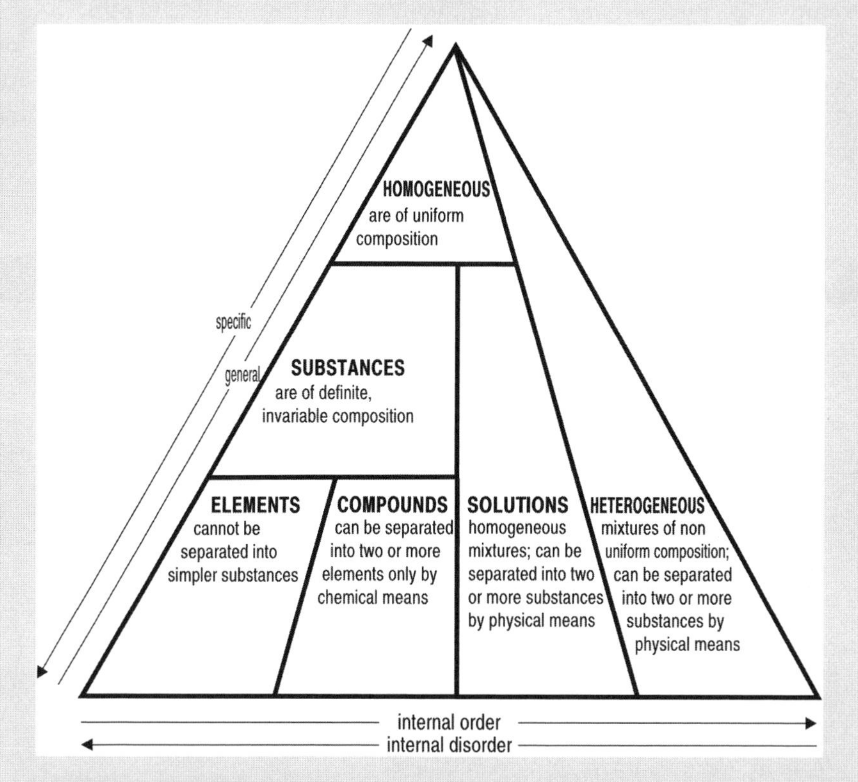

FIGURE 1 Classification of matter. Matter, everything that has mass and takes up space, is made up of very small particles known as "atoms." Particular kinds of matter that have uniform composition and properties throughout are known as "substances." Substances may occur alone (separate from) or mixed with other substances. Gold and lime, for example, often occur separate from others in nature: gold as gold nuggets and lime as limestone rocks. Mixtures of two or more substances may be homogeneous or heterogenous; seawater, wine, fruitjuice, and some alloys are homogeneous mixtures, with no visible separation between their components. Seawater, for example, is a solution (a homogeneous mixture) of various salts in water. Granite, brick, pottery, cement, and wood are heterogeneous mixtures of various substances.

1.2. MINERALS AND MINERALOIDS

Minerals are the most abundant type of solid matter on the crust of the earth; they are homogeneous materials that have a definite composition and an orderly internal structure. Minerals make up most of the bulk of rocks, the comminuted particles of sediments, and the greater part of most soils. Over 3000 minerals have been identified, and new ones are discovered each year. Only a few hundred, however, are common; most of the others, such as, for example, the *precious stones*, are difficult to find (Ernst 1969). Table 3 lists common minerals and mineraloids. Many schemes have been devised for classifying the minerals. In the scheme presented in Table 4, minerals are arranged in classes according to their increasing compositional chemical complexity.

TEXTBOX 2

THE STRUCTURE OF MATTER: ATOMS AND MOLECULES

Atoms All matter is made up of very small units called *atoms*, the smallest material units having recognizable, characteristic properties, identical to those of the bulk. Atoms, which are the building blocks of the universe, can exist either alone or in combination with others. The *chemical elements*, for example, are forms of matter made up of identical atoms, each atom of an element exhibiting the same characteristic properties. When extremely strong forces are applied to matter, however, atoms may break up into smaller parts; this shows that the atoms are not single physical units, but composites having a complex inner structure of their own. Studying the processes by which atoms break up made it possible, during the early twentieth century, to understand the inner structure of atoms (Asimov 2002; Pullman 2001).

Atoms are made up of varying numbers of extremely small particles, generally referred to as *subatomic particles*, whose size is too small to see or measure even under large magnification. *Protons*, *neutrons*, and *electrons* are examples of atomic particles. The *protons*, which bear a positive electric charge, and the *neutrons*, which are electrically neutral, are concentrated in the atomic *nucleus*, located in the center of the atom, as shown in Figure 2. Since the protons have a positive electric charge, the atomic nuclei also bear a positive electric charge. *Electrons*, which are much smaller than the protons and neutrons, have a negative electric charge and move rapidly in pairs around the atomic nucleus. All the atoms of any one *element* have the same number of protons and electrons and are,

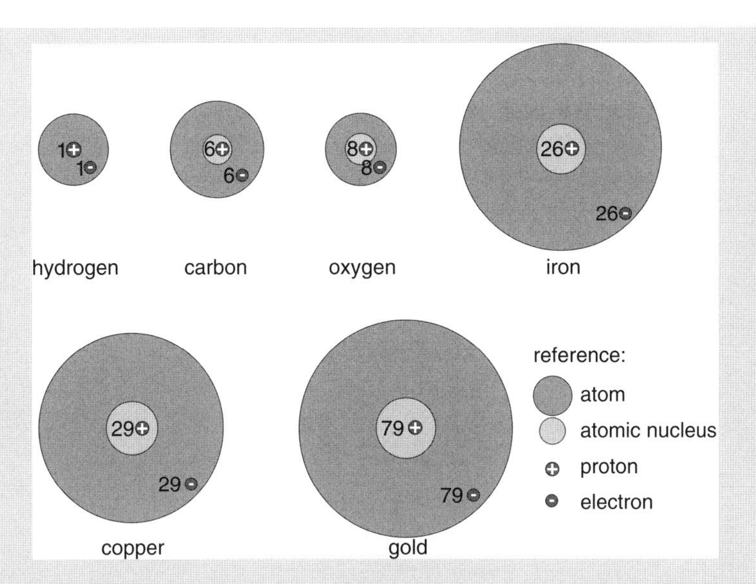

FIGURE 2 Elements and atoms. Matter that is made up of only one type of atom, and cannot be separated into simpler substances, is called an "element." Hydrogen, carbon, oxygen, iron, copper, and gold, for example, are elements. All matter on earth is composed of relatively few (only 92) different elements. The atoms that constitute any element are identical to each other but different from those of other elements. The atoms of all the elements, however, are built from basic particles known as "subatomic" or "elementary" particles, namely, protons, neutrons, and electrons. Most of the mass of each atom resides in the neutrons and protons, which occupy the central region of the atom known as the "atomic nucleus." The neutrons have no electrical charge and are, therefore, electrically neutral; the protons have positive electrical charges and the electrons, which surround the nucleus, have a negative electric charge equivalent to the positive charge of the proton. The figure shows diagrammatic representations of the atoms of several elements, including their nuclei and the number of protons and electrons in each atom. Atoms may combine with other atoms to form "molecules," and any form of matter made up of only one type of molecule is known as a "compound."

therefore, electrically balanced. Atoms of different elements, however, have different numbers of protons and electrons (Romer 1982; Asimov 1974).

Molecules Atoms of the same element or of different elements combine with each other into more complicated systems called *molecules*; *molecules* are the smallest units of chemical compounds, the basic components of the countless inorganic and organic compounds. A few molecules, such as, those of the chemical elements hydrogen, oxygen, and nitrogen, are com-

TABLE 3 Composition, Properties, and Ancients Uses of Common Minerals and Mineraloids

Common name	Composition	Physical properties				Main uses
		Color	Density (g/cm^3)	Hardness (Mohs scale)	Structure	
Minerals						
Alabaster	Calcium sulfate	Various colors	2.3	1.8	Crystalline	Lapidary
Alabaster (oriental)	Calcium carbonate	Various colors	2.5	2.8	Crystalline	Lapidary
Azurite	Hydrous copper carbonate	Blue	3.8	3.8	Crystalline	Ornamental stone, gemstone, blue pigment, building stone, making lime
Calcite (limestone, marble)	Calcium carbonate	White	2.7	3.0	Crystalline	Making glass
Corundum	Aluminum oxide	White	4.0	9	Crystalline	Abrasive
Diamond	Carbon	Colorless	3.5	10	Crystalline	Gemstone, abrasive, flux, and lapidary
Fluorite	Calcium fluoride	White	3.2	4	Crystalline	
Galena	Lead sulfide	Gray	7.5	2.5	Crystalline	Lead ore
Granite	Mixture of mica, quartz and fedspar	Various	2.7	6–7	Crystalline mixture	Lapidary, paving, building

Gypsum	Hydrated calcium sulfate	White	0.3	2	Crystalline	Building stone, making plaster of Paris
Common salt	Sodium chloride	White	2.2	2.5	Crystalline	Food products
Haematite	Iron oxide	Red	5.2	6	Crystalline	Iron ore, abrasive, red pigment, gemstone copper ore, green pigment
Malachite	Basic carbonate of copper	Green	3.6	3.6	Crystalline	Gemstone
Pyrite	Iron sulfide	Gold	5	6.5	Crystalline	Iron ore gemstone, making lithic artifacts
Quartz (flint, sand, etc.)	Silica	Colorless	2.7	7	Crystalline	Building material, making glass, gemstone
Turquoise	Hydrated copper aluminum phosphate	Blue	2.7	5.5	Crystalline	Gemstone
Mineraloids						
Amber	Fossil resin	Yellow to red	1.1	5	None	Gemstone, ornamentation
Opal	Hydrated silica	Varies	2	6	None	Gemstone
Obsidian	Natural glass	Black, brown	2.5	5.3	None	Making lithic artifacts

posed of two identical atoms bonded to each other. Most molecules however, such as those that make up all the inorganic and organic compounds, are composed of two or more different elements. The smallest molecules such as those of *common salt*, are made up of only two different elements (sodium and chlorine, in the case of salt); the largest, such as the molecules of the biological compounds *cellulose* and *DNA*, are made up of millions of atoms (Fine 1980; Fano and Fano 1973). Cohesive forces, known as *chemical bonds*, which arise from the sharing of electrons between atoms, hold the atoms together in the molecules (see Fig. 3).

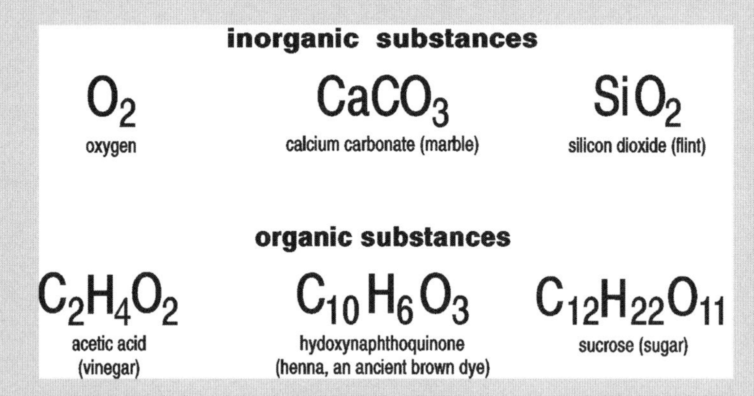

FIGURE 3 Molecules, compounds, and formulas. The atoms of the elements may combine with each other in an infinite number of ways to form molecules, and matter made up of only one type of molecules is known as a *compound*. The molecules are more or less complicated atomic structures that make up compounds, the main constituents of most matter on earth. Calcium carbonate, silicon dioxide, and sucrose, for example, are the compounds that make up marble, flint, and sugar, respectively. The molecules of calcium carbonate (lime), for example, are composed of atoms carbon, oxygen, and calcium; those of silicon dioxide (flint) are composed of silicon and oxygen, and the molecules of sucrose (a sugar) are composed of carbon, hydrogen, and oxygen. Within each molecule the atoms are linked to each other by electrical forces usually called "chemical bonds." The exact composition of the molecules and therefore of substances is expressed by "chemical formulas," a form of notation that conveys information on the types of atoms and their relative weight proportions in any specific substance. In the formulas, the chemical symbols of the elements indicate the types of atoms in the molecule and numerical subscripts show the relative number of each type of atom in the molecule. The chemical name and the formulas of well-known substances are illustrated.

TABLE 4 Chemical Classification of Minerals

Class	Compositional characteristic	Examples
Native elements	Single element	Native metals, native nonmetals
Oxides	Metal ion(s) + oxygen ion	Haematite (composed of iron oxide), a red pigment and an iron ore Corundum (composed of aluminum oxide), an abrasive silica (composed of silicon dioxide), common sand
Sulfides	Metal ion(s) + sulfur ion	Galena (composed of lead sulfide), a metal ore, from which lead and also silver (which occurs in galena as an impurity, are extracted Pyrite (composed of iron sulfide), an iron ore
Halides	Metal ion(s) + halogen ion	Common salt (sodium chloride), a component of animal diets Fluorite (calcium fluoride), a lapidary material and flux
Carbonates	Metal ion(s) + carbonate ion	Calcite (calcium carbonate), used for making lime, a building material Marble (calcium carbonate), building and lapidary stone
Sulfates	Metal ion(s) + sulfate ion(s)	Gypsum (calcium sulfate), raw material for making plaster of Paris Alabaster (calcium sulfate or calcium carbonate), a lapidary material
Silicates	Metal ion(s) + silicate ion	Granite (a metal silicate), used mostly for building

The (compositionally) simplest mineral class comprises the native elements, that is, those elements, either metals or nonmetals that occur naturally in the *native state*, uncombined with others. Native gold, silver, and copper, for example, are metals that naturally occur in a ductile and malleable condition, while carbon – in the form of either graphite or diamond – and sulfur are examples of nonmetallic native elements. Next in compositional complexity are the *binary minerals* composed of two elements: a metal or nonmetallic element combined with oxygen in the *oxides*, with a halogen – either fluorine, chlorine bromine, or iodine – in the *halides*, or sulfur, in the *sulfides*. The oxide minerals, for example, are solids that occur either in a somewhat hard, dense, and compact form in mineral ores and in rocks, or as relatively soft, unconsolidated sediments that melt at moderate to high temperatures. Most of the sulfide minerals are opaque, or nearly opaque, and have distinctive colors and colored streaks. More *complex minerals* are composed of three or more elements, such as the *carbonates* and *silicates*. The *carbonates* are easily identified because, when an acid is dropped on them, they effervesce, releasing carbon dioxide. The *silicates*, which have the most complex chemical composition among the minerals, are by far the most common minerals, making up about 25% of the known minerals and almost 40% of the most common ones in the outer crust of the earth.

TEXTBOX 3

THE STATES OF MATTER

All matter exists in three states familiar to everybody: the *solid*, *liquid*, and *gaseous* states. The actual state in which any particular type of matter occurs is determined by the amount of energy of the constituent atoms or molecules. The atoms or molecules of solid matter, such as, for example, minerals, rocks, ceramic materials, or metals, have little energy, so little that they cannot flow. Moreover, the atoms or molecules that make up solids are held tightly together in fixed positions by strong interatomic and/or intermolecular forces. Solid matter occupies, therefore, a fixed volume and retains its shape wherever it is located. When supplied with enough energy, however, most solids melt and become *liquids*. The atoms or molecules in *liquid* matter, such as water or oil, have sufficient energy to flow, but insufficient to rise out of a container. Liquids have, therefore, a fixed volume but not a fixed shape; they acquire the shape of the container in which they are placed. If supplied with enough energy, liquids boil and become *gases*. The forces that hold together the atoms or molecules of *gases*, such as air, carbon dioxide, or water vapor, are very weak; gases are therefore very fluid, much more fluid than liquids; they retain neither a shape nor a volume and may rise out of a container.

Changes of State The amount of energy in a material, which determines whether it is in the solid, liquid, or gaseous state (see Figs. 4 and 5), depends on its composition, temperature, and surrounding pressure. Different materials change from one state to another at widely different temperatures, although each substance (element or compound) does so at a specific and characteristic temperature and pressure. When iron, a solid at ambient temperature and pressure, is heated to a high temperature – namely 1538°C – it melts, turning into a liquid; and at 2861°C it boils, changing from a liquid to a gas. *Water*, on the other hand, the most abundant substance on the surface of the earth, is solid only at very low temperatures: ice (solid water) melts at 0°C and boils at 100°C. Because the temperature range of the liquid state is very narrow (only 100 degrees), water exists naturally on the surface of the planet in all three states, as ice, liquid water, and a gas (water vapor).

The melting points of mixtures and solutions depend on the nature and the relative amount of each component of the solution. They are, however, lower than those of the separate components. *Solder*, for example, an alloy of tin and lead, melts at 183°C, a much lower temperature than either of its components: tin melts at 231°C and lead, at 328°C.

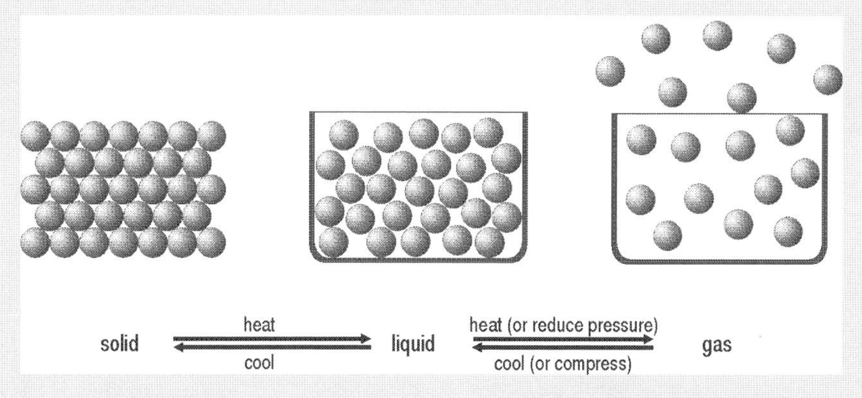

FIGURE 4 The states of matter. All matter – whether solid, liquid, or gaseous – is made up of atoms or molecules. The amount of energy of the atoms or molecules differs in each one the three states. In solids they have little energy; are tightly packed (usually in a regular pattern) and retain definite shapes. In liquids they have more energy than in solids, are closely packed together, but are not regularly arranged, sliding past each other; consequently, liquids keep no regular shape and acquire the shape of their container. The atoms or molecules of gases have much energy, move rapidly, are well separated from each other, and retain neither a regular arrangement nor a definite shape.

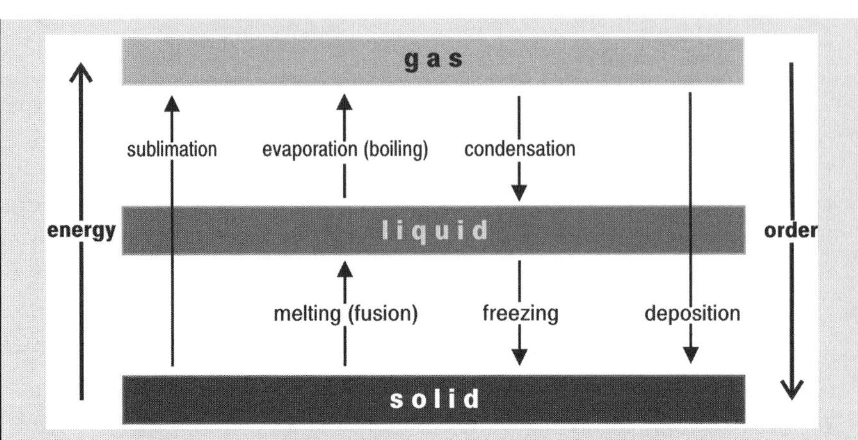

FIGURE 5 Changes of state. Changes in the amount of energy of the particles (atoms or molecules) that make up matter result in changes in the state of matter. The particles of solids, for example, have little energy and are held tightly together by attractive forces. When the temperature of a solid increases, however, the particles acquire energy and begin to move; as the temperature reaches a critical stage, known as the "melting point," the added energy breaks the attractive forces between the particles and the substance becomes liquid. A similar process takes place when the temperature increases still further: the liquid reaches another critical stage, the "boiling point," and becomes a gas. Also, as the temperature and the amount of energy of any form of matter increases, the particles move faster and the order in their spatial arrangement decreases. When the temperature is lowered, all this is reversed.

Any two samples of a particular mineral, whatever their source or place of origin, have the same basic composition and characteristic crystal structure; moreover, no two different minerals have identical chemical composition and crystal structure (see Textboxes 8 and 21). *Quartz*, for example, is a common and abundant mineral composed of *silicon dioxide*, a compound that occurs naturally not only as quartz but also in other crystal structures, known as *polymorphs* (polymorphs are minerals that have the same chemical composition but different crystal structure), some of which, listed in Table 23, have been used for a variety of purposes. The *crystal structure*, which is essential for the characterization of solid materials, is just one of a wide range of *physical properties*, that is, properties not involving chemical differences, which provide convenient criteria for characterizing and identifying solids.

TEXTBOX 4

THE PROPERTIES OF MATTER

All substances – elements as well as compounds – have characteristic and distinctive *chemical* and *physical properties* (see Table 5). The *chemical properties* are related to the changes in composition that substances undergo when interacting with one another. A chemical property of solid iron, a metal, for example, is that it interacts at ambient temperature with atmospheric oxygen, forming iron oxide, a dark brown material commonly known as *rust*. Copper, another metal, interacts at ambient temperature not only with oxygen but also with atmospheric moisture and carbon dioxide, forming copper carbonate, a green substance; copper carbonate, which covers copper surfaces exposed to the environment, often forms, on its outer surface, a layer of green "patina" on antique copper objects. Whether a substance reacts (interacts) or does not react with others is also a chemical property. A chemical property of gold, for example, is that it does not react with the components of the atmosphere or the soil or, for that matter, with practically any element or compound. Because of its low chemical reactivity, gold is known as a *noble metal*. Another noble metal is platinum.

TABLE 5 Properties of Substances

Physical properties	Color, refractive index, ductility, hardness, malleability, melting point, boiling point, density, thermal conductivity
Chemical properties	Composition; reactivity with other substances; stability to heat, radiation, and electricity

The *physical properties* of substances do not involve chemical changes. *Color* (see Textbox 17) and *crystal structure* (see Textbox 21), for example, are physical properties that are characteristic of a substance that serve to identify most substances. Other physical properties, such as *density*, *hardness* (see Table 3), *refractive index* (see Table 19), and *heat capacity* (see Table 101), are also useful for characterizing and identifying substances as well as distinguishing between different substances.

In addition to the minerals, there are also some rock-forming homogeneous materials that have neither the definite chemical composition nor the distinctive crystal structure characteristic of minerals. Such materials cannot, therefore, be considered as minerals and are known as *mineraloids*. *Obsidian*, for example, a natural material that has been widely used since prehistoric times for making lithic tools and decorative objects, is a mineraloid. Obsidian has neither a definite chemical composition nor a characteristic crystal structure and is not, therefore, a mineral. *Copal* and *amber* are other mineraloids that since antiquity have been treasured as semiprecious *gemstones*.

TEXTBOX 5

ELECTROMAGNETIC RADIATION; WAVES

Electromagnetic radiation (EMR) is the generic term used to refer to all forms of energy that, in the form of waves, travel through space and matter at very high velocity. *Visible light*, to which the eyes of humans and animals are sensitive, and *radio and television waves*, which provide much of the electronic information available today, are the most familiar forms of electromagnetic radiation. Less familiar, but no less important, forms of electromagnetic radiation include *infrared* radiation (also known as *heat waves*), *ultraviolet light*, *X-rays*, and *gamma rays* (see Textbox 13).

Waves

A *wave* is a form of movement, an oscillatory disturbance characterized by repetitive patterns in fixed time intervals, that propagates through space or matter without displacing mass, but energy (see Fig. 6). Ocean waves, sound waves, and electromagnetic waves are common examples of waves. There are two main types of waves: *longitudinal* and *transversal* waves. When the oscillation takes place along their axis of propagation, the waves are known as *longitudinal waves*; when the oscillation occurs across their axis of propagation, the waves are known as *transverse waves*. Sound waves are longitudinal. Water waves, in the seas and oceans, and the waves sometimes seen on the strings of musical instruments are transverse waves. Electromagnetic waves are also transverse waves.

All types of waves, whether longitudinal or transverse, can be accurately described by their *wavelength* and *frequency* values (see Fig. 6), which are mathematically related to each other by the expression $\nu\lambda = c$, where the Greek letter λ (lambda) is the wavelength, the Greek letter ν (nu) is the *frequency* of the wave, and c is the *velocity* of the wave.

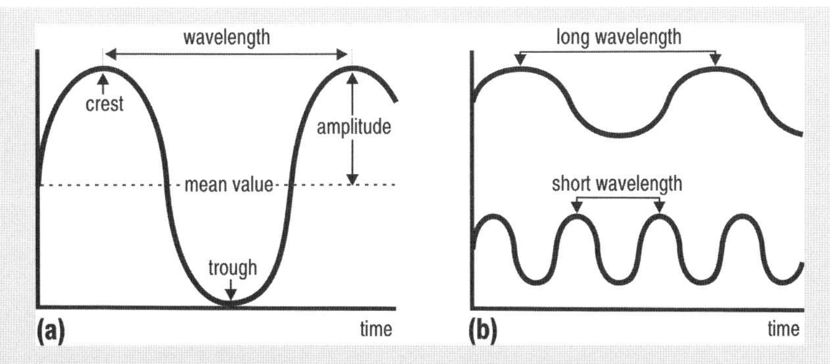

FIGURE 6 Waves. All types of waves are characterized by two features: amplitude and wavelength (or frequency). The amplitude, defined as half the linear distance between a crest and a trough **(a)**, is a measure of the maximum displacement of a wave. The wavelength is the distance on a straight line along a wave, from one crest (or trough) to the next **(a)**. The frequency is the number of times that a repeated event occurs per unit time; the shorter the wavelength of a wave, the higher is its frequency **(b)**.

Electromagnetic Waves *Electromagnetic waves* are created by moving electric charges. When an electric charge moves, it creates a *magnetic field*. If an electric charge *oscillates* (moves back and forth or up and down), its electric and magnetic fields change together, creating an electromagnetic wave. Different forms of electromagnetic radiation can be differentiated by their wavelengths, as listed in Table 6 and illustrated in Figure 7 (Bekefi and Barrett 1987). After inception, electromagnetic waves propagate through empty space (vacuum) or through matter, without displacing mass but displacing energy.

In a *vacuum* (empty space), all forms of electromagnetic radiation propagate at a velocity of 300,000 km per second, when propagating through air, water, or any kind of matter, they interact with the matter and their velocity is reduced. Differences in the manner of interaction between different forms of radiation and different types of matter generally reveal information on the nature and the constituents of matter.

The field of science that studies the interaction of electromagnetic radiation with matter is known as *spectroscopy*. Spectroscopic studies on the wavelength, the intensity of the radiation absorbed, emitted, or scattered by a sample, or how the intensity of the radiation changes as a function of its energy and wavelength, provide accurate tools for studying the composition and structure of many materials (Davies and Creaser 1991; Creaser and Davies 1988).

TABLE 6 Spectrum of Electromagnetic Radiation

Radiation	Typical sources	Wavelength (cm)	Frequency (Hz)[a]	Energy (eV)[b]
Gamma rays	Radioactive nuclei	below 10^{-9}	above 3×10^{16}	above 10^{5}
X-rays	Atoms	10^{-7} – 10^{-9}	$3 \times 10^{17} - 3 \times 10^{19}$	$10^{3} - 10^{5}$
Ultraviolet	Atoms	4×10^{-5} – 10^{-7}	$7.5 \times 10^{14} - 3 \times 10^{17}$	$3 - 10^{3}$
Visible	Hot bodies	$7 \times 10^{-5} - 4 \times 10^{-5}$	$4.3 \times 10^{14} - 7.5 \times 10^{14}$	2 – 3
Infrared	Hot molecules	$10^{-2} - 7 \times 10^{-5}$	$3 \times 10^{12} - 4.3 \times 10^{14}$	0.01 – 2
Microwaves	Electronic devices	10 – 10^{-2}	$3 \times 10^{9} - 3 \times 10^{12}$	10 – 0.01
Radio	Communication	above 10	below 3×10^{9}	below 0.001

[a] Hz = Hertz, unit of frequency – the number of times a wave oscillates per unit time – measured in cycles per second.
[b] eV = electron volt, unit of energy equal to the work done by an electron accelerated through a potenitial difference of 1 volt.

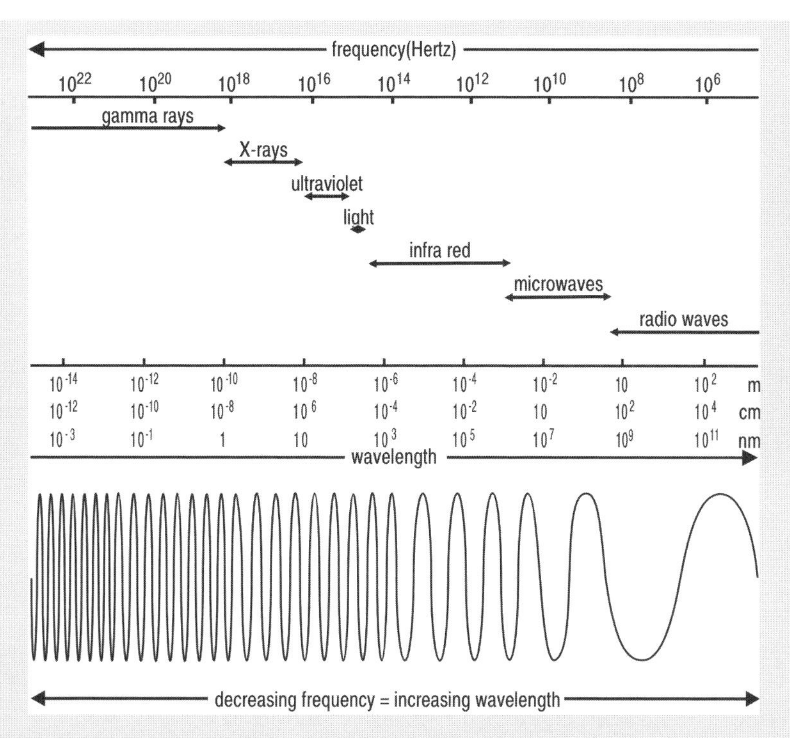

FIGURE 7 Electromagnetic radiation and the electromagnetic spectrum. The full range of electromagnetic radiation, extending from extremely long to very short wavelengths, is known as the "electromagnetic spectrum"; it includes the following well-defined regions: gamma rays, emitted by some radioactive isotopes or generated during nuclear reactions; X-rays, used for medical (radiography and the destruction of deceased cells), technological and scientific applications; ultraviolet radiation, which, from the part of the spectrum known as "ultraviolet A (UV-A)," enhances plant life and from another other part of the spectrum (UV-B) causes premature skin aging and skin diseases; visible light, which the human eye can perceive as colors ranging from violet (shorter wavelengths) to red (longer wavelengths); infrared radiation, associated with heat energy; microwaves, used in most forms of electronic communication; and long and extremely long wavelength radio radiation, also used in communication. Electromagnetic radiation with wavelengths shorter than ultraviolet rays has enough energy to cause atoms to lose electrons and become ions and is therefore known as "ionizing radiation;" long wavelength radiation, longer than those of ultraviolet rays, is "nonionizing."

Minerals, mineraloids, and metals are advantageous for specific applications because of one or more of their physical properties. The *softness* of soapstone, for example, makes this mineral useful for shaping and carving objects. The very opposite, the extremely high *hardness* of diamond and emerald, makes them resistant to wear and scratching and, therefore, valued as gemstones. It is also because of its extreme hardness that powdered diamond is the best abrasive material for rubbing, grinding, and polishing solids. Other extremely useful physical properties of materials are color and the refractive index (see Textbox 22). The *color* of many materials, such as the characteristic red of *minium* (a lead-containing mineral) and the green of *malachite* (a copper-containing mineral) makes them useful as pigment, while the high *refractive index* of diamond and emerald makes these two rare minerals glitter and sparkle and, therefore, desirable and highly valued as gemstones often set into jewelry.

TEXTBOX 6

ATOMS AND MOLECULES; CHEMICAL NOMENCLATURE AND FORMULAS

Elements and Atoms

The *chemical elements*, the building blocks of all matter, are made up of identical atoms (see Textbox 2). Very few elements, however, occur on earth singly, uncombined with others. Most are chemically combined with other atoms, either identical or different. Examples of elements that may occur singly are the metals copper and gold and the nonmetal sulfur, which are often found in their *native* form in the crust of the earth. The gaseous elements *oxygen* and *nitrogen*, the main components of the earth's atmosphere, occur as molecules composed of two identical, combined atoms; each atom in a molecule of oxygen or nitrogen is combined to another atom of the same element.

Most elements, however, are naturally combined with one or more other elements, forming compounds. Water and calcium carbonate (also known as *calcite*, *limestone*, or *marble*) are examples of common compounds formed by the combination of two or more elements; in water, one atom of oxygen is combined with two of hydrogen; and in calcium carbonate, one atom of carbon is combined with three of oxygen and one of calcium.

Molecules

Molecules, the smallest units of matter that have the properties of a substance, are made up of two or more atoms. The molecules of some chemical elements, such as oxygen and nitrogen, mentioned above, for example, are made up of two identical atoms. The molecules of compounds, that consist of two or more combined elements are made up two or more atoms of different elements bonded together (see Textbox 2).

The chemical and physical properties of single, uncombined atoms and those of the same atoms bonded (chemically linked) to other atoms in the molecules are basically different: single atoms are usually electrically uncharged, whereas when they are part of molecules, they bear a positive or negative electric charge of some measure. Atoms bearing an electric charge are known as *ions*. Positively charged ions, formed when atoms lose electrons, are known as *cations*. Negatively charged ions, formed when atoms gain electrons, are known as *anions*. Most atoms of such metals as copper and iron easily lose electrons when reacting chemically, and form cations. When the atoms of nonmetallic elements react, on the other hand, they generally gain electrons, and form *negative* anions. The anions formed by chlorine, known as *chloride*, and by sulfur, known as *sulfide*, are examples of common anions. Some nonmetallic elements – for example, nitrogen, carbon, phosphorus, silicon, and sulfur – form *complex anions*, which are made up of atomic groups of two, three, or more atoms. The *nitrate* anion, which is made up of one atom of nitrogen and three atoms of oxygen; the *carbonate* anion, made up of one atom of carbon and three of oxygen, and the *sulfate* ion, made up of one atom of sulfur and four of oxygen, are common examples of complex anions.

A few general rules are helpful for understanding the electric charges acquired by single as well as complex ions:

- Metals and hydrogen (a nonmetal) form positively charged cations.
- Nonmetals generally form negatively charged anions.
- Some nonmetals form complex (polyatomic) anions, which consist of a group of three or more atoms bearing a negative charge.
- Cations combine with anions to form compounds known as *salts*; when dissolved in water, the salts form solutions that conduct electricity.
- Solutions of nonionic compounds in water do not conduct electricity.
- Nonmetals can also combine with other nonmetals to form nonionic compounds (most biological substances are nonionic).

Chemical Notation and Nomenclature

An internationally accepted *chemical notation* makes use of *symbols* to represent elements and compounds, and advises on naming chemical compounds. In this notation, the elements are represented by one or two letters, many of which are drawn from the elements' Latin or Greek names. The number of atoms of an element in a molecule is represented by a subscript written after the symbol; thus Au (the first two letters of *aurum*, the Latin name for gold) represents an atom of gold; Cu (the first two letters of *cuprum*, the Latin name for copper), an atom of copper; and C (the first letter of carbon), an atom of carbon; O represents an atom of oxygen and O_2, a molecule of oxygen. The symbols listed below provide examples of the presently accepted form of chemical notation:

Ag: silver (*argentum*)
Au: gold (*aurum*)
C: carbon
Cu: copper (*cuprum*)
Fe: iron (*ferrum*)
Hg: mercury (*hydrargyrium*)
K: potassium (*kalium*)
N: nitrogen
O: oxygen
P: phosphorus
Pb: lead (*plumbum*)
S: sulfur
Sb: antimony (*stibium*)
Sn: tin (*stannum*)
Zn: zinc

The internationally accepted notation also establishes the method of expressing, with symbols, the composition of chemical compounds. These are expressed by chemical formulas consisting of the symbols of the component elements and numbers representing the relative amount of each elemental component in a molecule of a compound (Fox and Powell 2001). For any particular compound, the number of atoms of each element in a molecule of the compound is indicated by a subscript written after the symbol of the element; no subscript appears when there is only one atom of the element. Thus the formula of water, whose molecule consists of two atoms of hydrogen and one of oxygen, is H_2O, and that of carbon dioxide, whose molecule includes one atom of carbon and two of oxygen, is CO_2. The formula of glucose, $C_6H_{12}O_6$, means, therefore, that its molecule is composed of 6 atoms of carbon, 12 of hydrogen, and 6 of oxygen [since glucose belongs to the type of organic compounds known as the *carbohydrates*, which are, by definition, composed of a carbon and water; the formula for glucose is sometimes also written as $C_6(H_2O)_6$].

The chemical name of compounds composed of only two elements usually ends with the suffix *ide*. The chemical name for water, for example, which is composed of two atoms of hydrogen and one of oxygen, and whose chemical formula is H_2O, is, therefore, *hydrogen oxide*. The chemical name for common table salt, composed of one atom of sodium and one of chlorine, and has the formula is NaCl, is *sodium chloride*. Pyrite, an iron ore composed of one atom of iron (*ferrum* in Latin) and one of sulfur, has the formula FeS, and its chemical name is *ferrous sulfide*.

The name of compounds whose molecules contain three or more elements, usually including oxygen, ends in the suffix *ate*. Thus the chemical name for *limestone*, whose chemical formula is $CaCO_3$, is *calcium carbonate*. That of green vitriol, also known as *copperas*, whose formula is $FeSO_4$, is iron sulfate. The names and formulas of some chemical compounds related to archaeological studies are listed in Table 7.

TABLE 7 Common Substances and Their Chemical Formulas

Elements		Compounds	
Name	Symbol	Name	Formula
		Gas	
Oxygen	O_2	Carbon dioxide	CO_2
Nitrogen	N_2	Sulfur dioxide (an air pollutant)	SO_2
		Liquid	
Mercury	Hg	Water	H_2O
		Solid	
Copper	Cu	Iron oxide (rust)	Fe_2O_3
Iron	Fe	Lead sulfide (a lead ore)	PbS
Gold	Au	Benzenamine (magenta, a dye)	$C_{20}H_{19}N_3$
Sulfur	S	Sodium chloride (common salt)	NaCl
Phosphorus	P	Sucrose (sugar)	$C_{12}H_{22}O_{11}$

Early humans learned to recognize and appreciate some specific properties of minerals and mineraloids and, accordingly, put them to use. The hardness and the *conchoidal fracture* of flint and obsidian, for example, were recognized already during paleolithic times, and it is because of these properties that these two materials were found useful for making lithic tools. *Hematite*, a red-brown-colored mineral (composed of iron oxide), has also been appreciated since ancient times. Hematite occurs in the crust of the earth in two different forms: one consolidated and relatively hard and another powdery and loose. The consolidated, hard, and brilliant variety has been valued as a semiprecious gemstone, while powdery hematite has long been and still is widely used as a pigment for the preparation of paints, as well as a mild abrasive, for smoothing and polishing the surface of solid objects. Moreover, since the discovery of the smelting of iron, hematite, in all its varieties, has been used as a source from which the metal iron is extracted (see Chapter 5). Hematite is also just one example of a wide range of minerals that have been and are still used as mineral ores, raw materials from which metals are extracted (see Chapter 5). Other minerals, such as *clays* and *sand*, have been and still are used for making artificial (synthetic), human-made materials: *clays* for making pottery and *sand* for making glass. Table 4 lists minerals and mineraloids that, because of some specific property, have been used for many millennia for some specific applications.

1.3. ROCK AND STONE

Rock is any consolidated, compact, hard or soft natural material consisting of one or more conglomerated minerals. The rocks that make up the crust of the earth consist of either single minerals or natural cohesive aggregations of grains of minerals held more or less firmly together into a solid mass. Many rocks are mixtures of two or more different minerals and are, therefore, heterogeneous materials, while some, composed of only one mineral, and therefore known as *monomineral rocks*, are homogeneous materials. Examples of the latter are *marble*, which consists entirely of calcite (composed of calcium carbonate), and *alabaster*, which consists of consolidated plaster of Paris (composed of calcium sulfate).

TEXTBOX 7

MINERAL- AND ROCK-FORMING PROCESSES – CRYSTALLIZATION AND PRECIPITATION

Most natural solids, including minerals, are formed by one of two very different processes: *crystallization*, which takes place when a hot melt cools down slowly, or *precipitation*, which occurs mainly when a solid separates from a water solution.

Crystallization

When a hot melt cools down, the particles that make up the melt (that is to day, atoms, ions, or molecules) become rearranged, during the transition from the liquid to the solid state, into a regular, symmetric spatial arrangement, known as a *crystal structure*, which is regularly repeated throughout the bulk of the solid (see Textbox 21). Molten *magma*, for example, is a very hot, fluid mixture of liquid and gas formed deep below the crust of the earth. When magma emerges to the surface during volcanic eruptions (then known as *lava*), it cools down gradually. Depending on the chemical composition of the lava and the prevailing environmental conditions, and providing it cools down slowly, the atoms that make up lava become naturally grouped in orderly and symmetric patterns. As the melt finally solidifies, it is said to *crystallize*, the solid preserves the orderly and symmetric arrangement, known as the *crystal structure*, acquired while solidifying. The *crystal structure* of each substance is unique and characteristic, and provides a relatively easy way for characterizing minerals; the internal structure of several crystalline substances is shown in Figure 17.

When molten lava cools down rapidly, however, the orderly arrangement of the component particles into symmetric structures is curbed and the melt solidifies into an amorphous mass that lacks internal order. Such a substance is known as a *mineraloid*. Although mineraloids have a mineral appearance, they lack an internal orderly arrangement and are said to be *natural glasses* (see Textbox 27). *Obsidian*, widely used since prehistoric times for making tools and decorative objects, is a type of mineraloid. Another well-known mineraloid is *opal*, certain varieties of which have long been appreciated as gemstones (Middlemost 1985).

Precipitation

Precipitation, the other process by which minerals and rocks are formed, takes place from a solution. When solid particles separate out from a solution as the water evaporates, or as a consequence of cooling or of the biological activity of aquatic plants and animals, they are said to *precipitate*. The extremely small particles that precipitate bond with one another, forming larger particles that sink and create unconsolidated accumulations known as *sediments*. In time, as more particles precipitate, layers of sediments are formed. If this process continues over periods of time lasting millions of years, more layers accumulate; the higher-lying layers press down and compact the lower ones, a process that ultimately results in the formation of *sedimentary rocks*. Limestone (composed of calcium carbonate) and gypsum (composed of calcium sulfate) are examples of sedimentary rocks formed as a result of precipitation processes (Collinson and Thompson 1989; Ehlers and Blatt 1982).

Recrystallization

Extremely high pressures, shearing stresses, and/or high temperatures, although not sufficiently high to cause melting, may affect the internal structure of minerals and rocks and transform them into *metamorphic rocks*. The transformation (into metamorphic rock), known as *recrystallization*, a process that is seldom accompanied by changes in composition, but takes place merely through the formation of new crystalline structures and new textures (Herz and Garrison 1998; Raymond 1995).

Mineral or rock that has been naturally or artificially broken, cut, or otherwise shaped to serve some human purpose is known as *stone*. For millions of years, long before the discovery of the metals, humans used stone for making tools; erecting houses and monuments; building roads, floors, and roofs; sculpting statuary; and carving ornaments. It is not surprising, therefore, that stone is one of the most frequently encountered materials in archaeological excavations and one of the most often studied (Waelkens et al. 1992; Shadmon 1996).

1.4. STUDY OF ARCHAEOLOGICAL STONE

The study of archaeological stone is based mainly on chemical, physical, geologic, and mineralogic investigation; this entails identifying the composition of the stone, characterizing its structure and other physical properties, and elucidating the changes that it may have undergone since it was created or last used (Jeffery and Hutchison 1981). Additional, more specific studies of stone, however, are often also conducted in the framework of *petrology*, the science that deals with the origin, composition, and structure of rocks; and *petrography*, the classification and systematic description of rocks (Kempe and Harvey 1983). The main objective of analyzing archaeological stone, as well as most other archaeological materials, is usually to identify the rock unequivocally. This is not, however, the only objective, since the results of chemical analyses often also provide insights into any weathering processes that the stone may have undergone (see Textbox 45), as well as its provenance and chronology.

The Components of Materials. The composition of most materials – whether of natural origin, such as minerals, rocks, wood, and skin, or made by humans, as for example, pottery, glass and alloys – includes several kinds of components: *major*, *minor*, and *trace* elements (see Textbox 8).

TEXTBOX 8

COMPOSITION OF MATERIALS: MAJOR, MINOR, AND TRACE COMPONENTS

Substances prepared under carefully controlled conditions and using very pure chemicals, in a modern laboratory, for example, contain only the basic component elements, those that determine the actual composition and nature of the substances. Natural substances, whether of mineral or biological origin, and also most synthetic (human-made) substances contain, in addition to their main components, *impurities* foreign to their basic composition. Most impurities usually enter substances such as minerals, for example, in relatively small amounts, when the substances are created. Others, such as those in some rocks and the wood of trees, do so in the course of their existence. Once within a substance, impurities become an integral part of the host substance and impair the purity of the substance. Although they alter the actual composition of substances, impurities do not affect their basic properties.

Thus, three types of components can be distinguished in most substances, whether of natural origin or made by humans: *major*, *minor*, and *trace components* (see Table 8). The *major components*, also known as the *main* or *matrix components*, are those that determine the chemical nature and properties of a substance. The major components occur in the substance in high concentration, generally exceeding 1% of the total weight. In minerals and biological substances, for example, the major components are those that appear in the chemical formula that expresses their composition.

TABLE 8 Components of Most Natural Materials

Component	Relative concentration (%)
Major (matrix)	above 1
Minor	1.0 – 0.1
Trace	below 0.1[a]

[a]Parts per million (ppm) or parts per billion (ppb).

In practically all natural and in most synthetic substances there are, mixed with the major components, impurities in minor and trace amounts. *Minor components* occur in concentrations below 1% and down to about 0.1% of the total weight of a sample of the substance. Many additional impurities, usually referred to as *trace components* or *trace elements*, occur in host substances at extremely low concentration, generally below 0.1%; their concentration is generally expressed either as parts per million (ppm) or parts per billion (ppb) (1 ppm is equivalent to a one gram in one ton; 1 ppb, to one gram in one million tons). Minor and trace impurities do not alter the basic composition, nor do they affect most of the properties of substances, but they may change, even drastically, some of their physical properties. Trace impurities in otherwise colorless minerals, for example, often make the minerals highly colored.

A colorless mineral known as *corundum* (composed of aluminum oxide) is colorless. A red variety of corundum known as *ruby*, a precious stone, owes its color to impurities of *chromium* within the crystal structure of corundum. Blue and violet varieties of corundum are classified as *sapphires*, the blue being the result of iron and titanium impurities, and the violet of vanadium impurities within the corundum crystal structure. Another colorless mineral is *beryl* (composed of beryllium aluminum silicate); but blue *aquamarine*, green *emerald*, and pink *morganite*, are precious varieties of beryl including different impurities: aquamarine includes iron, emerald chromium and vanadium, and morganite manganese.

Impurities and Provenance

The nature and the relative amounts of the impurities in many natural and in some synthetic materials, are often characteristic of the geographic area where the materials occur or were made. This is of particular interest in archaeological studies, since determining the nature and the relative amounts of impurities in many materials allows one to determine their *provenance* (Maniatis 2004; Guerra and Calligaro 2003).

In any specific material, the *major* or *matrix* components (those that define and characterize the material) are always the same, regardless of how and where the material was created or where it is located. The matrix components of most sand, for example, are the same everywhere on earth, since sand consists of *silicon dioxide,* a compound in which silicon is bonded to oxygen in a well-known weight ratio: one part of silicon to two parts of oxygen. Pure silicon dioxide, however, is a colorless substance, while most sand on the surface of the earth or on the bottom of seas and oceans is colored. Its color is due to *impurities,* elements foreign to silicon dioxide, the main component of sand, that penetrated the silicon dioxide matrix in relatively small amounts, as minor or trace elements, when the sand was created. Much sand, for example, is yellow or red, colors that are due to small amounts of iron oxide within the silicon dioxide matrix; the color of black sand, on the other hand, is caused by minor or trace amounts of manganese dioxide within the silicon dioxide. It should be mentioned here that iron and manganese oxides occur as impurities not only in sand but also in many other natural materials such as clays and soils and also endow these materials with their characteristic (mostly dark brown) color. Moreover, iron impurities in clay and sand cause the characteristic color of human-made pottery and glass: red-brown in pottery and green in glass. Like sand, most rocks contain impurities, which may vary in number, from only a few to several dozens, and in concentration, from as little as a few parts per billion to as much as 1%. It is through the investigation of the nature, relative abundance, and properties of many minor and particularly trace elements that the *provenance* and *chronology* of archaeological rock and stone, as well as of other archaeological materials, can be studied.

1.5. CHEMICAL ANALYSIS OF ARCHAEOLOGICAL MATERIALS

Sampling archaeological materials for analytical purposes may sometimes be the most difficult stage in an analytical procedure (Bellhouse 1980; Cochran 1977). Since rock, ceramics, and cement are heterogeneous materials, obtaining a representative sample of them may be the most difficult step in a whole analytical procedure.

TEXTBOX 9

ANALYTICAL CHEMISTRY

Analytical chemistry is the branch of chemistry that deals with the composition of materials: whether they are single substances or mixtures, what their components are, and how much of each component they contain. Since it provides basic information on the nature and the composition of materials, analytical chemistry is of extreme importance in the study of archaeological materials (Pollard et al. 2007). Analyzing metallic objects, for example, reveals whether they are made of metals or alloys and which are the elemental components of the latter (see Chapter 5). Analyzing the composition of soils discloses their nature and may reveal whether and how human inhabitation affected their composition (see Chapter 8) (Ciliberto 2000; Pernicka 1996). Since the composition of materials quite often bears a relationship to their place of origin, chemical analysis also makes it possible to differentiate between materials from different regions. Moreover, as it discloses information on the components of materials from which objects are made, the analysis of archaeological materials provides clues for establishing their *provenance* and, at times, for differentiating between genuine objects and suspected (and even unsuspected) fakes or forgeries.

The nature and the relative amounts in which the components of materials have to be detected in different analytical studies varies greatly: from the identification and determination of the few major elements that make up a material, to the wide range, often in almost vanishing concentrations, of impurities. From a practical point of view and regardless of the objective of, or the type of information required from an analysis, most analytical procedures entail a sequence of three main operations:

1. *Sampling* a material
2. *Analyzing* and determining the nature of the constituents of the sample and their relative amounts
3. *Interpreting* the analytical results

Sampling

Sampling is the process of obtaining a small and representative portion from the relatively large amount of material that makes up an object or a feature in an object (Stoeppler 1997; Cochran 1977). Sampling homogeneous materials generally presents no particular problems. It merely entails separating a small portion from the whole. Sampling heterogeneous materials is, however, a complex task, at times the most difficult stage of an analytical procedure. Two simple guidelines need to be observed to ascertain that a sample from a heterogeneous material is unquestionably representative of the whole:

1. The sample should be as large as possible.
2. If the material sampled is made up of particles of different sizes, the large and small particles should be sampled separately and in the respective proportions in which they are present in the bulk. Moreover, each sample should contain as many particles as possible.

Sample Preparation A sample submitted for chemical examination is often in a form unsuitable for analysis. Generally it requires a process of preparation so as to bring it into such a condition that its characteristic chemical and physical properties and those of its constituents can be easily identified and evaluated. During the preparatory stage a solid sample may, for example, be dissolved or converted into a liquid or gas. Moreover, some of its components, which may be irrelevant for the purpose of the analysis and might interfere with an analytical procedure, are generally removed during this stage.

Analysis

A wide range of analytical techniques are today available for identifying and characterizing materials (Hancock 2000). Some, known as *qualitative techniques*, are designed to provide information only on the nature of the components of materials, that is, which components, elements, and/or compounds, make up a material (Masterton and Slowinski 1986). Most often, however, it is also essential to disclose precisely how much of each particular component there is in a material, and thus to reveal its exact composition. Such information is derived using *quantitative techniques* (Harris 2002; Jeffery et al. 1989).

The actual procedures by which the nature or the constituents of a material are qualitatively identified, quantitatively appraised, or both, constitutes the heart of the analysis; these are based, essentially, on identifying and measuring the magnitude of some property or properties of an element or compound in the material analyzed that makes possible their unequivocal identification. Fortunately, almost every element and compound has some characteristic chemical and/or physical properties that, under particular conditions, allow it to be distinguished from all others. This property may be a unique chemical reaction, a change in weight, volume, color, or wavelength of radiation absorbed or emitted, or any one of many others. What is important is that the property selected should be related, in a clear and unique way, to only one of the elements or compounds being analyzed, so as to make it positively identifiable. For

TABLE 9 Methods of Chemical Analysis and Their Uses

	Components or characteristics determined					
Analytical method	Single element	Multiple elements	Molecules	Surface composition	Crystal structure	Other
Gravimetry	✓	✓	–	–	–	–
Volumetry	✓	✓	–	–	–	–
Chromatography	–	–	✓	–	–	Separation
Electrochemistry	✓	✓	✓	–	–	✓
Electron probe microanalysis	✓	✓	–	✓	–	–
Mass spectrometry (MS)	✓	✓	✓	–	–	–
Nuclear activation analysis	✓	✓	–	–	–	–
Spectroscopy						
Emission spectroscopy	✓	✓	–	–	–	–
Absorption spectroscopy						
Infrared	–	–	+	+	✓	–
Fourier transform infrared	–	–	✓	–	–	–
Raman	–	–	✓	–	–	–
Ultraviolet	✓	✓	–	–	–	–
Visible	✓	✓	–	–	–	–
Auger spectroscopy	✓	✓	–	✓	–	–
Thermal methods	–	–	✓	✓	–	–
X-ray technology						
X-ray diffraction	–	–	✓	–	✓	–

historical reasons, the methods and techniques used in analytical chemistry are classified into two main groups: *chemical* (or *classical*) analysis and *physical* (or *instrumental*) analysis.

Chemical Methods of Analysis The *chemical methods* of analysis are those in which, in order to identify and/or measure the relative amounts of the constituents of a sample, use is made of characteristic and unique chemical reactions that alter the sample in a specific and unique way. Changes in the weight of the sample, for example, are studied using *gravimetric techniques*, while changes in the volume of solutions in which the sample is dissolved are studied using *volumetric techniques* (Harris 2002). When using *gravimetry*, samples are converted, by precipitation or combustion, to pure elements or compounds that, when weighed, provide an exact measure of the relative amount of each component in the sample. *Volumetry*, on the other hand, requires that samples be first dissolved in a suitable solvent so as to form solutions of known volume in which their components are dissolved in unknown concentrations. The concentration and, therefore, the relative amounts of these components in the sample are then determined by adding and accurately measuring the volume of solutions of specific reagents in precisely known concentrations until the reaction is complete. The relative amount of a component of a sample can then be calculated from the measured volume of solution added.

Physical Methods of Analysis The *physical methods* of analysis are based on measuring some specific physical property of a component of a sample that serves to ascertain its nature and/or its the relative amount in the sample. Such methods, most of which have been made possible only by modern developments in physics and electronics, make up the main source of analytical information on all substances at the present time (see Table 9 and Textbox 10) (Settle 1997).

The choice of the most appropriate technique for a particular analysis is determined by a number of considerations, the most important of which is the nature of the material to be analyzed. The size of the sample available, the accuracy required, the number of samples involved, and the speed with which results are required, however, also warrant attention. Sometimes, priorities may have to be established before deciding on the most appropriate method for the analysis of a particular sample.

TEXTBOX 10

PHYSICAL METHODS OF ANALYSIS FREQUENTLY USED IN ARCHAEOLOGICAL STUDIES

Most of the essential information on archaeological materials is derived, at the present time, using *physical methods* of analysis. This may include the qualitative or quantitative assessment of their composition, their provenance, the techniques used for their production, and their age. Some of the most widely used methods of chemical analysis based on physical principles are succinctly reviewed in the following paragraphs.

Spectroscopic Analytical Methods

Spectroscopy, the study of the interaction between matter and electromagnetic radiation, provides a wide range of analytical methods. The energy of all atoms and molecules is discrete (is said to be *quantized*); when atoms or molecules interact with (absorb or emit) radiation, a type of energy, they do so in a distinct manner characteristic of a given composition. This means that when they absorb or emit electromagnetic radiation, each type of atom and molecule does so at well-defined wavelengths. Detecting and measuring the spectroscopic characteristic of material provides, therefore, a powerful tool for the study of their composition and structure. Since the form of interaction of different types (infrared, visible, ultraviolet, X rays) of electromagnetic radiation with materials is different, spectroscopic studies also yield information on a variety of other characteristics of the materials (Davies and Creaser 1991; Creaser and Davies 1988).

Infrared spectroscopy, for example, studies the interaction of materials with infrared radiation and reveals information on their nature: detecting the wavelength at which a material absorbs infrared radiation and measuring the relative drop in the intensity of each wavelength absorbed provides information about their composition, the nature of the component molecules, and the concentration of particular molecules in a material or solution (Stuart et al. 1996). Another spectroscopic technique, known as *Raman spectroscopy*, is based on the scattering, by some substances of monochromatic light, mostly in the infrared, but also in the visible or near-ultraviolet ranges of the electromagnetic spectrum. Raman spectroscopy reveals information on the composition and structure of molecules in materials (Ferraro and Nakamoto 1994).

Atomic absorption spectroscopy and *atomic emission spectroscopy* are analytical techniques in which the wavelength of radiation absorbed or emitted by substances reveals information on the nature of the constituent atoms, while the intensity of the radiation absorbed or emitted is indicative of the relative amounts of specific atoms in a substance. *Atomic absorption spectroscopy* is based on the detection of the specific wavelengths of light depleted by atoms in a material from a continuous source of energy. *Atomic emission spectroscopy* rests on the detection of the characteristic wavelengths of the light emitted by atoms heated to high temperature (Young and Pollard 2000; Metcalfe and Prichard 1987).

Mössbauer spectroscopy is an analytical technique that, in archaeological ceramic studies, provides information on the condition and characteristics of the compounds of iron in pottery. Using the technique makes it possible to determine the relative amounts of the different (ferrous and ferric) ions of iron and hence to ascertain the firing conditions of the pottery at the time it was made. The technique involves irradiating a sample of pottery with gamma rays and then assessing the amount of radiation absorbed by the nuclei of the ions of iron within the pottery (Feathers et al. 1998; Béarat and Pradell 1997).

Nuclear magnetic resonance spectroscopy is a technique that, based on the magnetic properties of nuclei, reveals information on the position of specific atoms within molecules. Other spectroscopic methods are based on the detection of fluorescence and phosphorescence (forms of light emission due to the selective excitation of atoms by previously absorbed electromagnetic radiation, rather than to the temperature of the emitter) to unveil information about the nature and the relative amount specific atoms in matter.

X-rays Technology

Analytical *X-rays* techniques are used to characterize solids in a number of ways. X-rays penetrate solids opaque (impenetrable) to visible light but are slightly attenuated by matter. Irradiating a solid object with *X-rays* in *radiographic* equipment yields, therefore, an image, known as a *radiograph* of the internal structure of the object (Lang and Middleton 1997). Other X-rays-based analytical techniques frequently used in archaeological research include *X-rays fluorescence* and *X-rays diffraction* instruments. *X-rays fluorescence* (XRF) identifies the elemental constituents of a solid by the wavelength of electromagnetic radiation emitted by the atoms in the substance when excited by X-rays, while *X-rays diffraction* reveals the internal morphology as well as the atomic and molecular structure of substances (Jenkins 1999; Cullity 2001).

Spectroscopic analysis, in all its forms, is being strongly affected by the development of *lasers*, *microtechnology*, and *computers*. The high radiation intensities emitted by small lasers and the reduction in the size of samples made possible by microtechnological developments are being combined with the working proficiency of computers; such a combination provides new, small, portable, and extremely powerful instruments that can be taken to the field, and there yield highly sensitive analytical results, sometimes down to the level of detecting even single atoms and molecules.

Neutron Activation Analysis

Neutron activation analysis (NAA) is a technique for the qualitative and/or quantitative determination of atoms possessing certain types of nuclei. Bombarding a sample with neutrons transforms some stable isotopes into radioactive isotopes; measuring the energy and/or intensity of the gamma rays emitted from the radioactive isotopes created as a result of the irradiation reveals information on the nature of the elements in the sample. NAA is widely used to characterize such archaeological materials as pottery, obsidian, chert, basalt, and limestone (Keisch 2003).

Mass Spectrometry (MS)

In *mass spectroscopy*, sample molecules are ionized and the different masses of the ions formed are selected by use of an electric or magnetic field. In its simplest form, a *mass spectrometer* is an instrument that measures the mass-to-electric charge ratios of ions formed when a sample is ionized. If some of the sample molecules are singly ionized and reach the ion detector without fragmenting, then the mass-to-electric charge ratio of the ions gives a direct measurement of the weight of the molecule (de Hoffmann and Stroobant 2001).

Microscopy

The microscope is arguably the most widely used scientific instrument. Microscopes are used routinely in analytical chemistry and materials science, mineral exploration, and environmental science. The three most common types of microscopes, the *light microscope*, the *electron microscope*, and the *scanning microscope* are essential tools for the measurement of properties or for observing and measuring such qualities as the size and shape of particles, the texture and chemical composition of materials, and such physical properties as color, crystallinity, melting point, and refractive index (Eastaugh et al. 2005; Goodhew et al. 2000).

Separation Techniques

Complex mixtures, such as those that occur in ancient human remains or as residues in ancient vessels and tools, may need to be separated into their components before they can be analyzed. *Separation techniques* are analytical techniques used for separating and sometimes even identifying the components of chemical mixtures (Setford 1994). Two widely used separation techniques are *chromatography* and *electrophoresis*. *Chromatography* is based on allowing a gaseous or liquid mixture or solution, usually known as the *analyte*, to seep through a *stationary*, usually *solid medium* that separates between the components; the separation depends on the rates at which the different components of the analyte move through or along the stationary medium. If the moving mixture consists of gases, the technique is known as *gas chromatography* (GC); if it consists of liquids or is a liquid solution, it is known as *liquid chromatography* (LC) or *high-performance liquid chromatography* (HPLC) (Setford 1994; Miller 1988).

Hyphenated Analytical Techniques

Two or more combined analytical techniques, generally called *hyphenated* or *tandem techniques*, provide more information than that obtained from single traditional physical techniques, thus yielding more reliable results (Rouessac and Rouessac 2000). Many challenging analytical problems, for example, involve more or less complex mixtures that require, as an important first step, separation of their components. Some hyphenated techniques, therefore, couple a separation technique with an analyzing technique so as to separate and analyze complex mixtures. Such is the case with gas chromatography (GC), which is coupled with another separation and analyzing technique, mass spectroscopy (MS), to yield the *gas chromatography – mass spectrometry* (GC-MS) analytical technique. The combination of the two techniques allows a much finer degree of substance identification than does either technique used separately (Kitson et al. 1996; Scott 1997).

In many instances when analyzing archaeological objects, removing a sample may be aesthetically deleterious to the object, therefore rendering the removal of samples totally inadequate, or allowing the removal of extremely small samples so as not to affect the appearance of the objects. In such cases it is often necessary to turn to specialized techniques, such as nondestructive techniques, which do not require the removal of samples altogether, or to micro analytical techniques, for which extremely small samples are needed (see Textbox 11).

TEXTBOX 11

NONDESTRUCTIVE TESTING AND MICROANALYSIS

The terms *nondestructive testing* and *microanalysis* refer to techniques for examining the internal structure and analyzing the composition of objects that, because of aesthetic or other considerations, cannot be studied by conventional techniques.

Nondestructive Testing

Techniques of *nondestructive testing* (NDT) are material appraisal techniques that neither require sampling nor physically damage or impair the integrity of objects studied. Such techniques are invaluable for studying the constitution or the internal structure of unique archaeological objects. Most modern nondestructive techniques are based on the application of forms of *penetrating energy*, such as *acoustic waves* (sound and ultrasound waves) and *electromagnetic radiation*, as, for example, X-rays and gamma rays, as listed in Table 10 (Biro 2005; Shull 2002; Mix 1987).

TABLE 10 Techniques of Nondestructive Testing (NDT)

Technique	Characteristics detected
Optical (visual)	Cracks, color, finish, scratches
Electrical	Cracks, inclusions, voids
Magnetic	Inclusions, shape of internal parts
Radiographic	Inclusions, voids, materials variations, placement of internal components
Ultrasonic	Cracks, inclusions, voids, interfaces

Microanalysis

Microanalysis is the common name used to refer to a variety of techniques for identifying, characterizing, and evaluating minute amounts of materials. Some microanalytical techniques are scaled-down versions of well-known conventional or physical analytical techniques; others are specialized techniques that can be implemented only on extremely small samples. Table 11 lists the minimum size of samples required for microanalysis and the minimum amount of substance detectable by microanalytical techniques (Janssens and Van Grieken 2004).

TABLE 11 Analytical Techniques and Size of Sample Required

Analytical method	Minimum detectable (mg)	Sample size (mg)
Macro	10	over 100
Semimicro	1	about 10
Micro	0.01	about 1
Semiultramicro	0.001 (10^{-3})	0.01–0.1
Ultramicro	0.00001 (10^{-5})	0.0001–0.001 (10^{-4}–10^{-3})
Submicro	0.0000001 (10^{-7})	0.000001–0.00001 (10^{-6}–10^{-5})

Isotopes in Archaeology

The *isotopes* of many elements (see Textbox 12) provide important tools for obtaining information on many archaeological investigations; their use may enable researchers, for example, to determine the provenance (see Textbox 30) or the age of materials (see Textboxes 15 and 16), calculating ancient temperatures (see Textbox 47), or elucidating the nature of the ancient diets of human beings as well as of animals (see Textbox 59).

TEXTBOX 12

ISOTOPES

The basic material units of each chemical element are the almost identical atoms. Each atom of an element includes equal numbers of positive and negative electrically charged particles: positively charged *protons* in the nucleus and, surrounding the nucleus, negatively charged *electrons*. The nuclei of most atoms also include a variable number of *neutrons*, elementary particles that have a definite mass but are not electrically charged. Since the neutrons have mass, the nuclei and accordingly also the atoms of an element having different numbers of neutrons, have different masses. Any two atoms of any one element (which include in their nuclei an equal number of protons) may, therefore, have a different number of neutrons. The *atomic weight* of the two atoms is consequently different. Atoms of one element that have an equal number of protons but a different number of neutrons are known as *isotopes* of the element (see Fig. 8) (Putnam 1960).

All the *isotopes* of an element have the same number of protons in their nuclei and, therefore, they also have the same *atomic number*; consequently, they are chemically identical and indistinguishable from each other (the *atomic number* of an atom is the number of protons in its nucleus and determines the chemical properties of an element). Because they have different numbers of neutrons, however, each isotope of an element has a different atomic weight and, therefore, also slightly different physical properties.

The structure of some isotopes, generally referred to as *stable isotopes*, is immutable (see Textbox 13). Others, known as *radioactive isotopes* or *radioisotopes*, are unstable: in time they undergo what is known as

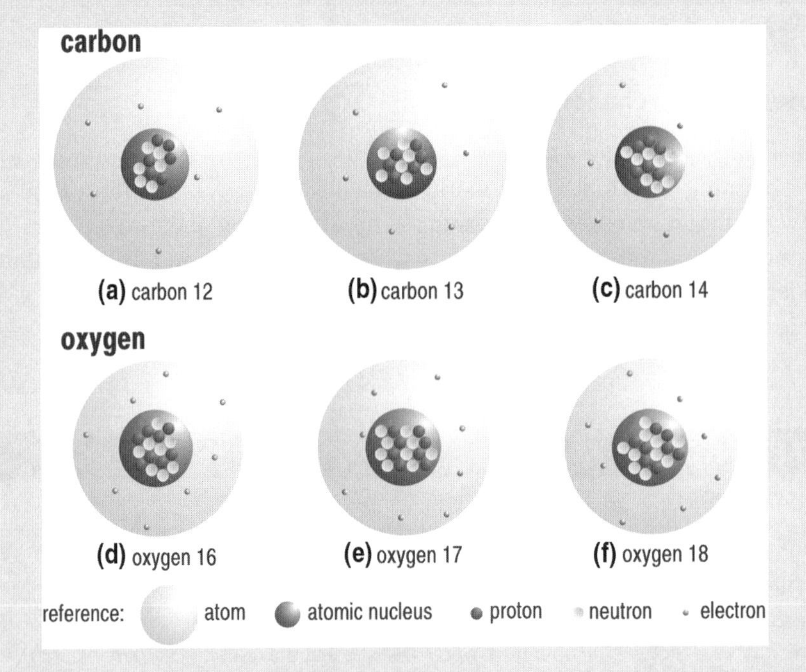

FIGURE 8 Isotopes; the isotopes of carbon and oxygen. Atoms of the same element can have different numbers of neutrons in their nuclei, and the different possible versions of each element are known as "isotopes." Carbon, for example, has three isotopes: carbon 12, carbon 13, and carbon 14. Carbon 12 **(a)**, the most abundant, has six neutrons; carbon 13 **(b)** has seven, and carbon 14 **(c)**, also known as "radiocarbon," has eight neutrons; carbon 14 is unstable and radioactive and is of use for dating archaeological finds. Oxygen has also three isotopes: oxygen 16, oxygen 17, and oxygen 18; the most common, oxygen 16 **(d)**, has eight neutrons, oxygen 17 **(e)** has nine and oxygen 18 **(f)** has ten neutrons. The weight ratio between the isotopes of oxygen in seawater, ice, and seashells, for example, provides a reliable indicator of past temperatures and climatic conditions.

radioactive decay processes, when they break down into smaller atoms. Moreover, the breakdown process is accompanied by the emission of various forms of *ionizing radiation* (see Textbox 13 and Fig. 9). Ionizing radiation can easily be detected and quantitatively measured with specialized instruments known as *radiation detectors*. Detecting and characterizing the radiation emitted by unstable isotopes makes it possible to qualitatively identify the isotopes, while measuring the intensity of the radiation makes it possible to quantitatively assess their amount.

Most naturally occurring elements are mixtures of two or more isotopes in which just one predominates. Isotopes can also be prepared

TABLE 12 Isotopes Used in Archaeological Studies

Isotope	Applications
Stable Isotopes	
Argon-40	Potassium–argon dating
Carbon-13	Provenance studies, study of ancient diets
Lead-206	Provenance studies
Lead-207	Provenance studies
Lead-208	Provenance studies
Neodymium-143	Provenance studies
Neodymium-144	Provenance studies
Nitrogen-15	Study of ancient diets
Oxygen-18	Paleotemperatures Provenance studies
Strontium-86	Provenance studies Study of ancient diets
Strontium-87	Provenance studies Study of ancient diets
Sulfur-32	Study of ancient diets
Sulfur-34	Study of ancient diets
Radioactive Isotopes	
Carbon-14	Radiocarbon dating
Cesium-137	Gamma-ray radiography
Cobalt-60	Gamma-ray radiography
Potassium-40	Potassium–argon dating
Thorium-232	Alpha recoil tracks dating
Thulium-170	Gamma-ray radiography
Uranium-234	Uranium series dating
Uranium-235	Uranium series dating
Uranium-237	Alpha recoil tracks and uranium series dating
Uranium-238	Alpha recoil tracks and uranium series dating

artificially. When some stable isotopes are irradiated with neutrons or with other energetic particles, they are converted into unstable, radioactive isotopes, some of which do not occur naturally on earth. The newly formed isotopes can then be identified and characterized using radioactivity detection methods. This has given rise to a significant branch of analytical chemistry known as *neutron activation analysis*, which is often used for the study of such materials as obsidian, glass, and pottery (see Textbox 10) (Neff 2000).

Many stable and unstable isotopes are used for studying processes relevant in archaeology (Katzenberg 2000). Carbon-14, also known as *radiocarbon*, an unstable isotope of carbon, for example, is of prime importance for dating archaeological finds (see Textbox 55). Carbon-13, a stable isotope of carbon, provides a tool for studying ancient diets (see Textbox 57). The heavy isotopes of oxygen, oxygen-17, and particularly oxygen-18 are of use for estimating past temperatures (see Textbox 47) (Wagner 2000). Isotopes frequently used in archaeological studies are listed in Table 12.

1.6. PROVENANCE OF ARCHAEOLOGICAL MATERIALS

The term *provenance*, or *provenience*, as the word is often also spelled, is used to refer to the geographic location, that is, the area or place from which a material or artifact originates. The main interest in the study of the provenance of archaeological materials and objects is twofold: determining the geographic source of archaeological materials, and studying the eventual distribution of such material to other areas and the routes through which they were traded (Hughes 1991). The study of provenance is based on the assumption that most materials have some detectable compositional characteristics that are uniquely related to their place of origin and particularly to the nature and the relative amount of the minor and trace elements at the place, which generally vary from place to place. To ascertain whether the composition of a material may provide relevant information on its provenance, it is necessary to ascertain three basic conditions:

1. Different samples of the material from a single source are of uniform composition.
2. There are compositional differences between samples of the same material from different sources.
3. Such differences can be unequivocally recognized and distinguished.

To be of value for reliable differentiation, such compositional differences should be greater between samples from different sources than between different samples from a single source. Most archaeological provenance studies have so far been related to stone, ceramics, glass, and metals. Compositional evidence on stone was used, for example, to prove that in neolithic times, large stones from Pembroke in Wales, in the British Isles, were transported a distance of about 200 miles to Wiltshire in England, where they were used to build a megalithic ring at Stonehenge (Newall 1959).

Some studies on the provenance of stone have been based on determining the weight ratios between different minor or trace elements or on the relative abundance and weight ratios of the stable isotopes of such elements as carbon and oxygen. Thus have been studied the provenance of some types of rock (Giauque et al. 1993; Waelkens et al. 1992), marble (Craig and Craig 1972), and pottery (Perlman and Yellin 1980; Millett and Cattling 1967). Establishing the provenance of metals presents a most difficult problem. With the exception of the native metals, that is, gold and silver, limited amounts of copper, and very little iron, most metals do not occur naturally as free metals but are combined with others, mainly as metal ores from which they have to be extracted by *metal smelting* procedures. Unlike stone, which is used "as is," metal ores undergo, during *metallurgical operations*, drastic chemical changes. As a consequence of these changes, there is usually little or no compositional correlation between the minor or trace element in mineral ores and in the metals derived from them. On occasion, however, the relative amount of the minor or trace elements in the ore and in the metal smelted from it bear some relationship and make it possible to elucidate the provenance of the metal (Cherry and Knapp 1991; La Niece 1983).

1.7. CHRONOLOGY OF ARCHAEOLOGICAL MATERIALS

Human evolution can be understood properly only if the time element is considered; establishing the chronology of past events and of ancient objects has been a preoccupation of historians since history was first recorded. In archaeology, in particular, dating ancient artifacts and structures is of fundamental importance, and an endless search continues for methods to accurately date material remains: to establish the time of occurrence as well as the chronological succession of past events. It has been only over the period of the last 5000 years (however, this date varies for different geographic regions on the surface of the planet) that written chronological data have become available. The nonliterate past of humans (*Homo sapiens*) left no records of the time of occurrence of events or the manufacture of objects during prehistoric times. The lack of chronological information led to the development of a number of methods for ascertaining, independently of written records, the age of materials and ancient objects and for establishing their chronological sequence (Hedges 2001; Taylor and Aitken 1997).

Chronology may be expressed in two forms: *absolute* and *relative chronology*. *Absolute chronology* depends on knowing the precise date of events or age of materials or objects. To sequence events or objects in absolute chronology implies organizing them in a certain order, for example, from oldest to most recent. *Relative chronology* is not expressed in specific dates but rather in a sequential relationship of events. *Relative dates* are therefore expressed in relation to those of other objects or events; they do not specify the precise age of objects or the time of occurrence of events. *Absolute dates* provide the precise age of objects and the time of occurrence of events, and are derived independently of those of other objects or events.

One method most widely used in archaeology for determining relative dates and elucidating chronology is *stratigraphy*, the study of stratified sediments and ancient remains. In any stratified sequence, higher strata were deposited later than lower ones. Provided there has been no disturbance of a stratigraphic sequence, therefore, remains in the higher strata are younger than those in the lower ones. This also determines the chronological sequence of objects buried in the different strata and therefore their relative chronology (Harris 1997). Another method for determining the relative date of tools is *seriation typology*, in which specific types of objects are classified according to the way in which some specific characteristic (e.g., their shape) undergoes successive gradual changes during the passing of time. Typological resemblance between objects may indicate that they are contemporary, although differences do not necessarily imply that they are not. Up to the mid-twentieth century serial types were usually constructed using a combination of empirical observation and intuition. Since then a mathematical approach and, since the mid-1990s, phylogenetic methods, borrowed from evolutionary biology, have been applied for the study of typology-based chronology (O'Brien and Lyman 1999; Dunnell 1996, 1971).

Already during the late nineteenth century a search began for techniques of dating involving the chemical and/or physical properties of materials (Carnot 1893; Middleton 1845). Some techniques, such as the *fluoridation of bone* (see Textbox 69), *geomorphology*, and *palaeomagnetism*, have been used for quite some time. Today, most dating methods are *radiometric*, based on the *half-lives* of *radioactive isotopes*, or the rate of cumulative changes to matter caused by *radioactive decay* processes. The possibility of determining *absolute dates* became possible only after the discovery of *radioactivity* and the process of *radioactive decay*, particularly since the 1950s, following the development of *nuclear* techniques. Since then, a wide and still increasing number of scientific dating methods have been and still are being developed (see Fig. 8) (Schwarcz 2002; Taylor and Aitken 1997).

TEXTBOX 13

RADIOACTIVITY

The *stability* of atoms – their property of being steadfast and remaining unchanged – is determined by the nature of their nuclei (see Textbox 12). Nuclei in which the number of neutrons is smaller than or equal to the number of protons are stable, while those in which the number of neutrons is larger than the number of protons are unstable. *Unstable nuclei* have a tendency to adjust the disparity between the number of neutrons and protons and become stable; they may do so by one of two processes, by *radioactive decay* or *nuclear fission*.

Radioisotopes and Radioactive Decay

Radioisotopes may occur in the earth naturally as *primordial radioisotopes*, formed when the planet was created, or be produced by natural or artificial processes. Most fast decaying primordial radioisotopes have long disappeared from the planet; since the earth originated about 4.5 billion years ago, such isotopes have decayed and reached a final, stable form. The relatively few primordial radioisotopes still extant in the earth today, therefore, decay very slowly. Among these are *potassium-40* and some isotopes of uranium, such as *uranium-235* and *uranium-238*, which are of use for dating archaeologically related minerals and rocks (see Textboxes 15 and 16).

Some radioisotopes are continuously being produced by the bombardment of atoms on the surface of the earth or in its atmosphere with extraterrestrial particles or radiation. One of these is *carbon-14*, also known as *radiocarbon*, which is widely used for dating archaeological materials (see Textbox 55). Many radioisotopes that are not primordial or are not created by natural processes are now produced artificially using specialized equipment; many of the "artificial" isotopes are of use for probing and analyzing materials.

The process of *radioactive decay* (also known as *radioactivity*) involves the ejection from a nucleus of one or more nuclear particles and ionizing radiation. *Nuclear fission* is a reaction in which the nucleus splits into smaller nuclei, with the simultaneous release of energy. Most radioisotopes undergo radioactive decay processes and are converted into different smaller atoms.

Radioactive Radiation

The *radiation* emitted when radioisotopes decay may be of one or more of three different types: alpha and beta radiation, which are streams of energetic particles, and gamma rays. *Alpha radiation* is a stream of bundled, positively charged particles known collectively as *alpha particles*; each alpha particle has a mass and charge equal in magnitude to two protons together with two neutrons. Alpha radiation travels very short distances (only a few centimeters) in air, as the particles lose their energy as soon as they collide with anything; they are therefore easily shielded by a sheet of paper or by human skin (see Fig. 9).

Beta radiation is a stream of negatively charged particles, known as *beta particles*, which have the same mass and electric charge as the electrons. Beta radiation travels in the air longer distances than alpha

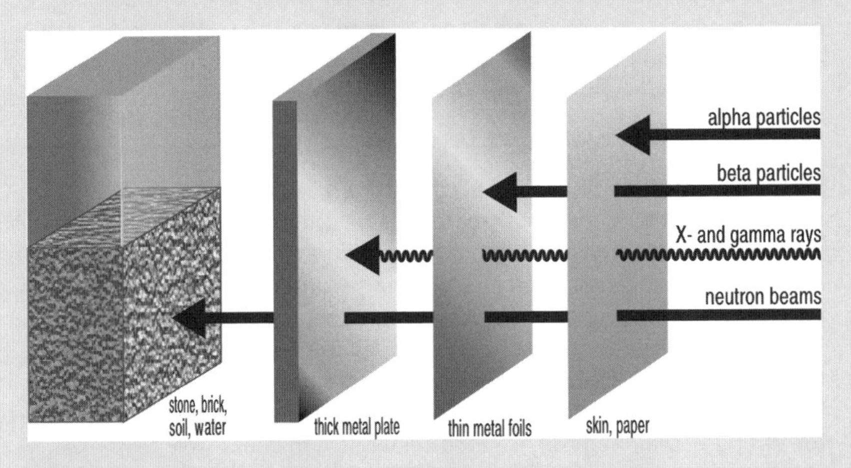

FIGURE 9 Penetrating radiation. Ionizing radiation (see Fig. 8) has sufficient energy to penetrate and interact with some types of matter and is therefore also known as "penetrating radiation." Gamma and X-rays, for example, travel many meters in air and readily penetrate thick layers of many materials, such as rock and metal. By extension, the term "penetrating radiation" is also used to refer to streams of energetic particles that penetrate and interact with matter. Streams of alpha particles emitted from some atomic nuclei undergoing radioactive decay, for example, known as "alpha radiation," have a very short range of penetration; alpha radiation travels only a few centimeters in air, but does not penetrate human skin or clothing. Streams of beta particles, which may travel several meters in air, are known as "beta radiation;" beta radiation penetrates human skin, other animal and vegetable tissues, and thin metal sheets, but not thick metal plates. Neutron beams, streams of uncharged nuclear particles, are the most penetrating form of particles. Neutrons (which can be generated with special equipment) penetrate very thick layers of most materials.

radiation (about one meter), passes through a sheet of paper, but is blocked by thin metal foils. Beta radiation is sometimes used to study archaeological problems as for example, to measure the thickness of gilding layers, to determine the relative amount of lead in lead glass, and to differentiate between different types of glass.

Gamma rays have no mass and no electric charge. They are highly energetic forms of electromagnetic radiation, similar to visible light or X-rays, but having very short wavelengths, the shortest wavelengths in the whole spectrum of electromagnetic radiation (see Textbox 5). Because of their high energy, gamma rays penetrate any type of matter, whether gaseous, liquid, or solid.

The wavelength of the gamma rays emitted by a radioisotope is unique and characteristic of the radioisotope; this makes gamma rays extremely useful for identifying radioisotopes. In *neutron activation analysis*, for example, gamma rays are used to characterize and determine the relative amounts of different elements in a sample (see Textbox 10). The attribute to penetrate solid matter makes gamma rays also a suitable complement to X-rays for radiographic purposes. Gamma rays emitted from a small sample of a radioisotope, for example, can be used for radiographing objects that would be difficult and often impossible to radiograph with X-rays (Hinsley 1959).

Radioactive isotopes that decay by the emission of alpha or beta radiation undergo a change in the nature of their nuclei and are converted into isotopes of other elements. The emission of gamma rays, on the other hand, does not change the nature of the nuclei of the radioisotopes from which the rays are emitted. Gamma rays are a form of dissipation of nuclear energy.

Some radioisotopes decay emitting only gamma rays, but many do so by the concurrent emission of beta and gamma radiation. The rate at which radiation is emitted from the nuclei of different radioisotopes varies considerably. Each radioisotope has a unique form of decay that is characterized by its *half-life* ($t^{1/2}$), the time it takes for the radioactivity of the radioisotope to decrease by one-half of its original value (see Textbox 14).

Radioactive decay is a stochastic process that occurs at random in a large number of atoms of an isotope (see Textbox 13). The exact time when any particular atom decayed or will decay can be neither established nor predicted. The average rate of decay of any radioactive isotope is, however, constant and predictable. It is usually expressed in terms of a *half-life*, the amount of time it takes for half of the atoms in a sample of a radioactive isotope to decay to a stable form.

TEXTBOX 14

HALF-LIFE

The concept of *half-life* applies to processes involving changes in which the rate of change of some quantity in a system depends on this quantity. The half-life is the time required, under specific conditions, for a quantity to decrease by one-half. After one half-life has elapsed, only 50% of the original quantity remains. During the second half-life, one-half of the remaining quantity vanishes, leaving one-quarter (25%) of the original. One-eighth (12.5%) is left after the third half-life, and so on. The decrease in the quantity follows a geometric regression, being continuously reduced by one-half, as illustrated in Figure 10. The half-life of specific materials is of use for dating the materials.

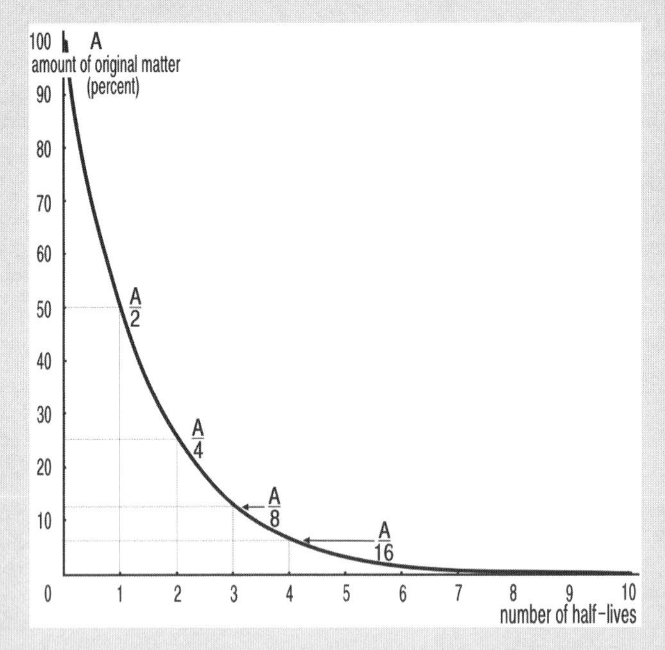

FIGURE 10 The half-life. It is impossible to predict when a radioisotope or an unstable substance (molecule) will decay or be decomposed. On an average, however, only half of any type of radioisotope or unstable substance (molecule) remains after one half-life ($A/2$); one-quarter will remain after two half-lives ($A/4$), one-eighth after three half-lives ($A/8$), and so on. The half-life is characteristic of every radioisotope and unstable molecule; that of radioisotopes is not affected in any way by the physical or chemical conditions to which the radioisotope may be subjected. Not so the half-life of chemically unstable molecules, which is altered by changes in temperature and by other physical and chemical conditions.

Isotopic Half-life

The half-life (t½) of a radioisotope is the amount of time it takes for that isotope to undergo radioactive decay and be converted into another. It is also a measure of the stability of the isotope: the shorter its half-life, the less stable the isotope. The half-life of radioisotopes ranges from fractions of a second for the most unstable to billions of years for isotopes that are only weakly radioactive. In the case of radiocarbon (carbon-14), for example, the half-life is 5730 years (see Fig. 61).

Chemical Half-life

The concept of half-life also applies to chemical reactions. The half-life of a chemical reaction is the time it takes for the amount of one of the reactants to be reduced by half. In some reactions the reaction rate is determined by the concentration of one particular reactant; as the reaction proceeds and the concentration of this reactant decreases, so does the rate of the reaction. This is the case for example, with amino acids, the components of proteins. *Amino acids* may occur in one of two different forms, the *l* and *d forms* (see Textbox 24). In living organisms, however, the amino acids occur only in the *l* form. After organisms die, the amino acids in the dead remains racemize and are gradually converted into the *d* form. Ultimately, the remaining amino acid, which is then known as a *racemic mixture*, consists of a mixture of 50% of the *l* form and 50% of the *d* form.

If the environmental temperature is constant, the racemization process takes place at a uniform rate, which is determined, at any time during the process, by the relative amounts of *l* and *d* forms of the amino acid can be measured. As the racemization proceeds and the concentration of the *l* form amino acid decreases, the rate of racemization gradually slows down. When there is a mixture of 50% of each of the *d* and *l* forms, the racemization process stops altogether. The half-life of the racemization of *aspartic acid*, for example, a common amino acid in proteins, at 20°C is about 20,000 years. This half-life makes it possible to date proteins as old as about 100,000 years. So far, however, the dates obtained with the technique have proved somewhat inconsistent, probably because of the difficulty in verifying whether the temperature of the amino acids has been constant.

TABLE 13 Radiometric Methods of Dating Used in Archaeology

Method	Measurement	Applicability	Reference
Methods Based on Measurement of Radioactive Radiation or Amount of Radioactive Isotope or Daughter Isotope in Materials			
Radiocarbon	Counting of beta radiation or amount of carbon-14	Organic matter	Textbox 52
Potassium argon	Relative amounts of parent- and daughter-isotope	Volcanic rocks	Textbox 15
Uranium series	Relative amounts of parent- and daughter-isotope	Calcareous rocks, shell	Textbox 16
Methods Based on Measurement of Cumulative Changes Caused by Radiation on Materials			
Fission track	Number of fission tracks in a solid	Obsidian, glass, slag	Textbox 26
Thermoluminescence	Amount of light emitted at high temperature	Pottery, burned flint	Textbox 24
Optically stimulated luminescence	Amount of light emitted when sample is illuminated with laser light	Sediments and soil	Textbox 24

Comparing the relative amount of a particular radioactive isotope of known half-life remaining in a sample of material with that of its decay products, or measuring the effects of the radiation of such an isotope on surrounding matter, for example, enables one to estimate the time that the material has existed. The most convenient isotopes for archaeological dating are those that are relatively common in nature and whose half-lives are compatible with archaeological dates. Table 13 lists radiometric methods of dating that are widely used in archaeological studies.

Potassium–Argon Dating

Potassium–argon dating is the only feasible technique for dating very old rocks that include potassium in their composition. It is based on the fact that one of the radioactive isotopes of potassium, potassium-40 (K-40), decays to an isotope of the gas argon, argon-40 (Ar-40). Since the half-life of potassium-40 is known, by comparing the relative amount of potassium-40 to that of argon-40 in a sample of rock, the date the rock was formed can be determined (see Textbox 15). Unlike the case in other radiometric dating techniques used in archaeology, the materials dated with the potassium–argon method, rocks, are directly related not to human activities but to geologic events. It is important, therefore, when using the potassium–argon method, that the association between the rocks dated and the human evidence to which the rock is related be carefully established. Bearing in mind this limitation, potassium–argon is the only viable method for dating very old, archaeologically related rocks such as hornblende, certain kinds of feldspar, lava, some natural glass, and types of clay (Walther 1997). A useful application of the technique in archaeological studies is, for example, to date volcanic tuff or lava flows overlying layers bearing evidence of early hominids and/or their activity; such dates provide evidence that the layers bearing archaeological evidence are older than the overlaying lava or tuff.

TEXTBOX 15

POTASSIUM–ARGON AND ARGON–ARGON DATING

Potassium–Argon Dating

Potassium (K) is one of the most abundant elements (2.4% by mass) in the earth's crust, where it occurs in three isotopic forms: *potassium-39, potassium-40*, and *potassium-41*. *Potassium-40*, which is radioactive, is the least abundant of the three isotopes, making up a very small fraction, just over 0.01%, of the total amount of potassium; one out of every 10,000 potassium atoms is radioactive potassium-40. The isotope decays by two different pathways: (1) 88% of the potassium-40 decays to *calcium-40*, and (2) the remaining 12% decays to *argon-40*, a stable isotope of argon also known as *radiogenic argon* (see Fig. 11). The decay of potassium-40 to argon-40 is the relevant decay pathway in the potassium–argon (K–Ar) dating method.

Potassium-40 decays, and as a consequence, in any mineral or rock containing potassium, the concentration of potassium-40 is gradually reduced, while, provided the rock is impervious to argon, the amount of argon-40 increases. Since the half-life of potassium-40 has been precisely determined (1.28×10^9 years), measuring the concentration of potassium-40 and argon-40 in minerals, rock, or volcanic ash enables one to determine their age (Faure 1986). Thus, if argon remains trapped within the minerals that make up rock or volcanic ash, *potassium–argon dating*, as the method is known, provides an invaluable tool for studying early evidence on human evolution. It should be noted, however, that potassium–argon ages are not directly related to human activity. They provide maximum time limits, indicating whether dated minerals, rocks or ashes are younger, or older, than the remains of the activity of humans. If human remains lie between two layers of volcanic deposits, for example, the potassium–argon dates of the layers provide a minimum and maximum age for the activity of humans. The method, which is particularly useful for dating mineral, rock, or ash rich in potassium, can be applied to mineral matter as old as the earth (4.5 billion years ago) and as recent as under 100,000 years ago (Walter 1997; Guillot and Cornette 1986).

Methodology

The determination of the potassium–argon age of rock involves the following experimental stages:

1. Sampling the rock
2. Isolating the potassium-40 and argon-40 in the sample from the rest of the mineral matter
3. Determining the relative amounts of these elements in the sample
4. Calculating the potassium–argon date of the mineral or rock

Experience shows that samples weighing 0.2–20 g are usually sufficient for dating by this method. The determination of potassium-40 is, at present,

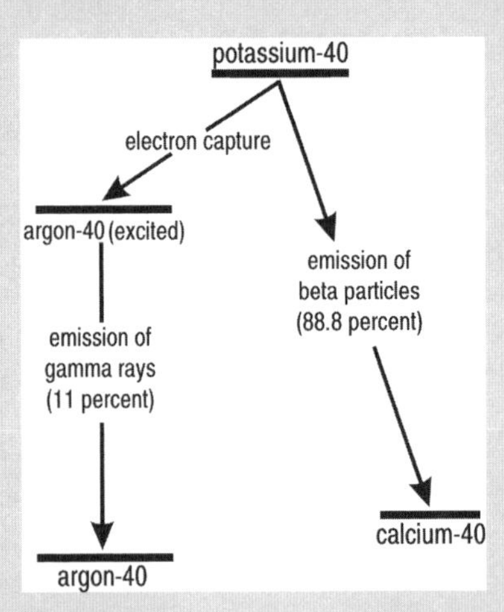

FIGURE 11 The decay of potassium-40. Potassium-40 may decay by one of two different radioactive decay modes: either by the emission of beta particles or by electron capture. The emission of beta particles is, however, its main mode of decay: almost 90% of radioactive potassium-40 decays by this mode to calcium-40; the other nearly 10% decays by electron capture, to excited argon-40. The diagram illustrates the two different decay modes. Only the decay by electron capture to excited argon-40, which is followed by the emission of gamma rays to argon-40, is the decay path used in the potassium–argon method of dating.

generally done using instrumental techniques, particularly absorption spectroscopy.

As for the determination of argon, since argon is a gas, it is extracted from rock by fusing the sample at a very high temperature (over 1600°C) and then, using a mass spectrometer, measuring the amount of gas released (Layer et al. 1987).

Knowing the total amounts of potassium-40 and of radiogenic argon-40 in a sample enables one to calculate first, their relative weight amounts and then, from these values, the age of the sample. The reliability of a potassium–argon date is generally expressed as a plus or minus value after the calculated date.

Relevance of the Technique

Not all potassium-bearing minerals are suitable for K–Ar dating. Some minerals are porous or permeable to gases and lose radiogenic argon; using the potassium–argon method for determining the age of such minerals therefore invariably yields inaccurate, young ages of no significance. Common minerals that can generally be dated using the potassium–argon method are listed in Table 14. There is practically no upper limit to the range of dates to which the method can be applied. Since samples of archaeological interest are relatively young, however, the lower limit of applicability of the method requires that samples suitable for dating be rich in potassium.

TABLE 14 Minerals and Rocks Suitable for Dating Using Potassium–Argon Technique

Minerals:
Adularia[a]
Anorthoclase[b]
Biotite[b]
Glauconite
Hornblende[b]
Lepidolite[b]
Leucite[a]
Plagioclase[a]
Rocks: basalt[b]

[a]Yields occasionally unreliable dates.
[b]Yields consistently reliable dates.

Errors

When a rock is formed from a lava melt, different grains of the same mineral within the rock solidify at slightly different times. The distribution of potassium-40 and consequently also of radiogenic argon is not homogeneous throughout any given rock; for the determination of reliable dates, therefore, the relative amounts of potassium-40 and argon-40 should be determined in the same sample. For practical reasons, however, the amounts of the two elements are generally measured in different samples of the same rock. It is important, therefore, to ascertain that every sample analyzed for dating should be truly representative of the composition of the whole rock.

One of the basic assumptions in the potassium–argon dating method is that the rock dated constitutes a *closed system*; this implies that it contains all the argon-40 generated in it, with nothing having escaped, and that it does not contain radiogenic argon other than that resulting from the decay of potassium-40. The largest errors in the method, however, are related to the occurrence of either an excess or a deficiency of argon-40 caused by geologic or weathering factors.

Argon–40/39 Dating

Relatively recent studies have led to the development of an additional dating method based on the use of argon-40, known as the *argon–argon* or, more specifically, the *argon-40–argon-39* dating method. Like the potassium–argon method, the argon–argon dating method is based on determining the relative amounts of the potassium and argon in a sample, but it uses a different methodology: the sample to be dated is irradiated with neutrons so as to convert potassium-40 to argon-39:

$$\text{Potassium-40} + \text{neutron} \rightarrow \text{Argon-39}$$

After the sample has been irradiated, heating the sample until it melts releases all the argon in the sample, including the isotopes argon-40, created by the decay of potassium-40, and argon-39, formed when the potassium-40 in the rock was irradiated with neutrons. Separating the two isotopes (argon-40 and argon-39) and finally determining the weight ratio between them provides values from which the age of the sample can be calculated (McDougall and Harrison 1999; Walter 1997).

The argon-40–argon-39 method thus provides a variation of the potassium–argon method; both methods yield the age of a sample by ascertaining the relative amounts of potassium-40 and argon-40. However, whereas in the potassium–argon method, the amount of potassium is determined directly in the sample, in the argon–argon method the potassium is instead converted to argon-39, and only then is the amount of the latter measured. It is claimed that the argon–argon method provides some advantages over the conventional potassium–argon method, such as the need for smaller samples and greater precision of the results, particularly when dating weathered rock.

Some of the most important applications of the potassium–argon dating method in archaeology have probably been for dating Pleistocene deposits whose age is reckoned in hundreds of thousands of years; the method may yield valuable information on extinct species, for example, by providing dates for the geologic beds in which their remains are found. Other valuable applications are for establishing a Cenozoic chronology and, most important, for providing a timescale for the evolution of humankind, from the earliest known hominids onward. Since it was first used, the potassium-argon dating method has made possible comparative studies on early humans over a timespan of several million years. The method have been used to date early hominids, including Java's Meganthropus and Olduvai Gorge's Zinjanthropus in East Africa (Evernden and Curtis, 1965).

Argon–Argon Dating

Another dating method based on the use of the decay of potassium-40 to argon-40 is the *argon–argon* dating technique (see Textbox 15). This technique is, however, applicable only to rocks containing substantial amounts of potassium, but makes it possible to date very young samples with relatively low accuracy (about 10%). Using the argon–argon method for dating pumice rock ejected from the Vesuvius volcano (during the eruption that destroyed Pompeii and Herculaneum during the year 79 C.E.), for example, made it possible to precisely date the eruption to 1929 ± 94 years ago, only seven years off from the true date (Renne et al. 1997).

Archaeologically Related Rock and Stone

Early in the development of humans, a variety of stones, such as flint, obsidian, basalt, marble, and turquoise, were put to multifarious uses; many of these stones were used for manufacturing implements, constructing buildings and roads, decoration and adornment or, after comminution into fine powders, as *pigments* and *abrasives* (see Table 15). *Flint* and *obsidian*, which were widely used for making implements, weapons, and decorative objects, are discussed in Chapter 2. Some others are described in the following pages. Their description and uses are followed by a discussion of studies addressed to characterize them and to elucidate the time of their use, their provenance, and the changes they have undergone since they were last used (Waelkens et al. 1992).

Limestone. *Limestone* usually occurs as a white or light-shaded sedimentary monomineral formed by the consolidation of calcite (composed of calcium carbonate) sediments. It is a soft rock that can easily be cut and shaped into any size and form, so it has been used, since early antiquity, for

TABLE 15 Archaeologically Related Stone

Stone	Common use	Compositional characteristics	Remarks
Igneous			
Andesite	Making tools	Complex metal silicates (about 60% silica)	Compact and tough; chips easily
Basalt	Building	Complex metal silicates (about 50% silica)	Dark-colored, fine-grained
Diamond	Precious-stone grinding and polishing	Pure carbon	High refractive index; hardest material on earth
Emerald	Gemstone	A variety of beryl	Very hard
Jade	Making jewelry and decorative objects	Metal silicates	Two varieties: jadeite and nephrite
Obsidian	Making tools and decorative objects	Natural glass	Extremely dense; sharp edges; conchoidal fracture
Sedimentary			
Alabaster	Making decorative objects	Calcium sulfate	Easily shaped and polished
Alabaster, oriental	Making decorative objects	Calcium carbonate	Easily shaped and polished
Flint	Making tools	Mainly silica	Very compact; sharp edges, conchoidal fracture
Gypsum	Building; making plaster of Paris	Calcium sulfate	Easily shaped and polished; decomposed by heat
Limestone	Building; making lime	Calcium carbonate	Easily shaped; decomposed by heat
Sandstone	Building	Mainly silica	Conglomerate of rounded silica fragments consolidated within calcium carbonate
Turquoise	Semiprecious gemstone	Hydrated phosphate of aluminum and copper	Carved with relative ease
Metamorphic			
Marble	Building; making statuary	Calcium carbonate	Easily shaped and polished
Quartzite	Making statuary	Mainly silica	Easily shaped and polished
Soapstone	Carving decorative objects	Complex silicate of magnesium	Greasy feel; easily carved

building and statuary. It is also the raw material from which lime cement is made (see Textbox 34), and an essential component in the process for recovering some metals from their ores and in the manufacture of glass. When added to iron ores during the iron-smelting operations, limestone reacts with the nonmetallic component of the ore forming a slag that can be easily separated from the molten iron. Limestone is also an important component of the most common type of glass, sodalime glass, in which it serves as a stabilizer that makes the glass resistant to dissolution by water and to decay.

TEXTBOX 16

URANIUM SERIES DATING

Uranium series dating is based on the decay characteristics of isotopes of uranium. The method takes advantage of the tendency of uranium dissolved in groundwater and seawater to precipitate together within calcium carbonate. The latter constitutes such minerals as calcite and aragonite and is a main constituent of bone, shell, coral, and speleothems (*speleothems* are mineral deposit formed in caves as stalactites and stalagmites by the evaporation of mineral-rich water). Ancient bone acquires most uranium after burial in the ground; shell, coral, and speleothems acquire it during their formation stages. All these can be dated by the uranium series dating method. The applicability of the method ranges from a few hundred to well over 500,000 years; Uranium series dating is particularly useful, therefore, for dating materials from the period beyond the useful time range of radiocarbon dating, that is, earlier than 50,000 years ago (Latham 2001; Schwarcz 1997).

Uranium, the heaviest element in the earth, is probably the best known of the radioactive elements. The use of compounds of uranium – for example, its natural oxide – dates back to at least the first century C.E., when uranium oxide was used to color ceramic glazes. But its use for dating is quite recent, beginning only in the last decades of the twentieth century. The atomic number of uranium is 92, which is also the number of protons in its nucleus. The number of neutrons in the nuclei of different uranium isotopes varies; therefore, uranium has several isotopes, three of which occur naturally: uranium-238 (which makes up 99.3% of the total amount of uranium), uranium-235 (0.7%), and uranium-234 (0.005%). Only the most abundant isotope, uranium-238, and the lightest, uranium-234, are used for dating. The two isotopes decay to form, ultimately, stable isotopes of lead; uranium-238 yields lead-206, while uranium-235 yields lead-207. The two do not, however, decay directly into stable lead isotopes; they do so through a rather long series of intermediate steps during which transient, radioactive daughter isotopes are formed. Uranium-238, for example, initially yields uranium-234, which then decays to thorium-230; only many more steps afterward is stable lead-206 finally formed. The half-lives of the transient daughter isotopes range from fractions of a second to hundreds of thousands of years. The longest-lived isotopes, including uranium-234, whose half-life is 250,000 years, and thorium-230, with a half-life of 75,000 years, are used to date prehistorical materials (see Fig. 12).

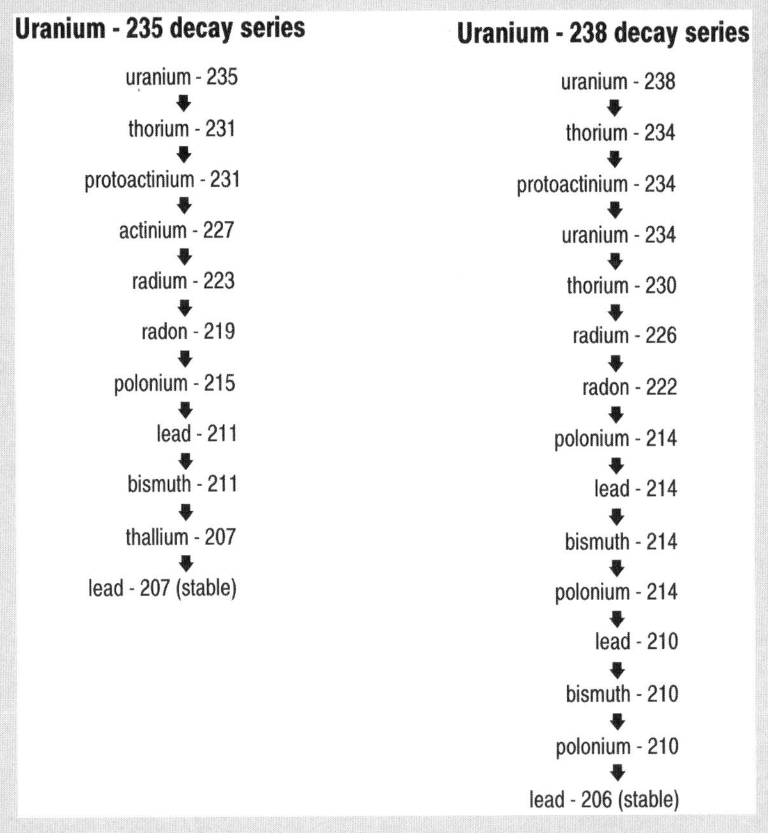

FIGURE 12 The decay series of the uranium isotopes. Most radioactive isotopes decay into stable ones. Some, however, decay into intermediate, unstable isotopes, over and over (repeatedly) in long "decay series," which terminates only when a stable isotope is formed; the stable isotope formed at the end of all the decay series is always an isotope of lead. Three decay series occur naturally in the crust of the earth: one beginning with the isotope thorium-232 and two others that begin with isotopes of uranium, namely, uranium-235 and uranium-238. The last two, which are of use for dating archaeological materials, are illustrated. Determining the length of burial time of objects that absorb uranium, such as bone, teeth, and eggshell, are typical applications of the uranium series method of dating.

The half-life implies that in any material where there is a certain amount of a decaying *parent isotope* there will be only half of the original amount after one half-life, the other half having decayed into a *daughter isotope*. In another half-life, half of the remaining parent isotope will decay, and the material will now consist of only one-fourth of the parent to three-fourths of the daughter isotope (see Textbox 14). Conversely, calculating the weight ratio of daughter to parent isotope in any given material makes it possible to calculate the length of time that has passed since the parent isotope started decaying, that is, since the material was created, as, in speleothems, or since the parent isotope was deposited within the material, as in buried bone.

Once uranium is incorporated into buried bone, shell, coral, or speleothems, the isotope uranium-235 decays, initially into the short-lived isotope (thorium-231) and then into long-lived protoactinium-231. Uranium-238, on the other hand, decays first into two successive short-lived isotopes (thorium-234 and protoactinium-234) and only then into a long-lived isotope, uranium-234 (see Fig. 12). The decay of uranium-235 to long-lived protoactinium-231 is used to date events up to 150,000 years in age; that of uranium-234 (derived from uranium-238) to thorium-230 is of use for dating events within the time range 1000–500,000 years.

Measuring the relative abundance of the successively formed isotopes is generally done using the mass spectroscopy technique, but it is also possible using nondestructive analytical techniques based on counting the alpha or gamma rays emitted during the decay of the isotopes. Dates obtained using the mass spectrometric technique have errors of ±0.5–1%, while with alpha ray counting the errors are much larger, ±5–10%. Comparing the relative amounts of isotopes from the decay of uranium-235 and uranium-238 makes it possible to verify the accuracy of the age estimated using the half-life of one of them.

Marble. The word *marble* is used as the common name for two types of *monomineral rocks*: one derived from *limestone* and therefore composed of calcium carbonate, the other derived from *dolomite* and composed of calcium magnesium carbonate. Extremely high pressures and heat during past geological times modified the structure of both limestone and dolomite, compacting them into a characteristic crystal structure. Most marble is white; however, minor and trace amounts of metallic impurities cause the formation of stains in a variety of colors, hues, and patterns, or of colored marble.

Marble occurs in many locations on the crust of the earth, so that many types of marble were known to the ancients. White Pentelic marble, from

Mount Pentelikus in Attica, Greece, and Carrara marble from Carrara in Italy, were probably the best known in classical antiquity. Red-mottled Siena marble, from Tuscany in Italy, and Tecali marble from Mexico are other well-known examples of widely used types of marble. Because it is easily shaped and polished, marble has been widely used since early antiquity for building and statuary. It is hardly surprising, therefore, that the characterization of archaeological marble, the determination of its provenance, and the study of ancient marble trading patterns have attracted and continue to attract much attention (Getty Museum 1990). A wide range of analytical techniques have been used to study, for example, the provenance of marble (Maniatis 2004). Some of these techniques include analysis of trace elements (Mathews 1997), thermoluminescence and electron spin resonance (Armiento et al. 1997), and determination of the ratios of stable isotopes of oxygen and carbon (see Fig. 13) (Walker and Matthews 1988; Craig and Craig 1972).

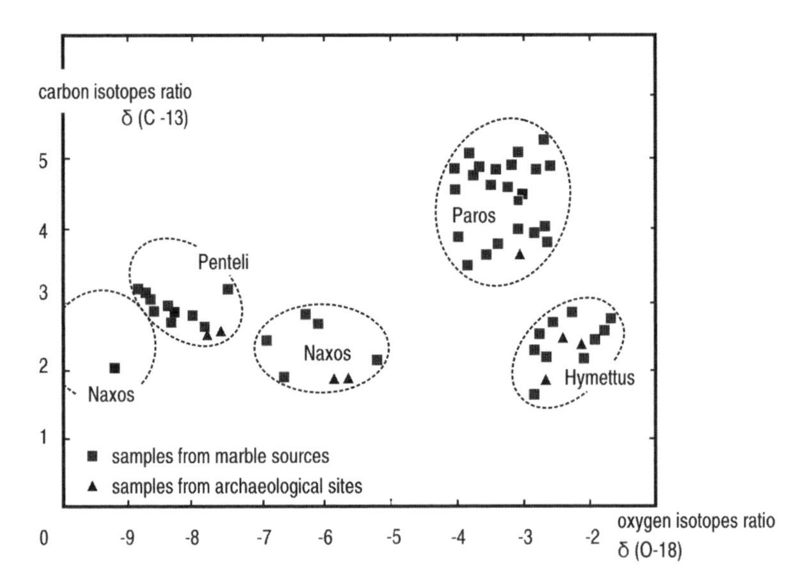

FIGURE 13 Carbon and oxygen isotopes in marble. Measuring the weight ratios between the stable isotopes of carbon and oxygen (i.e., of carbon-13 to carbon-12 and of oxygen-18 to oxygen-16) in marble often enables one to differentiate between different sources of the rock. Having determined such ratios (represented by δ in the graph) in samples from well-known ancient marble sources from the Greek Aegean Islands and from archaeological marble made it possible to identify the sources of the marble. With only one exception, that of marble from the Naxos quarries, these ratios are specific for each source and different from isotopic ratios in other sources (Craig and Craig 1972).

Quartzite. *Quartzite* is a very compact, exceptionally hard and tough metamorphic rock derived from sandstone. It consists mainly of rather large crystals of *quartz* (composed of silicon dioxide) naturally cemented by secondary quartz. Most varieties of quartzite contain over 90% quartz, and in some cases the quartz content exceeds 95% of the total weight of the stone. The color of most quartzite is white or light yellow, but if it contains iron oxide impurities it is red, while other metal oxide impurities may cause the rock to display patchy color variations. Quartzite is very hard, which makes it difficult to quarry. Nevertheless, because of its strength and resistance to weathering, it has occasionally been used for construction, sculpting statuary, and ornamentation.

During the fourteenth century B.C.E., for example, the ancient Egyptians used quartzite for making statuary, stelae, and other constructions. Among these is a famous group of monumental statues known by the plural name *Colossi of Memnon*, near Thebes (see Fig. 14). It is known that one of the Colossi was damaged in antiquity and repaired during Roman times by order of the emperor Septimus Severus. The *provenance* of the quartzite from which the Colossi of Memnon were made was studied using neutron activation analysis to analyze samples removed from the sculptures as well as from a number of quartzite sources from different geographic sites. Comparing the analytical results revealed that the quartzite used to make the statues most probably originated from a quarry quite a distance away from the location of the sculptures: almost 700 km down the Nile from Thebes. The quartzite used for the Roman repair, however, seems to have come from quarries nearer Thebes, upstream the river Nile (Bowman et al. 1984; Heizer et al. 1973).

Soapstone. *Soapstone*, also known as *steatite*, is a very soft metamorphic rock that feels greasy to the touch and can be easily carved and shaped. It consists mainly of the mineral talc (composed of hydrous magnesium silicate), often mixed with small, varying amounts of *magnetite* (composed of iron oxide), and/or *chlorite* (a complex silicate of aluminum, iron, and magnesium). Large outcrops of soapstone occur in many areas on the surface of the earth. Because of the ease with which it can be shaped, soapstone has been widely used since prehistoric times to manufacture small articles such as bowls, vessels, beads, and other decorative objects. It is for this reason that identifying the provenance and the routes of trade of ancient objects made from soapstone has been a subject of investigation since the late nineteenth century. Soapstone from different sources, and even from single sources, generally has variable composition, making it rather difficult to elucidate the provenance of soapstone. Some studies seem to indicate, however,

FIGURE 14 The Colossus of Memnon. The photograph shows one of two massive statues (each about 20 m high) that once flanked the entrance to the mortuary temple of Amenhotep III, a pharaoh of the eighteenth dynasty of Egypt, on the west bank of the river Nile near Thebes. The statues were sculpted during the fourteenth century B.C.E., each from a single block of quartzite. One of the statues was damaged in antiquity during an earthquake and, since then, it produces at sunrise a musical sound. The ancient Greeks related this sound to the mythological king Memnon calling to his mother. The present condition of the statues is not good, mainly because quartzite is prone to weathering.

that determining the nature and relative amounts of trace elements in the rock may contribute to distinguishing between samples from different sources and to the provenance of archaeological soapstone (Truncer et al. 1998; Kohl et al. 1979).

1.8. PIGMENTS

Pigments are intensely colored and finely powdered solids used (mainly in paints) to impart color to other materials. Since early antiquity most pigments have been prepared by mechanically grinding solid raw materials esteemed for their specific color; most are derived from minerals, although some are of vegetable and animal origin, derived from the bark of trees, the shells of nuts, and the bones and horns of animals (Eastaugh et al. 2004). A few artificial, human-made (synthetic) pigments, known by the generic names *frits* and *lakes*, have also been manufactured since antiquity (Barnett et al. 2006; Koenig and Metz 2003) (see Textbox 20).

TEXTBOX 17

COLOR AND COLORANTS – PIGMENTS AND DYES

The meaning of the word *color* varies in diverse fields of interest. In the behavioral sciences, for example, color refers to the psychophysiological perception of the appearance of objects or light sources; in chemistry, to substances that have certain optical properties (to exhibit color), and in physics, color refers to a specific property of visible light (Orna 1976). Different colors (i.e., different psychophysiological sensations) are perceived when light from, say, a blue sky, a red rose, or a green leaf falls on the retina of human eyes. Color may be perceived directly from a light source, where it originates, as, with a burning substance that emits colored light, or reflected from colored objects, as with the blue sky or the red rose.

In chemistry, color is generally associated mainly with *colorants*, intensely colored substances that may be either inorganic or organic, of natural origin or artificially made (Billmeyer and Saltzman 1981). Colorants that are insoluble in most liquids, particularly water, and that generally (not exclusively) are of inorganic origin, are known as *pigments* (see below, this chapter) (Lewis 1988). Organic colorants soluble in water and other solvents are known as *dyes* (see Chapter 15) (Hallas and Waring 1994). The color of pigments as well as of dyes is determined by the characteristic range of light wavelengths that they reflect. It is this property to reflect light of specific wavelengths (colors), that makes the pigments and dyes suitable for coloring other materials (McLaren 1986).

In physics, color is an optical property of electromagnetic radiation that is visible to the human eye and has wavelengths ranging between about 410 and 770 nanometers (see Textbox 22). White, colorless light is a mixture of light of all wavelengths within this range. If a beam of white light falls on a transparent glass prism, it is dispersed into a sequence of colored bands, each band characterized by a narrow range of wavelengths, as illustrated in Figure 15.

An object that reflects only part of white light (between, say, 540 and 600 nm) appears yellow. Yellow light can also be generated by the combination of orange–red and green, the two colors adjacent to yellow in the spectrum of white light, or by the combination of all colors except blue, the *complementary* color of yellow (*complementary colors* are pairs of colors that, when mixed with each other, produce another color; all colors have complementary colors).

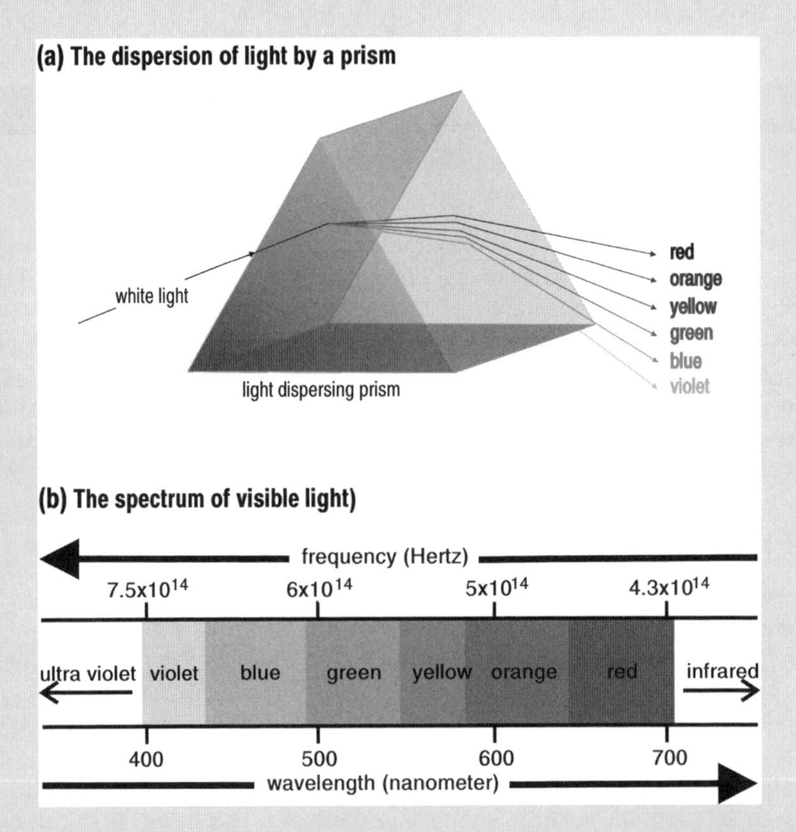

FIGURE 15 The spectrum of visible light. Visible light is the only form of electromagnetic radiation that humans can see. When all the wavelengths of the visible light spectrum in a beam of radiation strike the human eye at the same time, white light is perceived. White is not an actual color, but a combination of all the colors of the visible light spectrum. This can be seen when a beam of white light passes through a prism. Each one of the different colors that make up the beam has a different wavelength and interacts differently with the material from which the prism is made. As it passes through the prism, each color bends differently and the original white beam is spread out into a range of colors as in a rainbow **(a)**; violet has the shortest wavelength (and highest frequency) and red, the longest (and lowest frequency) **(b)**.

Two main properties render a material valuable for producing a pigment: that it reflects light of a particular color and that it covers and hides underlying surface imperfections, a property known as *hiding power*. It is also these two properties that make pigments ideal materials for making *paints*, preparations used to decorate and protect surfaces and obliterate their imperfections (see Textbox 18). Paints have been made and used since very early antiquity (Colombo 1995; Ashok 1993).

TEXTBOX 18

PAINT AND PAINTING

Paint

Paints are colored fluid preparations that are applied as thin coatings to the surface of solid materials where they *set*, serving decorative and/or protective purposes. After a paint is applied and *sets* (it is usually said that it "dries"), it forms a solid, stable, and generally durable coating that imparts its color and often also protects the underlying surface. Humans have been using paints for decorative purposes for well over 20,000 years. It was apparently only during the Middle Ages that it was recognized that the coating formed after paints set also shields the underlying surface from the surrounding environment; since then, paint has been applied specifically also for protective purposes (Gottsegen 1987; Doerner 1984).

Generally, although not exclusively, paints are mixtures of at least, three basic components: a *pigment*, a *medium*, and a *binder*. The *pigments* are extensively discussed elsewhere in this chapter. The *medium*, as its name suggests, provides the material agent in which the other components of the paint (the pigment and the binder) are either dissolved or suspended. The medium also determines the *fluidity* (the flowing characteristics) of the paint and it is therefore instrumental in the manner that the paint is applied and the way it sets. The *binder*, also known as the *film former*, is the component that, as the paint sets, forms a solid film that binds the particles of pigment to each other and to the painted surface. Most binders are fluids that solidify to form more or less transparent films. Binders used in antiquity included mainly carbohydrates (such as vegetable *gums*) and proteins (such as animal *glues*) (see Textbox 55). The use of drying oils (and particularly of linseed oil) as binders (see Chapter 10), which became increasingly widespread after the fifteenth century, can be traced no further back back than the sixth century C.E. (Wehlte 2001; Masschelein-Kleiner 1985).

Painting

The most common technique of painting, used since antiquity, entails the application – usually with a brush but also with a swab, a stick, or the fingers – of a coating of fluid paint to a surface. *Watercolor* painting, for example, entails the use of transparent paints made up of a mixture of pigments, water, and a gum, such as gum arabic. In *tempera* painting, the pigment is mixed with eggyolk (egg tempera) and often, also with water; in *oil* painting the paint is made by mixing the pigment with a drying oil, usually linseed oil in Europe and tung oil in eastern Asia in the past. Oil paint takes several days to set ("dry") into a compact and shiny paint coating (the oil does not really dry but sets by polymerizing into a solid layer that holds the pigment and adheres to the substrate). Less common painting techniques used in the past include *fresco* and *wax* (or *encaustic*) painting (Shirono and Hayakawa 2006).

Fresco *Fresco* (the Italian word for "fresh") painting entails mixing dry pigments with fresh, wet plaster of Paris or lime plaster, which serve as both medium and binder. After the mixture is applied as a coating to a surface (mainly walls, ceilings, and sometimes floors), the plaster dries and sets (hardens), binding the pigment to the surface, which remains permanently colored. The fresco technique employing plaster of Paris was used in India during prehistoric times. In Europe, where fresco painters seem to have preferred mainly lime plaster, the technique was used in Crete, as early as the fifteenth century B.C., to decorate extant Minoan buildings. Later, the Greeks and then the Romans widely expanded the use of fresco painting (Danti et al. 1990; Procacci and Guarnieri 1975).

Encaustic *Encaustic* or *wax painting* is a painting technique in which *beeswax* serves as both the paint's medium and the binder. *Beeswax*, the wax made by the common bee, is a solid at ambient temperatures that melts at about 65°C. For encaustic painting, pigments are mixed with previously heated, molten wax and the mixture is then applied to the surface to be painted, usually by one of two different techniques: *cestrum* (stump, in Latin) *encaustic*, in which the tool used to apply the paint is a short piece of wood, and *cauterium* (spatula in Latin) *encaustic*, in which a hot spatula serves to apply the paint mixture to the surface being painted, on which it hardens and becomes fixed on cooling. Encaustic painting was practiced by the ancient Egyptians in Fayum, for example, where the *cauterium* technique, in particular, was used for painting ancient portraits (Doxiadis 1995). After the sixth century C.E., the encaustic technique seems to have been almost forgotten in most areas of Western civilization (Mattera 2000; Laurie 1978).

TABLE 16 Natural Pigments

Color	Name	Composition
	Inorganic Pigments	
Black	Pyrolusite	Manganese dioxide
	Graphite	Carbon
White	Chalk (also known as whiting)	Calcium carbonate
	Gypsum	Calcium sulfate
	Kaolin	Aluminum silicate
Yellow	Orpiment	Arsenic trioxide
	Ochers and siennas	Natural earths, mixtures of silica and iron oxides
Red	Red ocher	Mixture of silica and iron trioxide
	Vermillion, or cinnabar	Mercury sulfide
Green	Malachite	Basic copper carbonate
	Chrysocolla	Copper silicate
	Green earth	Mixture of aluminum silicates
Blue	Mountain blue or azurite	Basic copper carbonate
	Ultramarine or lapis lazuli	Mainly a mixture of silicate and sulfide of sodium and aluminum
	Glaucophane	Sodium magnesium aluminum hydrosilicate
	Organic Pigments	
Black	Charred bone, charred wood	Mixture of charcoal, calcium phosphate, and calcium carbonate
	Soot	Mixture of charcoal and ash charcoal
White	Burned bone	Mixture of calcium phosphate and calcium carbonate
Brown	Tree bark	Mostly a mixture of cellulose and lignin
	Nut shells	Mostly a mixture of cellulose and lignin

The pigments can be classified according to a variety of criteria, such as their color (blue, red, green, etc.), composition (inorganic or organic), origin (natural or human-made), or other characteristics (Lewis 1988; Feller 1986). Common pigments used since antiquity are listed and classified, according to their color, in Table 16; some of those most commonly found in archaeological sites are discussed in the pages that follow.

White Pigments

The Latin term *terra alba* (white earth) is often used indiscriminately to refer to such white minerals as *chalk*, *gypsum*, and *kaolin*. *Chalk*, also known as *whiting* (composed mainly of calcium carbonate), and *gypsum* (composed of calcium sulfate) were the most commonly used white pigments of antiquity. *Kaolin* was used only in limited geographic regions where this primary clay was abundant or easily available. Chalk and gypsum rocks are widely distributed on the surface of the earth. Before being used as pigments, they are crushed and ground into fine powders. *Kaolin*, also known as *China clay*, generally occurs naturally mixed with unwanted contaminants, from which it is separated before use (see Chapter 4).

Probably one of the first artificially made pigments was *bone white*. Preparing bone white is usually done by first burning bones, generally in open fires, until all the organic material in the bones is burned away, and then crushing the product into a white-grayish, slightly gritty powder. Almost any sort of bone would suffice for making bone white, although many ancient recipes particularly recommended the use of specific types of bone such as from the wings of birds or the legs of domestic animals. White bone is composed mainly (85–90%) of calcium phosphate mixed with calcium carbonate (13–9%), minor constituents making up the rest.

Lead white (composed of highly poisonous basic lead carbonate) is another white pigment known since ancient times; its use seems to have started during Greek or Roman times. Lead white has excellent hiding power, that is, it effectively obliterates underlying imperfections on the painted surface; it also weathers well, remaining in a very good condition for long periods of time. These properties are unequaled by either chalk or gypsum. Although it occurs naturally, artificially made lead white seems often to have been preferred. Artificial lead white was made by exposing lead to the corrosive effect of vapors of vinegar (a natural solution of acetic acid) and carbon dioxide. The places where it was manufactured were reported in many catalogs and lists of pigments written since ancient times and on into the nineteenth century. Despite its toxicity, of which they may have been unaware, the ancient Romans used lead white not only as a white pigment but also as a cosmetic face powder (Gettens et al. 1967).

Black Pigments

The black pigments of antiquity fall into one of two categories: *mineral blacks* and human-made *vegetable* and *animal blacks*. The most common mineral blacks are *pyrolusite* and *graphite*. *Pyrolusite*, also known as *manganese black* (composed of the dioxide of manganese) has been widely used not only as a black pigment but also as a glass decolorizer (see Chapter 3). *Graphite*, also known as *plumbago*, can be either black or dark gray. It is composed of elemental carbon, as it is one of the two naturally occurring mineral allotropes of this element (see Textbox 19). Artificial, human-made black pigments have been derived mostly from either vegetable or animal matter. *Animal blacks* are made by charring (partly burning) either bone or ivory. To prepare *bone black*, bones are burned in an oven or closed pit in which there is a reducing atmosphere, that is, a restricted supply of air and therefore a dearth of oxygen.

TEXTBOX 19

ALLOTROPES

A number of chemical elements, mainly oxygen and carbon but also others, such as tin, phosphorus, and sulfur, occur naturally in more than one form. The various forms differ from one another in their physical properties and also, less frequently, in some of their chemical properties. The characteristic of some elements to exist in two or more modifications is known as *allotropy*, and the different modifications of each element are known as its *allotropes*. The phenomenon of allotropy is generally attributed to dissimilarities in the way the component atoms bond to each other in each allotrope: either variation in the number of atoms bonded to form a molecule, as in the allotropes oxygen and ozone, or to differences in the crystal structure of solids such as graphite and diamond, the allotropes of carbon.

Oxygen and Ozone

Oxygen has two allotropes, ordinary *oxygen* and *ozone*. Its allotropy is due to variations in the number of atoms bonded to each other in a molecule. The molecules of the most abundant allotrope, common *oxygen*, the gas that makes up over 20% by volume of the atmosphere of the earth, consist of two atoms of oxygen bonded together (O_2); those of *ozone*, also a gas, are made up of three bonded atoms of oxygen (O_3) as illustrated in Figure 16. The formation of ozone on earth is initiated mostly by ultraviolet radiation from the sun, which, on entering the stratosphere, splits oxygen molecules into separate atoms. Following the split, the separate atoms of oxygen recombine back into molecules, mainly of oxygen, but also of ozone. It is for this reason that most of the earth's ozone occurs naturally in the stratosphere, in the ozone layer, at a height between 15 and 30 km above the surface of the earth (see Fig. 82).

In the lower layers of the atmosphere, ozone occurs only in trace quantities. Some ozone, however, is also created in the lower atmosphere by lightning, during electric storms, accounting for the "smell of rain" that often arises when lightning strikes. Usually, a short while after lightning strikes, most of the ozone thus created begins to gradually revert back to ordinary oxygen. The small amounts of ozone that remain in the atmosphere constitute an active air pollutant and an aggressive oxidizing agent that greatly contributes to the decay of much organic matter. Archaeological materials that are attacked by ozone include skin and leather, textile fibers, dyes, and the remains of dead organisms.

Carbon: Graphite and Diamond

The allotropy of carbon is due to variations in the crystal structure of the element. There are three allotropes of carbon: graphite, *diamond*, and

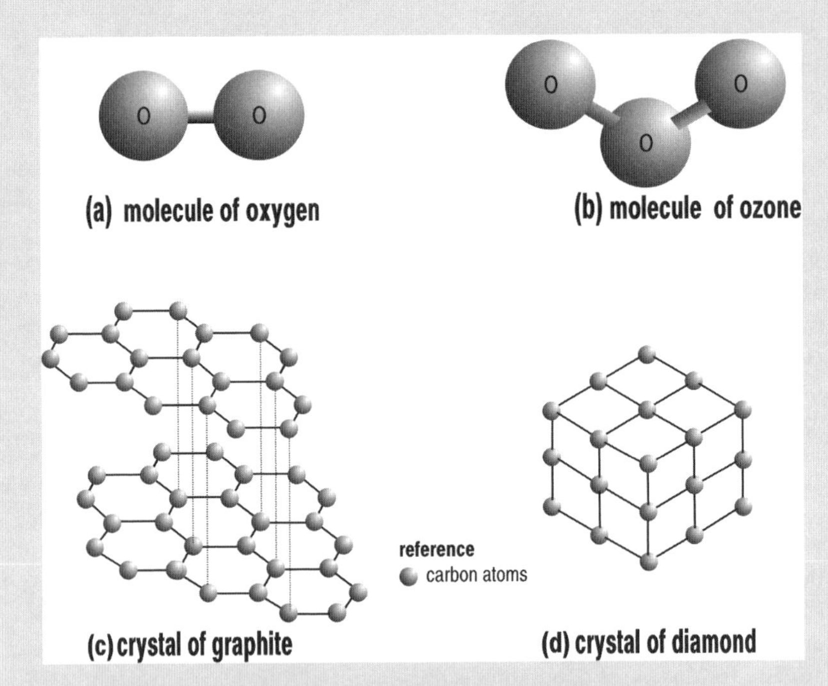

FIGURE 16 Allotropes. Allotropes are structurally different forms of the an element. Oxygen, for example, occurs naturally in two allotropic forms: oxygen and ozone. The molecule of oxygen **(a)** consists of two bonded atoms of oxygen, while that of ozone **(b)**, which is extremely active in the decay of ancient organic materials and particularly of dyes and textiles, is made up of three bonded oxygen atoms. Carbon also occurs naturally in two allotropic forms: graphite and diamond. In graphite **(c)** the carbon atoms are arranged in planar sheets, stacked parallel and loosely bonded to each other, which easily flake apart. In diamond **(d)** the atoms of carbon are arranged in a three-dimensional cubic pattern, which makes it extremely strong and hard.

fullerene. Only the first two, graphite and diamond, occur in nature; *fullerene*, which is artificially made by humans, was discovered quite recently, at the end of the twentieth century. *Graphite* occurs naturally as a dark gray, greasy-feeling mineral, which has been used, since antiquity, as a black pigment. Its greasy feel is due to the crystal structure of graphite: each carbon atom in the graphite structure is attached to another three, forming large, flat sheets of hexagonally bonded carbon atoms. The sheets have smooth surfaces that can easily slide and move relative to one another and also provide the greasy feeling. *Charcoal*, often found in archaeological excavations, consists of a porous mixture of graphite and partially carbonized organic compounds. It is formed when organic matter, particularly wood, burns in an atmosphere lacking or devoid of oxygen. In *diamond*, a gemstone, each carbon atom is attached to others within the ordered crystalline, tetrahedral structure. Since the carbon atoms are relatively small and the bonds between them are very strong, diamond is very hard, the hardest material known. It is because of its hardness, for example, that diamond powder has long been used as an abrasive for drilling holes and polishing surfaces. The crystals of diamond are also very tough and resistant to pressure; if tapped in just exactly the right place, however, they cleave along crystal faces.

Tin and Tin Pest

Another element that exhibits allotropy because of variations in the crystal structure is tin. The common allotrope is *tin metal*, also known as α *(alpha) tin*, which is stable at ambient temperatures. The other allotrope, which generally occurs as a gray powder and is known as β *(beta) tin*, but also as *tin pest*, is formed only at very low temperatures: when tin cools down to temperatures below –18°C, the ordinary allotrope, α tin, is converted to β tin, and the transformation is irreversible under ordinary temperatures. Tin objects exposed to temperatures below –18°C in very cold regions of the world, for example, are generally severely damaged when part of the tin converts to tin pest. In extreme cases, when exposure to low temperatures extends for long periods of time, the allotropic conversion may result in the transformation of tin objects into heaps of gray β-tin powder.

Black pigments of vegetable origin have generally been made from various kinds of charred plant matter, mostly wood, but also leaves or seeds; the *charcoal* formed during the charring process is then washed, to remove soluble matter, and finally ground to powder. Over 95% of well-burned charcoal black consists of elemental carbon; the remainder is ash, composed of salts and minerals in the original vegetable matter.

Red Pigments

The best-known and most widely used ancient red pigments were *red ocher* and the mineral *vermilion*. The *ochers* and *siennas*, also known by their more general name, *natural earths*, are a group of mixtures of clay, silica, and iron oxides. Their color, which may vary from yellow or light brown to dark brown and even to orange and red (see discussion on red ocher, below), is due mainly to iron oxides that occur either as *hematite* (composed of ferric oxide), or as *limonite* (composed of hydrated ferric oxide). Often the natural earths also contain small amounts of black *pyrolusite*, which makes them darker, and/or calcium and barium carbonates, which make them lighter. There is no sharp compositional difference between ochers and siennas, in either iron content or color; it is customary, however, to classify natural earths of lighter color and lower iron oxide content as ochers, and those of darker color and slightly higher iron oxide content as siennas. The intense red of red ocher, for example, is due mainly to a relatively high proportion of hematite (which is red) in the mixture. Deposits of red ocher abound in many places on the surface of the earth; only a few, however, are considered red enough and suitable for use as pigments. Red ocher has excellent hiding and staining powers, and it probably was one of the preferred red pigments of antiquity.

Vermilion and *cinnabar* are two bright red, toxic minerals that share an identical composition (they are both composed of mercury sulfide) but have different crystal structures. Two kinds of vermilion are known: one of natural origin and another made artificially. Finely ground natural vermilion may vary in hue from red to liver-brown and even to black. Artificial vermilion was made from mercury and sulfur; the method of preparation seems to have been developed by the Chinese and was introduced into Europe only during the eighth century C.E. (Gettens et al. 1972).

Red lead, also known as *minium*, is a bright red pigment with excellent hiding power. It is composed of lead tetroxide and occurs naturally as a mineral. Much of the minium used in the past as a pigment, however, seems to have been artificially made by oxidizing at high temperature, about 500°C, either metallic lead or the mineral litharge (composed of lead monoxide). To prepare minium, one of these materials, either the metal or litharge, was heated in a well-aired atmosphere, to ensure an abundant supply of oxygen, until the product acquired a desired red color. Red lead has been identified in wall paintings from China and Central Asia. It seems that in Egypt red lead was used only since Greco-Roman times; in latter times it became the favorite pigment of Persian and Byzantine illuminators (Fitzhugh 1986).

Another artificial red pigment is *madder lake*, an artificial, human-made pigment (Textbox 20). Madder lake is obtained when madder dye (see Chapter 14) is deposited on the surface of powdered kaolin or chalk grains. Preparing madder lake was, and still is, such a complicated process that even today some manufacturers are unwilling to disclose either its formulation or details on the actual method of preparation. A pink color in paintings on the plaster in tombs from the Greco-Roman period was identified as consisting of madder dye on a gypsum base.

TEXTBOX 20

MAN-MADE PIGMENTS: FRITS AND LAKES

Frits and *lakes* are two basically different types of human-made synthetic, *pigments*. *Frits* consist of colored glass, *lakes* of white or transparent painted powders. *Frits* are made by heating together, to high temperatures, a mixture of two or more minerals (including at least silica and lime) until they melt and form a colored glass (see Chapter 3). After cooling, the resulting solid is usually ground into a fine powder; it is only then that it becomes known as a frit, which can be used as a pigment. *Egyptian blue* is an example of a frit known since early antiquity (see Chapter 7).

Lakes are made by conferring a more or less permanent color to the surface of the grains of white or transparent powders insoluble in water. The powder is almost invariably of inorganic origin, for example, gypsum, chalk, or kaolin. To prepare a lake, any one of these materials, in powdered form, is treated with a water solution of a dye of a wanted color; the dye is either adsorbed on the surface of the powder grains, where it becomes more or less firmly held, or it reacts (chemically) with the surface of the grains. In either case, the result is a colored powder, insoluble in water that has the same coloring properties as natural pigments (Kirby 1977, 1988). Lakes were prepared in antiquity from many dyes. Some were valued for their bright colors, although most faded rapidly under sunlight. *Madder lake*, for example, derived from madder dye, and *lac*, derived from the red lac dye, have been widely used for painting since ancient times (Harrison 1936).

Yellow Pigments

Three main types of natural yellow pigments were known to the ancients: *yellow ochers* and *siennas* (see descriptions above) and *orpiment* (a highly poisonous, soft, lemon-yellow mineral composed of arsenious sulfide). Orpiment was identified in Egyptian paintings at Tell-Amarna, in Egypt, and it is mentioned in the writings of the Roman writers Vitruvius and Pliny. In the past orpiment was used not only as a pigment but also for the removal of hair from animal hides before the tanning processes (see Chapter 11). Another artificial yellow pigment, generally known by its Italian name *giallolino,* is composed mainly of basic lead antimoniate. It seems that giallolino was first prepared during the seventh century C.E. by heating a mixture of tartar emetic (composed of antimony potassium tartrate), common salt (sodium chloride), and lead nitrate; cooling the molten mixture resulted in the formation of a yellow solid that, when crushed into a powder, produced the pigment.

Green Pigments

A variety of mineral ores, mostly copper minerals such as *malachite* and *chrysocolla,* were probably the most used green pigments in the past. Various green minerals derived from metals other than copper, such as *green earth* (see below) were used in confined regions.

Malachite is a bright green mineral (composed of basic carbonate of copper), which is widely distributed in nature. The consolidated form of the mineral has also been appreciated since antiquity as a semiprecious gemstone. Powdered malachite was used as a pigment by the ancient Egyptians from predynastic times, later by the Romans in illuminated manuscripts, and by the Chinese in paintings from the ninth century C.E. *Chrysocolla,* also a copper mineral (composed of hydrated silicate of copper), may range in color from bright green to bright blue. It is often found naturally mixed with other minerals of copper, such as malachite and azurite.

Green earth, best known by the Italian name *terra verte,* is probably the main green pigment not derived from copper. Terra verte is a mixture of two minerals, *caledonite* and *glauconite,* both complex silicates of aluminum, calcium, iron, magnesium, and potassium silicate (Grissom 1986).

The most widely used artificial green pigment was *verdigris* (composed of basic acetate of copper). Exposing either metallic copper or copper minerals to vinegar for a few weeks, while allowing free access of air, results in the formation of a green solid composed of basic acetate of copper. Drying the solid and then crushing and powdering it produced verdigris.

Blue Pigments

Mountain blue, ultramarine, and *glaucophane* were the most widely used natural ancient blue pigments. *Mountain blue* is made by crushing and reducing to a powder the bright blue mineral *azurite* (composed of a basic carbonate of copper) that occurs in only a few regions of the world from where it was and still is exported in the form of massive rocks. *Ultramarine* is the name of the dark blue pigment obtained when lapis lazuli is crushed and powdered. Lapis lazuli, a rare rock of gemstone quality, is a mixture of minerals that include *calcite, muscovite,* and *lazurite* (composed of sodium calcium aluminum silicate sulfate). The source of most lapis lazuli used in the ancient world seems to have been mines in the Badakhshan region of Afghanistan, from where it has been continuously exported for over six millennia (von Rosen 1990). *Glaucophane,* a blue mineral (composed of sodium magnesium aluminum hydrosilicate), was used by the Greeks as a blue pigment as early as the seventeenth century B.C.E.

Egyptian blue is an artificial blue pigment that was made in the ancient Middle East, particularly in Egypt, where it is still made and used. Egyptian blue should not be confused with another, also blue or greenish-blue, *Egyptian faience,* which is made from much the same raw materials (see Chapter 7). *Egyptian blue* is made by first preparing a mixture of sand, natron, and copper filings and then heating the mixture to about 850°C. The raw materials melt at this temperature, reacting with each other to form a definite chemical compound (composed of copper calcium tetrasilicate). As the melt cools, it does not crystallize but solidifies into a *frit* (see Textbox 20), which is then powdered (Tite 1986).

Another artificial pigment, a particular light blue, is *Maya blue,* which was made by the Maya Indians in pre-Columbian Mexico. Maya blue is not a frit, however, as is Egyptian blue, but probably a lake. Although its precise components are still unknown, it may have been prepared by thoroughly mixing a clay (such as attapulgite or palygorskite) with blue indigo dye and then heat-treating the mixture before use (Reyes-Valerio 1993).

1.9. ABRASIVES

Cutting, grinding, and shaping stone, and in particular burnishing and polishing the surface of stone as well as metals, requires the use of *abrasive* materials that are harder than the solids to be cut, ground, burnished, or polished. Sapphire and ruby, two very hard gemstones, for example, can be cut or polished only with the assistance of diamond powder, an abrasive that is harder than sapphire or ruby. Diamond is the hardest material

TABLE 17 Natural Abrasives

Excellent (hardness above 7 on Mohs scale)	
Diamond	Garnet
Corundum	Emery
Average (hardness 5.5–7 on Mohs scale)	
Basalt	Quartz
Feldspar	Sandstone
Inferior (hardness below 5.5 on Mohs scale)	
Clay	Limestone
Alumina	Magnesia

known, and diamonds can be cut, ground, or polished only using diamond powder.

Abrasives are hard materials that, in powdered form, are used to cut or alter the dimensions, shape, or condition of the surface of solid bodies by abrasion, that is, by scraping away or rubbing the solid material. The abrading effectivity of any abrasive is determined mainly by its *hardness* (see Textbox 23), although it is also affected by its *toughness* (the resistance of materials to fracture under stress), as well as by the shape of its grains (Jacob 1928). All the abrasives used in antiquity were hard and tough powdered minerals such as *flint*, *corundum*, *emery*, and *garnet*. There is evidence, for example, that as early as the fifteenth century B.C.E. Egyptian crafters used abrasives when carrying out grinding and polishing operations (Peltz and Bichler 2000). A convenient way of classifying abrasives is by hardness and the type of minerals from which they are derived, as shown in Table 17.

1.10. GEMSTONES

Minerals, rocks, and a few consolidated materials of organic origin that, when cut and polished, acquire visual appeal and are also durable and relatively rare are known as *gems* or *precious stones*. *Gemstones* have been used since time immemorial for ornamentation and personal adornment, their use representing a sign of social position and wealth. The interest of human beings in gemstones can be traces so far back that the trail is lost in the dim reaches of time. Probably the earliest evidence of working gems is from Spain and southern France, where over 70,000 years ago stone, ivory, and horn were carved and polished. Since that time there has been a continuous and growing interest in gems. The transportation of gems through trade routes seems to be well over 5000 years old; lapis lazuli, probably from Afghanistan, for example, was then being traded in western Asia and Egypt (von Rosen 1990). Emeralds seem to have been favored during the Hellenistic period, while amethyst, sapphire, and diamond became relatively popular during Roman times. Many of these gemstones, set in jewelry, have survived to the present day (Hackens and Moucharte 1987).

Most gemstones are minerals; only a few are other inorganic or organic materials. Out of an estimated 3000 minerals known on the crust of the earth, however, only about 100 are valued as precious and semiprecious gemstones. When extracted from the crust of the earth, most gemstones have a rough surface and irregular shape, and to reveal and enhance their beauty they have to be cut and polished. Minerals of gem quality are formed and found in the most varied environments. Diamonds, for example, occur mainly among igneous rocks but also in sand or gravel deposits. Sapphires and rubies are usually found within metamorphic rocks. *Jade* is, in itself, a type of metamorphic rock. Turquoise and opal are formed in sedimentary rocks as a result of groundwater seeping through and dissolving rocks; when the water evaporates, the dissolved solids finally precipitate to form the stones.

TEXTBOX 21

THE STRUCTURE OF SOLIDS: AMORPHOUS AND CRYSTALLINE MATERIALS

Solid materials can be classified, on the basis of how their constituent particles (atoms, ions, or molecules) are arranged in space, into two groups: *amorphous* and *crystalline solids*. *Amorphous solids*, literally "solids without shape," such as obsidian and glass (see Chapters 2 and 3), are devoid of any regular internal arrangement. Their component particles are distributed at random within the bulk of the solid and lack short- or long-range geometric order. In *crystalline solids* such as table salt and quartz, the component particles occupy definite, fixed positions; they are arranged in a characteristic, symmetric, and repeating pattern that extends from one edge of the solid to the other. A crystalline solid is therefore a solid characterized by an ordered, regular, and symmetric three-dimensional arrangement of its component particles (see Fig. 17).

Crystalline Solids

Crystalline solids have a regular geometric shape bound by plane surfaces that intersect at characteristic angles. Their shape results from the arrangement of the particles (atoms, ions, or molecules) within the crystals, in an orderly pattern and connected by bonds in a repeating layout. Essentially, all crystalline solids are made up of basic units cells containing a small number of particles, thus known as *unit cells*, repeated in a regular way, again and again, throughout the solid. The geometric, regular arrangement of the unit cells is known as the *crystal form* and the surfaces that bound a crystal are known as the *facets*.

The most characteristic and unique property of crystalline solids is however, neither the shape of their crystals nor the relative size of the crystal faces, but the angle between any pair of crystal facets. For any substance, the angle between the crystal facets is constant and invariable, regardless of the overall shape or size of the crystals. Under some circumstances a substance may form short, wide crystals, while under others, the

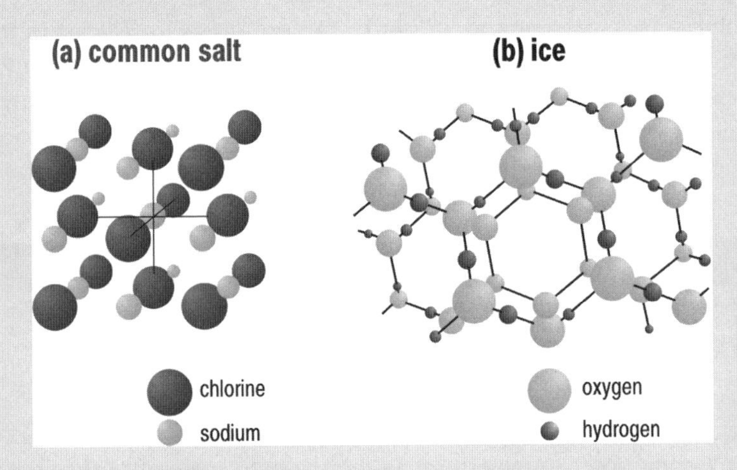

FIGURE 17 Crystalline solids. A "solid" is a relatively compact state of matter in which the component atoms are close together, forming a rigid structure that has a definite volume and shape. There are two main types of solids: crystalline and amorphous solids. In "crystalline" solids the atoms are arranged in orderly, regular and repeating patterns known as "crystals." The arrangement of the atoms within the crystals determines many of the chemical and physical properties of the solids. The smallest repeating pattern in a crystal is a "unit cell," a geometrically regular arrangement of atoms held together by electrical forces known as "bonds." All minerals are crystalline solids. In the crystals of sodium chloride (common salt) (a), for example, the atoms of sodium and chlorine are interlocked in a cubic spatial arrangement in which the unit cell is made up of four atoms of sodium and four of chlorine. In ice (b), also a crystalline solid, each atom of oxygen in the unit cell is surrounded by four atoms of hydrogen. The extremely small unit cells, invisible to the naked eye, combine with each other to form visible crystals. One small grain of common salt, for example, contains about 10^{18} (1,000,000,000,000,000,000) unit cells. In amorphous solids, such as obsidian and glass, the atoms that make up the material are randomly oriented relative to each other, lacking any long-range order.

crystals of the same substance may be long and narrow. The angle between any single pair of crystal faces, whether the crystals are short and wide or long and thin, will be the same in all the crystals of any substance, regardless of their dimensions.

All minerals and many solid substances of inorganic as well as organic origin exhibit characteristic crystal structures. The crystal structure of the *primary minerals*, for example, is formed when hot magma cools down; as the temperature of the magma decreases, the component atoms of minerals become spontaneously ordered in unique, symmetric arrangements. When the mineral finally solidifies, the solid preserves this symmetric arrangement in its characteristic crystalline structure. In *secondary minerals* the crystals are formed as a result of the integration, at a specific temperature and pressure, of small crystallites that came out of solution and precipitated (Goldman 1991).

The Study of Crystals

The structure of crystalline materials is studied using *crystallographic techniques*. Based mainly on the use of penetrating radiation, such as X- and gamma rays but also on other physical techniques, *crystallographic studies* reveal the exact position of atoms within solids. Such studies thus clarify the relationship between the structure and the properties of crystalline solids. Graphite and diamond provide illuminating examples of how the extremely different properties of these two substances are determined solely by the spatial arrangement of the atoms of a single element, carbon (see Textbox 19).

Gemstones can usually be identified and characterized by a number of physical properties; if they are minerals, mainly by their crystal structure, but also by their density, hardness, color, and other physical properties. All gemstones of any one type of mineral have the same crystal structure (see Textbox 21), but the crystal structure varies from one type of gemstone to another.

Mineral gemstones that have the same basic chemical composition, that is, are composed of the same major elements and differ only in color, are considered as variations of the same mineral species. As gemstones, however, minerals that have the same composition and crystalline structure but exhibit different colors are classified as different gemstones. *Beryl*, for example, a mineral (composed of beryllium aluminum silicate), includes a pink variety, known by the gemstone name of *morganite*, and also a well-known green variety, *emerald*. Table 18 lists and classifies, by composition and color, gemstones that have been appreciated since antiquity.

TABLE 18 Gemstones

Gemstone and Varieties	Composition	Color
	Metal Silicates	
Adularia	Potassium aluminum silicate	Varies
Beryl	Beryllium aluminum silicate	
aquamarine		Green-blue
emerald		Brilliant green
Jade		
jadeite	Sodium aluminum silicate	Mainly green, but also other colors
nephrite	Calcium magnesium silicate	
Quartz	Silica	
agate		Banded or patterned
amethyst		Purple
jasper		Varies
onyx		Black and white bands
rock crystal		Transparent
Topaz	Aluminum fluorosilicate	Pink to red
Tourmaline	Complex borosilicate	Variable
Turquoise	Hydrous copper aluminum silicate	Light blue
Zircon	Zirconium silicate	Variable
	Metal Oxides	
Chrysoberyl	Berylium aluminum oxide	
cat's eye		Variable
alexandrite		Red/green
Corundum	Aluminum oxide	
ruby		Red
sapphire		Any color but red
Spinel	Magnesium aluminum oxide	Variable
	Unrelated Composition	
Diamond	Carbon	Usually colorless
Garnet	Mixture of several minerals	Variable
Opal	Variable	Flashing
	Organic Origin	
Amber	Fossil resin	Yellow//brown/red
Coral	Calcium carbonate	Red/pink
Ivory	Mixture of calcium carbonate	Creamy white
Jet	Fossilized wood	Black
Nacre	Aragonite + organic compounds	Mainly white; iridescent
Pearl	Aragonite + organic compounds	Mainly white; iridescent

The color of most gemstones is due to either chemical impurities within their crystal structure or defects in the arrangement of the atoms that make up crystals [the word *impurity* is used here to refer to atoms, generally, although not exclusively, for metals, which occur in extremely small (trace) amounts, within the crystal structure of the gemstone (see Textbox 8)]. As a result of the impurities absorbing or reflecting light incident on them, gemstones display particular and characteristic colors. Pure *corundum*, a mineral of gem quality (composed of aluminum oxide), for example, is colorless when the only elements in its crystals are aluminum and oxygen. Even if minimal amounts of aluminum atoms in the corundum crystal are replaced by those of other metals, however, the mineral exhibits other colors. *Sapphire*, for example, is a variety of corundum in which some of the atoms of aluminum have been replaced by those of titanium or iron. If the replacing atoms are of chromium, the gemstone is deep red and is then known as *ruby* (O'Donoghue 1970; Herbert Smith and Phillips 1962).

TEXTBOX 22

LIGHT: ITS NATURE AND PROPERTIES

Light is electromagnetic radiation that is visible to the human eyes. The range of light wavelengths extends between approximately 410 and 770 nanometers, an extremely limited and narrow range when compared with the entire spectrum of electromagnetic radiation (see Textbox 5). All light originates in *excited atoms*, that is, atoms having more energy than do regular atoms. Excited atoms are formed when regular atoms gain energy from external sources, as, for example, when heated or struck by other particles. Atoms heated to high temperatures, above about 450°C, for example, acquire energy, get excited, and become *incandescent*, releasing the added energy in the form of light. The emission of light reveals the loss of previously acquired, excess energy. Light acquired from the sun is used by plants for carrying out the photosynthesis process (see Textbox 54) and by animals and humans to see. Since it originates in atoms, light is also extremely useful for a range of physical and chemical studies on matter, particularly for the identification and characterization of elements and substances.

The Nature of Light

Light has a dual character – it may be described either as a *wave motion* or as a *stream of moving particles*. Since light is a form of electromagnetic radiation, its properties, as with all forms of electromagnetic radiation, are those of a *wave motion* (see Textbox 5). In some cases, however, light behaves as a *stream of particles* moving along a straight line. Thus, although light can be accurately described as a wave motion, it can also be described as a stream of small particles, known as *photons*. Regardless of how it is described, whether as a wave motion or as a stream of moving particles, light carries energy, and the amount of energy it carries determines its color.

Light, as do all forms of electromagnetic radiation, moves very fast. In a *vacuum* (in empty space, devoid of any matter) light travels at the highest velocity that anything can move in the universe, 300,000 kilometers per second; when it travels through any kind of matter, however, its speed is reduced. This happens because in empty space there is nothing to obstruct the path of light and its traveling velocity is maximal; when it passes through any type of matter, however, light interacts with particles that it encounters in its path and is slowed down. The interaction of light with different forms of matter gives rise to various phenomena, such as those of *reflection*, *refraction*, *absorption*, *dispersion*, *scattering*, *polarization*, and *luminescence*.

Light Reflection and Refraction

When a beam of light reaches a surface delimiting between two different materials, such as air and water, glass and water, or any other pair of materials, part of the light beam is *reflected*, that is, it is turned back from the surface. Provided the second material is transparent, the part of the beam that is not reflected passes through the surface, enters the material and changes its path. Such a beam is said to be *refracted*.

The angle between the path of an incoming beam and the *normal* (the line perpendicular) to the surface is known as the *angle of incidence*. A reflected beam makes the same angle to the normal as the incoming beam but on the other side of the normal. When a beam of light is reflected from a smooth surface, all of the light is reflected in the same direction. If it is reflected from a rough surface, however, the beam is said to be *diffracted* (split) into many reflected beams going in many directions. This is because the normal at each spot on the rough surface points many different ways (see text below).

As for the *refracted* beam, the change in its direction is due to a change in the velocity of light as it passes from one medium to another. When a beam of light passes from a fast medium (such as a vacuum or air) to a slower medium (e.g., a crystalline solid), the change in velocity causes the beam to bend toward the normal (line perpendicular) to the boundary between the two media (see Fig. 18). The *index of refraction*, also known as the *refractive index*, the numerical value of the ratio between the speed of light in a vacuum to the speed of light in a material, provides

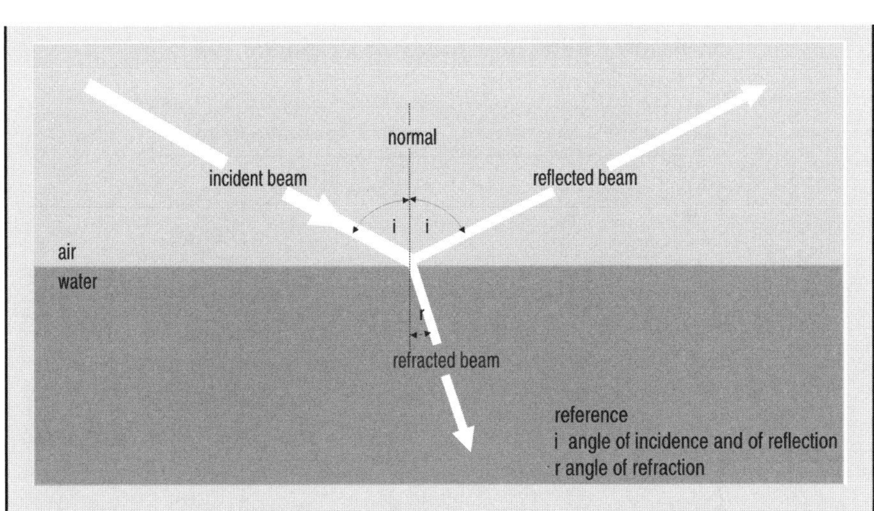

FIGURE 18 The reflection and refraction of light. Objects can be seen by the light they emit or, more often, by the light they reflect. Reflected light is light thrown back from a surface; it obeys the law of reflection, which states that the angle of reflection equals the angle of incidence (*i*) (the angle of incidence is the angle an incident beam makes with an imaginary line perpendicular to the reflecting surface). Light that passes from one medium to another as, for example, when traveling from air into water, is known as "refracted light." On changing media (from air to water in the example above), the speed of the light changes and consequently, in almost every case the angle of refraction (*r*) is different from the angle of incidence.

a measure of the extent to which a beam of light changes direction as it passes from a vacuum to a material. The index of refraction is characteristic for each material and different from that of other materials.

Measuring the index of refraction of a material is generally done using an instrument known as a *refractometer*. Typical values of the index of refraction of gemstones and other materials are listed in Table 19. The relatively high index of refraction of diamond, for example, accounts for its distinct brilliance and multicolor display; the index of refraction of rutile, a gemstone composed of titanium dioxide, is even higher than that of diamond, and rutile is indeed more brilliant than diamond.

Light Dispersion

White, ordinary light consists of a mixture of wavelengths. When a beam of white light refracts at a surface, the paths of the beams of different wavelengths *refract* to slightly different angles. After *refraction*, a beam of white light is, therefore, spread into many beams of diverse colors. The greater the index of refraction of the substance, the higher the dispersion

TABLE 19 Refractive Index of Common Archaeological Solids

Material	Refractive index (ρ)
Amber	1.55
Beryl	1.57–1.60
Calcite	1.48–1.66
Diamond	2.41
Ivory	1.51–1.55
Glass	
Soda-lime glass	1.50–1.55
Crystal (lead glass, flint glass)	1.60–1.85
Ice	1.33
Ivory	1.54
Obsidian	1.48–1.51
Opal	1.41–1.45
Pyrite	above 1.81
Quartz	1.54–1.55
Ruby	1.76
Rutile	2.65–2.69
Sapphire	1.76–1.78
Topaz	1.60–1.65
Turquoise	1.62–1.65
Water	1.31

of the light and the wider the spread of the different-colored beams (see Fig. 15a). The spreading of a beam of white light into beams of different wavelengths and of different colors is known as the *dispersion of light*. When a beam of white light passes through a transparent prism, for example, the white light is dispersed into different colors ranging in wavelength from about 400 nanometers (violet) to 770 nanometers (red), known collectively as the *visible light spectrum*, shown in Figure 15b.

Most materials exhibit specific colors because they absorb certain wavelengths (colors) from white, ordinary light. A red object exposed to white light, for example, appears red because atoms on its surface absorb all the other colors in the beam and reflect only red. If transparent materials contain coloring materials, such as dyes or pigments, they absorb the characteristic color of the coloring material.

Light Scattering

When a beam of light strikes small particles in its path, the light is said to be *scattered*: the particles send off the light incident on them in many directions. Part of a beam incident on a translucent material, for example, is scattered by particles or air bubbles within the material and only the remainder of the beam passes through. It is the scattering of light that makes it impossible to see through translucent materials. Opaque materials scatter and absorb all the light incident on them, blocking its passage.

Light Interference

When the paths of two light beams cross at a given spot, they are said to *interfere* with each other; they either add light to or subtract light from each other. If the two beams add together to form a stronger, more intense beam, providing brighter light than either beam separately, the phenomenon is known as *constructive interference*. If the beams subtract from each other, they give rise to a weaker, dim beam, or even to a region of darkness. This latter phenomenon is known as *destructive interference*. The interference of white light, for example, results in spectral colored fringes. It is the interference of light that causes the iridescent colors often seen in decayed archaeological glass.

Light Polarization

The electric field that creates the waves of a beam of ordinary light *oscillates* (varies in strength) in many directions perpendicular to that of the path of the beam (see Textbox 5), and ordinary light is said to be *unpolarized*. The phenomenon known as *light polarization* is caused by oscillations (variations in the strength) of the electric fields that make up a wave of light. If the direction of the oscillations points in just one direction the light is said to be polarized. Polarized light is used to characterize amino acids and also to date animal remains (see Textbox 66).

Luminescence

Materials gain energy either when heated or when irradiated with some form of energetic radiation, such as ultraviolet light or gamma rays. The higher the temperature or the amount or radiation, the higher the energy gained. When heated to high temperature most materials lose part of the energy gained by emitting light; at about 450°C rocks, minerals, and metals, for example, become *incandescent*, spontaneously losing part of the energy gained and shining brightly.

Many forms of nonmetallic matter can also lose energy they gained by processes other than becoming incandescent. Some materials previously heated or irradiated with some form of energetic radiation may even lose energy long before becoming incandescent and do so by a process known as *luminescence*. Two forms of *luminescence* are generally distinguished, *fluorescence* and *phosphorescence*. If the luminescence is emitted at the time of irradiation but ceases to be so immediately after the irradiation is discontinued, the luminescence is known as fluorescence. The fluorescence phenomenon is often used in archaeological studies to distinguish between genuine and fake materials (see Chapter 18). If the luminescence persists and continues after irradiation, it is known as *phosphorescence* [phosphorescence can sometimes be seen at night, at sites where bones have been exposed to intense solar (ultraviolet) radiation]. A particular form of luminescence, known as *thermoluminescence*, is emitted when some insulating materials, such as minerals and pottery, are heated below the temperature required for *incandescence*. Thermoluminescence is used to date archaeological pottery and other materials (see Textbox 24).

Other optical properties of gemstones, which also determine their beauty and other characteristics that make some of them unique, include the way they *disperse light* incident on them (see Textbox 22), their *refractive index*, which is unique to, and characteristic of every type of gemstone and is often used for their identification (see Textbox 22), and their *luster*, *adularescence*, *asterism*, and *brilliance*.

Luster, the gleam or brilliance of a solid surface in reflected light, can be either metallic or nonmetallic. The luster of gemstones is nonmetallic, and its characteristics are determined by the translucence and clarity of the mineral. Luster can be dull, glistering, shining, or superior. Other ways of describing luster include one or more of the following terms:

Greasy – when it has an oil-coated look
Pearly – when its iridescence is that of pearls
Resinous – when it has a resinlike look
Silky – when it reflects light in a silk–fiber manner
Vitreous – when it has a glassy appearance

Adularescence refers to changes in the color of the luster when an illuminated gemstone is turned, such as a change from white to pale blue and vice versa, resulting in a remarkable flash across the surface of the stone; changes of this type are particularly characteristic of adularia (a rare gem variety of the mineral feldspar orthoclase), although of other gemstones as well. *Asterism*, whereby light reflected from the stone appears in the form of a radiating star, is characteristic of such gemstones as ruby and sapphire. *Asterism* occurs when light reflected from the gemstone is concentrated into several rays that intersect with one another creating six-point stars, as in some rare gemstones. *Brilliance,* the property of gemstones to appear luminous, refers only to faceted gemstones; brilliance is due to the interaction between light beams incident on and reflected from the gemstone. Nonfaceted gemstones are never brilliant, although they are lustrous.

If the color and light dispersion characteristics determine their beauty, the durability of gemstones, that is, their resistance to being broken or abraded, therefore preserving their beauty, is determined by their *hardness* (see Textbox 23). Few minerals are valued as gems unless they can last in an unimpaired condition, unbroken and lustrous for a long time. Only very hard minerals, rated 7 or more on the Mohs hardness scale (see Table 20), wear well as gems. Quartz, for example, rated 7 on the scale, cannot be scratched with sharp steel instruments or with glass. Diamond, the hardest substance known, rated 10 on the Mohs scale, can scratch any other mineral but is not itself scratched by any material (Anderson 1974).

TEXTBOX 23

HARDNESS

Hardness is the property of solid materials to withstand abrasion, wear, penetration, or deformation by external forces. Some materials, such as the mineral talc, are so soft that rubbing them with the fingers breaks the bonds that hold them together and they become powdery; others, such as soapstone and marble, are somewhat harder but can be shaped by carving or abrasion. Still others, such as topaz and diamond, are extremely hard and are virtually unaffected by external penetrating or deforming forces. The particles that make up these hard materials are very strongly bonded to each other. In diamond in particular, the atoms of carbon are so strongly bonded to each other that diamond is the hardest natural material known; it cannot be scratched by other materials.

The hardness of any solid can be assessed with testing equipment that measures the relative ease with which a surface can be either scratched or penetrated by another solid of known hardness. A long-used classification of the hardness of minerals, the *Mohs scale of hardness,* is based on the *scratch test*: ten rather common minerals are arranged in the scale in order of their increasing relative hardness and listed in a scale varying from 1 to 10 (see Table 20). Each mineral in the Mohs scale scratches those with lower hardness numbers but does not scratch higher-hardness minerals. If a

TABLE 20 Mohs Scale of Hardness

Mohs hardness	Mineral	Common test	Relative hardness
1	Talc	Easily scratched by fingernail	1
2	Gypsum	Scratched by fingernail	3
3	Calcite	Scratched by copper	9
4	Fluorite	Scratched by mild steel	21
5	Apatite	Scratched by hard steel	48
6	Feldspar	Scratched by glass	72
7	Quartz	Scratched by glass	100
8	Topaz	No simple test	200
9	Corundum	No simple test	400
10	Diamond	Scratches all materials	1500

mineral or any other solid, for that matter, can be scratched by feldspar but is not scratched by apatite, for example, its hardness is between 5 and 6.

Other methods of determining the hardness of a material include a variety of "penetration" tests that yield hardness values measured in scales known as the *Brinell*, *Rockwell* (B or C), and *scleroscope scales*. These scales provide reliable hardness values for most materials, including ceramics, glass, metals and alloys, and wood (see Table 21). Unfortunately, as can be seen in the table, the various tests provide somewhat different hardness values for the same materials.

TABLE 21 Hardness: Conversion of Numerical Values between Different Hardness Scales

Brinell	Rockwell (C)	Rockwell (B)	Scleroscope
780	70	–	106
700	66	–	102
600	58	–	81
500	51	118	68
400	42	113	55
300	32	107	43
200	14	94	29
100	–	60	–

Cutting and Polishing Gemstones

The hardness, index of refraction, and transparency of gemstones determine the way they are cut and polished. Since ancient times gemstones have been shaped in two main styles: (1) as faceted gems, with many plane polished sides called facets, or (2) as nonfaceted, rounded, and polished gems, known as cabochons.

Some Archaeological Gemstones

Diamond. *Diamond,* the most brilliant, hardest, stiffest, and least compressible natural material and the most highly valued of all gemstones, consists of crystallized carbon. Its name is derived from the Greek word *adamastos,* meaning indestructible, and diamond is indeed chemically inert to most acids and alkalis. If diamond is heated to high temperatures in an oxidizing atmosphere, however, it burns, combining with oxygen to form carbon dioxide. The transparency of diamonds extends through the whole range of wavelengths of infrared, visible, and ultraviolet radiation. Their brilliance is due to a very high refractive index and a high dispersion of light (see Textbox 22). Gem diamonds are colorless and clear if composed only of carbon. Trace impurities within the crystal structure produce, however, diamonds in all the colors of the rainbow. Boron impurities, for example, which give diamonds a bluish color, and nitrogen, which adds a yellow cast, are common trace impurities. Since antiquity, diamonds have been appreciated for their beauty; and powdered diamonds have been used as abrasives for smoothing, cutting, carving, and drilling other gemstones and hard materials.

Emerald. *Emerald* is a brilliant, rich green, rare variety of the mineral beryl, which has been widely appreciated since antiquity as a gemstone. It is a relatively hard stone, 7.5 on the Mohs scale, harder than quartz, but not as hard as sapphire. Pure beryl (composed of beryllium aluminum silicate) is colorless, and the green of emeralds is due to trace amounts of chromium in the crystals (Ward 1994; Webster 1955). The name of the gemstone is derived from the French *emeraude,* which in turn goes back via Latin (*smaragdinus*) to the Greek root *smaragdos,* meaning simply "green gemstone." The provenance of emeralds set in a in a large variety of artifacts, dating from Gallo-Roman times (the first century B.C.E.) to as late as the eighteenth century, has been studied, using the weight ratio between the isotopes of oxygen in the gemstone as an indicator of its provenance (see Textbox 47). The study revealed that most of the analyzed emeralds from this long period of time originated from deposits that supposedly had been discovered as recently as the twentieth century. It seems that emeralds from Pakistan and Egypt, for example, were being traded already in antiquity by way of the Silk Route. Moreover, emeralds from these two sources and from Austria were the only ones from which European and Asian traders obtained gem-quality emeralds. It was only during the sixteenth century that a new source appeared and a new trade route for emeralds evolved; emeralds mined in Colombia were initially traded via Spain to Europe, the Middle East, and India, and later on directly via the Philippines to India (Giuliani et al. 2000, 1998).

Jade. *Jade* is the name shared by two usually green but distinct minerals that have been appreciated for many millennia. One is *nephrite* (composed of hydrous calcium, magnesium, and iron silicate), the most abundant in nature and apparently the earliest known form of jade. The other is *jadeite* (composed of aluminum sodium silicate), the most valued form of jade; jadeite is also known as Chinese jade, although it is found naturally in Burma and not in China (see Table 22). Although nephrite and jadeite are similar in appearance, they differ not only in composition but also in their index of refraction, petrographic microstructure, and X-rays diffraction (XRD) pattern. Thus, determining any one of these properties makes it possible to differentiate between the two. Jadeite is usually green, but white and red varieties are also common. It is remarkably tough and hard to carve or polish, the polished surfaces having a "greasy feeling" to the touch (Desautels 1971; Laufer 1912).

In ancient times the Chinese buried jade with their dead in the belief that the stone would prevent decomposition of the corpse. During the Han Dynasty (third century B.C.E. – third century C.E.), for example, pieces of jade were placed in each of the body orifices to ward off decay. It is ironic, therefore, that in contact with the corpse jade decayed and its hard, polished surface became damaged by chalky blemishes (Gaines and Handy 1975).

TABLE 22 Jade Minerals

Mineral	Color	Composition	Specific gravity	Hardness (Mohs scale)
Jadeite	Mostly green, but also white and red	Silicate of aluminum and sodium	3.3–3.5	6.5–7
Nephrite	Yellow, brown, green, blue, red, lavender	Silicate of calcium, magnesium and iron	2.9–3.2	6 –6.5

Ruby and Sapphire. *Ruby* and *sapphire* are "sister stones": both are gemstone forms of the mineral *corundum* (composed of aluminum oxide). Pure corundum is colorless, but a variety of trace elements cause corundum to exhibit different colors. Ruby is red corundum, while sapphire is corundum in all colors except red. The red in rubies is caused by trace amounts of chromium; the more intense the red color of a ruby, the more chromium it contains. The blue in *sapphires* is caused by titanium and/or iron impurities (Garland 2002; Hughes 1997).

Turquoise. *Turquoise* is an opaque, usually blue, mineral (composed of hydrated aluminum copper phosphate) that has been widely prized as a semiprecious gemstone for well over 8000 years, having been used to make beads, rings, and other decorative objects. Deposits of turquoise are formed when water flows and dissolves part of rocks rich in aluminum and then evaporates. The blue color of the stone is due to the copper; iron impurities give rise to shades of blue-green and even to green turquoise. Turquoise deposits occur in only a few and limited arid regions of the world, such as the Sinai Peninsula in the Middle East and the southwest of North America. There seems to have been a wide network of transportation and trade of turquoise from the places where turquoise was extracted to where it was worked and eventually used. Identifying the distribution routes of ancient turquoise has attracted much attention, and a number of investigations have been centered on establishing the distribution of turquoise from various mining regions. In one such study, concerned with turquoise beads in North America, differences in the nature and concentration of impurities in the gemstone were used to distinguish between turquoise from pre-Columbian sites in the midwest of the United States and some parts of Mexico. Most of the turquoise analyzed seems to have originated from mines in the southwest of North America, from where it was distributed, through well-established trade routes, to the workshops where it was worked (Harbottle and Weigard 1992; Sigleo 1975).

LITHICS
FLINT AND OBSIDIAN

2

Lithic artifacts, made for utilitarian or decorative purposes in a variety of shapes, were used by ancient humans for many millennia. Only after the discovery of the properties of metals did these gradually replace stone as the main material for making artifacts. Studying the properties of stone and of lithic artifacts is important, therefore, for tracing the early stages of human development (Edmonds 2001; Andrefsky 2005, 1998). Traditionally, the main guidelines for such studies have been typological: the artifacts, tools and ornaments are classified according to their shape into types, and the frequency of occurrence of the different types are investigated, scrutinizing the attributes and the variability of each type. There are, however, other criteria for studying lithic artifacts. The composition and physical properties of the stone from which artifacts are made, as well as alteration of the surface of the artifacts by human activity, provide important complements to typological analysis. Flint, a variety of the mineral quartz, and obsidian, a natural glass, are the two types of stone most widely used for making lithic artifacts.

2.1. QUARTZ AND FLINT

Flint is a hard and easily split variety of the mineral *quartz* (composed of silicon dioxide), which occurs not only as flint but also in a wide range of other varieties. Some of these exhibit different colors and colored patterns and have characteristic crystalline structures, while others are amorphous (see Textbox 21). In all its varieties, nevertheless, the hardness of quartz is very high, being graded as 7 on the Mohs scale (see Textbox 23). Almost all varieties are either transparent or translucent and display a distinctive luster. These properties made quartz an attractive material for making ornamental

Archaeological Chemistry, Second Edition By Zvi Goffer

and decorative objects and small artifacts. Structurally, three main forms of quartz are generally distinguished: macro-, micro- and cryptocrystalline (or amorphous) quartz. *Macrocrystalline quartz* occurs as relatively large crystals, large enough to be seen by an unaided eye. Rock crystal and amethyst, for example, are macrocrystalline varieties of quartz, valued as minerals of gemstone quality. *Microcrystalline quartz*, examples of which are the semiprecious gemstones jasper and agate, are made up from extremely small crystals that can be made visible and differentiated only under very high magnification. Flint belongs to the *cryptocrystalline*, also known as the amorphous variety of quartz. The extremely small crystallites in cryptocrystalline quartz are too small to be distinguished even under high optical magnification. Another common feature of all the cryptocrystalline varieties is a very compact structure that lacks cleavage (they cannot be split along planes of weakness intrinsic to the structure of crystals). A system of nomenclature in which the different forms of quartz often encountered in archaeological contexts are consistently identified, classified, and described is given in Table 23 (Sax and Middleton 1992).

Flint (also known as *silex*) was used for making lithic tools because when fractured by concussion, it breaks with a conchoidal fracture and the broken parts have smooth, sharp edged curved surfaces that resemble the shape of the shell of a bivalve. Flint frequently occurs in the crust of the earth embedded in chalk or other varieties of limestone; another occasional form of occurrence is as large flat slabs, only about 20 cm thick but usually very long, sometimes reaching a length of about 2 m (Sieveking and Bart 1986; Shepherd 1972). Silica, the main component of flint, usually makes up at least 98% of its mass. Interspersed within the silica there are usually pores, that may be filled with water, which makes up the remaining 2% of the weight of flint. If the water filling up the pores contains dissolved minerals, these may be characteristic of the geologic surrounding where the flint was formed. Determining the nature and the relative weights of the dissolved minerals may be useful for establishing the provenance of flint. Flint tools are shown in Figure 19.

When it consists of only silica and water, flint is basically colorless. Impurities within the stone, however, render colored varieties. Thus, much flint is tan, beige, or jet black with a brown horny appearance (the latter color is caused mainly by iron impurities), but there are also gray, pink, and even red varieties. If dark flint is heated to temperatures above 450°C, the outer layer changes to white. Flint that underwent such a heating process is known as *burned* or *calcinated flint*, a misnomer probably used because the white surface appears to have changed into lime (Luedtke 1992; Sieveking and Bart 1986). Burned flint is easier to work than is the natural mineral, a property

TABLE 23 Quartz Minerals

Variety	Luster	Light transmittance	Color	Other features
Macrocrystalline Quartz				
Amethyst	Vitreous	Transparent or translucent	Purple	Banded or uneven
Cat's eye, tiger's eye	Banded	Translucent	Yellow/yellow-brown	Banded
Ferruginous quartz	Vitreous	Transparent or translucent	Red/yellow-brown	Inclusions of iron and asbestos
Rock crystal	Vitreous	Transparent	Colorless	Conchoidal fracture
Rose quartz	Vitreous	Transparent or translucent	Pink/red-rose	Usually cracked
Smoky quartz	Vitreous	Transparent or translucent	Pale/black	Often banded
Microcrystalline Quartz				
Agate	Waxy	Translucent	Banded	Irregular, sometimes circular bands
Carnelian	Waxy	Translucent	Red/brown	Many forms, colors, and shapes
Chalcedony	Waxy	Transparent or opaque	Various	Splintery fracture
Moss agate	Waxy	Translucent	Various	Inclusions
Sardonyx	Waxy	Translucent	Banded	Whitish and brown-red bands
Cryptocrystalline Quartz				
Chert	Vitreous to dull	Translucent	Like flint	Splintery fracture
Flint	Waxy to dull	Translucent	Various	Conchoidal fracture
Jasper	Waxy to dull	Opaque	Various	Smooth fracture

Source: Sax and Middleton (1992).

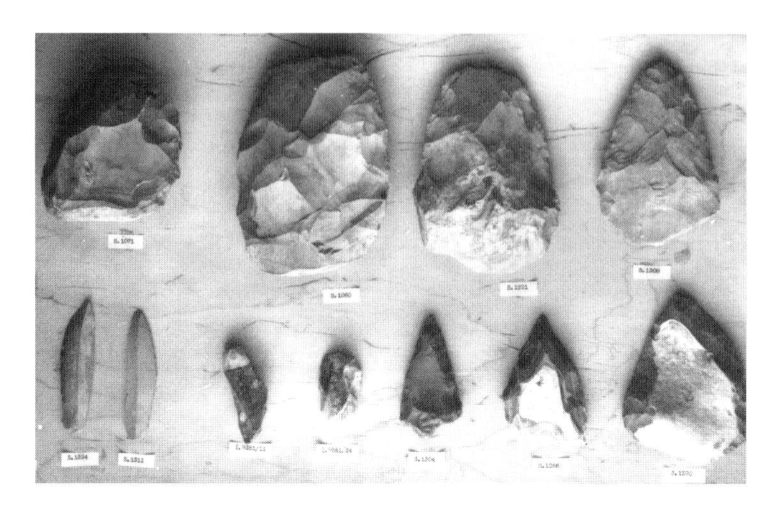

FIGURE 19 Flint tools. Axes, scrapers, and knives. Flint is a hydrated form of cryptocrystalline silica that occurs naturally as irregular nodules in chalk deposits. It is colorless and translucent when pure, but opaque and often colored when it contains impurities. When struck, flint breaks with a conchoidal fracture and the fragments formed have smooth, sharp edges. It is for this reason that prehistoric humans used flint to make tools. Since it also sparks when struck, flint was also used, until the eighteenth century, for lighting fire.

that was known and utilized by lithic toolmakers since very early times. Some varieties of flint split and break apart when heated at a fast rate. Therefore, in order to burn flint so that the product is suitable for making tools, the flint has to be heated gradually. The time when flint was heated to a high temperature and thus converted into burned flint can be dated using the *thermoluminescence dating* technique (see Textbox 24), and the dating of burned flint has received some attention (Aitken 1985; Bowman 1982).

TEXTBOX 24

THERMOLUMINESCENCE AND OPTICALLY STIMULATED LUMINESCENCE DATING

Thermoluminescence is the emission of light by moderately heated electrically insulating solids. Many minerals, rocks, and ceramics, when heated to a high temperature but below that required for them to become incandescent, emit light. The emission of light is related to the structure and previous environmental conditions of the solids. All solids on the earth's surface are continuously irradiated by energetic electromagnetic radiation, such as solar ultraviolet other forms of extraterrestrial radiation, and gamma radiation from radioactive elements (potassium-40, uranium, and thorium, for example). Some of these forms of radiation, known as *ionizing radiation*, such as gamma and X-ray, interact with the atoms that make up the solids, converting them to *excited atoms*, atoms having more energy than regular atoms. The greater part of the energy that the atoms acquire is soon released back to the surrounding environment, generally as heat, and most excited atoms soon return to their regular condition. Some of the added energy, however, may become "trapped" in imperfections or "defects," as imperfections within the structure of solids are usually termed, and not be released back. Over long periods of time, as more ionizing radiation interacts with a solid, more energy is trapped in the defects. Under normal conditions, that is, if the solid remains at the rather constant ambient temperature that prevails in most of the earth's outer crust, the added energy remains trapped within the defects indefinitely, while more energy is continuously being trapped. Only under special circumstances, as when a solid is heated to a relatively high temperature (above about 300°C) but below its melting point, is the trapped energy released, in the form of visible light. As the solid loses the trapped energy, the excited atoms in the solid lose the added energy (as light) and return to the condition of regular atoms. The light emitted as a result of heating is known as *thermoluminescence* (TL). A graphic representation of the thermoluminescence released by a heated solid, as a function of the temperature is illustrated in Figure 20 and is known as a *thermoluminescence glow curve*.

The thermoluminescence emitted by any type of solid is regulated by two factors: (1) the sensitivity of the solid to ionizing radiation; (2) the total amount of ionizing radiation energy that the solid acquired in the past. This implies that the total amount of energy trapped in the a solid is determined not only by the length of time the solid is exposed to radiation, but also by its sensitivity to trapping the radiation energy. Measuring two parameters, therefore, the sensitivity of a solid to ionizing radiation and the amount of thermoluminescence it emits when heated, enables one to date the solid, that is, to determine the time elapsed since it was last at high-temperature. If instead of using a regular heater, a beam of laser light is used to heat the sample and to stimulate the release of luminescence, the process is known as *optically stimulated luminescence* (OSL).

From a practical perspective, determining the time elapsed since the solid was at high temperature by thermoluminescence or optically simulated luminescence requires four successive stages:

1. Removing and preparing a representative sample of the solid.
2. Heating the sample with a conventional heat source or a laser and measuring the intensity of the luminescence emitted by the sample.
3. Relating the intensity of the thermoluminescence emitted to the amount of ionizing radiation absorbed by the solid since it was last hot. This is usually done by irradiating the sample with a well-calibrated source of ionizing radiation, then measuring the luminescence it emits after the irradiation, and finally comparing this value with the value measured in stage 2.
4. Calculating the average annual amount of ionizing radiation to which the solid was exposed over time. With the data at hand, the following relatively simple formula yields the time elapsed since the solid was last at a high temperature:

$$\text{thermoluminescence age} = \frac{\text{total amount of accumulated radiation}}{\text{amount of radiation accumulated per year.}}$$

Both the thermoluminescence and optically stimulated luminescence techniques are suitable for dating past heating events over a timespan ranging

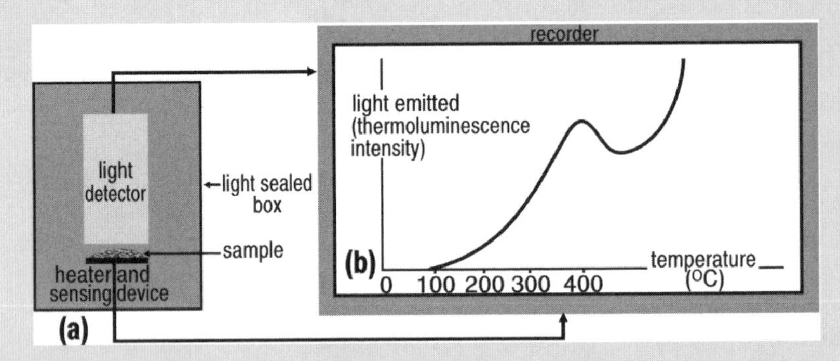

FIGURE 20 Thermoluminescence dating. Thermoluminescence (TL) is light emitted when electrically nonconductive solids, such as stone or pottery, are heated to high temperature although below their incandescence point. The intensity of the light emitted is related to the amount of energy the material absobed when exposed to surrounding ionizing radiation. The thermoluminescence emitted when a stone or a piece of pottery is heated therefore indicates the length of time since the solid was last heated to high temperature. An experimental setup, required to heat up a sample so that thermoluminescence is emitted and can be detected, is illustrated **(a)**. A recorded graph **(b)** usually provides the range of temperatures at which the material is heated and the intensity of the light emitted at each temperature. From these two values, the length of time elapsed since the material was at high temperature can be calculated.

from a few hundred to several hundred thousand years. This makes these techniques applicable for dating archaeologically related natural materials such as stone and sediments, human-made ceramics, and some types of glass. Ceramics and glass are fired during manufacture at temperatures exceeding 600°C. Such high temperatures are sufficient to release energy trapped in the "defects" of minerals components of these materials. Thus, ceramic-firing and glass-melting operations set the thermoluminescence "clock" of these materials to zero time and make it possible to date them using the technique. Animal remains (bone and shell) that underwent thermal events have also been dated using the thermoluminescence technique (Troja and Roberts 2000).

It should be mentioned, however, that although thermoluminescence dating has already been in use for about half a century, the dates derived with the technique are not yet accurate enough to be relied on in archaeological investigations. Thermoluminescence-derived dates are thus still considered to be in their developmental stage; they are used mainly for testing the authenticity of archaeological finds, not yet to determine their exact age (Aitken 1998; McKeever 1988).

2.2. OBSIDIAN

Obsidian is a dense volcanic solid often formed in lava flows where the lava cools so quickly that crystals cannot grow. Although it has a rather definite chemical composition, obsidian lacks the regular crystal structure characteristic of the minerals and is classified as a mineraloid, more specifically, as a natural glass. Like flint, when obsidian is struck it also breaks with a *conchoidal fracture* and the broken pieces have sharp edges; thus, obsidian was the type of stone most widely used for making lithic tools in areas where flint was not available. Chemically, obsidian consists of a mixture of metal silicates, a composition it shares with the mineral rhyolite, which also originates from molten magma. Rhyolite, a crystalline material, is formed when erupted magma cools down gradually and, while solidifying, its component atoms become regularly ordered in both short- and long-range geometric arrangements. Obsidian, on the other hand, is formed when the magma cools down rapidly, does not crystallize, and remains as a *supercooled liquid*, that is, an amorphous, hard, brittle glass (see Textbox 27). Often, microscopic crystals and gas bubbles locked within the glass render much obsidian opaque and usually gray or black. Colored obsidian, in a variety of colors, albeit not common, is also known. Red and brown varieties, for example, owe their color to inclusions of iron oxides. A semitransparent, smoky variety of obsidian, known as *marekanite* was valued as a semiprecious stone by some North American Indians (Doremus 1994; Shackley 1998, 1977).

Obsidian artifacts have been made for a variety of uses, and large numbers of artifacts, as well as of obsidian flakes (the waste produced when making the artifacts), are often found in many prehistoric sites. For this reason a number of early investigations centered on identifying some physical or chemical properties of obsidian that may provide information about its provenance. Most of the physical properties vary, however, in an irregular manner; the density and refractive index of obsidian from different sources, for example, overlap and provide inconclusive results regarding their place of origin. These properties and others that do not contribute to differentiating between different types of obsidian are therefore of very limited value for their identification. The chemical composition of obsidian from different sources is not identical, however. Although the major components of most obsidian are quite similar, there usually are great differences in the nature and relative amounts of the minor and trace elements they contain. Obsidian from single volcanic outflows, for example, generally is of homogeneous trace element composition, but different outflows are not (see Table 24). These differences make it possible to identify obsidian from a single source and to distinguish between obsidians from different sources. This is important in archaeological studies, since it is well known that obsidian, in the form of rather large cores, and in some cases finished obsidian artifacts, were widely traded already during prehistoric times (see Fig. 21). The cores or finished artifacts were generally transported from relatively few obsidian outflows in areas of volcanic activity in America, Asia, and Europe to the many sites where they were put to use (Shackley 1998, 1997).

Obsidians from different sources (outflows) are chemically distinguishable from one another; it is reasonable to deduce, therefore, that identity between the composition of obsidian objects or wastes and a particular

TABLE 24 Obsidian: Typical Elemental Composition

Component	Percent
Silica	63–76
Alumina	11–14
Sodium oxide	3– 6
Potassium oxide	3– 5
Iron oxide	1– 8
Others	below 1

source indicates that the material used to make the objects, or rejected as waste, was derived from that source (Neff 2000; Glascock et al. 1998). Based on this assumption, a number of studies have revealed quite clearly the distribution of obsidian tools in relation to the sources from which obsidian cores were obtained (see Table 25a) (Dixon et al. 1968).

In the Mediterranean Sea and Middle East area, for example, there are obsidian outflows only in Italy, in some islands in the Aegean Sea, and in Turkey. Artifacts made of obsidian, however, are widely distributed over much of this vast area. Chemical analysis of many of these artifacts has shown that most of the obsidian used to make them originated in one or another of the outflows mentioned, but also in far-distant places such as Armenia and Iran. Plotting on a graph the concentration of selected elements in samples from obsidian sources against that in samples from sites where it was used, enables the identification of the source of the samples (see Fig. 22). Moreover, this type of analysis also makes it possible to trace the routes through which obsidian (and most probably other goods) were traded in antiquity (Renfrew and Dixon 1976).

Also in the Mediterranean Sea region, obsidian artifacts are frequently found at prehistoric sites in northern Italy and southern France. Most of the

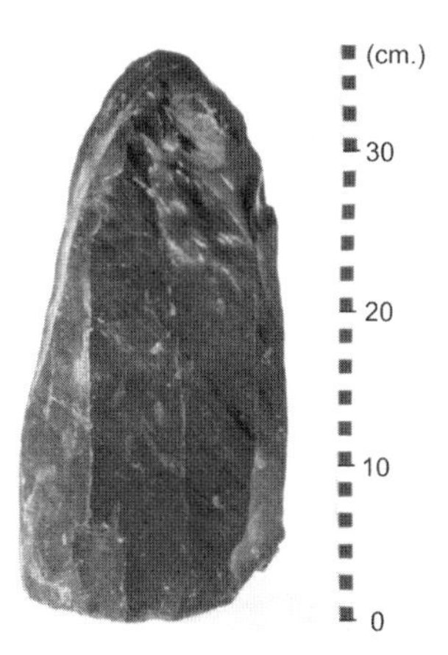

FIGURE 21 Obsidian core. Obsidian is a mineraloid formed when volcanic lava solidifies quickly so that there is not enough time for crystals to grow. When lava pours into a lake or ocean, for example, it cools down swiftly and remains in an amorphous solid condition, having the glassy texture characteristic of obsidian. Obsidian was used by ancient people to make pounding and cutting tools, such as hammers, axes, knives, and arrowheads. Natural obsidian resources are limited, so that the stone was traded in the form of massive cores as the one illustrated, which were relatively easy to transport over long distances. From these low bulk cores a large number of small items could be broken away and shaped into tools.

TABLE 25a Analytical Techniques Used to Characterize Archaeological Obsidian

Technique	Remarks	References
Electron microprobe analysis	Nondestructive	Moens et al. (2000)
Inductively coupled plasma emission spectroscopy (ICPE)	Destructive, although very small samples are required; major and trace elements can be determined	Tykot and Young (1996); Heyworth et al. (1988)
Mössbauer spectroscopy	Nondestructive	Scorzelli et al. (2001)
Neutron activation analysis	Nondestructive, but samples remain radioactive for some time after analysis	Asaro et al. (2002); Grimanis et al. (1997)
Proton-induced gamma emission spectroscopy (PIGE)	Nondestructive	Gratuze et al. (1993)
Strontium isotope analysis	Requires a very small sample	Gale (1981)
X-ray fluorescence spectroscopy (XRF)	Wavelength-dispersive XRF is generally destructive; not so energy-dispersive XRF	Giauque et al. (1993)

obsidian at these sites has been identified as originating in the islands of Sardinia, Lipari, Palmarola, and Pantelleria and, to a much lesser extent, from outflows in the Carpathian Mountains of central Europe. The identification of island obsidian on mainland sites points to the maritime navigational capabilities of people from the Neolithic period during approximately 6000–3000 B.C. (Tykot 1996).

Dating Obsidian

Freshly exposed surfaces of obsidian, such as those created when obsidian breaks or is flaked, react with environmental moisture (i.e., water), and the product of the reaction forms a thin layer of water-rich obsidian on the obsidian bulk. The surface is said to become *hydrated* while the underlaying bulk remains unaltered, as it is affected by neither the water nor other weathering processes (see Textbox 25). Microscopic studies have shown that the thickness of the hydrated layer depends on the relative amount of the water

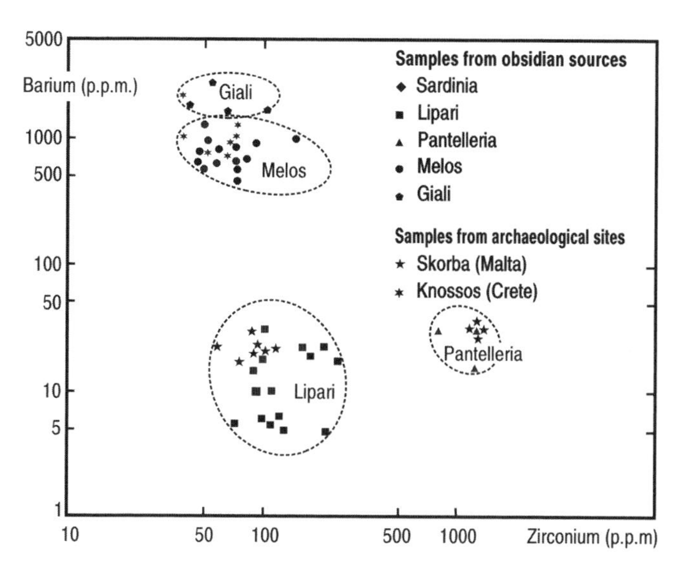

FIGURE 22 Obsidian in the eastern Mediterranean Sea area. Studying the relative concentration of trace elements in obsidian makes it possible to identify the obsidian and to determine its provenance. Determining the relative amounts of barium and zirconium in ancient obsidian tools and in samples from different sources of the natural glass, for example, made it possible to identify the provenance of obsidian used in eastern Mediterranean Sea area sites (Renfrew and Dixon 1976).

it contains: the more water in the obsidian, the thicker the hydrated surface layer becomes. With the passing of time and exposure to humidity, the exposed surface continues to react with water, the hydration process continues, and the hydrated surface layer gradually thickens. The thickness of the layer can be used for dating the time when the obsidian surface was first exposed to water or humidity (Liritzis et al. 2006; Ambrose 2001).

TEXTBOX 25

THE HYDRATION OF OBSIDIAN

The natural composition of obsidian includes very little water, generally less than 0.1%. When new obsidian surfaces are created, either by the natural breakdown of obsidian bulks or by human activity, the exposure of a new surface to humidity in the air or to water brings about a process known as *hydration*: the surface *adsorbs* (takes up) water and becomes *hydrated*. The hydrated layer may contain as much as over 10 times the amount of water in the interior, bulk obsidian.

Once initiated, and provided the surface continues to be exposed to the environment, the process of hydration continues at a slow, but measurable rate. The adsorption of the water is accompanied by changes in the physical properties of the obsidian. The *refractive index* of the obsidian, for example, is altered as it becomes hydrated. If the obsidian was subjected to alternative wet and dry periods, successive hydrated layers are formed on the surface. The differences in refractive index between the bulk and the hydrated layer (or layers) creates an interface between the bulk and the hydrated layer, and between the layers, that stands out sharply when observing a cross-cut section of obsidian under a microscope (see Fig. 23). Thus the thickness of the hydrated layer, or layers, can be measured.

The rate at which the hydration process proceeds, that is, the amount by which the layer grows in thickness per unit time (e.g., per year), is deter-

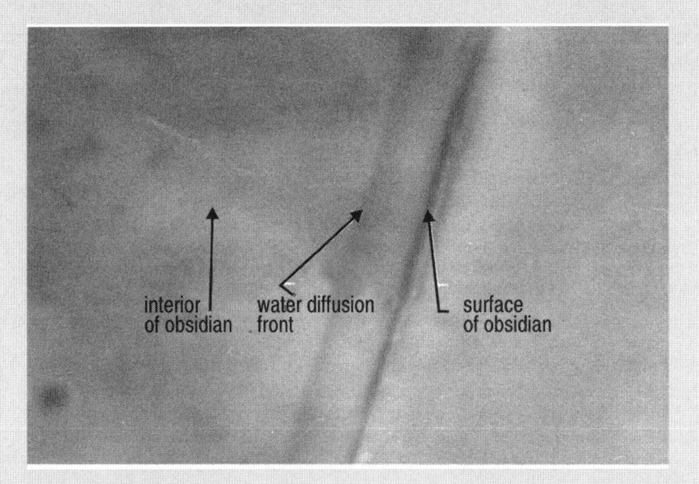

FIGURE 23 Hydration layer in obsidian. When obsidian is broken into two or more pieces, new surfaces are created. As a new surface is exposed to the environment, water (from atmospheric humidity, rain, or the ground) penetrates the surface; gradually, the water diffuses into the bulk and forms "hydrated obsidian," that is, obsidian containing water. With time, the thickness of the "hydration layer," as such a layer is known, gradually increases; the rate of increase is affected by such factors as the vapor pressure of the water in the atmosphere, the environmental temperature, and the composition of the surrounding environment as well as of the obsidian. If the hydration layer reaches a thickness of 0.5 microns or more, it becomes discernible under a microscope, the thickness can be measured, and the age of the surface calculated. The microphotograph shows an hydration layer on obsidian.

mined by a number of factors. These include the actual composition of the obsidian, the prevailing environmental temperature and relative humidity of the surroundings, the nature of the soil and of the groundwater if the obsidian is buried, and/or the extent of the radiation if exposed to solar radiation. It has been shown that the rate of increase in thickness of the hydration layer with time can be expressed by a mathematical equation. Specifically:

$$T = kt^{1/2}$$

where T is the thickness of the hydrated layer, t is the time since the surface was first exposed, and k is a mathematical constant dependent on all the factors mentioned before, which affect the rate of the hydration process.

In simple terms, the meaning of the equation is that by measuring the thickness of a hydration layer on the surface of a piece of obsidian or of an obsidian tool, it is possible to calculate when the surface was first created and became exposed to the environment (Stevenson et al. 2000; Friedman and Smith 1960).

Dating a substantial number of samples ranging in age from about 200 to 100,000 years has shown that the error of the calculated ages may be as high as 10%. Such a large margin of error is due, apparently, to the uncertainty of two factors that affect the technique: the temperature history at the site of burial, which determines the rate at which the hydrated layer grows, and the accuracy in measuring the thickness of the layer. The methodology of the technique, as presently used, may be in need of improvement before it is widely accepted as a reliable archaeological dating tool (Anovitz et al. 1999).

The rate at which the surface of obsidian reacts with water is regulated by the surrounding temperature; the hydration process is steady at constant temperature, slows down as the temperature is lowered, and accelerates at higher temperatures. Thus, provided the past temperature at the site where a piece of obsidian has been lying is known or can be estimated, measuring the thickness of the hydration layer makes possible dating the time when its surface was created. Moreover, since at equal and steady temperature the hydration layer grows at a constant rate, obsidian artifacts having layers of equal thickness are of similar age, contemporary with each other; those with thicker layers are older, and those with thinner layers are younger (Ambrose 2001; Friedman et al. 1997).

Another technique that has been used for dating obsidians, albeit only those that were heated in the past, is the *fission track* technique (see Textbox 26) (Fleisher et al. 1975). Fission track dating was used, for example, to date the time when an obsidian knifeblade found at a cave in Elmenteita, in Kenya, was heated (see Fig. 24). Since the shape of the blade was distorted, it was deduced that the blade had been heated to a temperature sufficiently high to partially melt it.

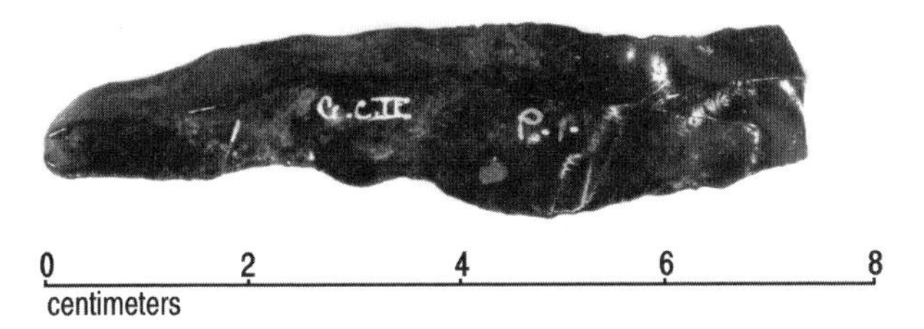

FIGURE 24 Obsidian knife. A knife made of obsidian in Mesolithic times and found in Elmenteita, Kenya, which was heated in the distant past. The time of heating (3700 ± 900 years ago) was determined by the fission tracks dating method (see Fig. 25).

TEXTBOX 26

FISSION TRACK DATING

Uranium is a heavy element that has a number of isotopes (see Textbox 16). Minerals and rocks as well as human made materials such as ceramics and glass often contain trace amounts of uranium as impurities. The most abundant isotope of this element, uranium-238, is radioactive and most of it decays into thorium-234 by the emission of alpha particles:

$$\text{Uranium-238} \rightarrow \text{Thorium-234} + \text{alpha particle}$$

One in every 2 million atoms of uranium-238, however, decays through a different process, known as *spontaneous fission*. This is an explosive occurrence during which, without any external stimulation, the isotope splits into two fragments of approximately equal mass that fly apart at great velocity, while a large amount of energy is released. When atoms of uranium-238 decay by spontaneous fission in electrically nonconducting solids such as minerals, ceramics, and glass, the rapidly moving fragments of the split atom collide with and disrupt other atoms in their path. While doing so, they plough out, in the surrounding solid, submicroscopic trails of damage, known as *spontaneous fission tracks*. The tracks are extremely small (about 0.0000003 mm wide and 0.015 mm long), too minuscule to be visible even when viewed under an ordinary microscope. They can, however, be seen under an electron microscope, which provides much greater magnification: if a freshly cut and polished surface of an uranium containing nonconducting solid is viewed under an electron microscope, the tracks are seen where atoms of uranium-238 disintegrated near the new surface (Fleischer et al. 1975, 1965a, 1965c).

For any uranium-containing solid, the number of uranium-238 atoms that have disintegrated increases with age, as does the number of fission tracks. When most solids are heated to a sufficiently high temperature, the tracks are erased, although the extent of the erasure varies: in some solids the tracks are erased only at very high temperatures, just below the incandescence temperature of the solid; in others, the tracks are erased at lower temperatures, known as the *healing temperature of the solid*; in another few, the temperature required for erasure is so low that most of the tracks are gradually erased even at ambient temperature. Determining the uranium content of a solid in which the tracks are erased only at high temperatures and counting the number of tracks in the solid provide a way to calculate its age.

Practical difficulties, related to viewing and counting the extremely small tracks electron microscopically, however, might render this dating method too cumbersome to be useful. Fortunately, the tracks can be enlarged by a relatively simple chemical procedure: treating a surface of the solid with an *etching* agent (an acid solution that preferentially dissolves those regions of the solid surrounding the tracks) renders the tracks visible under an ordinary microscope. Moreover, repeated but thoroughly controlled etching off such a surface enlarges the tracks enough to be easily countable under a regular optical microscope, as shown in Figure 25 (a). This procedure makes the fission track method a valuable method for dating archaeological materials and tools (Westgate et al. 1997; Wagner and van den Haute 1992).

Having enlarged and counted the tracks, all that is required to determine the age of a suitable solid by the *fission track method* is the determination of two values:

1. The track density, that is, the number of tracks per unit surface area
2. The concentration of uranium in the solid

The track density can be easily determined on a newly cut, polished, and etched surface by counting, under an optical microscope, the number of etched tracks in a measured area of the solid. The uranium concentration can be determined by a number of analytical techniques. Following these measurements, the age of the solid can be calculated, to a good approximation, using a mathematical relationship (Wagner and van den Haute 1992; Nishimura 1971). The *fission track age* thus determined measures the time elapsed since a solid was first formed, or since it was heated to a temperature sufficiently high to erase previously created fission tracks.

Uranium impurities in most natural solids generally occur in very low concentration. The density of tracks is, therefore, extremely low, usually too low to be measured. Thus, the fission track method is generally appli-

(a) fission tracks in obsidian (microphotograph)

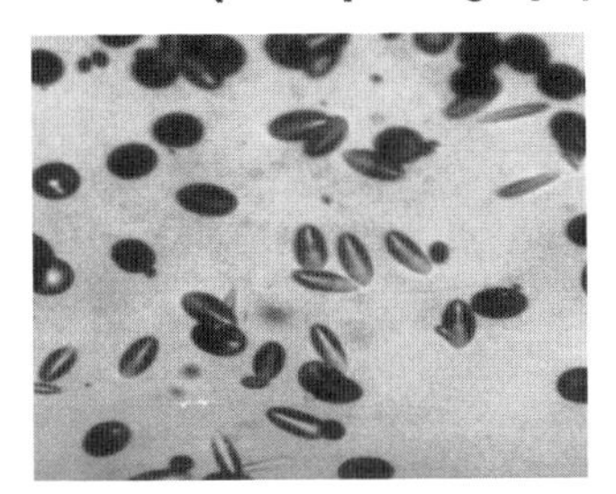

(b) the formation of fission tracks

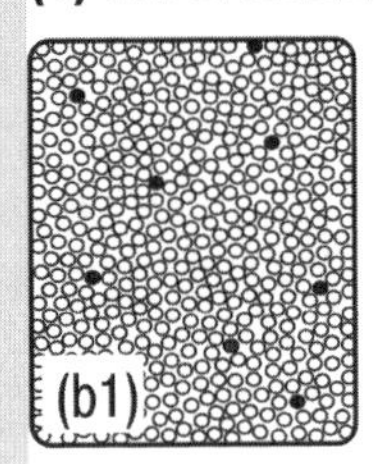

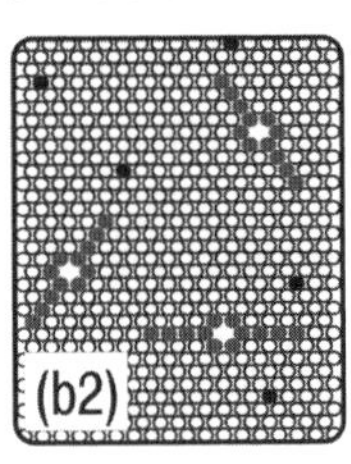

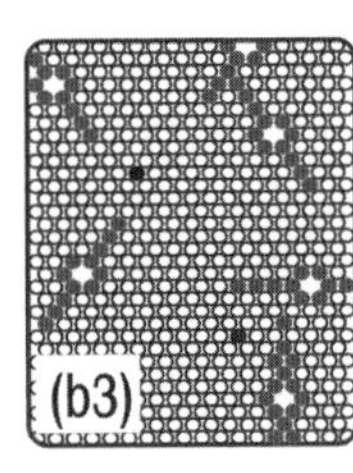

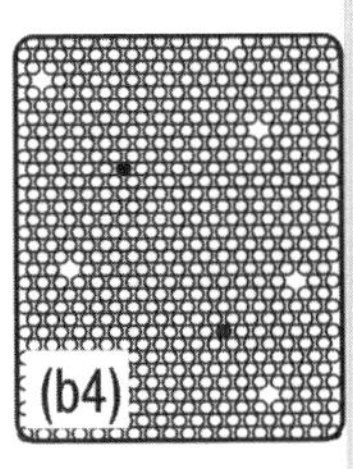

reference: ○ matrix atoms • uranium-238 • dislocated matrix atoms

FIGURE 25 Fission tracks. Fission tracks are extremely small trails in electrically nonconducting materials such as minerals, or synthetic (human-made) ceramics and glasses, which contain even the smallest amount of uranium impurities. In the microphotograph **(a)** can be seen fission tracks that were etched by a chemical process on a polished surface of obsidian. Melts, from which many solids are formed, do not record fission events **(b1)**. Only after a melt solidifies are fission tracks recorded and preserved, their number increasing with time **(b2)** and **(b3)**; the latter, which is the oldest, has more tracks. Fission tracks are thermally unstable: at temperatures higher than ambient, the tracks fade and may totally disappear **(b4)**. Since the number of fission tracks in materials such as archaeological obsidian, pottery, and glass increases over time, counting the number of fission tracks in these materials provides a way for determining when they were last at high temperature.

cable for dating mineral and rocks that are very old (at least 300,000 years old); however, "young" minerals and rocks with a relatively high uranium-238 concentration can also be dated. If uranium-containing stone were heated in the past, either intentionally or accidentally, when it was used to make tools or artifacts, or if tools or artifacts made from the stone were heated after they were made, the old fission tracks created before heating would have been erased. Provided such stone contains a relatively high concentration of uranium-238, it is possible, using this method, to date the time of heating. The error in the dates determined by the fission track method is, even under favorable circumstances quite high, usually on the order of 10%.

Heating a material to a high temperature erases fission tracks previously formed. Tracks detected in the Elmenteita knife were created, therefore, after the knife was heated. Examining a small polished and etched area on the surface of the knifeblade, disclosed only a small number of fission tracks; counting them revealed that the knife had been heated about 3700 ± 900 years ago; the wide range of uncertainty of the date, nearly 25%, it was claimed, is due to the low number of tracks in the obsidian (Fleischer et al. 1965b). The fission track method has also proved effective for dating quite a number of other archaeological finds, some of them listed in Table 25b.

TABLE 25b Fission Track Dating of Obsidian

Site	Reference
Anatolia	Bigazzi et al. (1994)
Ecuador	Miller and Wagner (1981)
Ecuador and Colombia	Poupeau et al. (1996)
Hadar Formation, Ethiopia	Aronson et al. (1976)
Olduvay Gorge, Tanzania	Fleischer et al. (1965a, 1965b, 1965c)
Zhoukoudian, China	Guo Shilun et al. (1990, 1980)

2.3. USE WEAR ANALYSIS

The ancient use of lithic tools is generally determined from an analysis of their shape. The latter is frequently assessed by analogy to modern equivalents, or by making experimental replicas and using them for different tasks.

Use may also be determined by examining the way the tools wore when used and by analyzing residues on their surface.

Lithic tools are altered by use and wear, which results in the formation, on the tool surface, of patterns of abrasion marks and layers of residues (Astruc et al. 2003; Semenov 1964). Repeated use of the tools for such tasks as slaughtering animals or cutting wood for example, leaves on their surface characteristic wear striation grooves and layers of different residues, such as blood remnants of animal tissues or vegetable sap, cellulose, or other vegetable tissues. Moreover, the interaction of the residues with the stone from which the tools are made results in the surface of the tools acquiring a distinctive and characteristic luster or *surface polish*, such as *bone polish* or *hide polish*. In sites where there was no washing away by flowing waters, scouring by moving particles, or consumption by biological activity, the wear pattern, residues, and/or surface polish have been preserved and their study is often significant for elucidating the tasks for which the tools were used in the past (Grace 1996; Plisson and Mauger 2001). Moreover, correlating between the shape and pattern of wear marks, the composition of remaining residues, and/or the surface polish on the tools is often indicative of distinct human activities (Hurcombe 1992; Hayden 1979).

SAND
GLASS, GLAZE, AND ENAMEL

The term *sand* is generally used to refer to natural accumulations of small rock particles, measuring between 0.05 and 2.0 mm in diameter, which are widely distributed over the surface of the earth. Most sand particles were once part of massive rocks that were partly or entirely broken down and comminuted by weathering processes (see Textbox 45). Following their formation, the particles are generally moved by environmental agents such as wind and flowing water, which may transport them over great distances from their place of origin. The smallest particles are separated out and generally moved away by transportation processes before settling down and becoming part of sedimentary deposits. The larger ones do so in places near the source rocks. Usually, the farther the sand particles are transported from the source rocks, the more wear they undergo and the rounder and better sorted they become, with fewer variations in size.

The composition of the particles is related to that of the source rocks. *Quartz sand* [composed of silica (silicon dioxide)], which makes up the most common variety of *silica sand*, is derived from quartz rocks. Pure quartz is usually almost free of impurities and therefore almost colorless (white). The coloration of some silica sand is due to chemical impurities within the structure of the quartz. The common buff, brown, or gray, for example, is caused by small amounts of metallic oxides; iron oxide makes the sand buff or brown, whereas manganese dioxide makes it gray. Other minerals that often also occur as sand are *calcite, feldspar* and *obsidian Calcite* (composed of calcium carbonate), is generally derived from weathered limestone or broken shells or coral; *feldspar* is an igneous rock of complex composition, and *obsidian* is a natural glass derived from the lava erupting from volcanoes; see Chapter 2.

Throughout the ages humans have used sand for many and the most varied uses. Some river sand, known as *auriferous sand*, contains native gold nuggets and has long been used as a source of the precious metal; see Chapter 5. Sand has been used to make *molds* in which metals are cast (see Textbox 38) and fine, mostly silica sand, as an *abrasive* to shape, smooth, or polish hard solids such as of stone and metals. The major use of sand for the last three millennia, however, has been for the preparation of *building cements* (see Chapter 4) and, since the manufacture of glass became widespread, for *making glass*.

3.1. GLASS, GLAZE, AND ENAMEL

It seems that sometime during the third millennium B.C.E., somewhere in the Middle East, it was discovered that when a mixture of silica sand and soda is heated to relatively high temperatures, the mixture fuses to form glass. Few other human-made materials are derived, as is glass, from such common and abundant raw materials. Its remarkable physical and chemical properties made glass, already in antiquity, one of the most useful and ubiquitous materials in many areas of the world.

Hot, molten glass is thick and cohesive; it can be shaped, and, as it cools down, it hardens while keeping its shape. Solid glass is extremely tough, withstands compression better than steel, is impervious to liquids, and is resistant to chemical attack. All this makes glass useful for making utilitarian artifacts, such as containers for solids and liquids, as well as ornamental and decorative objects (Tite et al. 2002; Tait 1991).

Glaze and Enamel. *Glaze* and *enamel* are forms of glass used to coat other materials: glaze, mostly for coating ceramics, *enamel*, to coat metals and alloys with thin, protective, and/or decorative coatings. Both are pliable liquids when hot, and as they cool down they solidify into amorphous, rigid, and impervious solids (see Textbox 27). The differences between glass, glaze, and enamel are not in their composition, but mainly in the way they are used: glass is generally shaped into freestanding objects, independent of any backing or supporting material; glaze and enamel are applied as coatings to the surface of artifacts made from other materials (Paynter and Tite 2001; Bimson and Freestone 1992).

Because of the similarities in their composition and chemical properties, glass, glaze, and enamel are usually referred to as *glassy* or *vitreous materials*. The distinct and specific use of each one of them determines, however, a number of technical differences between them. To facilitate shaping and working glass objects, for example, it is usually desirable that the melting point of glass be as low as possible and that it exhibit plasticity over a wide temperature range. For glazing and enameling, however, these properties are virtually irrelevant. The *coefficient of linear expansion*, a measurement of how much a material expands or contracts in length when heated or cooled per degree of temperature (see Textbox 76), is relatively unimportant for glass. For glazing and enameling, however, differences between the coefficient of linear expansion of the coating (glaze or enamel) and the coated ceramics or metals may lead to breakup of the coating, known as *crazing*; in extreme cases when the difference is great, the glaze or enamel may even come apart and be detached altogether (Shafer 1976). Table 26 summarizes the properties required from glass, glaze, and enamel.

TABLE 26 Properties Required from Glass, Glaze, and Enamel

	Material	
Property	Glass	Enamel or glaze
Ease to work	Low melting point plasticity over a wide range of temperatures	Determined by properties of substrate
Transparency	Usually desirable	Seldom desired since opaque layer obliterates underlying surface defects
Coefficient of linear expansion	Unimportant	Differences between substrate and coating layer lead to *craze* and peel

TEXTBOX 27

SUPERCOOLED LIQUIDS; GLASS AND VITREOUS MATERIALS

When a molten material cools down slowly and reaches its melting point, it solidifies by a process known as *crystallization*; the amorphous, fluid melt turns into solid, rigid crystals having a characteristic and distinctive internal structure, symmetry, and shape (see Textboxes 3 and 21). This is how for example, all primary minerals in the earth's crust, are formed. If the melt cools down rapidly, however, the material does not crystallize: as it cools its *viscosity* (its property to flow) gradually increases; as it cool still further, it turns rigid, solidifies over a range of temperatures (not at a sharp melting point) and remains internally amorphous, lacking internal structure (Feltz 1993).

Such an amorphous solid is also known as a *supercooled* liquid, a *glass* (Fig. 26), or *vitreous material*. At ambient temperature supercooled liquids have the characteristic of solids: they are hard, rigid, and generally also brittle, appearing as solids. Their constituent atoms are arranged in space so that they have short-range order (their positions relative to each other are quite regular and organized); their arrangement lacks, however, the periodicity and symmetry characteristic of crystalline solids (Plumb 1989; Zachariasen 1932). Obsidian, a naturally occurring mineraloid, and human made glass, glaze, and enamel are all examples of vitreous materials (see Chapters 2 and 3).

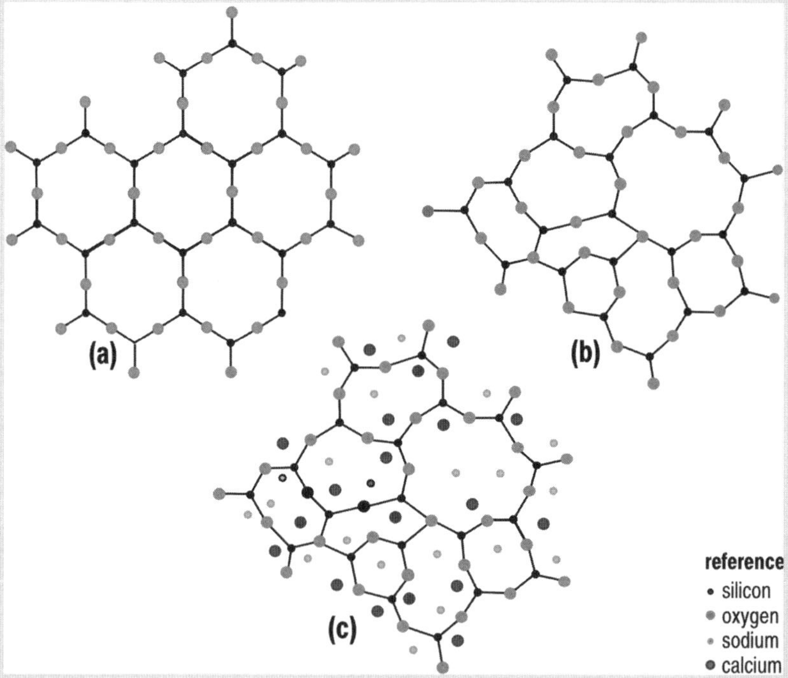

FIGURE 26 Glass. Glass is a supercooled liquid. Although it is distinctly different from liquids as well as from solids, it combines some of the properties of both of them. The figure shows a two-dimensional representation of the atomic structure of three different but related solids: crystalline quartz **(a)** composed of silicon dioxide, in which the component silicon and oxygen atoms are linked to each other in a regular, symmetric arrangement characteristic of crystalline solids; amorphous silica **(b)**, which has a disordered array of atoms, without a definite internal structure; and glass **(c)**, in which metal ions (sodium from the glass modifier and calcium from the stabilizer) are entrapped within a disordered array of atoms, similar to that of amorphous silica.

3.2. GLASS

If silica sand, a crystalline material (composed of silicon dioxide), is heated until it melts and then cooled rapidly, it hardens into *silica glass*, whose composition is identical to that of the original sand, although it lacks crystalline structure and properties. The temperature required for making silica glass is however, very high: silica melts above 1700°C, more than 500°C above the melting temperature of most types of ordinary glass and far beyond the temperatures attainable in antiquity. Tools made from silica glass, such as prehistoric hatchets and knives, and workshop debris found in the Sahara, provide evidence of the early human interest in the use of this material. Since chunks of natural silica glass do occur in the outer crust of the earth, it is reasonable to recognize that such tools were made from gathered natural silica glass (Clayton 1998; de Michele 1998).

Soda Glass

The addition of a relatively small amount of such substances as soda, potash, or borax, all known as *fluxes* or *fluxing materials*, lowers the melting temperature of silica sand from over 1700°C to below 1000°C, a temperature feasible in ancient furnaces (see Textbox 28).

TEXTBOX 28

FLUX

A *flux* (or *fluxing substance*) is a solid that, when added in minimal amounts to another solid, promotes melting of the other solid. In other words, a flux facilitates melting of solids by causing the melting temperature of a mixture of flux and a solid to be lower than that of the solid alone. Even in the ancient past, it was realized that the addition of small amounts of some materials to a large mass of a solid lowers the melting point of the latter. Adding a flux to minerals or ores that regularly melt at very high temperatures (above about 1300°C, too high to be reached in the furnaces and ovens available at the time, for example), results in the mixture melting at about 1000°C, readily attained in well-ventilated ancient furnaces. The use of fluxes thus made it possible to perform manufacturing processes that would otherwise have been impossible at the time. Fluxes have been and still are widely used in such manufacturing processes as *glassmaking* and metal *smelting*. Adding a flux to the raw materials of glass or to ores being smelted reduces the melting temperatures required for glassmaking or for metal smelting, making these processes easier to perform and more economical (since less fuel is required for their completion).

Fluxes are also particularly important in *soldering* and *brazing* operations, where they lower the melting point of the metals or alloys at the joining site. Moreover, they are effective in cleaning the surface of the joined parts, rendering them receptive to the solder or brazing metal. *Rosin* and *borax* are examples of fluxes used since antiquity for metal joining operations. *Rosin* is generally used mainly for soldering and *borax*, for brazing.

The activity of a flux is due to the *cations* (positively charged ions) that it bears. When heating a mixture of a solid and a flux, the cations of the flux penetrate the atomic lattice of the solid, breaking up *bonds* (links) and thus lowering the temperature at which the solid melts. In glassmaking, for example, the cations of the flux break up bonds between silicon and oxygen atoms in silica sand, the main component of glass; in extractive metallurgy they break up bonds in the silicate minerals, the main components of many metal ores, lowering the temperature at which the mixture melts. *Soda*, *potash*, and *borax* are examples of fluxes that have been widely used as fluxes (it is worth noting that these substances have also been used as cleansers, inhibitors of decay processes (e.g., during mummification), and mordants for dyeing.

Soda

Soda (composed of sodium carbonate) was acquired in antiquity either in the form of *natron*, or, when prepared, as soda ash. Natron is a natural mixture of sodium bicarbonate, sodium carbonate, lesser amounts of common salt, and sodium sulfate, and some organic matter. It occurs in a few places in the world, such as in dry lakebeds in desert regions, in Egypt and Siberia, for example. From these few sources, natron was traded and transported to many others in the ancient world, where it was used (von Lipmann 1937; Lucas 1932).

Potash

Potash (composed of potassium oxide), also a flux, was mainly used as a glass modifier. It was generally introduced into the glass melt in the form of either *pearl ash*, composed of potassium carbonate, *vegetable ash*, one of the main constituents of which is potassium carbonate, or *saltpeter*, a mineral composed of potassium nitrate.

Soda as well as potash have also been made, since early antiquity, by burning weeds until only their ash remains – thus known as either *soda ash* or *potash ash*. The ash may also contain as much as 5% of sodium or potassium carbonate. *Kelp*, a large seaweed of the order *Laminaria*, and *barilla* plants, of the genus *Salsola*, which grow on many seashores, have been used to produce soda ash and potash ash, respectively: burning kelp yields soda ash, while burning barilla may yield either soda ash or potash ash; burning *salsola soda* plants yields soda ash, while burning *salsola kali* yields potash ash.

Borax

Borax, another fluxing material, is a mineral (composed of hydrated borate of sodium), which has also been known since antiquity. There seems to be no evidence however that it was used in the past as a flux; it served mainly as a mordant, for dyeing textiles. Until the eighteenth century borax was apparently procured only from a lake in Tibet, whence it was exported to the Near East and Europe. Sources of borax in Asia Minor and in Tuscany, Italy, were discovered only in much later times.

Rosin, an Organic Flux

Rosin is an organic flux that has long been used for soldering. It is a yellow, transparent, and relatively hard *resin* secreted from wounds in the trunks of coniferous trees. Rosin is insoluble in water, and its exact composition and structure are as yet unknown.

The addition of a flux results not only in a mixture of silica and flux having a lower melting temperature than that of the silica, but also in the melt being less viscous, flowing more easily than silica (*viscosity* is a measure of the resistance of fluids, liquids, and also gases, to flow; fluids with high viscosity flow more slowly than do those with low viscosity). As a consequence of its relatively low viscosity, the hot molten mixture of silica and flux, a type of early glass, can be shaped with relative ease.

Soda (composed of sodium oxide) and *potash* (composed of potassium oxide) are relatively easy to obtain or to make, work, and handle. Although they do not occur naturally, they could be quite easily prepared in the past by burning some plants (see Textbox 28). Both these oxides were used as fluxes for making early types of glass articles, particularly small vessels and decorative objects.

Glass made from only two components, sand and soda or sand and potash, is generally known as *soda-glass* or *potash glass*. Because soda and potash are very soluble in water, high humidity, rain and flowing waters slowly but gradually dissolve them from within the glass. Moreover, their dissolution leads to gradual disintegration of the glass, so that most ancient objects made from soda or potash glass have vanished or are in an advanced stage of decay.

Sodalime Glass

Lime (composed of calcium oxide) is a very effective *stabilizer* of glass (a *stabilizer* is a substance that, when added to others, makes the mixture more stable and resistant to deterioration). The addition of lime to a soda glass or potash glass melt during the glassmaking stage produces a glass composed of three components: sand, soda, and lime (or sand, potash, and lime). Such a glass is stable to dissolution by water and is, therefore, resistant to weathering and more durable than soda glass or potash glass. *Limestone* (composed of calcium carbonate) can also be used, instead of lime, as a glass stabilizer. If added in place of lime to mixtures of sand and soda or sand and potash when glass is being made, limestone first decomposes to lime (see Textbox 34); the lime then combines with the other ingredients of the melt to form a three-component glass, known either as *sodalime glass* or *potashlime glass*.

At relatively high temperatures, above 1000°C, the viscosity of molten glass is low and it flows easily; as it cools down, however, its viscosity gradually increases until the melt turns into a hard, rigid mass. This is an important characteristic of glass; when molten glass cools down, it does not solidify at a well-defined temperature, as do crystalline solids (see Textbox 3): it stiffens gradually, over a rather broad range of temperatures known as the *hardening range* of glass. At the higher-temperature end of the hardening range the glass has low viscosity, flows easily, and collapses into a shapeless mass if an attempt is made to shape it; at the lower temperatures of the hardening range it is sufficiently dense and viscous to be worked and shaped with relative ease before it hardens. The lower-temperature end of the hardening range is, therefore, also known as the *working range* of the glass. Similarly, when solid glass is heated, it does not melt at a definite temperature but melts gradually over a range of increasing temperatures known as the *softening range* of the glass. These thermal properties are of extreme technical importance, since only when molten glass is highly viscous can it be shaped, by either casting, drawing, or blowing. The addition of lime to a glass melt not only stabilizes the glass, but also alters and extends the range of temperatures at which it can be shaped.

The Components of Glass. Silica, the main component of glass, is usually referred to as the *glass former*. Soda and potash, the fluxing materials that, when added to silica, lower the melting point of the mixture and alter the viscosity of the molten glass, are known as the *glass modifiers*. Lime, which stabilizes and improves the chemical durability of the glass, is known as the *glass stabilizer*: The terms *former, modifier*, and *stabilizer* simply define the respective functions of the three components of glass.

TABLE 27 Ancient Sodalime Glass: Typical Composition

Glass component	Relative amount (%)
Silica (glass former)	60–75
Soda (modifier)	30–15
Lime (stabilizer)	15– 8
Iron oxides (generally added unintentionally)	10–less than 1

Sodalime glass, made of sand, soda, and lime, has been, with only slight compositional variations from place to place or time to time, the most common type of glass made and used throughout the world since its discovery and until the present time. Although an infinite number of variations in the relative proportions of the three components is possible, a balance between them is essential. An excessive amount of modifier (soda) makes the glass chemically unstable and easily soluble in water. Glass that contains too much stabilizer (lime), tends to *devitrify*, that is, to change from the vitreous to the crystalline condition, and becomes very susceptible to *weathering* and *decay*. Most types of ancient as well as modern sodalime glass thus contain relatively uniform proportions of the components, generally about 75% sand, 15% soda, and 10% lime (see Table 27).

In addition to the three main components of glass (silica, soda or potash, and lime), other materials have been, intentionally or accidentally, added to the mixture of raw materials when making glass. The addition of *magnesia* (composed of magnesium oxide), for example, which is often naturally mixed with limestone, decreases the solubility of glass in water and therefore increases its durability. Also the addition, mostly unintentional in antiquity, of *alumina* (a mineral composed of aluminum oxide), which is a natural component of some sands, improves the chemical durability of glass and its working properties during shaping and forming operations.

Potash Glass

Glass made of sand, potash, and lime, known as *potash glass*, hardens faster and at a higher temperature than does soda glass. Because of these properties, potash glass was much less widely used than soda glass for making molded glass objects (Brill 1979; Olin and Sayre 1974).

Lead Glass

Lead oxide is a glass modifier that yields a very special type of glass known by a variety of names, including lead glass, flint glass, lead crystal, and crystal glass. Although the intentional addition of lead to glass has a long history, the use of lead oxide as a modifier in early antiquity seems to have been rather limited. Lead oxide was generally introduced to the glass melt during the manufacturing process either directly, as one the natural oxide minerals of lead, such as the mineral *litharge* (composed of lead monoxide) and *red lead* (composed of lead sesquioxide), or indirectly, as one of the natural salts of lead, as, for example, *white lead* (composed of basic lead carbonate) and *galena* (composed of lead sulfide). During the heating process required for glassmaking, all these were converted to lead oxide. Lead glass, as well as lead glaze have a low softening temperature and high density. They also exhibit a characteristic brilliance and, when struck, emit a metallic ring, properties that have been much appreciated since ancient times, particularly by the ancient Romans, for making high-quality decorative objects.

Colored Glass

Most of the properties of glass are determined by the nature and relative amounts of the glass former, modifier, and stabilizer used for making it. Some of its properties, however, are determined by the minor and trace constituents that together usually make up less than 5% of the total mass of glass. The color of glass, for example, is generally due to the presence, whether intentional or otherwise, of relatively small, often trace amounts of metal oxides. Heating a mixture of pure silica, soda (or potash) and lime, for example, yields glass that is essentially colorless. The addition to such a mixture of oxides of some metals, even in very small amounts, causes the glass to become colored. *Iron oxide*, for example, the almost universal and ubiquitous coloring agent of practically all ancient as well as modern ordinary glass, produces a typical and characteristic, usually unintentional, greenish glass. When yellow-brown sand, the color of which is due to iron impurities, is used for making glass, the iron in the sand causes the glass to acquire a green tint. The tint can be easily seen, even in apparently colorless ordinary glass, when a sheet of glass is viewed edgewise. Its occurrence is due to an optical effect caused by iron ions. The iron impurities occur in glass as ferrous and ferric ions (see Textbox 6); the ferrous irons impart to the glass a blue tint, while ferric ions make it yellow. The combined effect of the two ions is the ubiquitous *bottle green* of ordinary glass. Less than 1% of iron oxides is generally sufficient to give glass the characteristic green tint that was in antiquity, and still is, regarded as aesthetically deleterious to most types of glass.

As early as Roman times, it was found that the green coloration due to iron impurities could be offset by adding to the glass, while still in the molten state, relatively small amounts of the mineral *pyrolusite* (composed of manganese oxide). Manganese ions, which impart a deep violet color to the glass, optically mask the green imparted by the iron, modifying it to a less noticeable gray. Since it serves as a glass *decolorizer*, pyrolusite is also known as the *glassmaker's soap*. The ubiquity of iron impurities in the components of glass and particularly in sand is sometimes turned to good use. Melting glass under mildly reducing conditions (created when fuel burns with a restricted supply of oxygen) increases the proportion of ferrous ions and results in the glass acquiring a bluish tint. Pyrolusite, on the other hand, is often used not only as a glass decolorizer but also as a coloring agent. Under particular glass-melting conditions the addition of small amounts of pyrolusite produces amethyst (violet) colored glass. Much ancient glass was also colored by the addition of relatively small amounts of other metal ions. The addition of highly oxidized copper, in the form of cupric ions, for example, makes the glass appear bright blue, whereas mildly oxidized copper, in the form of cuprous ions, colors it red. Table 28 lists metal ions used in ancient glassmaking to produce colored glass.

Thus, most of the color in glass is produced by metallic ions derived from minerals, usually in relatively small amounts within the glass structure. Already in antiquity there was awareness of the effects of some other metals on the color of glass. It was known then, for example, that even relatively small amounts of specific metals, included as minerals in the components of a glass melt, could produce colored glass; particular *metalliferous minerals* were therefore added to glass melts so as to produce special and

TABLE 28 Glass Coloring Ions

	Coloring metal		
Glass color	Metal	Ionic form	Glass furnace environment
Black	Manganese	Mn^{2+}	Reducing
	Copper	Cu^{+}	Reducing
Red	Copper	Cu^{+}	Reducing
	Gold	–	Reducing
Pink	Manganese	Mn^{4+}	Oxidizing
Yellow	Uranium	U^{4+}	Oxidizing
	Silver	–	Reducing
Green	Copper	Cu^{2+}	Oxidizing
	Iron	Fe^{2+}	Reducing
	Chromium	Cr^{3+}	Oxidizing
Blue	Copper	Cu^{2+}	Oxidizing
	Cobalt	Co^{2+}	Reducing
Violet	Manganese	Mn^{3+}	Reducing

sometimes highly appreciated colors. During Roman times, for example, the addition of small amounts of silver mineral was used to produce a characteristic yellow glass, and of extremely small amounts of gold to make one of two types of glass: either displaying a variety of red hues or exhibiting dichroic properties [*dichroism* is the display of two different colors, one when a material (such as glass) is viewed with reflected light and another when viewed with transmitted light]; see discussion below (Weyl 1999).

Silver was also used for *shading* colored glass, that is, for grading from light to dark the color of glass used for making *stained glass* (see text below). The process of shading glass with silver entails three working stages: (1) applying to the surface of colored glass, usually with a brush, a coating of varying thickness of a paste consisting of a powdered silver mineral mixed with a clay binder and water; (2) heating the coated glass to about 600°C in a *reducing* atmosphere, to cement the coating to the glass; and (3) cooling the now shaded glass at a controlled rate. Above 600°C the ions of silver in the coating layer are first oxidized to silver oxide, which, since the glass is partially molten at these temperatures, migrate and penetrate into the glass. Then, under reducing conditions the silver ions are reduced to metallic silver, which darkens the glass. The variation in the thickness of the coating, and therefore in the amount of silver, creates shades of varying intensity.

Stained Glass. *Stained glass* has been made since the Middle Ages. A stained glass work consists of relatively small pieces of glass of different colors and shapes assembled together within lead frameworks arranged in figurative displays. Stained glass has to be distinguished, incidentally, from *painted glass*: while stained glass is made from colored glass, painted glass is made from transparent, colorless glass, to the surface of which a coat of more or less transparent but colored paint is applied.

Opaque Glass

When molten glass cools down, it solidifies as a supercooled, amorphous liquid that lacks internal structure. In some instances, however, whether deliberately or accidentally, while the glass solidifies and remains amorphous, one or more of the glass melt components crystallize. This results in the formation of myriad small crystallites that remain suspended within the molten glass at first and, as the glass cools down, within the solid glass. These crystallites *scatter light*; that is, they deflect light incident on them from its path and prevent it from passing through the glass (see Textbox 22). The intensity of light scattering depends on the relative amount of the crystallites within the glass; the more crystallites there are in the glass, the more light is deflected and does not pass through the glass. As a consequence, the glass may appear *cloudy, smoky, opalescent* (displaying a "milky" variety of fine luminous colors), or *opaque*. All this was (intuitively) known in antiquity and was used to produce a variety of different types of glass. *Opaque glass* (and also *opaque glaze* and *enamel*) does not transmit any light incident and is, therefore, impenetrable to sight. Another variety, known as *opal glass*, which is somewhat translucent and iridescent, was especially made to imitate the gemstone opal (Vendrell et al. 2000; Turner and Rooksby 1961, 1959).

Two substances whose glass-opacifying properties where recognized quite early are antimony oxide and tin oxide. Antimony oxide was probably added to molten glass as the mineral *stibnite* (composed of antimony sulfide). In ancient Syria and Mesopotamia, incidentally, stibnite was used not only as a glass opacifier but also as a cosmetic, for darkening the skin around the eyes. When stibnite is added to molten glass, it reacts with other components of the melt to form metal antimoniates (compounds composed of antimony, oxygen, and one of several metals). Thus, depending on the actual nature of the metals in the melt, several metal antimoniates may be formed. Calcium, for example, forms colorless calcium antimoniate, which crystallizes on cooling and produces *white opal glass*. If there is lead in the glass melt, however, yellow lead antimoniate is formed, which produces a *yellow* variety of *opal glass* (Wainwright et al. 1986). It seems that during the fourth century C.E. the use of antimony oxide as a glass opacifier faded and was gradually replaced by tin oxide, which later became the glass opacifier almost exclusively used in Europe and the Middle East (see Table 29). Tin oxide occurs naturally as the mineral *cassiterite*. Its use as an opacifier in Europe seems to have began only between the second and fourth centuries C.E. (Turner and Rooksby 1961, 1959). In more recent studies, however, it is claimed that tin oxide was already used as a glass and glaze opacifier in Mesopotamia, probably as early as the sixth century B.C.E. (Mason and Tite 1997). An Islamic technology of making opaque glaze, based mainly on the use of tin oxide as the opacifier, was developed during the eighth century C.E. and continued

TABLE 29 Opacifiers in Ancient Glass

Opacifier	Color of glass	Example
Tin oxide	White	upper Egypt, second century B.C.E.
Lead tin oxide	Yellow	Koban, Russia, ninth century C.E.
Calcium antimoniate	White	Turkey, eleventh–tenth century B.C.E.
	Blue	Nimrud, seventh century B.C.E.
Lead antimoniate	Yellow	Rhodes, second century B.C.E.

to be widely used in Egypt and Mesopotamia until the tenth century, when it spread to all of the Islamic world and, eventually, also to Europe (Kleinmann 1986).

3.3. GLASSMAKING

The basic process for making glass, although not the actual technology, has changed little since antiquity. Over the centuries the technology has advanced, being continuously improved and refined beyond recognition (Bray 2000; Neri 2000). Six main manufacturing stages are involved in the glassmaking process:

1. Selecting the raw materials
2. Comminuting and mixing the raw materials
3. Heating and melting the mixture
4. Fabricating, that is, forming and shaping objects
5. Annealing the objects
6. Finishing

The *batch*, as the mixture of the raw materials required for making glass is called, consists almost invariably of the same basic ingredients: the glass former (sand), a modifier (mostly soda or potash but other suitable materials as well), and a stabilizer (lime). Usually, the batch also includes pieces of broken glass from previous batches or glass waste, known as *cullet*. When required, *decolorizing* and/or *coloring* agents and *opacifiers* are also added to the mixture of raw materials. Some of the basic components of the batch are not generally available as required. Modifiers such as soda or potash and stabilizers such as lime (i.e., the oxides of sodium, potassium, and calcium) are rarely available in nature as such. Thus other, easily available, compounds of these metals, such as their naturally occurring carbonates and/or hydroxides, are used instead; during the heating and melting stage, these compounds are converted into oxides. Table 30 lists the raw materials most frequently used for making glass.

After comminuting and thoroughly mixing the raw materials, including cullet, the mixture is gradually heated. The cullet softens and melts first, at a temperature lower than most of the other mixture ingredients; it provides a fluid through which the still solid grains of the other ingredients move around, mix with each other, and, as the temperature is steadily increased, gradually melt. Only above 1000°C does the entire mixture melt into a highly viscous liquid. The temperature of the molten batch is then further increased

TABLE 30 Main Components of Common Glass

Constituent	Raw material (Mineral)	Chemical name
Former	Sand	Silica
Modifier	Soda ash	Sodium oxide
	Natron	Sodium carbonate
	Salt cake	Sodium oxide
	Potash	Potassium oxide
	Litharge	Lead oxide
	Minium	Lead oxide
Stabilizer	Limestone	Calcium carbonate
	Burned lime	Calcium oxide
	Hydrated lime	Calcium hydroxide
	Dolomite	Magnesium carbonate

so as to reduce its viscosity, dislodge and remove bubbles of entrapped gasses, and melt extraneous minerals that melt at slightly higher temperatures and may cause stains or streaks within the finished glass. Chemical and physical changes at this high temperature result in the conversion of the heterogeneous mixture of raw materials into a homogeneous mixture that has the properties of molten glass. After some time (usually several hours) of further mixing, lowering the temperature to about 1000°C results in the melt becoming increasingly viscous and suitable for *glass forming* operations. These operations, including, *blowing* and *casting*, for example, have to be completed in a relatively short period of time; depending on the actual composition of the melt and the working conditions, the forming operations may extend from as little as a few seconds to as much as several minutes. This is the period of time during which molten glass is within a *working range* of temperatures (900–600°C for ordinary sodalime glass) at which it has a viscosity appropriate for shaping.

Forming molten glass into objects may be done using a variety of techniques, including *drawing*, *blowing*, *casting*, and *pressing*. The description and discussion of these techniques is, however, outside the scope of subjects discussed in this book, and the interested reader is referred to one of many works on the subject (see, e.g., Newton and Davison 1997; Thorpe and Whiteley 1955).

When cooled down after forming operations, glass objects are subject to *internal stresses* generated during the handling and shaping operations. Such stresses may be harmful and cause damage to the glass, including cracking and even the spontaneous breakage of finished objects. The stresses are usually removed by a thermal process known as *annealing*, which alters their internal structure (see Textbox 29). Glass objects are annealed by first heating them to a temperature just below the softening point of the glass, keeping them at that temperature for a short while (usually a few minutes) during which time the internal stresses are relieved, and finally cooling to room temperature at a relatively slow and carefully controlled rate.

TEXTBOX 29

THERMAL OPERATIONS: ANNEALING, TEMPERING, AND SINTERING

Annealing, *tempering*, and *sintering* are thermal processes known since antiquity for altering the physical properties of such materials as metals, alloys, and glass by heating them under carefully controlled conditions. *Annealing* removes the internal stresses and reduces the hardness of solids; *tempering* reduces brittleness and develops desired strength characteristics, and *sintering* causes conglomerates of small particles, often of different materials, to coalesce into coherent masses.

Annealing

Newly made glass objects are often subjected, during hot shaping operations, to an uneven distribution of heat. When hot glass objects cool down to ambient temperature the ill-balanced heating conditions may result in the generation of internal stresses; these, may be detrimental and cause cracking, breakage, and even the spontaneous disintegration of glass made objects. *Annealing* removes the internal stresses of materials such as glass and some metals and alloys and reduces their hardness.

The annealing process entails heating the material to a well-established temperature, maintaining that temperature for a specified length of time, and then cooling it down at a carefully controlled rate. Annealing metals and alloys, for example, reduces internal stresses caused either by uneven heat distribution during metallurgical procedures or by working (such as hammering) operations. Annealing also softens metals and alloys, so that they can be worked with relative ease. Shaping metallic objects therefore, often entails subjecting them to repeated cycles of hammering and annealing.

Tempering

The *tempering* process entails heating a solid (generally a metal or an alloy) to a temperature just below its melting point and then cooling it rapidly. In the past, tempering was used exclusively for hardening metals and alloys, especially steel. In more recent times, since the twentieth century, however, some glass objects are also tempered. Tempering a metal or an alloy entails a sequence of two thermal cycles: (1) heating the metal or alloy to a high temperature (below its melting point); (2) *quenching* (cooling it rapidly) by immersion in a cold liquid. Water, oil, and other liquids such as blood or urine were frequently used in the past for quenching; urine, in particular, was widely favored for quenching steel. Following quenching, and to complete the tempering process, the metal or alloy is heated again to a well-specified high temperature, but lower than that reached when quenching, and is finally cooled down, at a controlled slow rate, to ambient temperature.

The tempering process changes the internal structure and the physical properties of the solid. Steel, for example, which can be put to many different uses, is often required to have widely differing properties such as varying degrees of hardness, strength, or toughness. To satisfy these varying specifications, the internal structure of the steel is modified by tempering it under varying conditions. Some tempering processes may require heating to distinct and precise temperatures for different lengths of time, or quenching in different liquids.

Sintering

Sintering is a process for causing powdered materials or mixture of materials to become a coherent mass by heating but without melting; it requires heating the powder to a temperature just below the melting point of a material or the main constituent of a mixture, which then undergoes three different but concurrent processes: (1) joining of the small powder particles, often of different composition; (2) recrystallization; and (3) removal of porosity.

Sintering processes have been used since antiquity to alter the properties of ores and ore concentrates, powdered metals and alloys such as bloomery iron, ceramics, and glass. *Bloomery iron*, for example, was sintered in a forge, and the operation was often referred to as *forge sintering*.

Other operations on solid glass objects, such as reshaping, joining, cutting, drilling, engraving, grinding, and/or polishing, are intended to make them useful for particular tasks or give them a desired appearance. Such operations may be performed when the glass is either hot or cold. *Reshaping* and *joining*, for example, are usually done when the glass is hot, while cutting, drilling, engraving, grinding, and polishing are normally carried out at room temperature.

3.4. ANCIENT GLASS STUDIES

Most of our knowledge on the composition of ancient glass comes from two types of resources: ancient glassmaking recipes and other written information, and the analysis of ancient glass. In cuneiform tablets from Nineveh, in ancient Mesopotamia, for example, there are recipes for the manufacture of plain and colored glass and glaze (Oppenheimer et al. 1970). It is the analysis and study of the properties of ancient finds, however, that provides the best information on the raw materials, the manufacturing technology and the characteristics of ancient glassy materials. Chemical analysis provides a most direct way of learning about the composition of ancient glass, the knowledge of which often clarifies archaeological problems and is also of value in the history of glass technology. One of the difficulties encountered when analyzing ancient glass, however, is the compositional similarity of glass from different periods and different sources. Much ancient glass is of the soda or sodalime varieties, and new analyses that merely confirm this fact do not contribute new knowledge. Relevant studies usually expose compliance to known compositional patterns or differentiating features recognizable in well-defined groups of objects. Studying the results of systematic chemical analyses of glass makes it possible, for example, to evaluate the ancient raw-materials selection processes or the making of particular types of glass. Moreover, to fully appraise ancient glassmaking technology, it is often necessary to replicate ancient methods using equipment available in antiquity (Schuler 1963, 1962, 1959). Most analysis of glass is now done using relatively fast instrumental techniques, such as those listed in Table 31.

The analytical results of the major and minor components of glass are customarily expressed not as a weight proportion or percentage of the elemental components, as is usual with analytical results, but as a weight proportion of the metal oxides that make up the glass. Thus, when expressing the composition of glass, it is an accepted practice to specify all or most of the following components: silica (silicon dioxide), soda (sodium oxide) or potash (potassium oxide), lime (calcium oxide) and/or magnesia (magnesium oxide), manganese oxide, and iron oxides. All these are expressed as a percentage of the total mass of glass. Trace elements, on the other hand, are usually expressed as parts per million (ppm) or parts per billion (ppb) of the total mass. Close examination of much published analytical work reveals some generalizations on the composition of soda-lime glass, the most common type of ancient glass:

- *Silica*, the universal glass former, generally makes up 60%–70% of the total mass of glass; only in very few cases does the amount of silica exceed 70%.

TABLE 31 Analytical Techniques Frequently Used to Study Ancient Glass

Analytical technique	Sample analyzed	References
Atomic absorption spectroscopy (AAS)	Bulk	Schreiner et al. (1998)
Emission spectrography	Bulk	Sayre and Smith (1974)
X-ray diffraction	Trace	Perez y Jorba and Dallas (1984)
X-ray fluorescence (XRF)	Surface	Schreiner et al. (1999)
Neutron activation	Bulk	Hughes et al. (1991)
Electron spectroscopy for chemical analysis (ESCA)	Surface	McGovern and Michel (1991)
Ion beam spectrochemical analysis	Surface	Lanford (1986)
Auger emission spectroscopy	Surface	Ciliberto and Spoto (2003)
Scanning electron microscopy (SEM)	Surface	Jembrih et al. (2000)
Electron microprobe (EMPA)	Surface	Senn et al. (2001)
Particle-induced X-ray emission spectroscopy (PIXE)	Surface	Malmqvist (1995)

- *Soda* and/or *potash* usually occur in a higher proportion than in modern glass. Although the relative amounts of these components may vary widely, the upper limit of their concentration is usually about 30% of the total mass of glass. Soda generally predominates over potash.
- *Lime* and *magnesia* usually make up about 10%, that is, similar to, or slightly higher than their concentration in modern glass; magnesia usually occurs below the range 2–5%.
- *Alumina*, widely used in modern glassmaking, was probably not added deliberately in the past. It was, however, often introduced as an impurity, together with the other raw materials; usually it occurs in low concentrations, below the range 1–5%.
- *Iron* may occur in a wide range of concentrations varying from trace amounts to a maximum of about 10%. It was also not intentionally added to the glass mixture, but was probably introduced as an integral constituent of sand.

The main source of silica in antiquity, as today, was sand and, very occasionally, crushed rock or pebbles. The most common modifier was soda, which was obtained from natron lakes, as in ancient Egypt, for example, or from vegetable ash. Most lime was derived from limestone, although some lime could also enter the mixture of glass raw materials together with soda from plant ashes, as mentioned previously, or with sand, in which it occurs as shell detritus in sea sand.

Most ancient Egyptian glass is of the sodalime variety; specimens from as early as the twentieth century B.C.E. are certainly of this type. The constituents of such glass are similar to those of modern ordinary glass, although variations in the proportions of the major components are great. Much ancient Egyptian glass has also a high magnesia content. Sodalime glass was produced in Mesopotamia as early as the middle of the third millennium B.C.E. Many Mesopotamian samples have a high magnesia content, and the glass should be more appropriately called a sodalime–magnesia type of glass. Much Egyptian glass also has a high magnesia content and thus a high magnesia to lime ratio. This ratio is usually much higher in Mesopotamian glass. On occasions, these differences have been used to distinguish between glass from the two regions. An uninterrupted continuation of a glassmaking technology has been attributed to the fact that much Islamic glass has a composition similar to that made as early as during the second millennium B.C.E. in the Middle East. Roman sodalime glass from Italy, however, does not seem to bear any resemblance to the same type of glass made in either Mesopotamia or Egypt.

Some Special Types of Ancient Glass

Cameo Glass. *Cameos* are decorative carvings on colored stone, such as onyx, shell, or multilayered glass, exposing a background of a different color. The technique of making cameos was known to the ancient Egyptians and widely practiced in the Roman Empire during the years just preceding and following the birth of Christ. The tradition of making cameos continues to this day (Goldstein et al. 1982); the most common type of glass used for carving cameos, which is known as *cameo glass*, is usually made up of two – although occasionally as many as five – differently colored, bonded layers. In cameos made of two-layered glass the carving is designed on the top layer, which is mostly opaque. The backing layer often consists of a more or less transparent glass of a different color from that at the top (Mommsen et al. 1997a).

A study of the composition and characteristics of an ancient vase made of cameo glass provides enlightening insight into this ancient glass technology: the vase is made of two-layered glass, exposing figures carved in an opaque white layer on a dark blue backing layer. Analysis of the glass of the vase showed that the composition of the blue (backing) layer conforms to that of ancient sodalime glass. In addition, however, the backing layer also contains low concentrations of phosphorus oxide, magnesia, and potash and, in still lower concentrations, oxides of cobalt, copper, iron, and manganese.

FIGURE 27 Millefiori glass. Mold-cast cups made of millefiori glass during the second century C.E. Millefiori glass (the Italian word *millefiori* means a thousand flowers), is a type of mosaic glass characterized by flowerlike patterns. It is produced by first heating bundles of thin glass rods of different colors until the rods fuse together. The fused bundles are then pulled thin and sliced cross-sectionally while still hot, so as to produce small disks with flowerlike designs. When the cool cross-sectional disks are arranged side by side and heated until incipient fusion, repetitive flowerlike patterns, as those in the illustration, are formed.

Its blue color is due primarily to the cobalt and copper oxides. The composition of the opaque top layer is very similar to that of the colored glass, but without the coloring oxides. It includes relatively large amounts of calcium antimoniate (an opacifier) and sulfur trioxide. These two compounds most probably indicate that the opacifier was added to the glass melt as the mineral stibnite (composed of antimony sulfide). At high temperature, when the glass was made, stibnite reacted with other components of the glass melt, forming calcium antimoniate and sulfur trioxide. It is probable, therefore, that the two layers were made from a common, ordinary type of sodalime parent glass. The colored backing was then colored blue by the addition of oxides of cobalt and copper. The opacity of the top layer resulted from the addition to the glass melt of stibnite. When the glass cooled down during manufacture, calcium antimoniate, derived from stibnite, crystallized, forming myriads of crystallites, which make the glass opaque (Turner 1959).

Dichroic Glass. *Dichroism* is the property of some materials to exhibit two different colors: one when the material is viewed with *light reflected* from them, the other when viewing *light transmitted* through them. Several ancient types of glass, mostly Roman but also some Islamic, exhibit dichroism. The chemical analysis and physical examination of some of these glass specimens has helped to clarify the source of dichroism. The Lycurgus cup, dated to the fourth century C.E., at present in the British Museum, is an example of ancient Roman glass that exhibits dichroic properties. The cup has an opaque pea-green appearance when viewed in reflected light, but it is seen as a deep magenta in transmitted light. Analysis of a small chip from the cup revealed that the cup is made from ordinary sodalime glass, similar to most other glass of the same period, to which some additional components were intentionally added. These include minute particles of metallic silver and gold that cause two optical effects: the selective absorption and the scattering of light. Together, these two effects create the dichroism exhibited by the glass (Barber and Freestone 1990).

Lead Glaze. *Lead glaze* is usually made by fusing a mixture of a mineral of lead, such as litharge, and sand. Melting the mixture requires relatively low temperatures, and making either transparent or opaque lead glaze involves simple working conditions. Also its application to a surface is relatively simple, since lead glaze has a low tendency to *crawl* (draw back on application to a surface and thus leave unglazed surface areas on the finished objects) during application. Moreover, finished surfaces are highly brilliant and do not *craze* (crack as a result of internal tensile stresses) easily. All these properties made lead glaze one of the most appreciated and widely used glazes on ceramic objects (Tite et al. 1998). Transparent as well as opaque varieties of the glaze seem to have been first made in the Occident during Roman times, although it may have already been made in China some centuries earlier. Its use grew widely in the Islamic world between the ninth and eleventh centuries and since then it continued to be used until quite recent times. Because of the high toxicity of lead compounds, however, the use of lead glaze has been banned in most counties since the twentieth century.

The Provenance of Glass

Lead Isotopes in Glass. The weight ratios between the isotopes of lead in objects made from lead–glass can be used to differentiate between lead minerals from different sources and to determine the provenance of the lead mineral used for making the glass (see Textbox 30). In the decoration of Egyptian glass vessels, for example, the weight ratios of the lead isotopes are similar to that in lead minerals from some local mines. This has been interpreted as indicative of the pigment used to make the decoration probably being locally mined lead minerals. Also the weight ratio between the isotopes of lead in ancient Greek glass panels of different colors seems to reveal that the lead used for making the differently colored glass came from different sources, probably from outside Greece (Brill et al. 1990). There are limitations, however, to the application of the lead isotope ratios technique for classifying or establishing the provenance of lead glass. The most important one is that although the isotope ratios in minerals are usually characteristic of a given mining region, they are rarely uniquely so. Minerals from geologically similar environments occasionally have identical lead isotope ratios, even though the regions themselves may be widely separated. This limits the applicability of lead isotopes to study the provenance of lead in glass and in lead-containing objects in general.

TEXTBOX 30

THE ISOTOPES OF LEAD

Natural *lead*, a metallic element, is a mixture of the following four isotopes: *lead-204*, *lead-206*, *lead-207*, and *lead-208*. Only lead-204 is a *primordial* isotope of *nonradiogenic* origin; all the others are *radiogenic*, each isotope being the end product of one of the radioactive decay series of isotopes of thorium or uranium, namely, *uranium-238*, *uranium-235*, and *thorium-232*; the decay series of the uranium isotopes are listed in Figure 12:

lead-204 (1.4%), a primordial isotope
Uranium-238 → lead-206 (23.8%)
Uranium-235 → lead-207 (22.2%)
thorium-232 → lead-208 (52.6%)

The four isotopes, as those of any element, have the same chemical properties. The four are not, however, uniformly distributed in the earth's crust: the occurrence of three of them, in minerals and rocks, is associated with the radioactive decay of isotopes of thorium and uranium. In most minerals and rocks the relative amounts (or the isotopic ratios) of the isotopes of lead (often expressed relative to the amount of stable lead-204) are generally within well-known ranges, which are independent of the composition of the mineral or rock; they are, however, directly related to the amounts of radioactive thorium and uranium isotope impurities in them.

Minerals and rocks of similar composition but of different geographic or geologic origin generally include different relative amounts of thorium and uranium impurities; after generally long periods of time they also include, therefore, different relative amounts of the isotopes of lead. The relative abundance of the isotope is generally characteristic of the geographic or geologic origin of each rock or mineral.

The relative amounts of the isotopes of lead in materials and objects thus provide a useful tool for differentiating between the place of origin of the minerals or rocks from which the objects are made or derived. They provide, therefore, a tool for determining not only the *provenance* of minerals and rocks but also of their use as raw materials for making man made materials; among these are metallic lead and its alloys, other metals, such as silver and copper, with which lead is often associated in nature, as well as materials that include lead in their composition such as lead-pigments, and lead glass, glaze and enamel, in which lead may occur in a wide range of concentrations (Gale and Stos-Gale, 2000, 1996).

Oxygen Isotopes in Glass. Oxygen is one of the basic elementary constituents of all the minerals used for making glass; it is a main component of sand, soda, and lime, and accounts for about 45% of the total weight of most glass. Oxygen occurs naturally as a mixture of three stable isotopes: oxygen-16, oxygen-17, and oxygen-18. Variations in the relative abundance of these isotopes in minerals from different sources and in glass have been studied as probable differentiators, to distinguish between minerals from different sources and, in the case of glass, to distinguish lead minerals used as raw materials for making the glass (see Textbox 47). Analyses of ancient glass from different regions reveals considerable differences in the relative abundance of the stable oxygen isotopes; these differences have been used for classifying glass, and Figure 28 illustrates a classification method that may provide answers to specific provenance problems. Attention should be drawn, however, to the fact that mixtures of entirely different raw materials may occasionally result in the haphazard formation of glass with identical oxygen isotopes composition (Brill et al. 1990).

3.5. THE DECAY OF GLASS

Glass, a supercooled liquid, is in a *metastable state*, that is, an apparently stable condition that may be perturbed by external conditions and undergo unpredictable changes, so that the supercooled liquid may be converted to a solid. When glass is made from a well-balanced mixture of former, modifier, and stabilizer, it is remarkably stable. Environmental changes may,

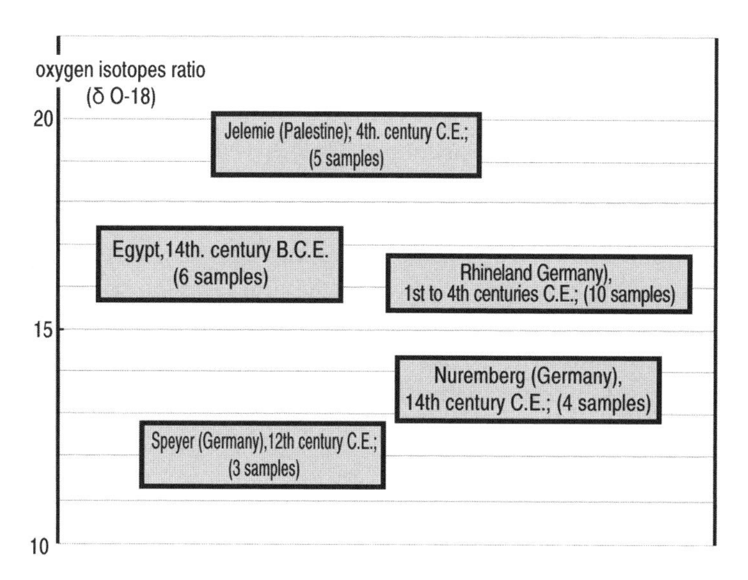

FIGURE 28 Oxygen isotopes in ancient glass. Oxygen, which consists mainly of a mixture of the stable isotopes oxygen-16 and oxygen-18, is one of the main components of most types of glass. In any sample of glass there is a definite, measurable weight ratio between the two isotopes. This weight ratio may reveal the provenance of the glass. Since the ratio is very small, a mathematical transformation is used to turn it into a more easily managed number known as the "δ O-18 value." The illustration shows the value of δ O-18 in glass from different geographic regions (see Fig. 46).

however, cause the glass to crystallize, or, as the condition is known, to *devitrify*, that is, to lose its vitreous (glassy) properties.

Glass exposed to the environment or buried in the soil under dry conditions, even for long periods of time, is usually stable and undergoes very little devitrification or decay. The more humid the environment or burial site, the more easily glass decays and the more extensively it devitrifies. Extended periods of alternating dry and wet conditions may result in periodic decay effects and the formation of devitrified layers, first on the outer faces and then throughout the bulk of the glass (see Fig. 29). The chemical decay of glass often starts when its alkaline components, soda, potash, or lime, are *leached* by water from the surrounding of the glass (*leaching* is the process of extraction of the soluble components of a solid by their dissolution, usually in water but also in mild acids). The tendency to, and the extent of decay of glass are determined mainly by its composition, the environmental temperature and humidity, and/or the surrounding water conditions at the location sit of the glass. Salts in, and the pH of groundwater, and even microorganisms with which glass is in contact, alter its rate of decay.

Various stages in the decay of glass have been defined: *dulling*, which entails the loss of clarity and transparency, is the simplest; *frosting*, the formation of a network of small cracks on the surface follows; *strain cracking*, the occurrence of small cracks running in all directions, is a more advanced form of decay that may result in the partial or total disintegration of glass (Newton and Davison 1997; Griffiths 1980). Frosting and strain cracking take place particularly when water is abundant: the water leaches from the glass most of the soda and potash and part of the lime, leaving behind only thin layers of hydrated silica. The final stages of such a decay process may result in the glass becoming just a residue of generally separate, flaky, highly porous, layers of hydrated silica displaying a sugarlike appearance that eventually totally disintegrates (Geilmann 1956).

FIGURE 29 Crusts of weathered glass. The deterioration of glass and the formation of weathered layers under humid or wet conditions is a rather complex process. Seasonal variations of temperature and of the amount of environmental moisture may provide the trigger that initiates the glass weathering process. Still, the final product of the process are opaque layers composed mainly of hydrated silica (silica is the main component of glass). In the figure, a microphotograph of a cross section of weathered glass, the hydrated silica layers remaining after many years of weathering can be clearly distinguished. It has been suggested that, in areas with alternating dry and wet yearly seasons, one weathering crust is formed each year. Thus counting the number of weathering layers on a glass surface may reveal the length of time the glass was exposed to wet or humid conditions.

FIGURE 30 Decaying glass. When glass is exposed to humid or wet conditions for long periods of time, weathered glass layers are formed that may become detached from the glass body. When exposed to light, the layers display colorful iridescent effects characteristic of ancient weathered glass. The iridescence is due to light interference effects arising from the front and back surfaces of the weathered layers. The illustration shows the weathered surface of a glass perfume flask made during the Byzantine period (fifth century C.E.).

Partial leaching and devitrification, although basically damaging, often enhance the appearance of old glass; they give origin to *iridescence*, the display of rainbow-like variegated colors when illuminated old glass is moved or turned. Iridescence generally occurs on ancient glass that has been recurrently exposed to seasonal variations, in some climates in yearly cycles, of temperature and humidity. Such cyclic processes result in the formation, on the outer, exposed faces of the glass, of very thin layers of decay products (composed mainly of hydrated silica); the thickness of single layers has been measured and found to vary in the range 0.3–15 microns. Light incident on these layers causes *interference* between beams reflected from their front and back surfaces and gives rise to the variety of colors often seen on ancient glass objects (see Fig. 30) (Laubengayer 1931; Raman and Rajagopalan 1940).

SECONDARY ROCKS
BUILDING STONE, BRICK, CEMENT, AND MORTAR

There is no reliable evidence about the first materials used by humans for the construction of dwellings, nor about the early techniques used for building them. It is reasonable to assume, however, that the first dwellings may have been tents made from the branches and leaves of plants, rushes, reeds, and/or animal skins. Developments in the form of settlement may have brought about efforts to build strong, lasting living quarters that eventually led to the use of masonry and, later still, of bricks consolidated with mortar. The convenient general term *building materials* is used here to represent natural and human-made materials whose principal use is for the construction of buildings, fences, roads, and other such structures. This includes natural materials such as *stone* and *mud* as well as human-made *mortar* and *brick*. Building stone, mud brick, and the different mortars used to consolidate and cover them are the main subject discussed in the following pages. Burned brick, a ceramic material, is discussed in Chapter 7 (Wright 1998; Davey 1961).

4.1. BUILDING STONE

Much building material has been derived from two *monomineral sedimentary rocks*: *gypsum* (composed of hydrated calcium sulfate) and *limestone*, which consists of *calcite* (composed mostly of calcium carbonate). Freshwater and seawater contain dissolved calcium carbonate and calcium sulfate. Most limestone and gypsum are formed when, as a consequence of the evaporation of water, calcium sulfate and calcium carbonate precipitate out of the water solutions as either gypsum or limestone. Limestone is also formed as a result of the activity of living organisms. Many sea- and freshwater animals, such as snails, clams, and corals, as well as some water plants, draw

Archaeological Chemistry, Second Edition By Zvi Goffer

calcium carbonate out of the water and use it: the animals to make their shells and the plants their inorganic solid matter. When the organisms die, their living matter decays. The calcium carbonate that constitutes the shells, corals, or plants, however, remains in the form of small particles that gradually accumulate as sediments. Over long periods of time, such sediments become consolidated into beds of cohesive limestone deposits (Boynton 1980).

Both gypsum and limestone are widely distributed throughout the earth's crust. They are mostly white, as are pure calcium sulfate and calcium carbonate. Because they often contain impurities, many gypsum and limestone rocks are colored, usually off-white, but occasionally they display intense colors, bands, and speckles. They are excellent thermal insulating materials, and although they are strong, they are relatively soft and can be cut easily, without splitting. These properties makes them ideal for building sheltering constructions, sculpting statuary, carving ornaments and engraving epigraphs.

Gypsum and Alabaster

Gypsum, one of the more common sedimentary rocks in the earth's crust, occurs in varieties that differ from each other in texture rather than in composition. In most of its varieties gypsum is very soft and can be scratched even with a fingernail. One massive and fine-grained variety of gypsum, known as *alabaster*, much appreciated for designing delicate decorative objects, has been used particularly for carving and ornamentation.

There is some confusion regarding the use of the term *alabaster* in archaeological contexts. Although alabaster (see Fig. 31) is a fine-grained form of gypsum (composed of hydrated calcium sulfate), much of the stone known as *ancient alabaster* from ancient Egypt, for example, where it was widely used for decorative work, consists not of gypsum but of calcite (composed of calcium carbonate) (Lucas 1962). Ancient alabaster also known as *oriental alabaster*, *Egyptian alabaster* or *onyx marble*, therefore, is a variety not of gypsum but of limestone. The appearance, carving, and sculpting properties of ancient alabaster are, however, very similar to those of ordinary alabaster. The two types, alabaster and ancient alabaster, can, however, be easily distinguished from each other by some simple tests. Alabaster (a variety of gypsum) is so soft that it can readily scratched even by a fingernail; its hardness on the *Mohs scale* is 1.5–2 (see Table 20). Ancient alabaster, whose hardness is about 3 on the Mohs scale, is too hard to be scratched in this way. Moreover, since ancient alabaster is composed of calcium carbonate, it effervesces when in contact with any acid, whereas gypsum alabaster is unaf-

FIGURE 31 Alabaster. A small reconstructed cosmetic cup made from ancient (or oriental) alabaster (seventh century B.C.E.) found at Horvat Radum, Israel. The surface of one of the joined parts is coarse and rough because of part dissolution by groundwater. The two parts were found in slightly different locations at the site, but only one of them was affected by water and its surface disintegrated. Two basically different forms of alabaster have been employed, since time immemorial, for carving statuary and ornamental objects. In mineralogy the term "alabaster" refers to a massive, fine-grained and translucent variety of gypsum (composed of calcium sulfate). This type of alabaster, often called "true alabaster," is very soft and is easily scratched with a fingernail. The other type, known as "ancient alabaster," also translucent, but somewhat harder than "true alabaster," widely used in the Middle East, for example, is a variety of marble (composed of calcium carbonate).

fected by acids. The stones can also be distinguished from each other, therefore, by a chemical test: just a drop of any acid on the surface of ancient alabaster will dissolve part of the stone while carbon dioxide is evolved from the affected area; the acid will not have any effect on ordinary alabaster.

Limestone and Marble

Limestone varieties differ greatly from one another in their texture and the impurities they contain, and consequently they also differ in color. The color of limestone may vary from white (when it contains practically no impurities) to off-white and even to intensely colored. Minor inclusions within the limestone structure are often of silica, usually in a concentration below 5%, as well as feldspar and clay in still lesser amounts. Many types of limestone also include embedded fossils. Much limestone deposits in the outer crust of the earth are altered during geologic *metamorphic processes* that involve mainly pressure and heat but also liquids and gases. *Marble*, for example, a metamorphic rock derived from calcium carbonate, is white when composed only of this substance; colored metal ions and other impurities impart to marble a wide range of colors such as red, yellow, and green and also give rise to speckled and banded varieties (Pacifici 1986). The beauty of marble has long been highly valued, particularly for buildings monuments and sculpting statues. The generally preferred type of marble for these uses is translucent marble, in which incident light penetrates the surface of the marble and is then diffracted. Translucent marble quarried from Mount Pentelicus in Greece, for example, was used to construct the Parthenon, and from Carrara, in the Apennine Mountains of Italy, to build monuments.

4.2. CEMENT

A profusion of terms is used to describe the various stony bonding materials used in building for consolidating and/or covering stone and brick or grains of sand; some of these are *cement*, *concrete*, *mortar*, *plaster*, and *stucco*. The short discussion that follows, which is relevant in archaeological contexts, may differ somewhat in terms of those used in building technology.

TEXTBOX 31

CEMENT

Cement is the generic name for materials, or mixtures of materials, used for bonding, that is, for holding together separate units or parts of a unit without the aid of mechanical devices such as nails or bolts. A cement, generally in a somewhat fluid and pliable condition when recently prepared, adheres to and joins solid objects or parts of objects into cohesive units. Included under the category of cements are an almost countless variety of materials and mixtures ranging from the inorganic *building cements*, metallic *solders*, *brazing metals*, and organic *adhesives*. *Building cements*, such as mud, gypsum cement, and lime cement, serve for joining together and concealing stones and bricks. *Solders* and *brazing* metals are used to join metals and alloys (see Textbox 44). The organic *adhesives*, generally referred to simply as *adhesives*, are preparations of organic substances and water used to join wood, skin, and leather to other materials of vegetable or animal origin (see Textbox 56).

Practically all the inorganic building *cements*, such as mud, lime, and gypsum cement, are prepared as mixtures of at least three components: a *binder*, a *filler*, and a *medium*, the latter always water. The *binder* (or *binding material*) is a substance that acts cohesively, adheres to surfaces and, when dry, is strong and resistant to external forces and to weathering. The most widespread building binders are clay in mud cement, lime in lime cement, and gypsum in gypsum cement. The *filler*, also known as the *aggregate*, is a comminuted solid that enhances the mechanical properties of the dry binder. In practically all types of building cements some type of sand, but occasionally also comminuted stone, sea animal shells, twigs, dung, or straw are used as *fillers*. Inorganic cements are therefore, *composite materials*, that is, mixtures of two or more solids that when mixed together have better mechanical properties, particularly strength and toughness, than the separate components (see Textbox 32).

TEXTBOX 32

COMPOSITE MATERIALS

Composite materials, or simply *composites*, are mixtures of two or more solids in which each component enhances the mechanical properties of the other(s). A composite is stronger and tougher than any of its components. Although composite materials consist of mixtures of two or more different materials strongly bonded together, each material in a composite remains a separate chemical entity (Chawla 1998b). Many composites exist in nature, wood and bone being the two most common and most abundant natural composites. Wood, which combines cellulose fibers and lignin, is a much stronger and tougher material than are cellulose fibers or lignin alone (see Chapter 9). Bone consists of a network of collagen fibers entwined with crystals of carbonated hydoxyapatite, a blend that has even greater compression strength than reinforced concrete (see Chapter 15).

Most composites are made up of two basic ingredients generally known by the generic names *matrix* (or *binder*) and *filler*. The matrix as well as the filler may be of either inorganic or organic origin. In freshly prepared human composites, the *matrix* is often a fluid or soft, pliable material that solidifies into a hard, compact mass encasing the filler. Wet clay or lime, hot bitumen, and resin, all of which are fluid when freshly prepared or when hot, are examples of binders that have been extensively used for making composites. The *filler* may be either in the form of particulate matter as, for example, sand, gravel, and seashells, or as fibrous matter, such as vegetable tissues (leaves, twigs, and straw) or the dung of herbivorous animals.

A freshly prepared mixture of a fluid or soft binder with a solid filler is generally pliable and can be easily worked and shaped. As the binder solidifies, either because of drying, as when the binder is a cement, firing at high temperature, when the binder is clay, as in ceramics, or cooling down below the melting point, when the binder is bitumen or an alloy, the composite becomes increasingly hard until it sets into a strong, tough material, stronger and tougher than its separate components.

Humans have made composite materials for thousands of years. Examples of composites that have been widely used since early antiquity include *adobe* (mud cement), *calcareous* and *gypsum cements*, *mud brick* and *burned brick*, *pottery* and most other ceramic materials, and some alloys. A block of dry mud, for example, is easily broken by bending, that is, by applying a mechanical force on its surface. Straw, on the other hand, is very resistant to stretching, although very fragile and labile to creasing and crinkling. When mud is thoroughly mixed with straw and the mixture is left to dry, the product is *mud brick*, a strong, tough composite material, resistant to most mechanical forces. The mixture of mud and straw also makes an excellent building cement for joining stones or mud bricks. Other ancient, well-known building cements such as lime and gypsum cements are also composite materials; they consist of a coarse mixture of sand, gravel or small stones with wet lime or gypsum (see Textboxes 33 and 34). When any of these cements dries and sets, it becomes extremely strong and resistant to compression and bending, much stronger than any one of its separate components (de Wilde and Blain 1990).

Mortars are cements used for bonding together masonry units, such as stones or bricks. When a cement is used to conceal masonry, as a more or less smooth covering on walls, for example, it is referred to as *plaster*. A very fine plaster, known as *stucco*, is made of very thin sand or finely comminuted marble. Freshly prepared plasters and stuccos are spread on consolidated masonry to form more or less uniform and smooth layers; stucco also provides a smooth and often flat outer coating.

Concretes are cements containing a large proportion of gravel. *Hydraulic cements* are cements that *set* (harden) in wet environments, as required when building structures submerged in water. Like all other cements used in ancient times, hydraulic cements were also composite materials in which one particular component, such as *pozzolana* in ancient Rome (see text below), endowed the cement with the property of setting in wet environments (Gani 1997; Akroyd 1962).

Mud

Probably the oldest known building cement, *mud* has been widely used since time immemorial, and its use still continues in many areas of the world.

Building mud, a composite material, is easily prepared by mixing clay or clayey soil with fibrous matter, such as straw or dung of herbivorous animals, and sufficient water to obtain a plastic, pliable mass. In ancient Egypt, for example, mud was made by mixing clayey soil with sand, chopped straw, and sufficient water so as to make the mixture pliable and suitable either for use as mortar or stucco or for making bricks.

Mud Brick. *Bricks* are small, usually rectangular blocks of sun-dried or fired mud or clay, used in building. Making bricks is probably one of the most ancient technologies: mud bricks (see Fig. 32) have been made for at least 6000 years by molding a mixture of wet mud and fillers such as straw, leaves, or dung into rectangular units of regular size that are then dried in the sun. Mud bricks, also known as *sun-dried bricks* or, using the Spanish name, *adobe bricks*, are quite hard, compact, and stable. After drying in the sun, if mud bricks are fired at relatively high temperatures, generally below 850°C, *fired bricks* are obtained (see Chapter 7) (Prandtstetten 1980; Brown and Clifton 1979, 1978).

Digging, collecting, and mixing the raw materials and then molding and drying the shaped mud are the main stages involved in making sun-dried brick. The dug-out clay, sometimes exposed for some time to the atmosphere

FIGURE 32 Mud brick. Mud brick, first century C.E., Jerusalem, Israel. Mud brick, also known as "adobe brick," is a type of brick used in dry, hot climates. It is made by mixing soil, which consist mainly of clay, with some form of fibrous matter such as straw, leaves, or animal dung and sometimes, also comminuted stone. Adding water (generally just below 20%) to such a mixture results in a workable mixture that is then poured into regular and usually rectangular molds and left to dry under the sun. After the water evaporates, a strong and durable composite material is formed in which the clay particles are bound to the fibrous matter or/and gravel. Mud brick is still used, as of old, in many developing areas of the world.

so as to make it more suitable for shaping, is then mixed with fillers and water. The mixture constitutes a *composite material* that prevents cracking and breakup of the brick while drying and makes the dry, finished bricks strong tough, and resistant to weathering, and to external forces (see Textbox 32). The addition of water makes the mixture pliable and easily molded into the required shape (Lynch 1993; Gurke 1987).

Sun-dried bricks can be used only in relatively dry climates, since high humidity brings about their softening and rain may turn them into a fluid, flowing mass. They were widely used in Mesopotamia some 6000 years ago, and similar ones are still being used today in areas such as Mexico, Egypt, and central Africa, where the climate is dry and warm (Clifton 1977). Bricks were often also made from river mud mixed with straw and sometimes even with sand. A mud brick bearing the cartouche of Ramses II was found within the walls of an Egyptian temple together with many bricks stamped with the names of his predecessors; the cartouche and other inscriptions were stamped while the brick was still damp and soft.

When building walls with mud bricks, each layer of bricks was consolidated with a layer of moist mud of a composition similar to that used for making the bricks. Often, the consolidated wall was covered with a relatively smooth layer of plaster also made from the same mud. Drying of the mortar and the covering layer makes the wall a solid mass of a dry and mechanically strong composite material, mud, as found, for example, at Giza and in the step pyramid at Saqquara, in Egypt. One of the components of the mortar from this pyramid is comminuted limestone, which was used as a filler instead of sand; the relative amount of the comminuted limestone in the mud varied over a relatively wide range, from as little 3% to as much as 55% of the total amount of mortar (Lucas 1962). The ancient Egyptian word for sun-dried brick was *debet*; it is quite possible that this word was reintroduced into modern nomenclature in the Spanish word *adobe*, now widely used to refer to mud cement made of a mixture of mud, additives, and water (Prandstetten 1980; Brown and Clifton 1978).

Bitumen

The Babylonians and Assyrians cemented stone slabs as well as bricks with *bitumen* (which is sometimes also referred to as *asphalt*), a mixture of a black or brown natural *organic material* with a pitchy luster and comminuted mineral. Bitumen is a *thermoplastic solid* material; when heated above 50°C, it softens and becomes a thick, viscous liquid that reverts to a solid on cooling. It is composed of a natural mixture of *hydrocarbons* (organic substances composed of carbon, hydrogen, and oxygen) that occurs in natural deposits on the surface of the earth or in bituminous or asphalt rock, from which it is extracted. When used as a building material, to join and consolidate or to cover stone or brick, bitumen is applied only after it is heated and turns into a hot, viscous liquid. In the molten state it penetrates into the pores, cracks, and holes of stones and bricks, cementing and consolidating them together and providing an effective and lasting cementing and/or covering material. Bitumen is highly waterproof; it is unharmed by most acids and salts, and thus prevents weathering of the structures it covers (Serpico and White 2000a; Marschner and Wright 1978).

Lime and Gypsum Cements

The two other main types of human-made building cements, *lime cement* and *gypsum cement*, have been and still are used in many areas of the world. Both these cements require quite elaborate thermal procedures for producing their main components, which are *slaked lime* in lime cement and *plaster of Paris* in gypsum cement. Making them involves the calcination of an appropriate type of stone, a process that has been practiced since prehistoric times. Slaked lime is made by the calcination of limestone; plaster of Paris, by the calcination of gypsum (see Textbox 33) (Coburn et al. 1990; Lea 1962).

TEXTBOX 33

CALCINATION AND ROASTING

Calcination and *roasting* are two thermal operations that entail heating minerals to high temperatures (above 650°C) so as to alter their composition and render them friable.

Calcination

Calcinating a mineral removes its volatile components, such as water or carbon dioxide and leaves an usually crumbly solid residue. Calcinated secondary minerals such as limestone are the basic components of building cements, and in extractive metallurgy operations they facilitate the smelting of metals. Calcinating *limestone* (composed of calcium carbonate), for example, drives away carbon dioxide, leaving a solid, friable residue of *quicklime* (composed of calcium oxide):

$$\underset{\substack{\text{limestone} \\ \text{calcium carbonate}}}{CaCO_3} \xrightarrow[\text{(over 650°C)}]{\text{heat}} \underset{\substack{\text{quicklime} \\ \text{calcium oxide}}}{CaO} + CO_2\uparrow$$

The major uses of *quicklime* are as a component of ordinary glass, as a flux in metal smelting operations and (mostly), for making building *cement* and *mortar* (see Textbox 34).

Likewise, the calcination of gypsum drives away water. Calcinating gypsum (composed of hydrated calcium sulfate) causes that part of the water combined with the calcium sulfate to evaporate, leaving a solid, friable residue usually known as *plaster of Paris* (composed of calcium sulfate hemihydrate); *plaster of Paris* is used as a *cement* and *mortar* as well as an efficient *casting material* (see Textbox 35).

$$\underset{\substack{\text{gypsum} \\ \left(\text{hydrated calcium sulphate}\right)}}{2CaSO_4 \cdot 2H_2O} \xrightarrow[(150°C)]{\text{heat}} \underset{\substack{\text{plaster of Paris} \\ \left(\text{calcium carbonate hemihydrate}\right)}}{2CaSO_4 \cdot \tfrac{1}{2}H_2O} + 3H_2O$$

Roasting

Roasting, another heat-induced process, entails the oxidation of minerals and/or ores so as to make them suitable for further use. Metal rich minerals and ores are generally roasted to make them suitable for smelting the metals they contain. When minerals composed of metal sulfides, for example, are heated while an abundant supply of oxygen is maintained, the metal sulfides are converted to metal oxides, which are more convenient for use in metal smelting. Roasting galena, composed of lead sulfide, a major ore of lead, for example, results in the conversion of the lead sulfide to lead oxide, from which the lead can be easily extracted by relatively simple smelting operations:

$$\underset{\substack{\text{galena} \\ \text{(lead sulfide)}}}{2PbS} + 3O_2 \overset{\text{roast}}{=} \underset{\substack{\text{lead} \\ \text{oxide}}}{2PbO} + \underset{\substack{\text{sulfur} \\ \text{dioxide}}}{SO_2 \uparrow}$$

Roasting pyrite, an iron ore composed of iron sulfide, results in the oxidation and decomposition of this compound to volatile sulfur dioxide and the formation of iron oxide, which can be smelted with relative ease into iron:

$$\underset{\substack{\text{pyrite} \\ \text{(iron sulfide)}}}{2FeS} + 3O_2 \overset{\text{roast}}{=} \underset{\substack{\text{iron} \\ \text{oxide}}}{2FeO} + \underset{\substack{\text{sulfur} \\ \text{dioxide}}}{SO_2 \uparrow}$$

Although slaked lime as well as plaster of Paris unmixed with other materials are relatively good building cements, they have little mechanical strength. Adding a filler to them results in the formation of composite materials with significantly improved strength, toughness, and resilience. Thus the two cements have been and still are generally prepared as mixtures of either lime or plaster of Paris with some type of filler. The most widely used filler has been and still is *sand*, although comminuted stone and mollusc shells, and occasionally also ceramic waste material, have often been used as fillers, instead of sand.

Preparing the actual cement mixture entails, in both cases, mixing the product resulting from the calcination or roasting of a rock with a given amount of filler and water, so as to produce a wet, semifluid, pliable mass. Freshly prepared cement adheres to stones and bricks and, after *setting*, forms a relatively hard, consolidated solid that is mechanically strong, stable, and resistant to weathering and decay.

TEXTBOX 34

LIME CEMENT; QUICKLIME, SLAKED LIME

Calcinating limestone (composed of calcium carbonate) removes its volatile component (carbon dioxide) and results in the formation of *quicklime* (composed of calcium oxide) (see Textbox 33).

$$\underset{\substack{\text{limestone}\\\text{(calcium carbonate)}}}{CaCO_3} \xrightarrow{\text{above } 600^\circ C} \underset{\substack{\text{quicklime}\\\text{(calcium oxide)}}}{CaO} + CO_2 \uparrow$$

This process, usually carried out in a kiln and at a temperature well above 600°C, seems to have been practiced as early as the Stone Age (Gourdin and Kingery 1975). Quicklime is a basic component of calcareous cement. Before the cement can be prepared, however, it is essential to *slake* (disintegrate and break up) the quicklime by the addition of water: water reacts with quicklime to form *slaked lime*, composed of calcium hydroxide:

$$\underset{\substack{\text{quicklime}\\\text{(calcium oxide)}}}{CaO} + H_2O \rightarrow \underset{\substack{\text{slaked lime}\\\text{(calcium hydroxide)}}}{Ca(OH)_2} + H_2 \uparrow$$

When the excess water (that is, the amount of water in excess of that required to bring about the formation of slaked lime) evaporates, the slaked lime sets into a hard, stony material in two consecutive stages. The first stage lasts only a few hours after the addition of water to the quicklime; during this stage the excess water evaporates, which results in the slaked lime setting into a solid, relatively hard and stable mass. The second stage is a much longer one, often lasting many years after evaporation of the water; during this time the already dry and hard lime absorbs carbon dioxide from the surrounding atmosphere and is thus converted back to calcium carbonate, that is, to limestone:

$$\underset{\substack{\text{calcium hydroxide}\\\text{(dry slaked lime)}}}{Ca(OH)_2} + CO_2 \rightarrow \underset{\substack{\text{calcium carbonate}\\\text{(fully set lime cement)}}}{CaCO_3} + H_2O$$

TEXTBOX 35

GYPSUM CEMENT (PLASTER OF PARIS)

When *gypsum*, a sedimentary mineral (composed of hydrated calcium sulfate) is calcinated, most of its volatile constituent, water, evaporates and is therefore removed. The friable material remaining after the product of the calcination process cools down to ambient temperature is commonly known as *plaster of Paris* (composed of calcium sulfate hemihydrate) (see Textbox 33).

Plaster of Paris has long been used as a casting material, a cement, and a mortar. If mixed with water, plaster of Paris forms a very soft and pliable mixture. After a very short time, lasting only 5–8 minutes, the wet, pliable mixture *sets*, that is, it hardens into a stable, firm solid. The setting process entails the incorporation of water molecules (a process known as *hydration*) into the calcium sulfate hemihydrate and the consequent formation and crystallization of hydrated sulfate of calcium. In other words, when water is added to plaster of Paris, the two combine, again forming gypsum, which soon crystallizes into a hard solid mass:

$$\underset{\substack{\text{plaster of Paris}\\\text{(calcium sulfate hemihydrate)}}}{CaSO_4 \cdot \tfrac{1}{2}H_2O} + 1\tfrac{1}{2}H_2O \rightarrow \underset{\substack{\text{gypsum}\\\text{(hydrated calcium sulfate)}}}{CaSO_4 \cdot 2H_2O}$$

The crystallites of gypsum thus formed grow entangled and interlocked with one another, and the solid mass becomes relatively hard, strong, stable, and durable. Moreover, when plaster of Paris sets, it expands, growing linearly in size by about 0.3%, a property that makes it suitable for *casting*, producing sharp, accurate casts. When plaster of Paris is to be used as a mortar or building cement, a *filler* such as sand or comminuted stone is generally added during the preparatory stage. As the cement dries, the grains of sand or comminuted stone (the *filler*) become embedded in the newly created crystallites of gypsum (the *matrix*), making up a rigid, strong, and durable *composite material* (see Textbox 32) (Coburn et al. 1990).

When using lime cement, the actual mixing of slaked lime with sand or other fillers, was in the past, and in many places still is carried out, in the proportion of one part slaked lime to three parts filler. Sufficient water is then added to the mixture to form a thick-consistency paste. Since the chemically active component of the cement is slaked lime, the cement sets into a hard, consolidated mass, by two successive but different processes: first by drying, that is by losing water, a process that may last for a few hours or days. The setting process continues by the slow absorption, by the lime, of atmospheric carbon dioxide, resulting in the conversion of the slaked lime into calcium carbonate. The ultimate result of the setting of the cement is the formation of a coherent mass of a composite of calcium carbonate crystals embedding the grains of the filler (Ashurst 1991; 1983). As the fresh cement loses water and sets in the days and weeks following application of the cement, only a small part of the slaked lime in the cement is converted into calcium carbonate. Completion of the reaction is extremely slow and often remains incomplete for very long periods of time. Minute traces of still unreacted quicklime, for example, were found during the twentieth century in lime plaster from the Egyptian pyramids and from the walls of the palace at Knossos, Crete, where they had been exposed to the atmosphere for several millennia.

Hydraulic Cements. To build constructions submerged in a sea, lake, or river, it is necessary to use *waterproof cements*, generally known as *hydraulic cements*, which harden even in the presence of excessive amounts of water. In the past, such cements were prepared by heating a mixture of limestone and a considerable amount of clay or other powdered siliceous material. At high temperature (above 650°C), the quicklime, formed when the limestone is calcinated, reacts with the silica and alumina in the siliceous additives and forms new complex silicates and aluminosilicates of calcium, which are waterproof and set even under water. In Egypt and in India, for example, hydraulic cements were made firing heaps of a mixture of lime and clay at temperatures well above 600°C. In ancient Mesopotamia, ash was added to the lime to make a hydraulic cement that was used for plastering the interior walls of water reservoirs. The Romans made two different types of hydraulic cement; one consisted of a mixture of freshly prepared wet lime cement and oil. The best and the most widely used Roman hydraulic cement, however, was made by mixing slaked lime (about one part), volcanic ash, which consists of siliceous minerals (two parts) and sand (six to nine parts). Such a mixture, known as *pozzolana cement*, was widely used for the construction of port facilities and other underwater building projects. A study of the cement used during Roman times to construct harbor quays at Caesarea on the Mediterranean Sea and on the shores of Lake Tiberias in ancient Israel confirmed the high strength and durability of pozzolana cements (Oleson 1988; Malinovsky et al. 1961).

4.3. STUDY OF ANCIENT CEMENTS

There are various reasons to study the composition of *ancients cements*. The actual composition of a cement, for example, provides information on its nature, the technology used for making it, and the provenance of its components (Middendorf et al. 2005). It may also elicit differences between the nature of an original cement used for building and that used for later repairs (Streicher 1991; Jedrzejewska 1990). Most analytical work concerning ancient cement in the recent past has been based mainly on the use of optical microscopy and classical analysis techniques. Sometimes, such studies are complemented with information derived by instrumental techniques (Bläuer-Böhm and Jägers 1997).

ORES
METALS AND ALLOYS

The *metals* are a group of chemical *elements* having in common physical and chemical properties that make them unique among the wide ranges of materials available to humans. They have a *metallic luster*, their smooth surfaces reflect light incident on them, they are *ductile* and *malleable*, when subjected to external forces they are deformed rather than shattered, and when struck many metals produce a *metallic ring*. It is quite probable that it was the luster and the plasticity (ductility and malleability) of metals that first attracted human attention. Moreover, when metals are mixed and heated together with other metals or even with some nonmetals above their melting point, they form new materials, known as *alloys*, which also have metallic properties, although the physical properties of the alloys differ considerably from those of their component metals.

The Greek poet Hesiod (eighth century B.C.E.) identified four successive ages in the legendary prehistory of human beings: the golden, silver, bronze, and iron ages. The Roman historian Lucretius (first century B.C.E.), wrote: "The earliest weapons were the hands, nails and teeth. Then came stones and clubs. These were followed by iron and bronze, but bronze came first." Today it is generally accepted that the first metals known to humans were those occurring in the *native state*: gold, copper, and meteoritic iron. There is, of course, no way of knowing which one was used first. During the early neolithic period, probably about 12,000–10,000 years ago, human were already acquainted with some metals, mainly *gold* and *copper*. For many millennia, while relying mostly on stone as the main raw material for making artifacts, they also shaped small objects from lumps of gold and copper they found on the surface of the earth. In time, other metals were recognized and put to use.

Archaeological Chemistry, Second Edition By Zvi Goffer

The Old Testament, for example, mentions six metals: *gold, silver, copper, iron, tin,* and *lead*. The ancient Hindus also knew about these six metals, which were listed as well by Arabian chemists of the eight century and by European alchemists of the thirteenth. From the repeated reference to these six metals in ancient literary sources, it has been inferred that the discovery of other metals came at a much later date. Nevertheless, a few other metals, such as *mercury* and *zinc*, were known to ancient people in some regions of the world, thus bringing the total number of metals known in antiquity to the eight listed in Table 32 (see Fig. 34 for melting points of these metals). Still others, such as *antimony, nickel,* and *platinum*, were known only in rather confined geographic areas (Tylecote 1976, 1962; Raymond 1986).

TABLE 32 Metals of Antiquity

Metal	Chemical symbol	Abundance in earth's crust (ppm)	Density (g/cm^3)	Melting point (°C)	Hardness (Mohs scale)
Copper	Cu	70	8.9	1,083	2.5–3
Gold	Au	0.005	19.3	1,063	2.5–3
Iron	Fe	50,0000	7.9	1,536	4 –5
Lead	Pb	16	11.4	326	1.5
Mercury	Hg	0.5	13.6	−39	(Liquid)
Silver	Ag	0.1	10.5	960	2.5–3
Tin	Sn	40	5.8	232	1.5–1.8
Zinc	Zn	132	7.1	419	2.5

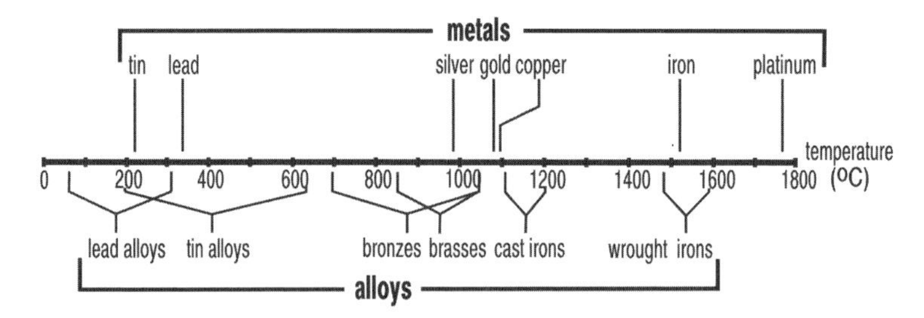

FIGURE 34 Melting point of the metals and alloys of antiquity. Heating techniques to attain temperatures higher than about 1500°C were developed as recently as the late nineteenth century. Only metals and alloys melting below 1500°C were, therefore, smelted in antiquity.

TEXTBOX 36

METALS AND ALLOYS

The 92 chemical *elements* that occur naturally in the earth can be divided into two main groups: *metals* and *nonmetals*. Although the distinction between the two is not always sharp and clear, it can be said that over 70 of the 92 elements are metals; among the fewer than 22 remaining nonmetals, six are known as *metalloids*, which have properties that fall between those of metals and nonmetals (see Appendix I).

In common parlance, the term *metal* is used to refer to two different types of *metallic materials*: *metals* and *alloys*. The *metals* are chemical elements; each metal (e.g., copper, iron, and gold) is composed of only one type of atom. The *alloys* are mixtures that have metallic properties. All alloys include two or more elements in their composition; some are made up of two or more metals, others of one or more metals mixed with one or more nonmetals. Bronze, for example, is made up of two metals: copper (60–85%) and tin (40–15%); *steel* includes iron, a metal (98–99.97%) and carbon, a nonmetal (2–0.03%). Metals and alloys share many common properties:

1. They are *malleable*, that is, they can be shaped by applying an external pressure, as when hammering a lump of metal or alloy into different shapes or into thin sheets; they are also *ductile* (can be drawn into wire), although cast iron is brittle.
2. They have a distinctive metallic *luster*, although some nonmetals have a similar luster.
3. They are good conductors of electricity and heat.
4. All the metals and therefore the components of the alloys combine with nonmetals to form *compounds*. In the compounds thus formed, the metals always form *cations*, which have a positive electric charge, and the nonmetals form negatively charged anions. The compounds formed by the metals with oxygen, for example, known as *metal oxides*, include a positively charged metal (the cation) and oxygen, a nonmetal that has a negative charge (the anion). Since the metal oxides generally react with water to form *basic substances*, the metal oxides are said to be *basic* (or *alkaline*).

Relatively soon after ancient humans recognized the metals and their special properties, they also discovered ways to make alloys. Some alloys were produced in antiquity directly, by the smelting of ores that include two metals in their composition or mixtures of ores of different metals. *Arsenical copper*, *bronze*, and *brass*, for example, three alloys of copper widely used in antiquity, were sometimes prepared from ores or mixtures of ores that incorporate two metals in their composition. It seems that in some regions of the world arsenical copper was derived from ores that include in their composition copper as well as arsenic. Some ancient bronze and brass seem to have been produced also by smelting together ores of copper and tin or of copper and zinc, respectively. *Cast iron*, as a number of alloys of iron and carbon are known (see Table 33), was formed mostly when smelting iron, as a result of the combination of hot iron with carbon, the main component of wood and charcoal used as fuel in the smelting furnace.

TABLE 33 Ferrous Alloys

	Composition (%)	
Alloy	Iron	Carbon
Cast iron		
Pig iron	96.5–95.8	3.5 –4.2
White iron	98.2–97.0	1.8 –3.0
Malleable iron	98.0–96.4	2.0 –3.6
Gray iron	97.5–96.2	2.5 –3.8
Ductile iron	97.8–95.8	3.2 –4.2
Steel		
Mild	above 99.7	0.03–0.3
Medium	99.7–99.2	0.3 –0.8
Hard	99.2–98.0	0.8 –2.0

Many of the so-called metals used in everyday life are actually not metal but alloys. Pure gold, for example, is too soft a metal for most practical purposes and is therefore generally alloyed with other metals. When alloyed with copper or silver, for example, it forms relatively hard and durable alloys from which most of the so called "gold objects" are made. In informal, common language, however, the gold alloys are usually, albeit erroneously, referred to as "gold." Also, most objects said to be made from "silver" or "iron" are actually made from silver- or iron alloys and not from the pure metals (see text below). Making alloys usually entails adding, to a molten *base metal*, as the main component of an alloy is usually referred to, other metals or nonmetals, thoroughly mixing the resulting melt, and then cooling it under controlled conditions to cause the alloy formed to solidify.

The alloys are often classified according to different criteria. One convenient classification is based on the most characteristic method used for shaping alloy-based objects. Examples are *cast alloys* and *wrought alloys*. *Cast alloys*, such as cast bronze, cast brass, and cast iron, are generally shaped into objects by *casting*: pouring the molten alloy into a *mold* and then letting the melt cool and solidify so as to acquire the shape of the mold. *Wrought alloys*, such as, wrought iron and several alloys of silver and gold, on the other hand, are shaped into objects by mechanical means, as by hammering at temperatures well below their melting point.

Another way of classifying alloys is based on the nature of the *base metal*, the major constituent of the alloy. Since iron is the most widely used metal, its alloys, known as the *ferrous alloys*, are often listed separately. Table 33, for example, lists alloys that include iron as the base metal. Alloys of other metals are listed in Table 34.

TABLE 34 Nonferrous Alloys[a]

	Composition (%)					
Alloy	Copper	Zinc	Tin	Lead	Silver	Nickel
Copper alloys						
Brass						
Common	65	35				
Red	82–85	10– 7	3– 5	5– 3		
Roman	60	39	1			
Bronze						
Chinese	78		22			
Roman	58–60	34–12	42–40			
Jewelers'	89	9	2			
Leaded	80		10			
Statuary	65–85	below 5	30–10			
Coin	92–95		5– 4			
Bell metal	75–80	below 1	25–20			
Mirror						
Chinese[b]	16	73	11			
European	66–88	8– 1	11–26			
Paktong	42–58	46–34				12–8
Tin alloys						
Pewter						
Common			85–95	15– 5		
Medieval[c]		below 2	80	19		
Roman			70	30		
Silver alloys						
Coin silver	25– 5				75–95	
Shibu-ichi[d]	51–86				49–14	

[a]See also Table 48.
[b]Also contains antimony (about 8%).
[c]Also contains bismuth (about 6%).
[d]Also contains traces of gold.

Few metals occur in the earth's crust uncombined with others. Those that do, such as gold, silver, mercury, and some copper, are known as the *native metals*. Together with some *native nonmetals* such as carbon and sulfur, the native metals make up the relatively small group of *native elements*, naturally occurring masses of single elements that were recognized and put to use by humans in quite early times. Most metals occur in nature combined with nonmetals in the form of mineral and rock deposits, and from those minerals they are extracted by means of a variety of *metallurgical* techniques.

5.1. NATIVE METALS

The few metals that naturally occur uncombined, in a relatively pure state, do so mainly in the form of *lumps* or *nuggets*. The obvious luster of gold nuggets on the surface of the earth could hardly have been ignored by prehistoric humans; it is highly probable, therefore, that because of its natural glitter, native gold was the first metal to be noticed and recognized. Native copper was also relatively abundant in antiquity in some areas of the world. Even today, deposits of native copper are still occasionally discovered in some places. Since copper is affected by, and reacts with, weathering agents and pollutants, native copper lumps are seldom lustrous; they are mostly covered by a surface coating of dark green or reddish brown copper products, usually known as *patina* (Coghlan 1951). Native iron occurs only as grains of microscopic size that would not have been recognized by ancient humans. But large masses of iron, known as *meteoritic iron*, have been brought to the earth's surface by meteorites from outer space. Silver, too, occurs in some places as relatively large nuggets, although mostly in veins too deep below the surface of the earth to have been accessible to the ancients.

The native origin of metals used in prehistoric sites for making objects can often be revealed by chemical analysis and microstructural studies. If quartz and iron are found in gold, for example, they are often a good indication of the native origin of the metal, since such impurities do not occur in smelted gold. *Native copper* is generally free of impurities, a characteristic that makes it distinguishable from the smelted metal. *Meteoritic iron* is generally characterized by the presence of phosphorus and is sometimes naturally alloyed with nickel. Among the oldest known human-shaped metal objects are native copper beads found in northern Iraq, probably dating from the beginning of the ninth millennium B.C.E. (Smith 1965).

5.2. METALLIFEROUS ORES

The overall abundance of most metals in the upper crust of the earth is rather small (see Table 32). At such low concentrations, separating them from other associated matter could be quite a monumental task. Fortunately, natural geochemical processes have resulted in the uneven distribution and abundance of the metals on the outer part of the planet. As a result of geologic and natural weathering processes, metals occur in high concentration in geologic beds usually referred to as *metalliferous ores*, metal-bearing minerals, valuable enough to be mined (Brimhall 1990; Ixer 1990). In the *ores*, the metals are combined with other elements, mainly with oxygen, sulfur, carbon, or silicon. Common ores of iron and tin, for example, are minerals composed of the oxides of these metals, such as *magnetite* (composed of iron oxide) or *cassiterite* (composed of tin oxide). Copper, lead, mercury, silver, and zinc, and also some iron, often occur in minerals in which the metal is combined with sulfur (in the form of metal sulfides) and even with other nonmetals. It is from these ores that practically all metals have been recovered, since antiquity (Brimhall 1990).

Between the fifth and third millennia B.C.E., while already familiar with the native metals, humans began to realize that there was a relationship between metals and metalliferous ores. It was probably at this time that the process of *smelting*, for the production of metals from metalliferous ores, was discovered. In retrospect, it is impossible to pinpoint with any degree of certainty precisely when, or where, the smelting process was first practiced. Nor is it clear how the process was discovered. What is certain is that the discovery of the smelting processes brought about the gradual replacement of native metals by metalliferous ores as the main sources of metals. By the end of the third millennium B.C.E., nearly all the metals that could be recovered from metalliferous ores had been smelted and put to use (Maddin 1988; Forbes 1997a).

5.3. MINING

The procedures necessary for excavating and removing metalliferous ores from their earthy surroundings developed quite early into an organized *"mining technology."* Ores on or near the earth's surface required relatively easy mining techniques, while those far beneath the surface, could be extracted only after digging, sometimes deep underground. Still others ores had to be extracted from the bottom of lakes or rivers (Shepherd 1993; Craddock 1990). The ores exploited in early antiquity were probably easily recognizable because of the color, shape, and size or other characteristics of the rocks, boulders, or pebbles from which the metal could be recovered by relatively simple smelting procedures. This confined mining centers to the relatively few geographic regions where such ores were abundant. On the island of Cyprus, for example, which served as an important copper-mining center in antiquity, many of the large copper ore deposits are easily recognizable by their mostly greenish color. Also in Europe much tin was mined, for many centuries, from easily recognizable ore deposits in Britain, which was therefore referred to as the "island of tin."

5.4. ORE DRESSING

Metalliferous ores are generally mixtures of metal-rich minerals and varying amounts of worthless other minerals. After mining, ores therefore undergo a process known as *ore dressing*, intended to separate and concentrate the metal-rich minerals from the useless *gangue*, as the worthless minerals are called. Ancient ore dressing most probably entailed such operations as hand picking, pounding, screening, or washing the raw ore with running water, so as to omit or remove the generally less dense gangue. Ores containing as little as 10% of metal-rich mineral for example, could be separated during *dressing operations* into two fractions: the *ore concentrate*, containing as much as 60% of metal-rich mineral, and the *gangue*, waste material with little, if any, metal (Craddock 1980; Evans 1987). Some ore concentrates, those in which the relative amount of metal-rich mineral consisted of metal sulfides or metal carbonates, were then either *roasted* or *calcinated* so as to convert them into metal oxides (see Textbox 33).

5.5. SMELTING

To recover metals from *dressed mineral* ores, the latter are *smelted* (processed at high temperatures to convert them to metals) (see Textbox 37). Early *smelting* was probably initially accomplished in open hearth furnaces, as found in many ancient archaeological sites, consisting of a hole dug in the ground, which may have been lined with a layer of fire-resistant clay or stone. In most furnaces, *charcoal* was layered on top of the clay floor or stone floor, and then covered with a layer of metal ore. After charcoal is ignited, it burns, releasing *reducing gases* that react with the mineral and reduce the combined metal to free metal. The burning temperature in such a furnace is sufficiently high to melt the metal, thus creating a molten mass of dense metal topped by a lighter layer of waste products, known as *slag*. When the hearth is finally cooled, the generally brittle and glassy slag is broken off and removed, leaving an *ingot* of crude metal at the bottom of the hole. More complex smelting techniques make use of *crucibles*, ceramic pots into which the fuel and metal ore are placed. *Bellows*, usually made of animal skins, direct a draft of air at the fire, raising the temperature of the furnace where the crucible is located. The metal ingot produced by this technique has fewer impurities than does that in the open hearth. Metal ingots are usually cast into the shape of bars, rings, and other shapes suitable for transportation and trade.

TEXTBOX 37

METALS SMELTING AND REFINING

Smelting is a *pyrometallurgical* process based on the use of fire for extracting and recovering metals from metalliferous ores. It involves heating an ore to high temperature and then inducing in it chemical changes that ultimately result in the detachment and separation of the metal in the ore from the other elements with which it is combined. The final products of all smelting processes are a metal and waste; the latter, generally known as the *slag* or *dross*, contains the nonmetallic elements from the ore. The conditions required for any particular smelting process are determined by two chemical factors: the composition of the ore and, more specifically, the strength of the bond between the metal and the other elements in the ore. The stronger the bond between them, the higher the amount of energy required to separate between them and therefore the higher the smelting temperature. Most ores require smelting temperatures sufficiently high to melt the extracted metal as well as the slag, so that the end products of most smelting processes are a molten metal and molten slag. To lower the temperature of smelting and make the process economical in fuel, a flux is often added to the ore (see Textbox 28). The addition of flux is particularly noteworthy, since it lowers the smelting temperature of ores that otherwise melt at temperatures too high to be attainable in antiquity.

In practical terms, most smelting operations involve heating to high temperature, in a *smelting furnace*, an ore (often previously *calcinated* or *roasted*) or, if required to lower the smelting temperature, a mixture of ore and flux. When at a high temperature, the metal in the ore is chemically *reduced* to free molten metal, by gases generated when the fuel burns. The appropriate conditions for the reduction and separation of the metal from the slag are created when carbon atoms in the fuel (wood, charcoal, or dung in the past) react with atmospheric oxygen to form carbon monoxide (CO), a *reducing* gas:

$$\underset{\text{carbon}}{2C} + O_2 = \underset{\substack{\text{carbon monoxide} \\ \text{(a reducing gas)}}}{2CO}$$

Carbon monoxide reacts with combined metal in the ore and reduces it into uncombined molten metal. Smelting iron, for example, proceeds by the conversion of iron oxide, in the ore, to uncombined iron metal:

$$\underset{\substack{\text{iron oxide} \\ \left(\substack{\text{natural, calcinated} \\ \text{or roasted ore}}\right)}}{Fe_2O_3} + CO \overset{800°C}{=} \underset{\substack{\text{iron} \\ \text{(molten metal)}}}{2Fe} + 3CO_2 \uparrow$$

Similarly, smelting copper oxide, often derived from roasted copper sulfide ores, proceeds as follows:

$$\underset{\substack{\text{copper oxide} \\ \left(\substack{\text{derived from roasted} \\ \text{copper sulfide}}\right)}}{CuO} + CO \overset{\text{about } 1000°C}{=} \underset{\substack{\text{copper} \\ \text{(metal)}}}{Cu} + CO_2 \uparrow$$

Most smelting processes (see Fig. 33) result in the formation, in the smelting furnace, of a two-layer melt: a dense layer of molten *crude* (impure) *metal* at the bottom and a lighter layer of waste, the slag, forming a scum on top. After the furnace cools down and the products of the smelting process solidify, the slag is removed and discarded. The crude metal can then be used as such, or it may be further refined (Tylecote 1992; Bachmann 1982).

Crude Metals and Metal Refining

The crude metal produced in most smelting processes is quite impure. The impurities may be derived either from the ore, the flux, or the fuel; during smelting, parts of these combine with the hot, molten metal imparting to it deleterious properties. Metal *refining* is the process of entirely removing or reducing to acceptable levels the relative concentration of deleterious impurities in the metals. Some impurities, such as gold and silver, which are not necessarily deleterious, have sufficient value to make their recovery profitable. Still others, which are neither deleterious nor valuable, are frequently ignored and remain in the refined metal (Engh 1992).

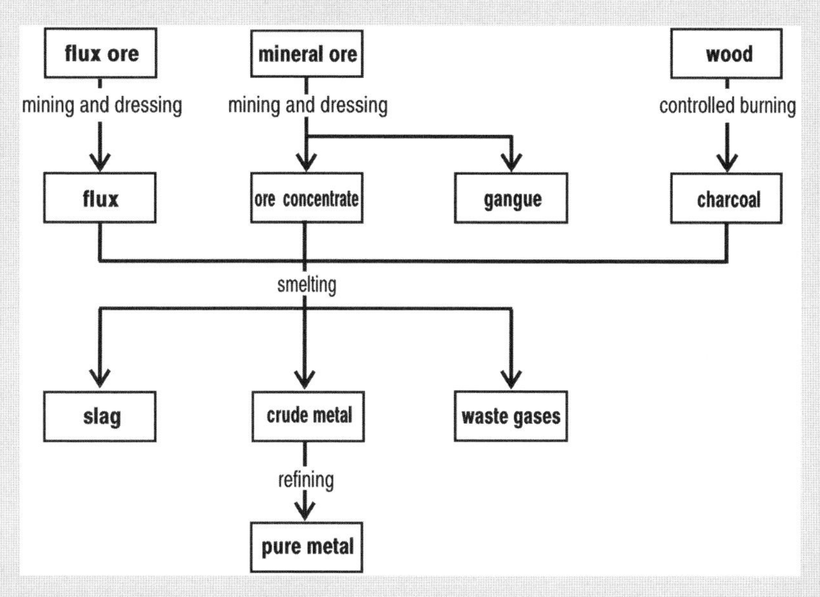

FIGURE 33 Smelting; flowchart. The chart shows, in diagrammatic form, the sequence of metallurgical processes required for extracting metals from their ores: from the initial mining of metalliferous ores to the final refining of the smelt metals.

Smelting Fuel and Metal Reduction

Wood and *charcoal,* both of which are combustible materials, were used as the *fuel* in the smelting furnace. Charcoal, in particular, was used in extractive metallurgy from very early times and its use as a fuel for smelting continued until quite recent times. It was only during the middle of the eighteenth century that it was gradually replaced by *coke,* a change that revolutionized the metal industry. It should be noted that *coal* from which coke is derived and *peat,* two of the most common industrial fuels during the eighteenth and nineteenth centuries, were apparently unknown as combustible materials to most ancient peoples; their use was confined to a few regions of the world, such as the Ruhr area in Germany and part of the British Isles (Forbes 1997a; Neuberger 1969).

5.6. METAL REFINING

Smelted metals, generally referred to as *crude metals*, usually contain a relatively large number of *impurities*. These include carbon from the fuel, silicon from the slag (formed during the smelting process), oxides of the smelted metal (resulting from accidental oxidation during the process), as well as a variety of other metals and nonmetals, mostly in trace quantities, from the ore. Methods for *refining* crude metals, freeing them of impurities, vary widely, depending on the specific metal to be refined and the nature of the impurities. Two general rules, however, apply to most refining processes:

1. The crude metal is molten before refining.
2. The refining process generally entails either the volatilization or the controlled oxidation of the impurities.

Some impurities are converted into gases that are then released from the hot molten metal; others are oxidized (converted into oxides), making up *slag*, the layer of useless matter on the upper surface of the molten, refined metal, from where it is removed.

Precious metals such as *silver* and *gold*, which are seldom oxidized even at high temperatures, are often refined by *cupellation*, a process for removing from them *base metal impurities* such as lead and tin, with which they are associated in many ores. Hot lead and tin are easily oxidized. In the cupellation process, a crude, impure precious metal is placed in a shallow cup or crucible made of *bone ash*, known as a *cupel*, and is then heated by a blast of hot air. At high temperatures, the base metal impurities are oxidized by oxygen in the hot air, and the oxides thus formed are absorbed by the porous bone ash. The Chaldeans are said to have been the first to have utilized (ca. 2500 B.C.E.) cupellation to remove lead and purify silver from lead–silver ores.

The smelting process seems to have been independently discovered in the Middle East, China, and Southeast Asia. The chronological sequence in which the different metals were brought into use was probably determined by the difficulty in winning them from their ores. Copper, for example, which melts at just above 1000°C, a temperature within reach in a vented wood or charcoal fire, was separated from its ores quite early. Iron, which melts at above 1500°C, a temperature that for a long period of human technological development was out of reach, was first produced at a much later time.

5.7. ALLOYS

Sometime after the discovery of processes for smelting metals, it became clear that some of their properties could be altered and in many cases improved by *alloying*, that is, by mixing metals with other elements. Some alloys made by mixing two metals, for example, were found to be harder or softer than the separate metals. Also the melting point of an alloy was often lower than that of its components, which made the alloys easier to work. Soon it was appreciated that many other properties of alloys, such as their strength, workability, and resistance to decay, were more suitable for required needs than were its components, and the manufacture and use of alloys become widespread (see Table 35).

The first alloys made by humans were probably those of copper, namely, *bronze* and *brass*, which were already being made during the Chalcolithic period (see Table 33). The most important, however, later became the alloys of iron, known as the *ferrous alloys* (from the Latin word *ferrum*, for iron). Since iron ores are one of the most abundant metalliferous ores on the crust of the earth, and its alloys are relatively easy to produce, ferrous alloys have been the most widely used alloys for the last three millennia (see Table 34).

TABLE 35 Alloying and the Mechanical Properties of Alloys

Property	Effect of alloying on base metal	Implication
Melting temperature	Lower	Alloy easier to cast and work than the base metal
Strength	Higher	Alloy stronger than the base metal
Ductility and malleability	Lower	Gain in strength more than offsets the loss in these properties

TEXTBOX 38

SHAPING MATERIALS: CAST AND WROUGHT OBJECTS

Subtracting matter, by cutting, chipping, or carving, for example, provides a way to shape solid materials into objects. Many objects are not shaped by the subtraction of matter, however, but are either *cast* or *wrought*: *cast* objects are shaped when the material used to make them, such as glass, plaster of Paris, or bronze, is in a fluid condition. *Wrought* objects, on the other hand, are made form malleable materials, almost exclusively metals or alloys, by applying an external force on them, by hammering for example, at suitable temperatures below their melting point.

Casting

Casting is the process of shaping objects by pouring into a *mold* (a receptacle of a given shape) a fluid that *sets* (solidifies) while retaining the shape imposed by the mold. The process of casting dates back many thousands of years. In antiquity casts were made from such varied materials as wet mud, plaster of Paris, and clay mixtures, as well as from hot molten glass, metals, or alloys (Gerin 1972). The *mold*, or *die*, is a hollow, "reverse" pattern of the object to be cast. Molds are either carved from a bulk of solid matter, such as stone or wood, or shaped from a soft and pliable material, such as, wet mud or plaster, which, after drying, hardens while preserving its shape. Casting hot molten fluids such as glass, metals, or alloys requires that the mold be made from a *refractory material*, stable at high temperatures, such as sand or clay.

The actual casting process entails filling the mold with the fluid casting material, then letting it set, and finally removing the finished cast from the mold. Casting materials set either at room temperature, or when cooled down from high temperatures. Wet mud and plaster of Paris, for example, which are fluid when mixed with water, set at room temperature when the water evaporates (see Textbox 33). Glass, metals, and alloys, which are fluid only at high temperatures, set when the cast cools down: glass as it cools down below its softening temperature, metals and alloy as they cool below their melting point (Bareham 1994). Ancient methods of casting can be broadly classified into three categories:

1. *Open-mold casting*
2. *Hollow sand casting*
3. *Lost wax* or *cire perdue* casting.

In *open-mold casting*, the shape of the desired object is carved as a "negative" depression or hollow into a suitable bulk of material that withstands the temperature required for the process; wood is often used for cold casting molds; stone, sand, and dry mud, for hot casting. The fluid casting material is then poured into the carved depression, where it is left to set and from which, when solid, it is finally removed.

Hollow *sand casting* is a relatively simple process for making cast metal objects that for thousands of years was the most widely used of all casting methods. For the most basic castings, the molds are made from ordinary silica sand mixed with water so as to keep the sand particles compacted together to maintain the required hollow-shaped form. The hot, fluid metal is then poured into the mold. After the hot metal has cooled down and solidified, the surrounding sand mold is broken down and discarded, and the cast object is removed. Hollow sand casting produces rough metal castings that require further refinement by either forging, mechanical manipulation, or both; hollow sand castings not further refined are readily recognized by the sandlike texture of the casting's surface.

Casting by the *lost wax* method entails initially creating a *core* of clay covered with a layer of *beeswax*, and modeling the outer layer of the wax in the exact pattern of the desired cast. Once the wax pattern is made, the sequence of operations listed below is generally followed (Feinberg 1983; Ammen 1979):

1. A mold is made from a refractory material to surround the wax pattern.
2. The mold is heated so as to melt and remove the wax.
3. The casting fluid (metal or alloy) is poured into the mold and allowed to cool slowly.
4. After the cooled cast sets the mold is removed.

Wrought Metals

Wrought metal objects, made mostly from iron or steel but also from other malleable metals and alloys, are shaped (at an appropriate temperature) by beating or hammering. *Wrought iron*, for example, is generally shaped in a *forge*, a special furnace in which the iron is heated to a suitable temperature before shaping (Bealer 1969). Some wrought metal objects are shaped by hammering the metal, in the form of thin sheets, into open *dies* (metal blocks of required shapes); a series of different dies are often used successively before a finished shape is produced (Sabroff et al. 1968; Smith 1967).

5.8. THE METALS AND ALLOYS OF ANTIQUITY

Copper

Copper (chemical symbol Cu, from the Latin name of the metal, *cuprum*), the metal that in Roman times was known as the Cyprian metal (since much of the metal came from Cyprus), is reddish brown, malleable and ductile, and can be easily shaped by cold- or hot-working techniques (see Fig. 35) (Scott 2002). Native copper occurs mainly in the form of boulders, nuggets, dendrites, and laminar outgrowths. It was certainly in its native form that copper was first recognized and used; for over five millennia since then, however, the bulk of copper has been derived from copper ores by a variety of smelting methods. Table 36 lists ores from which copper was and still is extracted.

Smelting such copper ores as the minerals *malachite* (composed of copper carbonate) or *chalcopyrite* (composed of copper sulfide) was a well-established metallurgical technology in many regions of the world before the end of the third millennium B.C.E. The smelting furnaces and smelting techniques used in each region seem to have been quite different, and characteristic for that region (Herbert 1984). In the Wadi Arabah, in southern Israel, and in the Sinai Peninsula, for example, ancient Egyptians smelted mainly

TABLE 36 Common Copper Ores

Ore	Composition
Atacamite	Copper chloride
Azurite	Basic copper carbonate
Malachite	Basic copper carbonate
Cuprite	Cuprous oxide
Tenorite	Copper oxide
Chalcocite	Copper sulfide
Chalcopyrite	Copper iron sulfide
Chrysocolla	Copper silicate
Vitriol	Copper sulfate

FIGURE 35 Copper vessel. Basket-shaped copper vessel from the Cave of the Treasure, Nahal Mishmar, Israel. Copper, one of the earliest used metals, has been one of the most important materials in the development of materials technology. Masses of native copper were being pounded into tools and ornaments as early as the tenth millennium B.C.E. Because of the relative ease with which it is recovered from its ores, its remarkable physical properties (high ductility, malleability, and thermal conductivity) and its resistance to corrosion, copper has been among the major metals in terms of the quantities consumed.

malachite and chrysocolla (composed of hydrated silicate of copper) ores as early as the third millennium B.C.E. In ancient Greece, where extensive information on the ancient metallurgy of copper has been preserved, roasting (see Textbox 33) was already being used at that time to convert copper pyrite ores (composed of copper sulfide) into copper oxide.

Mixing molten copper with other metals yields a variety of alloys, such as *bronze* when alloyed with *tin*, *brass* with *zinc*, and *arsenical copper* with *arsenic* (see Table 34 and text below). All these alloys have extremely good mechanical and working properties and have, therefore, been employed for applications requiring strength and hardness (West 1982).

Bronze. *Bronze* is probably the oldest alloy made on a wide scale. The term *bronze* refers not to a single alloy but to a relatively large class of alloys

FIGURE 36 Bronze mirror. A bronze mirror, Jerusalem, second century B.C.E.–first century C.E. Bronze, an alloy of copper and tin and sometimes also other metals, was widely used for casting tools, weapons, and decorative works since as early as the fifth millennium B.C.E. After molten bronze is poured into a mold, it expands as it cools and solidifies, filling every detail in the mold. Moreover, when it cools further, after solidifying, it shrinks slightly, preventing bronze casts from sticking to the mold. Because of these unique properties, bronze has been a preferred alloy for casting functional and decorative objects. When exposed to the environment, the surface of bronze undergoes chemical changes resulting from its interaction with the environment and develops initially a film and after some years an incrustation of corrosion products known as "patina."

composed mostly of copper (80–95%), tin (20–5%), and often also smaller proportions of other metals such as zinc and lead. Bronze is harder and has a higher tensile strength than copper, but is less malleable. Because it melts at lower temperatures than does copper, it is more suitable for casting operations (see Textbox 38). The ease with which bronze can be cast accounts to a great extent for the development of the metal casting technique (see Fig. 36) (Scott 2002). The melting point, color, and mechanical properties of bronze vary according to the amount of tin present in the alloy. The mechanical and thermal properties of the alloy, however, are affected not only by the composition of the alloy but also by *mechanical* and *thermal treatments*, since these alter its internal structure. Hammered bronze, for example, is under internal stresses and is quite hard; heating it, however, restores its ductility.

Bronze is strong and wear- and corrosion-resistant. Because of these properties, even in early antiquity bronze was put to innumerable uses, such as making statuary, small and large utilitarian objects, decorations and adornments, and weapons such as knives, and axes. Some bronze, containing 78–80% copper and 22–20% tin, is also *sonorous* and has been used for making bells; such bronze is therefore known as *bell metal*. Substituting part of the tin with zinc results in an alloy with improved workability, albeit at the cost of sacrificing a measure of strength. Statuary bronze, for example, which typically includes copper (about 90%), tin (over 5%), and smaller amounts of zinc (3%), and lead (1%), has been used since antiquity, mainly in Europe, for casting statues and ornamental objects that require smooth surfaces and fine details. The lead apparently increases the fluidity of the molten alloy and yields sharp castings (Born 1985).

Brass. *Brass* is a yellow, ductile, malleable alloy of copper and zinc. Most ancient types of brass are composed of 60–85% copper and 40–15% zinc and occasionally, also lesser amounts of other metals. Variations in the relative amounts of copper and zinc results in variations, sometimes pronounced, of the properties of the resulting brass. Brass is stronger and harder than copper, and, although its malleability and ductility vary with the composition, it is generally more malleable and ductile than pure copper. This is the reason that, since antiquity, brass alloys have been used for casting jewelry and other decorative objects. There are examples of brass objects dating from the first millennium B.C.E. The Greeks were well acquainted with it in Aristotle's time (the fourth century B.C.E.) and knew it as *oreichalcos*. It was only much latter, during Roman times, that brass was first used on any significant scale for making military and decorative objects and coins (Bayley 1998; Caley 1967).

Arsenical and Antimonial Bronze. The addition of *arsenic* or of *antimony* to copper produces hard alloys, which hold a cutting edge better and are less likely to bend in use than copper. Arsenic often occurs as an impurity in copper ores, and small amounts of arsenic in early smelted copper were probably not intentionally added or particularly noticeable in the final product. When the proportion of arsenic in the alloy reaches over 2%, however, the effects are quite noticeable, and ancient copper arsenic alloys, which are generally known as *arsenical copper* or *arsenical bronze*, were presumably intentionally made. In western Asia and in Europe, for example, copper arsenic alloys were regularly produced between approximately the middle of the fourth and the second millenniums B.C.E. (Lechtman 1991). The results of many analyses of ancient copper alloys seem to suggest that tin gradually replaced arsenic in copper alloys. On the reasons for this gradual change opinions vary. One suggested view is that the gradual cessation of the use of arsenic lies in the extreme *toxicity* of all arsenic compounds, particularly the arsenic oxides, that are readily formed when arsenic-containing minerals are processed. This view is challenged by another asserting that metal workers at the time learned to control the dangers associated with arsenical compounds (see text below).

Copper–Iron Alloys. Iron may enter copper during smelting operations, and many samples of ancient copper also contain varying amounts of iron. A study of a large number of samples of ancient copper and copper–iron alloys of different origin revealed that changes in the relative amount of iron may be indicative of changes in the smelting technology of the copper. Many early samples of copper, which were apparently made using rudimentary smelting technologies, contain relatively small amounts of iron. Others, made at later times and apparently using more advanced technologies, contain relatively higher amounts of iron, some as much as 30–40%. The iron was probably introduced deliberately to render the resulting alloys suitable for *minting* coins (Craddock and Meeks 1987).

Shakudo. *Shakudo* is a copper alloy used in Japan since the third century B.C.E. for making decorative and ornamental objects. In addition to copper it also includes gold (1–4%), silver (about 2%), and lead (about 1%). Occasionally gold replaced part or even all of the silver. Exposed shakudo surfaces acquire a layer of patina consisting mainly of the mineral *cuprite* (composed of cuprous oxide), which exhibits a characteristic shine. Varying amounts of gold in the alloy are said to have a marked effect on the color and the shine (Notis 1988; Oguchi 1983).

Iron

Iron (chemical symbol Fe, from the Latin name for the metal, *ferrum*), the most prominent of all the metals in the history of human technology, is a gray *base metal* that easily combines with oxygen and becomes corroded (Friend 1926). Its importance is most likely due to a number of factors:

1. Iron is one of the most abundant metals in the upper crust of the earth. It is the fourth mineral-forming element (after silicon, oxygen, and aluminum), constituting about 5% of the earth's crust (see Table 1). Large deposits of its ores are numerous, widely distributed, and easily accessible.
2. Iron ores are comparatively easy to smelt into metal.
3. As a result of factors 1 and 2, iron is the most accessible metal to produce.
4. Iron combines with carbon to form a remarkable series of extremely useful alloys known as *steels* (see Table 33).

Iron does not occur in nature as a native metal. Lumps of *meteoritic iron*, which fell to the surface of the earth from outer space, are often found, however. It has been argued whether the earliest iron used by humans was of meteoritic origin or smelted from ores (Piaskowsky 1988). Combined with other elements, iron occurs in a varied range of *ferruginous* (iron-containing) *ores* that are widely dispersed on the upper crust of the earth; some common iron ores often used for smelting are listed in Table 37.

The sporadic use of iron can be traced back as far back as nearly 6000 years ago, but its widespread application came at a much later date, later than that of copper. This is probably due to a number of factors, mainly the high melting point of the metal (1535°C) and, accordingly, the high temperatures required for smelting it. Temperatures above 1200–1300°C were hardly accessible in early antiquity. A rudimentary form of iron known as

TABLE 37 Common Iron Ores

Ore	Composition
Goethite	Iron oxide
Hematite	Iron oxide
Magnetite	Iron oxide
Limonite	Hydrous iron oxide
Siderite	Iron carbonate
Pyrite	Iron sulfide

bloomery iron was produced quite early and is still being produced in some underdeveloped areas of the world, without resorting to the high temperatures required for smelting. Producing bloomery iron is very simple: comminuted iron ores mixed with *fuel* (mostly charcoal) are piled up in crude furnaces, known as *bloomery furnaces*, which often consist of only a pit sunk in the ground. After igniting the fuel, the charcoal and carbon monoxide produced when charcoal or wood burn reduce the iron without melting. The result of the process are very *porous* (spongy) lumps of *iron bloom* covered with slag, which collect at the bottom of the furnace. After cooling, at the end of the process, the spongy iron bloom lumps, which are of no practical use until forged, are removed from the furnace; the slag is knocked off, and the iron is forged and shaped into objects (Avery 1982).

Smelting Iron. Molten iron is almost indistinguishable from the hot slag with which the metal is covered in a furnace during smelting operations. It is possible, therefore, that the difficulty in discerning between the layers of iron and slag may have hindered the spread of the iron-smelting process and also the use of the metal. Smelting technologies used since antiquity and until the eighteenth century entailed heating the ore mixed with some fluxing material (to reduce the temperature at which the ore melts) in close contact with the fuel (almost exclusively charcoal, which also makes up the metal reducing agent). Above about 1000°C, charcoal-derived gases formed during combustion, especially carbon monoxide, reduce the iron that sinks to the bottom of the furnace. The flux, together with the useless materials in the ore form a slag that floats on the molten iron. After cooling, a layer of *pig iron* solidifies, covered with a layer of slag. *Pig iron* is an impure and brittle form of iron, actually an *iron–carbon alloy*, which cannot be worked or shaped as such, but only after being remelt and cast into molds. It was used in antiquity as the main source of iron metal from which, after thermal and chemical processing, practically all forms of iron and steel were derived.

Ferrous Alloys. Many ancient objects allegedly made of iron actually consist not of the pure metal but of alloys of iron and carbon known by the generic name *ferrous alloys*. These can be broadly classified into two classes: *steel* and *cast iron*. *Steel* is the common name for iron–carbon alloys in which the relative amount of carbon ranges between 0.03% and 2%. If the relative amount of carbon in the alloy exceeds 2%, the alloy is known as *cast iron* (see Table 33) (Angus 1976; Wertime 1961). Steel is outstanding because of the mechanical properties that it acquires when subjected to heat treatment, which causes changes in its structure and physical properties (see Textbox

29). Many of the important properties of steel are related to the amount of carbon they contain. The malleability, toughness, and ductility of steel, for example, decrease with increasing relative amounts of carbon. All *cast irons*, with the exception of one – known as *malleable iron*, which can be worked and has properties similar to those of mild steel – are brittle (not malleable) when cold or hot; they cannot be forged but only cast, hence their common name, *cast iron*. Malleable iron is made stronger and more ductile by heat treatment (*annealing*), which removes the brittleness associated with cast iron and adds resistance to breakage under heavy impact or distortion (see Textboxes 29 and 38) (Angus 1976).

Dating Iron Alloys and Iron Slags. The carbon contained in ancient ferrous alloys makes it possible to date them by the radiocarbon dating method. From earliest times until just over 200 years ago, when coke was first used as a fuel, charcoal was the almost universal and exclusive fuel used for smelting and, hence, as an additive to iron, to produce steel and cast iron. Making charcoal requires the use of cut wood. Now, the carbon in all iron alloys is derived from the smelting fuel (see text above). The radiocarbon age of carbon in ancient iron and iron slags is, therefore, indicative of the time of death of the wood from which the charcoal was made and generally, but not always, of the time of manufacture of the iron. Table 38 lists radiocarbon dates obtained for iron samples (van der Merwe and Avery 1982).

Iron is easily corroded when exposed to the atmosphere (see text below), and few specimens of ancient metallic iron are recovered in archaeological

TABLE 38 Radiocarbon Dates of Ancient Iron and Iron Slags

Sample	Location	Historical date	Radiocarbon date
Iron Metal			
Bloomery iron	Scotland	83– 87 C.E.	1850 ± 80 B.P. (100 ± 80 B.C.E.)
Carburized steel	Yugoslavia	480– 180 B.C.E.	2130 ± 60 B.P. (180 ± 60 B.C.E.)
Cast iron	Honan, China	480– 220 B.C.E.	2380 ± 80 B.P. (430 ± 80 B.C.E.)
	Sian, China	221– 220 B.C.E.	2060 ± 80 B.P. (10 ± 80 C.E.)
	Saugus, Massachusetts	1648–1678 C.E.	350 ± 60 B.P. (1600 ± 60 C.E.)
Slag			
Bloomery iron	Kgolpone, South Africa	1430 ± 60 C.E.	530 ± 60 B.P. (1430 ± 60 C.E.)
Bloomery furnace	Matsepe, South Africa	1870 ± 60 C.E.	80 ± 60 B.P. (1870 ± 60 C.E.)

excavations. Sometimes, only stains in the soil remain at the site where ancient iron artifacts were once buried. *Slags*, the refuse of smelting, are, however, practically indestructible; they consist mainly of a mixture of silica and cinders of unburned charcoal. Iron slags are often found in large quantities in archaeological sites, and the charcoal in the cinders can be mechanically extracted and dated. Carbon in iron slag is therefore often used for dating (by the radiocarbon method) iron-related operations, establishing chronological sequences, and also as a surveying device in ancient iron-smelting sites (van der Merwe 1969).

Gold

Pure *gold* (chemical symbol Au, from the Latin word *aurum*, for gold) is a soft yellowish, *precious*, *noble metal*. Since it is a noble, inert metal, it does not react with most chemicals and is therefore always found in nature in the metallic state, as *native gold*. Moreover, gold is not affected by air pollutants, as it neither tarnishes nor corrodes when exposed to the environment. Much native gold occurs in quartz veins, sea gravel, or sand. From quartz veins it was extracted in the past by *amalgamation* with mercury (see text below) or by smelting and then refining, to obtain the gold in a high degree of purity. From alluvial sands it was, and it still is, recovered by placer methods, such as washing sand or gravel in a pan or dish: the gold particles remain and are collected in the pan or dish, while the sand is washed out by the running water. A quite detailed outline of the smelting process for the extraction of gold from quartz veins by the ancient Egyptians in Nubia (the Land of Gold) was written by the Greek historian Diodorus during the first century B.C.E. He describes how gold-rich quartz was comminuted down to very fine particles that were then smelted together with lead. Cupellation of the raw lead and gold alloy obtained from the smelting process removed the lead and yielded pure gold beads, which remained in the cupel (Cauet 1999).

Gold is the most malleable of metals, as it is deformed by hammering without breaking. It is because of its high malleability that gold can be made into very fine sheets known as *gold leaf*, which has been used since early antiquity for *gilding* other metals and wood and for illuminating manuscripts (see below) (Shimura 1988; Nicholson 1981). Gold is also very *ductile*, and can be drawn into wires without fracturing. *Gold thread*, made by winding very thin and narrow strips of gold around a fibrous core of cotton, silk, or other yarns, has long been used for making gold cloth (Járó 1990; Glover 1979). Gold and its alloys were widely used throughout the world even before the dawn of history: gold jewelry (see Fig. 37) and ornaments have been found in Stone Age tombs at many sites throughout the world.

Egyptian goldsmiths of the earliest dynasties were highly skilled gold artisans. Gold ornaments about 4600 years old from the royal graves at Ur, in Mesopotamia, indicate that a wide range of gold metallurgical techniques were being used at such early times. Many gold artifacts from the pre-Columbian period were also found in the Americas, such as in Eldorado, Colombia, and the Cocle province, Panama.

Gold Alloys. In common usage the word *gold* is generally used to refer not only to the metal but also to alloys in which gold is an important component (excepting the amalgams, which see below). Pure gold, also known as *fine gold*, is too soft a metal to be used for almost any purpose, as it is subject to easy deformation and rapid abrasion. To make it harder, as well as to produce less costly objects that resemble gold, the metal has generally been alloyed with others such as copper and silver, and these alloys are also usually referred to as gold. To differentiate between the metal and the various alloys, however, it is accepted practice to express the relative amount of gold in any specific gold alloy in *karats*. A *karat* is an ancient unit that expresses the fineness of gold alloys as parts by weight of gold in 24 parts of alloy; 24-karat gold is, therefore, pure gold; the relationship between a karat number and the relative proportion of gold in a gold alloy is shown in Table 39.

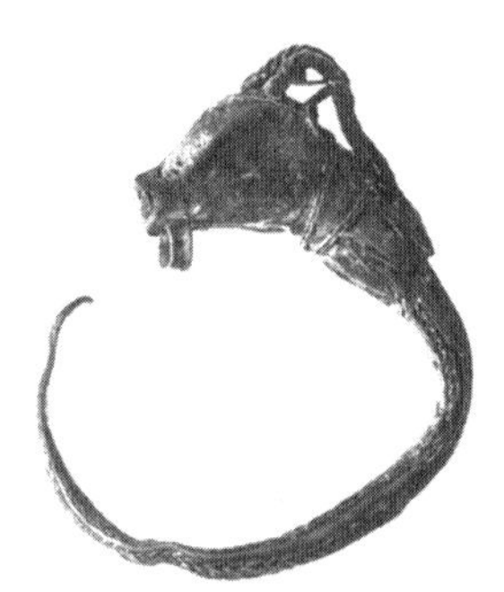

FIGURE 37 Gold earring. Elaborate gold earring featuring an antelope's head (seventh century B.C.E.) found at Tel Malhata, Israel. Throughout the development of human civilizations gold has been a symbol of prosperity and wealth. The metal and its alloys have been crafted in skillful ways to satisfy the wishes of rich patrons and royalty. Since it is a noble (chemically inert) metal that does not corrode, native gold nuggets were easily recognized in early antiquity. As early as the Stone Age ornamental gold plates were made by cold hammering native gold. Most gold, however, occurs in the crust of the earth as minute, shiny yellow flakes in "veins" (fissures, faults, or cracks in rocks) of minerals such as quartz or metamorphic rocks, or in alluvial deposits that originated from these sources.

TABLE 39 Karat Values; Relative Amount of Gold in Gold Alloys

Karat value	Relative amount of gold (%)
24	100
18	75
14	58
12	50
9	37.5

There is often confusion, incidentally, between the term *karat*, used to express the relative amount of gold in gold alloys, and *carat*, a unit of weight, equivalent today to 200 milligrams. The carat, also an ancient unit, is used to weigh gemstones, such as diamonds, emeralds, and pearls. To add to the confusion, in British usage the word *carat* has both meanings, a unit of weight relationship and of weight; also, in many texts the term *carat* as well as its abbreviations, *c.* and *ct.*, are used indiscriminately to refer to the purity of gold as well as to the weight of gems.

Gold Leaf. *Gold leaf* is gold in the form of extremely thin sheets, less than 0.0001 mm thick. Making gold leaf is done, to this day, by *gold beating*, an ancient metallurgical technique based on the great malleability of gold. *Gold beating* entails striking a bulk of gold, again and again, with a mallet. Continued beating (hammering), results in the gradual diminution of the thickness of the bulk and an accompanying growth in its surface area until the desired thickness is attained. Only pure gold or very rich gold alloys can be beaten until reduced to very thin sheets without breaking (Nicholson 1981).

Gilding. In addition to the use of bulk gold for making objects, much ancient gold was used for *gilding*, coating the surface of objects made from lesser materials with a very thin layer of gold leaf. The materials gilded varied from paper to leather to wood to ceramics and base metals such as silver, copper, and their alloys. Gilding seems to have been practiced since the end of the third millennium B.C.E. and probably earlier, using a variety of mechanical or chemical techniques, some of which are still practiced today (Oddy 2004; Drayman-Weisser 2000). *Mechanical gilding*, probably the earliest known gilding technique, entails two processing stages: first making gold leaf and then either attaching the thin leaves mechanically, by pressing and/or hammering, or sticking them with glue or gum to the surface to be covered. Common surfaces gilded by either of these techniques include paper, leather, ceramics, ivory, glass, and metals or alloys.

Chemical gilding methods used in antiquity included such processes as *fire gilding, depletion gilding,* and *dipping*. Other chemical gilding methods, as for example, electroplating, based on the use of electric currents, and vapor deposition, using evaporated metals, were developed only as recently as during the nineteenth and twentieth centuries (Budden 1991; Oddy et al. 1979). *Fire gilding,* also known as *mercury gilding* or *amalgam gilding,* was widely practiced in antiquity to gild metals; it is based on the property of mercury to evaporate from amalgams when heated, leaving behind the gold (see text below). In practice, the object to be gilded, generally made of copper, bronze, or brass, is first coated, using a brush or a swab, for example, with a thin layer of freshly prepared gold amalgam; heating the amalgam-coated object expels the mercury and leaves behind a well-adhered, thin layer of gold on the surface of the object. Fire gilding was the major technique for the gilding metals since 300 B.C.E. in China and 200 C.E. in Europe until the invention of electroplating in the nineteenth century (Anheuser 2000, 1999; Oddy 1991).

Depletion gilding, or *depletion plating,* as the process is also known, is a process that was widely used in antiquity for gilding sheets of alloys of gold and copper with a thin layer of gold. In contrast to the other gilding techniques, which require the use of gold from an external source, in depletion gilding the gold is initially an integral constituent of the original sheet of alloy (La Niece 1995). To gild a sheet of a gold–copper alloy by the *depletion gilding* technique, an object made of the alloy is subjected to a repeated sequence of *hammering, annealing,* and *pickling* operations, so as to reduce its thickness. *Hammering* reduces the thickness and shapes the sheet, but it also hardens the alloy and, in order to soften and make it malleable again, the sheet is annealed (see Textbox 29). Heating during *annealing* causes oxidation of the copper in the surface of the alloy and its conversion to copper oxide. Subsequent *pickling* of the sheet by immersion in an appropriate acid (vinegar or fruitjuice in ancient times) dissolves the copper oxide, leaving the surface slightly depleted of copper and enriched in gold. Repeating the sequence of hammering, annealing, and pickling many times until the sheet is modeled into a required shape removes enough copper and leaves on the surface a sufficiently thick and continuous gilding layer. The process was apparently used in places as far distant from one another as Mesopotamia, Greece, and Peru. The Greeks seem to have used depletion gilding to make Corinthian bronze, the most prized of all the copper alloys of classical times (Jacobson and Weitzman 1992; Lechtman 1984).

Dipping, a process also known as *flush gilding* or *wash gilding,* was used to coat the surface of objects made of base metals with a thin layer of molten gold. Copper and its alloys were gilded by dipping in pre-Columbian South

America and ancient Rome. The Romans, for example, used the dipping technique to give copper the brilliance of gold. The *dipping* process entails two successive working stages: (1) heating under reducing conditions (to prevent oxidation of the surface of the object to be gilded) and (2) dipping the hot object for a short while in a bath of molten gold or one of its alloys at a temperature lower than that at which the base metal melts. The molten gold or gold alloy thus wets and covers the object with a thin layer that, on removal from the bath and cooling, solidifies as a well-attached gilding layer (Scott 1983; Bergsoe 1937).

Gold Thread. *Metal threads* are thin threads of one of the precious metals, mostly gold but also silver, which are employed to make highly decorative textiles known as *gold* and *silver cloth*. Most used in antiquity to make gold or silver cloth were two types of metal threads: (1) very thin strips of a precious metal; (2) composite threads, made by winding very thin, narrow strips of a metal around a fibrous core of cotton, silk, or other yarn (Járó 1990; Lee-Whitmann and Skelton 1984). Gold and silver cloth (woven or decorated with metal threads) have always been identified as symbols of wealth and opulence and are among the most admired and precious antiquities. Only during the late twentieth century, when the spread of inexpensive, precious-metal-appearing threads and mechanized textile decorating methods became easily available to all, did the interest in textiles decorated with precious metal threads decrease.

Silver

Silver (chemical symbol Ag, from the Latin word *argentum*, silver) is a light gray, highly ductile, malleable, and very soft precious metal. The softness of silver makes the pure metal impractical for many applications. Some alloys with copper contain between 80% and over 90% silver and are much harder, and therefore such alloys have been used for most applications. As with the alloys of gold, in everyday usage the alloys of silver and copper are generally referred to as *silver*. Neither the metal nor its alloys are oxidized when exposed to the atmosphere; they easily react, however, with hydrogen sulfide, an air pollutant, forming a dark gray and a esthetically impairing surface layer known as *tarnish* (composed of silver sulfide) (see text below).

Native silver is rarely found in the form of nuggets, and it is probably for this reason that silver was recognized and came into widespread use at later times than did gold. Much native silver occurs as minute grains dispersed in small concentration in *argentiferous* (silver-rich) sands. Now, when silver is mixed with mercury, it forms *amalgams* (see below); the *amalgamation*

TABLE 40 Common Silver Ores

Ore	Composition
Argentite	Silver sulfide
Horn silver	Silver chloride
Galena	Lead sulfide; the main ore of lead and one of the most important sources of silver
Pyrargyrite	Silver sulfide – antimony sulfide

process has therefore been used since antiquity for extracting the native metal from argentiferous sands and from other ores in which it occurs native as very small grains.

Most silver occurs in nature combined with other elements, forming a variety of minerals, some of them listed in Table 40. These minerals have been the main resource from which the metal has been and still is recovered. Also some minerals of other metals, such as lead and particularly *galena* (composed of lead sulfide), contain relatively large amounts of *silver sulfide* impurities, and for several millennia galena deposits have also been exploited as major silver resources. Mines in Asia Minor (modern Turkey) are considered as one of the first sources of silver; the large-scale extraction of the metal seems to have begun there sometime during the third millennium B.C.E. and supplied much of the silver requirements of the Near East, Crete, and Greece.

After extraction from its ores, crude silver is generally refined by the process of cupellation, mentioned earlier. Since ancient times the main use of silver has been for making articles of value such as ornaments, decorative objects, jewelry, and coins. In Mesopotamia, much silver was used between the twentieth and fifteenth centuries B.C.E. to make decorative and ornamental objects. It seems that in Egypt, during the same period of time, the metal was scarcer and perhaps even more costly than gold (Hess et al. 1998; Mischara and Myers 1974).

Lead

Lead (chemical symbol Pb, from the Latin name for the metal, *plumbum*) is a gray, soft, ductile, and very poisonous metal, although its poisonous properties were probably unknown to the ancients. The metal has been used, particularly in China and India, since very ancient times. Lead is not found in nature in the native, metallic form, although tiny particles of the metal are occasionally encrusted in rocks. It is unlikely, therefore, that the metal would

TABLE 41 Common Lead Ores

Ore	Composition
Galena	Lead sulfide
Cerusite	Lead carbonate
Anglesite	Lead sulfate
Massicot (lead ocher)	Lead oxide
Minium	Lead oxide

had been known to early humans. Moreover, despite the fact that numerous minerals include lead in their composition, only a few, listed in Table 41, can be considered as lead ores to be exploited for the recovery of the metal; in the table, the minerals are listed in decreasing order of their importance as a resource of the metal.

Galena seems to have been the main mineral resource for most of the metallic lead smelted in antiquity. In addition to lead sulfide, the main component of galena, many galena ores also contain appreciable amounts of valuable impurities such as gold, silver, and copper, which made it economical to process the ore for their recuperation. Thus, the ancient metallurgic processing of galena was often designed to extract not only lead but also copper, silver, and gold.

Smelting lead could have been carried out in simple furnaces fueled with wood or charcoal and with an ample, regular supply of air. In Europe, before the fifteenth century C.E., for example, lead-smelting furnaces consisted of circular or rectangular stone structures with a clay floor, usually built into the side of a hill. The ore was placed in the furnace on top of the fuel, which, after being ignited, was fed with a forced air draft; under such conditions the lead was reduced to the crude, metallic condition. Silver or gold in significant quantities in the ore were generally recovered from the crude lead by the cupellation process.

TEXTBOX 39

LEAD SMELTING

When *galena*, a lead ore (composed of lead sulfide) is roasted in a well-ventilated, open furnace, part of the lead is oxidized by air oxygen to lead oxide and the sulfur to sulfur dioxide, which is released into the atmosphere (see Textbox 33):

$$\underset{\text{galena (lead sulfide)}}{PbS} + O_2 = \underset{\text{lead oxide}}{PbO} + \underset{\text{sulfur dioxide}}{SO_2\uparrow}$$

The lead oxide thus formed then reacts with galena still in the furnace to yield molten lead, while more sulfur dioxide is evolved:

$$2PbO + PbS = 3Pb + SO_2\uparrow$$

The molten lead flows to the bottom of the smelting furnace (which is usually made of clay) and from there, through a channel and an opening at the bottom of the furnace, collecting in a lead reservoir. Part of the sulfur dioxide formed during the reaction is further oxidized by still more air oxygen into sulfur trioxide:

$$\underset{\text{sulfur dioxide}}{2SO_2} + O_2 = \underset{\text{sulfur trioxide}}{2SO_3}$$

The sulfur trioxide reacts with lead oxide created during the roasting process (see above) to form lead sulfate

$$SO_3 + \underset{\text{lead oxide}}{PbO} = \underset{\text{lead sulfate}}{PbSO_4}$$

and the lead sulfate reacts with still more of the remaining galena to produce additional molten lead, which also drips to the bottom of the furnace and into the molten lead reservoir:

$$PbSO_4 + PbS = 2Pb + 2SO_2$$

On cooling, the lead solidifies as an *ingot* in the shape of the lead reservoir.

Freshly cast lead has a bright, silvery appearance. On exposure to the atmosphere, however, lead in the surface layer combines with atmospheric oxygen and carbon dioxide to form a dark, stable gray coating of mixed lead oxide and basic lead carbonate. This layer usually protects the metal from further oxidation and corrosion (see Fig. 38). Protected by a weathered surface layer, solid lead is stable to further corrosion. Lead is also very ductile and soft, being the softest metal known in antiquity. It is mainly because of these properties that lead was widely used for building, to make pipes and roofs, and in naval construction, for example. Solid lead flows, albeit very

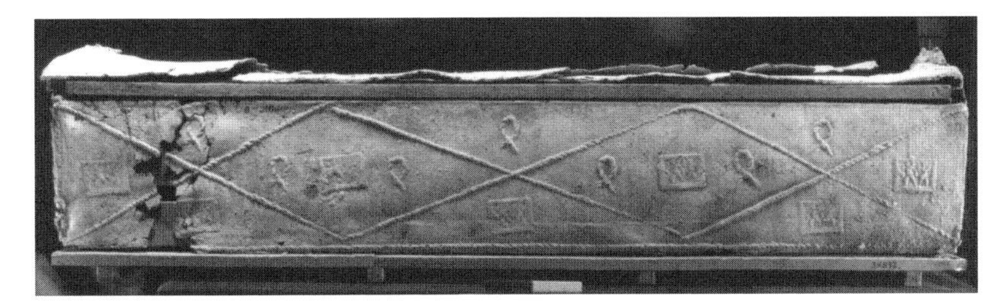

FIGURE 38 Lead coffin. Lead coffin (first–third centuries C.E.) from Jerusalem, Israel. Lead, widely used in many ancient civilizations, was one of the first metals to be recovered from its ores. Lead objects date back from as early as the seventh century B.C.E. In Mesopotamia molten lead was used to fasten bolts and shafts into masonry. In Syria it was made into rods used as currency, and in Greece it was cast into coins. During the Roman Empire the use of lead become so widespread that the health hazards caused by lead exposure are suspected to have been one of the factors affecting the fall of the Roman Empire. Since it is very resistant to corrosion, lead was also used by the Romans, for making coffins as the one illustrated.

slowly, at room temperature. Because of this flow, sheets of lead in ancient slanting roofs often develop, over long periods of time, a lower edge much thicker than at the top. In addition to its use as a metal, a great part of the lead produced in antiquity was mixed with other metals to make alloys such as *pewter* and *soft solders*. *Pewter*, an alloy of tin, lead, and some other metals, was used to make a wide range of utilitarian and decorative objects. Other lead alloys served for minting coins and sculpting statuary.

The ancient Egyptians used lead as early as the fourth millennium B.C.E. Later, the Romans developed and expanded the use of lead during the Roman Empire to impressive levels, so lead is often referred to as the *Roman metal*. The large-scale use of the metal most probably caused considerable environmental contamination of food, drink, and the atmosphere (Nriagu 1983).

Tin

Tin (chemical symbol Sn, from the Latin name of the metal, *stannum*) occurs as a native metal only as small, rare nuggets; it is very doubtful, therefore, whether native tin would have been noticed, never mind used, by ancient people. Nevertheless, tin was one of the earliest metals to have been produced. Tin ores occur in few places on the upper crust of the earth, mostly as the mineral *cassiterite* or *tin stone* (composed of tin oxide) from which most tin has been and still is extracted. *Tin stone* is a usually brown or black, high-density ore, whose dark color is due to impurities, mostly compounds of iron. It occurs mainly in two distinct forms: as a vein ore, within fissures in rocks, or as *stream tin*, in alluvial gravel. The bulk of the tin ore used in antiquity seems to have been drawn from alluvial deposits. Since it was indispensable for the production of several important alloys, mainly bronze and several solders, the trade in tin or in its ores played an important role in antiquity. Sometimes it was transported from far distant sources to the places where it was used (Muhly 1973).

Smelting tin in antiquity generally proceeded through the reduction, with charcoal, of the mineral to the metal. The crude metal obtained after smelting was often quite pure. If impure, it was refined by heating it in a furnace with a reducing atmosphere (Penhallurick 1986; Franklin et al. 1978). Tin is a gray–white, very soft metal that takes a very bright polish. It is not affected by exposure to air or water at ordinary temperatures, nor to water at temperatures up to its boiling point. Little tin was used in the metallic form in antiquity, although there seems to be evidence for its use in pre-Columbian America (Caley and Easby 1964).

Tinning. Since tin is not affected by exposure to air or water at temperatures up to its boiling point (above 2000°C), the metal has been used for protecting objects made from less stable metals and alloys. A thin layer of tin is, therefore, often applied to protect objects made from iron alloys or to decorate those made of copper or its alloys. The coating process is known as *tinning*, which is usually accomplished by dipping an object, for a short while, into a bath of molten tin. A different tinning procedure is used to coat just one surface, such as, for example, the internal surface of cooking vessels. The vessel whose internal surface is to be tinned is heated to a temperature sufficiently high to melt a small amount of tin placed on the surface (above 231°C, the melting point of tin), and the molten tin is then spread, usually with a cloth swab, while the object is kept hot (Hoare 1948).

The Allotropes of Tin; Tin Pest. Metallic tin may occur in three allotropic forms (see Textbox 19): the common form of tin, also known as *white tin* or *beta tin*, is stable at ambient temperatures; its stability extends between –18°C and 170°C; below –18°C tin is converted to a gray powdery allotrope, known as *alpha tin* or *tin pest*. A third allotrope, known as *rhombic tin*, is the form of tin stable at temperatures above 170°C. If ordinary white tin remains for extended periods of time at temperatures below –18°C, therefore, it is slowly converted to the gray, brittle, and powdery allotrope *tin pest*; the conversion is accelerated at still lower temperatures. Tin objects kept in regions of the world where extremely low temperatures (below –18°C) prevail, initially acquire a gray, powdery surface; if low temperatures persist for extended periods of time, however, the powdery tin crust grows gradually thicker until entire objects may eventually collapse and totally disintegrate (Kawohl 1988; Miles and Pollard 1985).

Zinc

Zinc (chemical symbol Zn, from the German name of the metal, *zink*, of obscure origin) is a bluish-white metal that is only slightly tarnished in moist air. High-purity zinc, which is brittle at ordinary temperatures but ductile and malleable between 100°C and 150°C, was apparently unknown until quite recent times (during the fifteenth and sixteenth centuries). Zinc does not occur naturally in the metallic state, but as a component of a large number of minerals, some of which have been used since antiquity for smelting the metal (see Table 42). *Calamine* (a mineral of complex composition composed of carbonate and silicate of zinc) seems to have been the main ore from which the metal was extracted for many centuries in India and China, long before it was introduced into Europe (Zhou 1996).

TABLE 42 Common Zinc Ores

Mineral	Chemical composition
Calamine	Zinc carbonate
Hemimorphite	Hydrated zinc silicate
Willemite	Anhydrous zinc silicate hydroxide

TEXTBOX 40

ZINC SMELTING

The ancient process for smelting zinc has not yet been fully understood. The available evidence about the way zinc smelting was conducted has been interpreted in different ways (Craddock 1998; Morgan 1985). One explanation suggests that before zinc was smelted from an ore as *calamine* (composed of zinc carbonate), for example, the ore was first calcinated to produce zinc dioxide while carbon dioxide evolved:

$$\underset{\text{calamine (zinc carbonate)}}{ZnCO_3} \rightarrow \underset{\text{zinc oxide}}{ZnO} + CO_2 \uparrow$$

Zinc dioxide is a friable material that could then readily be reduced to metallic zinc by carbon, the main component of the fuel:

$$2ZnO + C \rightarrow \underset{\text{zinc metal}}{2Zn} + CO_2 \uparrow$$

Zinc melts at a fairly low temperature, 419°C, and boils at 918°C, below the temperature at which zinc oxide is reduced to zinc. If zinc had been smelted in the way described, the metal would have distilled when formed. To prevent its reoxidation when in the gaseous condition, the atmosphere of the furnace would have had to be kept free of oxygen so that the metal could cool and condense. The earliest available information on zinc distillation, however, is from as late as the twelfth century C.E. in India; the suggested explanation advanced here does not, therefore, provide an answer as to how zinc was smelt before that time.

Ancient objects made from zinc are extremely rare, and it seems that practically no zinc was produced and used as a metal in antiquity. One of the few ancient objects that are known to be made of zinc is a roughly rectangular fragment of corroded zinc sheet found at the Agora of Athens (Farnsworth et al. 1949). Even before it was produced and recognized as a metal, however, zinc was being used for making alloys; it has also been claimed that mixed zinc–copper ores containing sufficient zinc might have been used to produce brass on smelting, although it is doubtful whether the process could have been accomplished in antiquity. When mixed with copper, zinc forms two alloys known in antiquity: *brass*, which usually contains 15–40% zinc, well known in Europe (see pg. 170), and *paktong*, which contains 33–46% zinc, in China. Paktong, a silver-colored alloy, also contains a small proportion of nickel (Gilmour and Worrall 1995; Needham 1980).

Mercury

Mercury (chemical symbol Hg, from the Latin name of the metal, *hydrargyrium*, liquid silver), previously also known as *quicksilver* is, at ordinary temperatures, a silvery white liquid metal that boils at 360°C. The metal is occasionally found in nature in the native state. Most mercury has been derived, however, from the red mineral *cinnabar* (composed of mercuric sulfide) that was also used in the past as a red pigment known as *vermilion* (see Textbox 41). The Greek philosopher Aristotle, writing in the fourth century B.C.E., described mercury, and Theophrastus, his disciple, gave details on a method used at that time for extracting the metal from its ore: rubbing native cinnabar with vinegar in a copper mortar with a copper pestle yielded the liquid metal (Hill 1774), an apparently very early example of a mechanical chemistry reaction.

TEXTBOX 41

MERCURY SMELTING

Smelting *cinnabar* a mercury ore (composed of mercury sulfide) is a relatively simple, two-stage heating process: first roasting of the ore and then reducing and boiling the metal. Heating the ore at a relatively low temperature, in an atmosphere rich in oxygen, roasts the mercury sulfide, which is converted to mercuric oxide:

$$\underset{\substack{\text{cinnabar} \\ \text{(mercuric sulfide)}}}{HgS} + O_2 \rightarrow \underset{\text{mercuric oxide}}{HgO} + SO_2\uparrow$$

Further heating results in decomposition of the oxide formed and the release of mercury vapor:

$$2HgO \overset{\text{heat}}{\rightarrow} 2Hg\uparrow + O_2\uparrow$$

The mercury vapor evolved was, in the past, directed toward cooled earthenware pipes over the furnace, where it condensed into liquid metal. From there it was then collected in special containers, together with considerable amounts of soot. The crude metal thus obtained was refined by further distillation.

Amalgams. When gold or silver is mixed with mercury, it forms alloys known as *amalgams*, which are soft when freshly prepared but harden after some time. If an amalgam is heated to about 360°C, the boiling temperature of mercury, the mercury in the amalgam evaporates, leaving behind the gold or silver with which it was amalgamated. The property of gold and silver to amalgamate was known in antiquity, and it was, and still is, used to extract gold and silver from some of their ores. When extracting gold from auriferous sands, for example, the sand was brought into close contact with mercury so that the gold combined with the mercury and formed an amalgam that can then easily be separated from the sand. Heating the amalgam, generally in pottery vessels, resulted in the evaporation of mercury while rather pure gold was left behind in the vessels. Such a process was described in ancient texts. Vitruvius (first century B.C.E.) and Pliny the Elder (first century C.E.), for example, describe how gold in auriferous sand was dissolved in mercury and the amalgam formed was periodically removed while more mercury was added to the sand. The mercury was then removed from the amalgam by heating, and pure gold globules obtained. The evaporated mercury was probably recovered by condensation on ceramic pipes. Gold amalgam was also used to gild objects made from lesser metals and alloys, such as copper, bronze, and brass; see above, discussion on fire gilding.

Platinum

Platinum (chemical symbol Pt, from the Spanish word *plata*, silver) is a very rare silver-colored metal; together with gold, platinum belongs to the group of metals known as the noble or precious metals. Platinum melts at a very high temperature, 1774°C; since the maximum temperatures which could be reached in ancient furnaces was lower than that, platinum could have been neither smelted nor melted in antiquity. It is even doubtful whether platinum was recognized as a distinct metal in the Old World. Still, two outstanding platinum finds were found in the Old World: (1) a casket found at Thebes, in Upper Egypt, dating from the seventh century B.C.E., which was shown to have been made of an alloy of platinum; and (2) a box made of copper inlaid with silver and platinum decorations (Berthelot 1901). In South America, Indians from Ecuador and Colombia made small artifacts from (probably natural) gold–platinum alloys before the discovery of America by the Europeans (Scott and Bray 1992; Bergsoe 1937).

5.9. DETERIORATION OF METALS AND ALLOYS – CORROSION

The environment has negative effects on most metals; thus, when metallic archaeological objects are eventually found, they are generally in an advanced state of decay. The decay of metals and alloys caused by the chemical action of gases and/or liquids in the environment is known as *corrosion*. *Corrosion* processes are natural destructive processes that result in the waste of most metals and alloys. The ultimate result of all corrosion processes is the reversion of most metals from the metallic condition in which they are used, to the chemically combined form in which they naturally occur in the crust of the earth. *Rust*, the reddish-brown corrosion product that forms on the surface of iron and steel exposed to the environment, for example, has the same composition as the iron ores from which the metal is extracted.

"Lay not up yourselves treasures on earth where moth and rust doth corrupt," states an admonition in the New Testament, implying that the inevitable course of corrosion was already familiar in ancient times. Not only iron and steel but most metals corrode. Unprotected surfaces of copper exposed to the atmosphere, for example, grow dull within a short time after exposure; the dullness is generally due to the formation of a coating created when the metal surface reacts with atmospheric components (oxygen, water vapor, and carbon dioxide) and pollutants (such as hydrogen sulfide in the atmosphere and salts carried by the wind). After more or less extended periods of time a layer of copper carbonate and copper sulfide, gradually growing in thickness, turns the initial dullness of the copper into an intensely green, characteristic coloration on ancient copper surfaces. Even in relatively unpolluted environments a copper surface will take on, within a few years of exposure, a green bloom and will eventually turn green all over. Copper roofs, for example, turn green in regions with relatively uncontaminated air within 50–100 years; in polluted urban air, however, the change takes place in less than about half that time. The corrosion of copper or bronze buried in soil, or immersed in water, is generally faster. Most metals corrode because in any natural environment all metals and alloys, with the exception of only the noble metals, react chemically and are converted to more stable compounds than the metals and alloys themselves. The process of corrosion is, therefore, natural and unavoidable (Scully 1990; North and Macleod 1987).

The reaction that takes place during corrosion processes can be expressed by the general equation:

$$\text{Metal or alloy} \quad + \quad \begin{matrix}\text{oxidizing elements}\\ \text{or compounds}\\ \text{in the environment}\end{matrix} \quad \rightarrow \quad \begin{matrix}\text{corrosion products,}\\ \text{stable to the environment}\end{matrix}$$

All metals and alloys, except for the noble metals, have a natural tendency to combine with oxidizing substances in their surroundings, particularly oxygen, and to be corroded. During metal-smelting operations, this natural tendency is reversed: through the application of large amounts of energy, the metals are recovered from their ores. But this is only a temporary situation; nature takes its unavoidable course, and in time, corrosion processes reestablish the equilibrium that exists in nature. All that can be done to preserve metals and alloys and prevent their return to the crust of the earth as metal oxides is to slow down corrosion processes for shorter or longer periods of time (Jones 1996; Scully 1990).

TEXTBOX 42

CORROSION

Corrosion is the process of gradual waste and degradation undergone by most metals and alloys exposed to weathering agents in the environment. The products of the process are chemical compounds in which the corroded metals are combined mainly with oxygen but also with other elements or ions, such as sulfur, carbonate, and sulfate. The composition of the *corrosion products* is often almost identical to that of the metalliferous ores from which the metals are extracted.

Corrosion processes are *electrochemical*, involving the transfer of electric charges between matter (see Fig. 39). A number of corrosion paths are recognized, *uniform corrosion* and *localized corrosion* being common in ancient metals. *Uniform corrosion*, for example, common in iron, copper, and their alloys, generally spreads over extended areas, forming a layer of corrosion products that often covers the entire surface of metallic objects. Uniform corrosion layers on copper, silver, and their alloys generally cover and seal the outer metallic surface from further contact with the environment and hence from further corrosion. Such a corrosion layer, generally referred to as *patina*, may be beneficial for archaeological objects made of these metals since it often prevents further corrosion of the underlying metal.

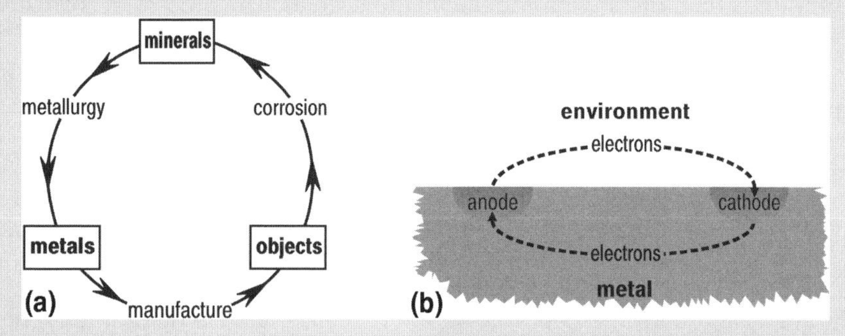

FIGURE 39 Corrosion. Corrosion is the process of gradual deterioration of metals and alloys as a result of their interaction with the environment. The corrosion process is a reversal of metallurgical processes, whereby metals are recovered from the minerals in which they occur in nature **(a)**. It is an electrolytic process, brought about by the passage of electric currents. Any metal or alloy contains sites in which there are slight local compositional differences. When such compositional differences are exposed to a humid or wet environment, extremely small electrolytic cells as the one shown in **(b)** are created; in each cell, an electric current drives the otherwise nonspontaneous corrosion reactions. In a surface undergoing corrosion there are millions of electrolytic cells.

Localized corrosion, another form of metal waste, generally occurs on small, confined areas of metallic bodies. Localized corrosion processes expand mostly by penetrating deep into the bulk of the metal, forming holes and cracks that may endanger the integrity of corroded objects (see Textbox 43).

Almost all new metallic surfaces exposed to the environment are sooner or later coated with a layer of corrosion products: metal oxides, sulfides, and carbonates, for example, are common corrosion products formed when a metal or alloy interacts with contaminants in the environment. If the layer is continuous and stable, as in uniform corrosion, it may conceal the underlying metal from further exposure and protect it from additional corrosion; if it is discontinuous, or chemically unstable, however, the metal surface below the initial layer of corrosion products remains in contact with the environment. Exposed to humidity and pollutants, the corrosion process continues, penetrating deeper into the metallic bulk and eventually resulting in its total destruction.

With the exception of the noble metals, which are stable and do not corrode under any circumstances under normal environmental conditions, all metals and alloys are corroded. Different metals and alloys, however, corrode in different forms, and even the same metal or alloy undergoes different corrosion processes under different environmental conditions.

Corrosion and Water

Water, whether as a liquid, moisture in the soil, or water vapor in the atmosphere, is essential for corrosion processes to take place. Under dry environmental conditions most metals and alloys are resistant to corrosion. The more humidity there is at a site, the more active are the corrosion processes. Some metals and alloys that are resistant to corrosion under dry conditions rapidly corrode under humid or wet conditions, particularly, in the presence of pollutants. Depending on their susceptibility to corrosion processes, the metals and alloys can be divided into three groups:

1. The *noble metals* (in antiquity only gold and in a very few instances platinum) and their alloys, which resist corrosion in all natural environments and under any environmental conditions.
2. Metals that are easily but uniformly corroded, such as silver and copper and their alloys. Some of the corrosion products of these metals and alloys are somewhat stable and serve as protective layers that protect the underlying metal from further corrosion.
3. Metals and alloys known as *base metals*, which corrode rapidly under most environmental conditions, such as iron and steel.

Following this system of classification, the metals known in antiquity can be listed in what is usually referred to as the *galvanic series of metals*: a listing of the metals and alloys arranged according to their stability to corrosion, as measured by their *electrochemical potential* a numerical magnitude used to express the tendency of metals and/or alloys to corrode (see Table 43). As can be seen in the table, at the top of the galvanic series are the noble metals, which have a low electrochemical potential, undergo little corrosion, and are stable under almost any conditions. The more reactive, easily corroded base metals have a high electrochemical potential and are listed at the bottom of the series. Among the metals of antiquity, gold, a noble metal that was the most chemically stable metal known at the time, has the lowest electrochemical potential. The least stable and most easily corroded metal, zinc, whose electrochemical potential is the highest, is last on the list.

TABLE 43 The Galvanic Series: Ancient Metals and Alloys

1. Platinum (the most stable metal; has the highest electrochemical potential)
2. Gold
3. Silver
4. Copper
5. Bronze
6. Brass
7. Tin
8. Lead
9. Lead–tin (solder) alloy
10. Mild steel; wrought iron
11. Zinc (least stable ancient metal; has the lowest electrochemical potential)

Under equal environmental conditions, the rate of corrosion of alloys is generally faster than that of their components. The tendency of bronze, an alloy of copper and tin, to corrode is greater than that of its separate (unalloyed) components, copper and tin. Moreover, once a corrosion process sets up in bronze, it also proceeds, under similar conditions, at a faster rate than in either copper or tin.

Cathodic Protection

When two objects made of different metals or alloys are in contact with each other in the presence of an *electrolyte* (a medium that provides a transport mechanism), an electric current flows between them. The direction of the current's flow is determined by the electrochemical potential of the metals in contact: the baser metal, having the higher electrochemical potential value, will be preferentially corroded, whereas the one having the lower electrochemical potential is more passive and will remain unaltered. The process whereby metals are preserved, known as *cathodic protection*, is often the dominant factor in the preservation of ancient metallic objects found in archaeological excavations. Metallic objects preserved by cathodic preservation are usually uncovered in excellent condition, next to the sometimes unrecognizable corroded remains of objects made from baser metals, which originally were in contact.

Other terms, such as *rusting* and *tarnishing*, have connotations similar to that of *corrosion*. *Rusting*, however, is only used to refer specifically to the corrosion of iron and its alloys; the word *rust* refers to iron oxide, the main product of the rusting process. *Tarnishing* is a form of corrosion that results in loss of the luster of polished surfaces of copper, silver, and their alloys. *Tarnish*, as the corroded layer on these metals and their alloys is generally known, is caused by atmospheric pollutants, such as sulfur or ozone, reacting with the surface atoms of the metal or alloy; the outcome of the tarnishing process is the formation of a layer of metal sulfides (if the pollutant is sulfur), or metal oxides (if the pollutant is ozone), which impair the appearance of the surface.

Corrosion of Ancient Metals and Alloys

The corrosion of metals and alloys generally starts at the surface with the formation of an outer layer, which may develop into a crust of corrosion products. If a crust is formed, it generally has a layered structure comprising two or more compounds: (1) an outer, rather stable, mineralized layer that often covers entirely the surface of the objects, and underneath, (2) a less mineralized, unstable, and chemically active layer. Some corrosion layers may also bind ugly and disfiguring earthy accretions.

Copper and the Copper Alloys. Copper and its alloys are relatively resistant to corrosion: dry, unpolluted air rarely affects them at normal temperatures; surfaces of the metal or its alloys exposed to polluted air, even under ordinary atmospheric conditions, however, are tarnished by pollutants such as hydrogen sulfide and/or carbon dioxide. Given sufficient time, the activity of the pollutants result in the formation of a usually green layer, known as *patina*, which coats and surrounds the bulk of the metal or alloy (see Fig. 40). If the patina is chemically stable, that is, if it is hard, is nonporous, and covers the entire surface of an object, it protects the underlying metal core from further corrosion. Such a patina consists mostly of basic copper carbonate with smaller amounts of the related blue mineral azurite, generally forming a fairly uniform green crust. Sometimes, the patina preserves details of the shape of objects and may, under certain conditions, enhance their appearance and add to their beauty. Naturally formed patina, which is often referred to as *noble patina*, is much esteemed as characteristic of antiquity and is often imitated and produced artificially (Craddock 1992; Lewin and Alexander 1967, 1968).

Most corrosion processes in copper and copper alloys generally start at the surface layer of the metal or alloy. When exposed to the atmosphere at ambient temperature, the surface reacts with oxygen, water, carbon dioxide, and air pollutants; in buried objects the surface layer reacts with the components of the soil and with soil pollutants. In either case it gradually acquires a more or less thick patina under which the metallic core of an object may remain substantially unchanged. At particular sites, however, the corrosion processes may penetrate beyond the surface, and buried objects in particular may become severely corroded. At times, only extremely small remains of the original metal or alloy may be left underneath the corrosion layers. Very small amounts of active ions in the soil, such as chloride and nitrate under moist conditions, for example, may result, first in the corrosion of the surface layer and eventually, of the entire object. The process usually starts when surface atoms of the metal react with, say, chloride ions in the groundwater and form compounds of copper and chlorine, mainly cuprous chloride, cupric chloride, and/or hydrated cupric chloride.

Under favorable environmental conditions, a chemical equilibrium is established between a corroded surface layer and its surroundings, which may lead to the preservation of the bulk of copper; thus ancient objects made

FIGURE 40 Patina. Patina is a colored (usually green) layer of corrosion products that frequently develops naturally on the surface of copper and copper alloys exposed to the environment. Since it is sometimes appreciated aesthetically and as a proof of age, patina is also developed artificially, by chemical means, as a simulated product of aging. Copper patina generally includes such compounds as copper oxides, carbonates, and chlorides. In bronze and brass patinas, these compounds are mixed with the oxides of tin and lead resulting from the corrosion of the other components of the alloys. In any particular patina there may be many layers, not necessarily in the order shown in the illustration.

from copper or copper alloy may survive burial for centuries without appreciable deterioration. The equilibrium is generally disrupted, however, when the objects are excavated and removed from the environment in which they have been buried: some compounds formed during the corrosion process but inactive while buried may become chemically active after removal from a site and exposure to a new environment, and cause the total destruction of the objects. When the compound active in the decay process is cuprous chloride, the decay process is known as *bronze disease* (see Textbox 43).

TEXTBOX 43

BRONZE DISEASE

Bronze disease is the name given to a form of corrosion of bronze and some other copper alloys, in which light blue-green outgrowths form on the surface (see Fig. 41). It is an especially obnoxious form of corrosion that particularly attacks ancient excavated bronze objects. Unless terminated by specialized treatment soon after excavation, bronze disease usually results in the complete corrosion and total destruction of the object.

Bronze disease generally appears first in the form of bulky, bright blue-green, powdery spots that disfigure the surface of objects made from the alloy. The spots are quite easy to loosen and remove, but even after they have been removed, when the objects are exposed to the environment, the process continues and the spots recur in a short time. The spots are formed wherever copper atoms in the bronze react with chloride anions from salts in the soil and/or in groundwater. In dry places the growth

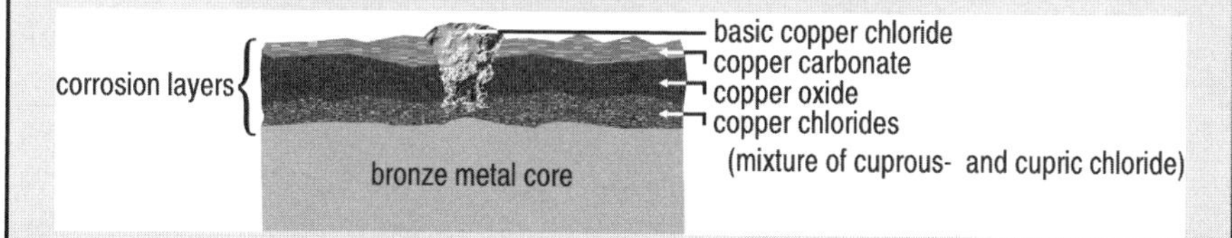

FIGURE 41 Bronze disease. Bronze disease, one of the most serious corrosive processes besetting recovered bronze antiquities, results from the interaction of one component of the patina on ancient bronze, namely, cuprous chloride with atmospheric oxygen, in a damp environment. Small spots of a light green powder (composed mainly of cuprous chloride) that grow rapidly on the surface of the patina are indicative that the bronze disease process is active. Unless the chemical activity of cuprous chloride is inhibited by some conservation procedure, bronze disease generally results in the eventual total destruction of bronze objects.

becomes active only after excavation, when objects are exposed to environmental humidity, since water provides favorable conditions for furthering the interaction between copper and chloride ions (Scott 1990).

The bronze disease corrosion process usually starts when copper atoms in bronze react with chloride ions under humid conditions, forming *cuprous chloride*, a bright blue-green, very unstable compound of copper:

$$2Cu + 2Cl^- = Cu_2cl_2$$

$2Cu$	+	$2Cl^-$	=	Cu_2cl_2
copper (in bronze)		chloride ions (in the soil ground-water, or seawater)		cuprous chloride

Following this initial reaction, a number of hypotheses have been advanced to account for the ensuing decay process. The explanation that follows is one of several that have been proposed (Gilberg 1988; McLeod 1981). Part of the cuprous chloride initially formed reacts with oxygen in the air to form cupric chloride:

$$2Cu_2Cl_2 + \tfrac{1}{2}O_2 = Cu_2O + 2CuCl_2$$

$2Cu_2Cl_2$	+	$\frac{1}{2}O_2$	=	Cu_2O	+	$2CuCl_2$
cuprous chloride				cuprous oxide		cupric chloride

The remaining cuprous chloride is oxidized and also reacts with environmental water (or water vapor) to form more cupric chloride, as well as basic cupric chloride:

$$3Cu_2Cl_2 + \tfrac{3}{2}O_2 + 3H_2O = 2CuCl_2 + CuCl_2 \cdot 3Cu(OH)_2$$

$3Cu_2Cl_2$	+	$\frac{3}{2}O_2$	+	$3H_2O$	=	$2CuCl_2$	+	$CuCl_2 \cdot 3Cu(OH)_2$
cuprous chloride						cupric chloride		basic cupric chloride

Some of the cupric chloride, in turn, reacts with additional copper in the alloy to form more cuprous chloride:

$$CuCl_2 + Cu = Cu_2Cl_2$$

$CuCl_2$	+	Cu	=	Cu_2Cl_2
cupric chloride				cuprous chloride

Unless interrupted by preventive conservation treatment, the sequence of reactions of the bronze disease process recurs again and again, until all the copper in the alloy is converted to copper compounds and objects made from the alloy turn to waste.

TABLE 44 Corrosion Products Identified in Archaeological Copper and Its Alloys

Corrosion product	Color
Oxides	
Cuprite (copper suboxide)	Red
Tenorite (copper oxide)	Gray to black
Carbonates	
Azurite (basic copper carbonate)	Bright blue
Malachite (basic copper carbonate)	Dark green
Chalconatronite (sodium copper carbonate)	Blue green
Chlorides	
Nantokite (cuprous chloride)	Pale gray
Atacamite (basic copper chloride)	Bright to dark green
Botallackite (copper hydroxychloride)	Green-blue
Sulfides	
Chalcocite (cuprous sulfide)	Black
Covellite (cupric sulfide)	Indigo blue
Chalcopyrite (copper iron sulfide)	Brass yellow
Sulfates	
Brochantite (basic copper sulfate)	Green
Connellite (copper chlorosulfate)	Bright blue

Bronze disease is one of the corrosion processes that probably causes most damage to excavated archaeological objects made from copper or its alloys. Soon after objects are excavated, bronze disease initially manifests itself as isolated green spots on the surface of the objects. The spots gradually grow in size, and then expand over the entire surface and deep into the bulk of the metal. Unless arrested by special treatment, objects with bronze disease are soon converted into a green powder. Other environmental agents, in addition to chloride ions, are also instrumental in the formation of corrosion products. Acids and naturally generated electrical currents in the ground, for example, play an important part in the conversion of copper and its alloys into minerals. Together, these and other factors cause the formation of a great variety of minerals; some of the minerals listed in Table 44, have been identified as products of bronze disease in ancient copper and its alloys (Gettens 1963).

Iron and the Ferrous Alloys. Iron exposed to a moist environment reacts with water and atmospheric oxygen to form *rust*, a brown, crumbly corrosion product (composed of hydrated iron oxide). Initially rust forms a surface layer that is usually held in much disfavor. If an ancient iron object is of any interest, the layer should, whenever feasible, be removed, any traces of rust eradicated, and the surface protected against further rusting. If the rust is not removed and the iron surface is not protected, the layer eventually flakes away from the underlying metal and exposes fresh iron to more water and oxygen; these, in turn, cause further rusting and after some time, depending on the environmental conditions at the site, the entire mass of iron is converted to rust.

Iron objects submerged under seawater, in which there is a free supply of oxygen, are easily corroded: initially the surface and eventually the bulk of such objects are fast transformed into rust. If and when submerged iron objects are removed from the water and exposed to the atmosphere, the rusting process is almost immediately accelerated; unless terminated by appropriate conservation procedures, the rusting may continue until the objects are entirely corroded and collapse. If submerged under water or in the seabed under *anaerobic* conditions (where there is little or no free oxygen), however, iron rusts slowly and may endure for long periods of time in a relatively well-preserved condition.

Silver and Silver Alloys. The surface layer of silver and its alloys is readily tarnished by hydrogen sulfide, an atmospheric pollutant. On occasions the tarnish may develop into a thick layer of black patina composed mainly of silver sulfide. Silver retrieved from the ground is often encrusted with a gray and stable layer, known as *horn silver*, composed of silver chloride formed as a result of the interaction of the metal with chloride ions in the groundwater. The formation of a thick horn silver layer surrounding a silver object, or its penetration into the bulk of the metal, is occasionally accompanied by strong forces generated by dimensional changes that can cause the deformation of buried silver objects.

Lead. Ancient lead objects excavated from the ground are often affected by ions in the groundwater and become covered by a white layer made up of a wide range of lead corrosion products (see Table 45). Sometimes, however, the early formation of a corrosion layer of *cerussite* (composed of lead carbonate), provides the metal with a protective coating that prevents the progress of the corrosion process.

5.10. STUDIES ON ARCHAEOLOGICAL METALS AND ALLOYS

The scientific study of ancient metals and the remains of metallurgical activities is of obvious importance from the history of technology standpoint; the conclusions and inferences drawn from such studies are often also helpful

TABLE 45 Corrosion Products Identified in Archaeological Lead

Corrosion product	Color
Oxides	
Massicot (lead monoxide)	Yellow
Platnerite (lead dioxide)	Brown to black
Carbonates	
Cerusite (lead carbonate)	Gray
Hydrocerusite (basic lead carbonate)	White
Chlorides	
Cotunnite (lead dichloride)	White
Phosgenite (mixture of lead chloride and lead carbonate)	White
Sulfates	
Anglesite (lead sulfate)	White
Sulfides	
Galena (lead sulfide)	Gray

for resolving practical archaeological problems. Analyzing ancient metallic objects and learning about their composition provides clues as to their nature and on metallurgical operations the metals used to make them underwent when recovered from the ores as well as during forming. Identifying the minor components of a metal or alloy is often useful for elucidating compositional variations made to a metal before using it for making different objects. The analysis of ancient ores and slags, on the other hand, provides information on the exploitation of specific ores for the production of certain metals and on ancient smelting techniques. Similarities and differences in the composition of a metal smelted at different locations may even provide guidelines to establish the provenance of the objects (Pernicka 2004).

Also the importance of the identification and analysis of *slags*, the waste materials of metallurgical processes, often the only evidence of previous metallurgical activity, cannot be overemphasized. Establishing the composition of slags makes it possible to determine the nature of the ores, fluxes, and reducing agents used in particular smelting processes and to establish whether metallurgical operations were carried out in a random fashion or under carefully controlled conditions. Of all ancient metals, the most intensively investigated so far have been copper and iron and their alloys. Much fewer studies have been done as yet, however, on other metal and alloys, although some research has been dedicated to the provenance of gold and silver, much of it related to the composition of coins (Buchwald and Wivel 1998; Bachmann 1982).

Ancient Metallurgy

The extraction of the common metals from their ores in antiquity was based mainly on relatively simple equipment and processes. Lumps of copper or iron ore, for example, that may have formed part of a ring of stones around an ancient domestic fire and become embedded in its embers, could have been reduced to metal. It is quite reasonable to conjecture, therefore, that some prehistoric campfire became, quite accidentally, the first metallurgical furnace. All that is needed to convert a campfire into a smelting furnace is a small depression in the ground to receive the molten metal. A furnace of this type is illustrated in Figure 42 (Gowland 1912; Killick 2001).

The study of slags from copper-smelting sites in Wadi Arabah in southern Israel provides an interesting insight into the technology of ancient copper smelting used there at the time. Some slags, as old as 4000 years, and others from later times, have remained undisturbed since the time when they were first made. The ores used for smelting at Wady Arabah were locally mined malachite, azurite, and/or chalcocite. After mining and ore dressing, the minerals were apparently crushed into fine granules and mixed with ores composed mainly of iron oxide or manganese oxide (a reducing agent) as well as with dolomite or limestone (a flux) and charcoal or wood (the fuel).

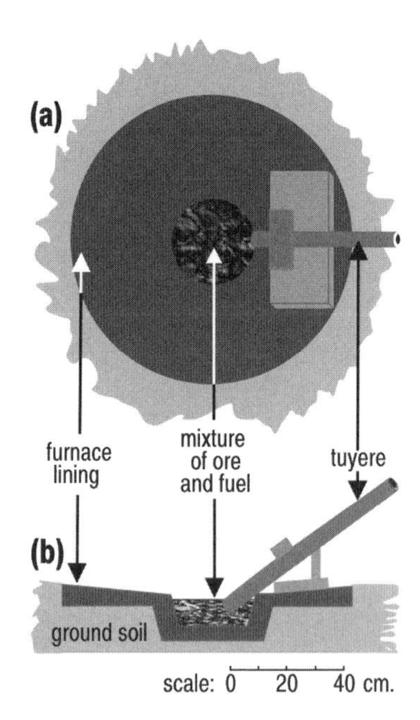

FIGURE 42 Ancient smelting furnace. A simulated ancient smelting furnace used to replicate ancient metallurgical processes; **(a)** plan; **(b)** cross section.

TABLE 46 Composition (%) of Copper Ingots from Wadi Arabah

Copper	Iron	Lead	Sulfur
89–92	9–5	0.2–1.2	below 0.1

From the evidence provided by the remains of ancient furnaces and the analysis of the slags, it seems that during the early second millennium B.C.E. crude smelting operations were conducted in very simple furnaces in which it was practically impossible to separate between the recovered metal and the slag. After smelting was completed and the furnace was cooled, the process yielded a mixture of copper and slag that could be separated only after breaking it up into pieces, by pounding, for example. The impure smelted copper pieces could then have been remelted and used as such or otherwise refined. A more elaborate smelting furnace and technique were reportedly developed later, during the twelfth century B.C.E., and its use continued throughout the Roman period, until Byzantine times. In this furnace the molten copper sank to the bottom of the furnace, where, after cooling, it solidified into bun-shaped ingots covered by a layer of slag. A typical copper ingot, at this time, weighed up to 15 kg and consisted of impure copper containing as much as 10% impurities (see Table 46). The slag was tapped, while still molten, into a waste pit.

Studying the structure of the slags provided an estimate of the smelting temperature, which seems to have been within the range 1200°C–1350°C. Reaching such high temperatures was made possible with the aid of a *tuyere*, a vent in the furnace through which air, and therefore oxygen, was injected into the burning fuel either by a natural draft or with bellows.

Arsenical Copper

Arsenical copper alloys were widely used in antiquity, and arsenical copper finds have been reported in such places, among others, as the Dead Sea area in Israel, the Cyclades Islands in the Aegean Sea, and South America (Renfrew 1967; Lechtman and Klein 1999). The compositions of some arsenical coppers are listed in Table 47.

It seems that making arsenical copper was characteristic of a transitional stage of technological development, the alloy apparently first replacing pure copper and then eventually being supplanted by bronze. It is possible that during the early Bronze Age it was realized that the use of arsenic-rich copper ores, or the incorporation of arsenic ores into copper ores smelting

TABLE 47 Ancient Copper Arsenic Alloys

		Composition (%)					
Origin	Sample	Copper	Arsenic	Antimony	Silver	Lead	Reference
Israel	Chalcolitic artifacts	96.0	3.5	0.2	0.1	0.3	Kay (1964)
		87.0	11.9	0.6	0.2	0.04	
		87.0	12.0	0.7	0.3	Trace	
Argentina	Arm band	90.7	3.8	2.9	N.A.	None	Fester (1962)
	Breast plate	96.9	1.2	0.5	N.A.	None	

mixtures, resulted in the production of alloys with particular, useful properties. Copper–arsenic alloys generally have properties superior to those of pure copper but inferior to those of bronze: they melt more easily, are harder, produce sounder castings, and are more suitable for subsequent forging than is copper. Alloys having an arsenic concentration of up to 8% are also easily worked, either when hot or cold, and are as hard as some bronze. The inclusion of arsenic into copper would, therefore, have improved the casting and working properties of copper.

The question arises as to whether these alloys were intentionally made, or were merely the result of the chance use of raw materials containing arsenic. The answer to this question can be found inferentially if some pertinent facts are reviewed. The arsenic content of native copper is generally low and the use of native copper–arsenic alloys can be ruled out as the only source of the alloy. Arsenic is naturally associated with copper mainly in two minerals, both of them composed of copper sulfoarsenates: *enargite* and *tennantite*. When either of these is metallurgically processed, it yields copper–arsenic alloys rich in arsenic. Some nonarsenical minerals of copper are occasionally associated with *arsenopyrite* (composed of iron sulfoarsenide) that when processed also yields substantial amounts of arsenical copper. The possible ways to deliberately produce copper–arsenic alloys relatively rich in arsenic would be either selecting copper minerals particularly rich in arsenic, such as those mentioned above, or adding to copper ores, during the smelting process, minerals of arsenic such as *orpiment* or *realgar* (both of them composed of arsenic sulfides). Modern experiments indicate that arsenical copper can be easily prepared when fusing copper with arsenopyrite (Earl and Adriens 2000). In a hoard of metal artifacts discovered at the Dead Sea, there are technical differences between type of artifact and its composition, as reflected in their analyses (see Table 47); all the ornaments, for example, which are made of arsenical copper, were finished off to close tolerances, whereas the tools, all made of unalloyed copper, were left rough. This may be so either because the tools are older, or made by different crafters than the ornaments.

The high toxicity of arsenic may have been what led, at a later stage, to the gradual discontinuation of the use of arsenic in copper metallurgy. It is now known, however, that toxicity had little influence on the choice of technological processes in early times. An interesting sidelight on arsenical copper technology is the fact that arsenical copper ores were not of common occurrence in the old world. The only known deposits of importance then known seem to have been in Armenia. This would suggest that the supply routes along which the raw materials could be brought from Armenia to the Mediterranean Sea area were already established as far back as Chalcolithic and early Bronze Age times.

Damascus Steel

Damascus steel, also known in the Western world as *Damascene steel,* is a special type of steel that was and is still used to make sward and knifeblades. Apparently, Damascus steel was first made in India, where it was known as *wootz* or *kuft,* and later (during the second century B.C.E.) it was developed in Persia. The name "Damascus steel" was used by the Crusaders to describe the steel used by sword smiths of Damascus, Syria, famous for their ability to hammer and temper the steel into fine blades. The sword blades made from the steel had a reputation for their exceptional properties, especially their toughness, the retention of their cutting edge, as well as for a particular and characteristic decorative pattern on their surface (Figiel 1991).

Metallographic and chemical studies have shown that the toughness of Damascus steel is due to a very high carbon content of the steel, while the retention of the cutting edge resulted from variations in the distribution of the carbon within the steel blade. Damascus steel blades also had a characteristic decorative, much appreciated pattern that was produced by a controlled process of etching the surface with an appropriate acid solution: the acid selectively dissolved iron from the surface layer, creating the decorative pattern (Wadsworth and Sherby 1980; Voss 1976). It has been suggested that some Damascus steel blades may have been made exclusively from steel smelted from selected ores, and that impurities in the ores, passed on to the metal during smelting, may have caused the formation of the characteristic etch patterns on the finished blades (Verhoeven et al. 1998; Verhoeven and Peterson 1992).

Granulation and Filigree

Granulation and *filigree* are two ancient techniques for decorating gold and silver surfaces with small metal granules (granulation) or fine threads (filigree) in more or less complicated, decorative patterns. Both techniques were practiced by the Sumerians as early as 2000 B.C.E. and later were widely used in Egypt, Greece, and Rome. At some time during the Roman Empire, however, both techniques were lost for many centuries to be only rediscovered in relatively recent times. (Duval et al. 1989; Wolters 1983, 1981). Granulation and filigree are remarkable in that they exemplify an exceptional ancient technical achievement. The usually complicated, decorative design patterns required a large number of metal joints to be made in very small areas, and it was not possible to hold together neither soldered nor brazed joints in position while others were being worked. Thus the regular methods of joining metals could not be used and required a special technique, often referred to as *fusion joining* or *colloidal hard soldering* (see Textbox 44).

TEXTBOX 44

METAL JOINING TECHNIQUES: SOLDERING AND WELDING

Joining together metallic parts into a steady single unit may be effected either mechanically or metallurgically. *Mechanical fastening* makes use of such devices as wires, bolts, staples, or screws, which do not join the parts into single, permanent units, but keep them more or less tightly fastened to each other. Joining metallic parts of similar composition into single, permanent units is generally done using *metallurgical techniques* such as by *soldering* or *welding*. *Soldering* involves the deposition, in the adjoining areas of the parts, of a hot molten filler alloy, known as a *solder*, which solidifies on cooling, adheres to each part and firmly joins them together. *Welding* does not require the use of solder. It is accomplished by heating the parts under reducing conditions (in the absence of oxygen) until their edges undergo incipient fusion. Then, with or without the application of pressure, the parts adhere to each other into one homogeneous, permanent single unit.

Soldering and Brazing

Filler alloys used for soldering or brazing melt at a temperature lower than that of the metal or alloy soldered. A *filler* fills in gaps between areas of the parts to be joined and, as it cools down, solidifies and adheres to each part, providing a permanent and stable join between them. Two soldering techniques are usually distinguished: *soft soldering* and *hard soldering*, also known as *brazing*). The two techniques are differentiated by the temperature at which the solder melts: solders used for soft soldering, obviously known as *soft solders*, melt at temperatures below 425°C; those used for hard soldering and known as *hard solders* or *brazing metals*, melt at higher temperatures. Soft soldered joints are generally weaker than brazed joints (Peter 2006).

All soft solders are alloys composed mainly of lead and tin mixed in proportions that vary from as little as 30% lead (and 70% tin), to as much as 98% lead (and only 2% tin). The brazing metals are mostly alloys of copper or silver; their composition, as well as that of other alloys used since antiquity for soldering, are listed in Table 48 (Zhadkevich 2004).

TABLE 48 Solders and Brazing Alloys

		Composition (%)			
Alloy	Melting range (°C)	Copper	Tin	Lead	Silver
Solder					
Ordinary	above 250		13–65	87–35	
Plumber's	above 215		50	50	
Tertiarium[a]	above 190		66.6	33.3	
Brazing bronze	above 800	50	50	50	
Brazing silver	above 650	15–50	5–40	33.3	80–10

[a] Latin name for a solder used in anciet Rome, where it was also known as *argentarium*.

Welding

Two welding techniques were practiced in antiquity: *forge welding* and *fusion welding*. In *forge welding* adjacent parts become joined when subjected to heat and pressure (the pressure was applied exclusively by hammering). As early as the tenth century B.C.E. around the Mediterranean Sea, metal parts, mainly iron, were joined using the forge welding technique.

For *fusion welding*, the parts to be joined are heated until they reach a temperature at which their contiguous edges melt and form a homogeneous piece. Fusion welding was used for producing *granulation* and *filigree* work. The finished objects produced by either technique show no visible evidence of the joining operation.

Fusion welding is possible because copper, a component of most gold and silver alloys, easily forms *eutectics*, alloys that melt at a temperature lower than any of the metals composing them (Duval et al. 1989; Wolters 1983, 1981). To attach granules or wires to a surface of gold or silver alloy (as in *granulation* and *filigree* work), the granules or wires are first coated with a wet layer of some adhesive, such as a gum or a glue preparation, and while still wet, they are distributed in a chosen pattern on the surface to which they are to be joined. After the adhesive dries and attaches the wires or granules to the surface, the assemblage is heated. As the temperature increases and exceeds about 250°C, the adhesive burns and is converted partly to carbon and partly to carbon dioxide, carbon monoxide, and water:

$$\underset{\text{(organic substances)}}{\text{gum or glue}} \xrightarrow[\text{over } 250°C]{\text{heat}} \underset{\text{carbon}}{C} + \underset{\text{carbon monoxide}}{CO} + CO_2\uparrow + H_2O\uparrow$$

As the assemblage is further heated and reaches a temperature of about 850°C, the carbon reacts with copper oxide on the surface of the alloy, reducing the copper to metallic copper, while the carbon is oxidized to carbon monoxide (it should be noted that practically all exposed copper surfaces acquire a thin layer of copper oxide formed by the oxidation of the metal when exposed to oxygen in the atmosphere):

$$\underset{\text{carbon}}{C} + \underset{\text{copper oxide}}{CuO} \rightarrow \underset{\text{copper}}{Cu} + \underset{\text{carbon monoxide}}{CO\uparrow}$$

The carbon monoxide prevents reoxidation of the hot copper. A further temperature rise to about 900°C results in the copper and gold (or silver) at the surface of the parts interacting to form a eutectic. The eutectic melts and runs freely, wetting the surface as well as the attached wires or granules. When the assemblage is finally cooled, the eutectic solidifies, firmly joining the wires or granules to the now decorated surface.

Coins

Of all the ancient metallic artifacts that have been left from antiquity, coins are among the most numerous. Since ancient times coins have generally been made from coinage metals or, mostly, from coining alloys, whose chemical and physical properties and economic qualities make them suitable to be used for this purpose. Until the twentieth century, gold, silver, copper, and their alloys were practically the only metals from which coinage was made. All these metals and alloys have the following properties:

1. They are physically and chemically stable.
2. Their intrinsic value is appreciated by humans.
3. They are easily workable, that is, they are ductile and malleable.
4. In the past, they were available in all regions of a country or political entity (see Fig. 43).

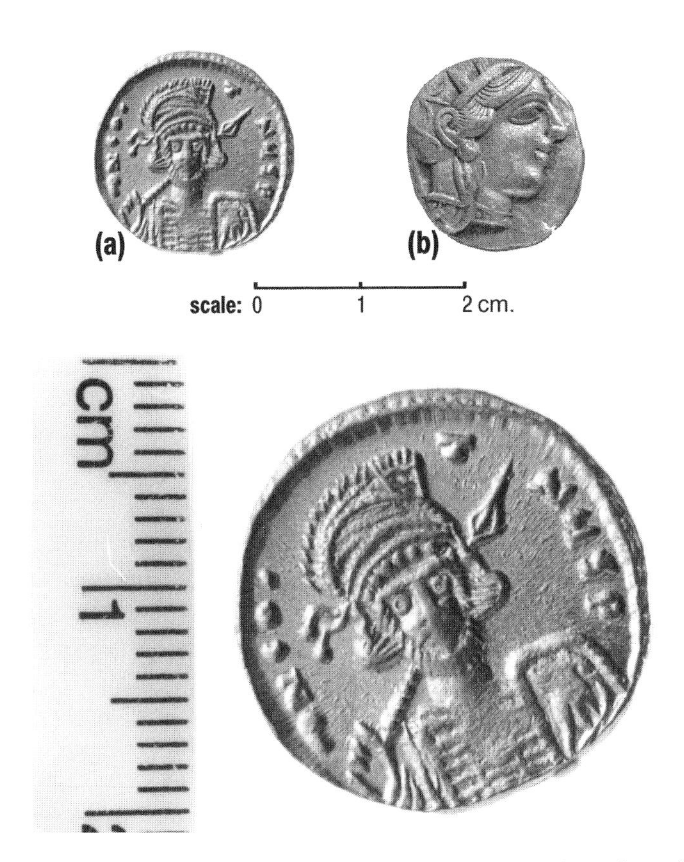

FIGURE 43 Coins. Early coins were made from precious metals or their alloys, had a consistent weight, and were warranted by a ruling central administration. The earliest coins known, minted in Asia Minor during the seventh century B.C.E., for example, were made from electrum, a naturally occurring alloy of silver and gold. They were minted for only about 50 years and were then replaced by coins made of synthetic (human-made) alloys of silver and of gold. Copper coins become common only since the fourth century B.C.E. Like the face values of most modern coins, the face value of copper coins was greater than the value of the metal, and therefore they were "token coins". The figure shows **(a)** gold Constantine IV solidus, 674–681 C.E., minted from a gold silver alloy containing 96–98% gold and **(b)** silver Athenian Tetradrachm, fifth century B.C.E., minted from a silver copper alloy including 95% silver.

The actual process of making coins, known as *coining*, may be carried in two different ways: either casting the molten metal, or working the solid metal by compression, that is, in antiquity, by hammering by hammering (see Textbox 38) (Cooper 1988; Oddy 1980). Although much numismatic research was conducted in the last few centuries, the main concern in numismatic studies involved the examination of inscriptions, denominations, and types (Cassey and Reece 1988). The composition of the coining metals and alloys is, however, of not less importance than their design. Studying the chemical composition of ancient coins can provide useful information on a variety of ancient technological aspects:

1. Composition usually reflects the level of development of the metallurgical technology and the economic trends at the time when, and at the place where the coins were struck.
2. Compositional variations in sequential coin issues, say, from year to year, may reflect differences in the source of the ores from which the coinage metals or alloys were extracted.
3. Compositional variations within a single coin issue point to variations in the quality control of the metallurgical processes.
4. Compositional differences of particular coins make it possible to recognize imitations and fakes.

The reluctance of museum curators and collectors to allow permanent damage to antiquities was, until not long ago, the main reason for the small amount of analytical work done on ancient coins. This was understandable since performing chemical analysis required removing a sample from the coin or damaging its surface, which meant either the destruction or defacement of, at least, a portion of a coin. More recently, however, a number of nondestructive methods of analysis such as neutron activation, X-ray fluorescence, and some techniques of surface analysis have been successfully applied to obtain information about ancient coins and the people and societies involved in their production (Carter 1993; Barrandon et al. 1977).

SEDIMENTS AND SOILS

Most of the rocks that make up the upper crust of the earth lie hidden beneath layers of *sediments*, unconsolidated accumulations of particles derived from the *weathering* of minerals and rocks (see Fig. 44 and Textbox 45) (Keller 1957). Once formed, the particles are either carried away or moved by the wind, rain, and gravitational forces into the seas and oceans or, before they get there, into depressions in the land. There they accumulate in a wide range of shapes and sizes (see Table 49) (Rocchi 1985; Shackley 1975).

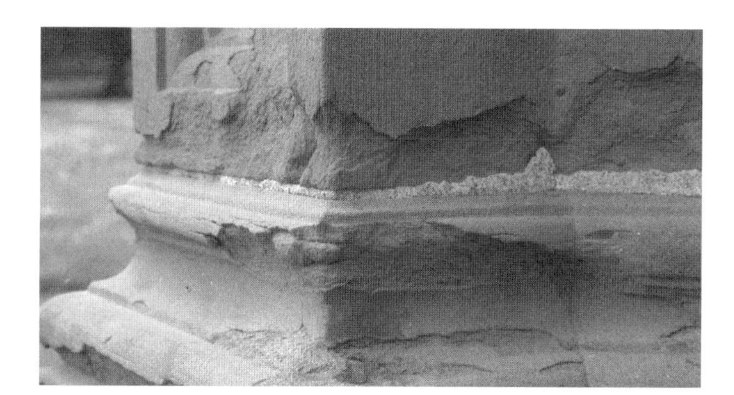

FIGURE 44 Weathering. A weathered sandstone column. Calcite (composed of calcium carbonate) is dissolved by rain and groundwater (see Textbox 73). When stone in which calcite is a main component as, for example, sandstone, limestone, and marble, is in contact with water for long periods of time, it is weathered and partly or entirely dissolved. Pollutants such as sulfur dioxide are fundamental in accelerating the weathering and dissolution process. When sulfur dioxide, for example, dissolves in rainwater, it forms sulfuric acid, a strong acid that, at ambient temperatures, rapidly dissolves calcium carbonate.

Archaeological Chemistry, Second Edition By Zvi Goffer

TABLE 49 Classification of Sediments by Particle Size

Sediment	Particle size (mm)
Clay	less than 0.002
Silt	0.002– 0.05
Sand	0.05 – 2
Pebbles	2 – 50
Boulders	50 –250

Most sediments are formed by accumulations of loose mineral particles. If there is living activity in their surroundings, however, the sediments also include debris derived from plants, animals, humans, and human-made materials. Because of differences in the size of the particles and in their specific gravity, sedimentary processes generally result in the *segregation* and *stratification* of the particles in layers whose texture, color, and even composition may vary from one layer to another. Studying the structure and composition of sediments discloses, therefore, information on the geologic past of the sites (Selley 1982). Many mineral ores rich in chemical elements important to human beings (such as copper and iron) often occur in highly concentrated, sedimentary layers. Also the occurrence of gold nuggets in some water streams is due to a segregation process. Since the specific gravity of gold is much higher than that of most common minerals, the nuggets are concentrated in places where water currents or water waves are strong enough to keep the lighter minerals suspended and in transport (see Chapter 5).

Like mineral and rocks, also archaeological materials exposed to the atmosphere, are weathered when exposed to the environment. If and when they become buried in the soil, they continue undergoing physical and chemical changes brought about by their interaction with groundwater, salts in the soil, and even underground biochemical activity. As a result of these changes, which are known collectively as *diagenesis*, some materials may be partially or entirely dissolved by groundwater. Others may become the receptors of new minerals that precipitate within their pores and crevices. Whatever changes they undergo, after long periods of time, diagenetic changes result in the ultimate conversion of buried materials into others that are gradually integrated in stratified sediments (see Textbox 46) (Wilson and Pollard 2002; Stein and Farrand 2001). Under unusual burial conditions, however, diagenetic processes may occasionally result in the preservation of the exact shape and appearance of animal and plant remains. Thus are formed *fossils,* in which dead biological matter is replaced by minerals (see Chapter 17) (Fortey 1991).

The study of the diagenesis of buried remains and the composition of sediments has greatly contributed to the interpretation of the changes that take place in buried archaeological remains (Hedges and Millard 1995a; El-Kammar et al. 1989). Of particular interest in this context is the study of the relative abundance of the isotopes of oxygen in archaeological sediments;

TEXTBOX 45

WEATHERING

Weathering is the deterioration and decay of solids on, or just under the surface of the earth by physical, chemical, and biological processes. All solids – from minerals and rocks in the natural land features to the remains of dead organisms, built structures, and man-made objects exposed to the environment – are weathered by the activity of such natural factors as wind, rain, snow and hail, groundwater flow, changes in temperature, solar radiation, and/or biological activity. It is because of weathering processes that archaeological objects and structures exposed to the atmosphere, buried in the ground or submerged under water, are always in a more or less advanced state of decay (Rolls and Bland 1998; Drever 1985).

Most weathering processes initially affect the external layers of solids. Soon after exposure to the environment, the surfaces of practically all solids are altered by the force of the wind or of flowing waters, the energy of the sun's ultraviolet radiation, or the activity of living organisms. Also, strong pressures, such as those activated by temperature changes on water in the pores and/or crevices in solids, result in the breakup and even in the bursting of large masses of matter into small pieces (see Fig. 45).

Physical weathering takes place whenever solids are exposed to radiation from the sun, abrasion by the wind or flowing waters, or are subjected to drastic changes in temperature. *Chemical weathering* involves changes in composition resulting from the interaction of the solids with atmospheric oxygen, air or soil pollutants such as ozone and carbon dioxide in the atmosphere, rain and groundwater, and substances dissolved in the water. The main factors that determine the *rate of weathering* of a material are its constitution and structure and the surrounding environmental conditiuons. Chemical weathering processes are most pronounced in regions where temperatures and rain precipitation are high (Ollier 1969).

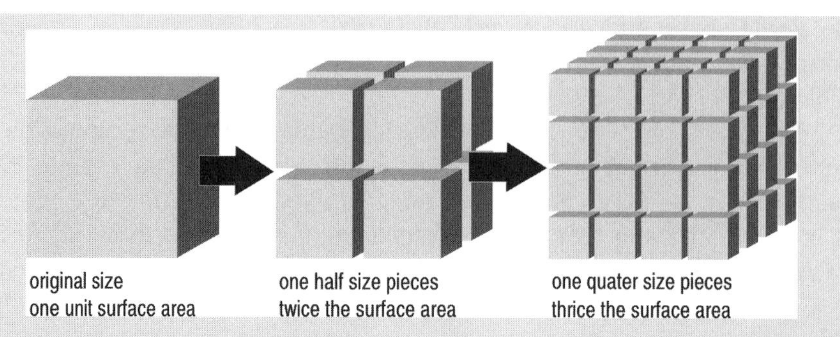

FIGURE 45 Freeze wedging. Freeze wedging, a form of mechanical weathering illustrated in the figure, is caused by the abnormal thermal behavior of water. The dimensions of most liquid and solid substances increase as the temperature increases and decrease as it is lowered. Water is at its smallest dimensions at about 4°C, expanding whenever it is heated above or cooled below this temperature (see Textbox 72). Thus, when solids such as minerals and rocks cool below 4°C they contract, while water in pores and crevices within the rocks expands; the opposite but concurrent dimensional changes generate forces strong enough to break apart the solids into fragments. The process, which is known as "freeze wedging," does not alter the composition of the weathered solid; as can be seen in the figure, however, its exposed surface area increases by a large factor and its exposure to further weathering is also enhanced. Moreover, since most chemical reactions begin at surfaces, the increase in the surface area caused by freeze wedging also accelerates chemical weathering processes.

TEXTBOX 46

DIAGENESIS

The term *diagenesis* is used in archaeology to refer to the chemical, physical, and biological changes affecting materials after burial, which ultimately result either in their dissolution, conversion into sediments, or lithification and formation of consolidated rock. All solids exposed to the environment are affected by weathering processes that initially induce changes, mainly to their outer surface layers (see Textbox 45). When buried, materials continue to be altered by physical, chemical, and biological processes. The changes resulting from these processes may include dissolution of part or all of the solids by groundwater, precipitation of new substances within the solids, and chemical reactions of some of their components with salts dissolved in groundwater. Such fundamental changes generally result in gradual alteration of the composition as well as of the porosity, permeability, density, and other physical properties of buried solids.

Diagenetic processes generally begin at the surface, pores, voids, and/or crevices and may continue for long periods of time, whenever one or more of the components of the solid become unstable vis-à-vis the environmental conditions. The way in which specific solids are altered, and the extent of the alterations is determined by their nature (their actual composition and physical characteristics), the prevailing environmental conditions (temperature and pressure), and the nature and amount of groundwater flowing around and through them (Reiche et al. 2003a; Wilson and Pollard 2002). A number of diagenetic processes are often differentiated, as detailed in Table 50.

The diagenesis of *bone* (a composite material consisting mainly of crystals of carbonated hydroxyapatite within intertwined collagen fibers), for example, seems to proceed initially via two parallel pathways: the relatively fast biodegradation of the organic components (the collagen fibers, which are used as food by microorganisms) and the slower decay and breakdown of the mineral component (carbonated hydroxyapatite) (Hedges 2002). The two paths are in no way exclusive of each other, although it is not clear to what extent they can proceed in isolation. Which is the predominating path depends on environmental conditions at the burial site.

TABLE 50 Diagenetic Processes

Process	Chemical and/or physical changes
Authigenesis	Crystallization between particles or in voids of sediments or material remains of new minerals introduced by groundwater
Bioturbation	Disruption of sediment layers by activity of living organisms
Cementation	Binding of particles of a material by new solids precipitated from groundwater
Compaction	Reduction of volume of pores and voids in materials by rearrangement of particles
Recrystallization	Reorientation of crystals that make up a material as a result of its dissolution in groundwater and subsequent reprecipitation
Replacement	Replacement of minerals by new minerals formed in situ
Solution	Dissolution of matter by groundwater

these made possible the development of two relatively new fields of investigation, known as *paleotemperatures* and *paleoclimates*, which, until not long ago, were unforeseen scientific disciplines. The knowledge of ancient temperatures and climates is important in archaeology, since awareness about their constancy or variation contributes to the understanding of the development of humans and their environment. The constancy of climate was, until the mid-twentieth century an accepted concept in archaeology and anthropology. It was only when geochemical and particularly isotopic techniques began to be applied to ancient temperature and climatic studies, that it became possible to estimate ancient temperatures with a certain degree of accuracy, and to investigate ancient climates (Hecht 1985; Wigley et al. 1985).

6.1. SEDIMENTS, OXYGEN ISOTOPES, AND ANCIENT TEMPERATURES

The shells of marine organisms such as corals and molluscs are composed mainly of calcium carbonate that the organisms create from substances they extract from water in their natural surroundings. Marine sediments are enriched in the isotope oxygen-18, and variations in the relative amounts of this isotope in the sediments is determined by the temperature prevailing at the time when the sediments were created (see Textbox 47). Also marine shells are enriched in oxygen-18, and the degree of their enrichment in this isotope is determined, as is that in marine sediments, by the temperature prevailing at the time when the shells were created. Thus, determining the weight ratios between the relative amounts of oxygen-18 and oxygen-16 in marine shells and other deposits of dead marine organisms reveals the temperature of the water at the time of shell formation. Moreover, since many sea sediments are derived from the shells of dead marine organisms, such sediments bear a record of the past isotopic composition of the oxygen of seawaters. The fluctuation in the weight ratio of these isotopes provides a kind of "isotopic paleothermometer" for establishing past variations in the temperature of the water (Bradley 1999; Aitken and Stokes 1997).

Experimental data derived from isotopic studies have greatly contributed to broadening knowledge on climatic trends during the last 100 million years and have also revealed detailed information on past climatic changes, such as those that took place during the glaciation cycles of the Ice Ages (see Fig. 46).

TEXTBOX 47

STABLE ISOTOPES OF OXYGEN AND ANCIENT TEMPERATURES

Oxygen, the most abundant element in all the outer layers of the earth, has three stable isotopes: *oxygen-16*, chemical symbol ^{16}O, the lightest, *oxygen-17* (^{17}O), and *oxygen-18* (^{18}O), the heaviest (see Textbox 12). The relative abundance of each isotope at any place or in any material varies, depending on whether the source of the oxygen is the atmosphere, water, or a mineral. Nevertheless, oxygen-16 is always the most common and most abundant. In the atmosphere, for example, 99.76% of the total oxygen is oxygen-16, only 0.20% is oxygen-18, and as little as 0.04% is oxygen-17. The existence of the three isotopes gives rise to three slightly different molecules of water, each molecule containing a different isotope of oxygen: $H_2{}^{16}O$, $H_2{}^{17}O$, and $H_2{}^{18}O$ (see Textbox 72). The relative abundance of oxygen-17 is so low, however, that in the discussion that follows it is ignored altogether; only oxygen-16 and oxygen-18 are considered.

Since the oxygen isotopes are of different weights, there are differences in the weight of the molecules of water that they form: water molecules including oxygen-16, are the lightest, while those with oxygen-18 are the heaviest. Whenever water evaporates, there is preferential evaporation of the lightest water molecules, those that include oxygen-16. In the remaining water (the water that did not evaporate), there is consequently, an increase in the concentration of water molecules containing the heavier isotope, oxygen-18. The degree of *oxygen fractionation* (the difference in the relative amounts of the oxygen isotopes in water vapor and nonevaporated water) is quite small, however. Still, depending on the temperature of evaporation, the weight ratio oxygen-16/oxygen-18, which is known as the *δ Oxygen-18 value*, in water vapor is a few percent higher than in sea water. The isotopic composition of water vapor in the atmosphere is, therefore, not the same as that of sea water; atmospheric water vapor carries more isotope oxygen-16 whereas sea water is enriched in the oxygen-18 isotope (see Fig. 46a).

Water vapor enriched in oxygen-16 is transported by wind in the atmosphere from the sea to land. When the water vapor condenses and precipitates as rain, snow, or hail, the water becomes rich in oxygen-16. Eventually the oxygen-16 rich water is incorporated into rivers, lakes, glaciers, and polar ice, which are, therefore, also rich in oxygen-16. Thus the isotopic composition of groundwater and the water of rivers, lakes, and glaciers is not the same as in seas and oceans.

Whenever atmospheric carbon dioxide dissolves in groundwater or in the water of rivers, lakes, and glaciers, it also reacts with the water to form mild carbonic acid, which renders the water slightly acidic:

$$CO_2 + H_2O \rightarrow \underset{\text{carbonic acid}}{H_2CO_3}$$

The carbonic acid thus formed is rich in oxygen-16. The mildly acid groundwater as well as the water of rivers and lakes, which is, therefore, also enriched in oxygen-16, dissolves limestone from surrounding rocks, to form calcium bicarbonate, which is soluble in water:

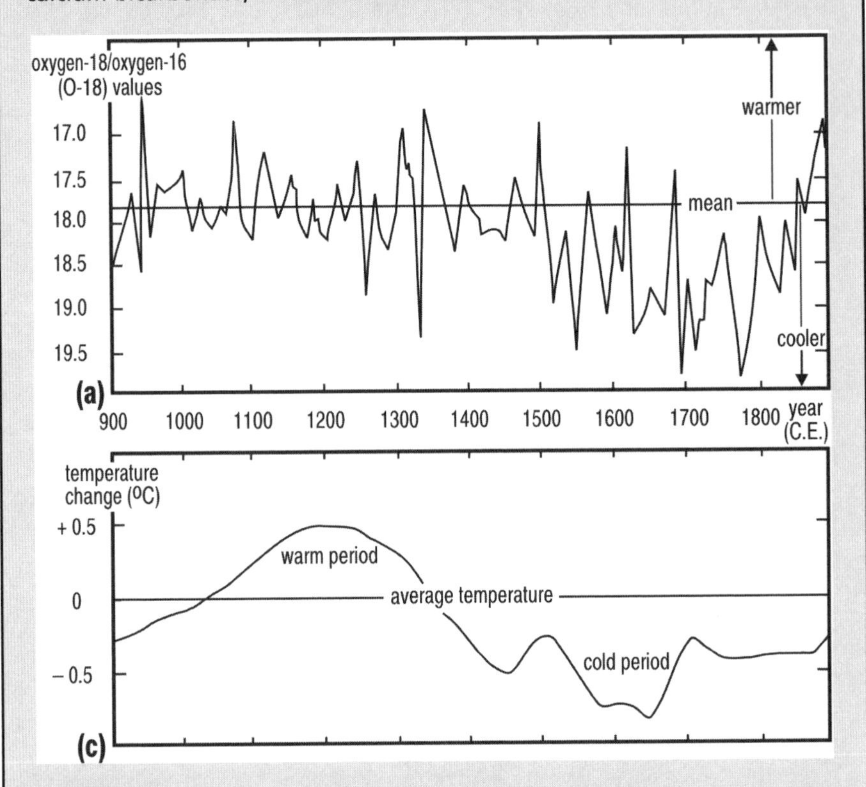

FIGURE 46 Oxygen isotopes and ancient temperatures. The relative amounts of stable isotopes of oxygen (oxygen 18/oxygen 16) in secondary minerals that include oxygen in their composition (e.g., calcite, composed of calcium carbonate) depends on two factors: the temperature at which the minerals were formed and the oxygen isotope composition of the water from which they originally precipitated. In any such mineral there is, therefore, a measurable ratio between the amounts of the oxygen isotopes. Its numerical value, which is known as the "δ O-18 value," is indicative of the temperature at which the mineral was formed. Oxygen isotope data from minerals and fossils can, consequently, be used to interpret climatic changes or fluctuations. The illustration shows the δ O-18 values determined in minerals **(a)** and the calculated temperatures (°C) **(b)** for the last 10 centuries in the region where the minerals were formed.

$$\underset{}{H_2CO_3} + \underset{\substack{\text{calcium carbonate} \\ \text{limestone}}}{CaCO_3} \rightarrow \underset{\text{calcium bicarbonate}}{Ca(HCO_3)_2}$$

Whenever natural water is heated and evaporates, some of the calcium bicarbonate reconverts to calcium carbonate and reprecipitates as limestone:

$$2Ca(HCO_3)_2 \rightarrow CaCO_3 + CO_2\uparrow + H_2O$$

It follows that, depending on the temperature, whenever limestone precipitates, it includes the oxygen isotope in which the water is rich; in sea waters, for example, the limestone is enriched in oxygen-18:

$$CaO + H_2C^{18}O_3 \rightarrow CaC^{18}O_3 + H_2O$$

while in groundwater and in the water of rivers and lakes it is enriched in oxygen-16:

$$CaO + H_2C^{16}O_3 \rightarrow CaC^{16}O_3 + H_2O$$

Calcium carbonate is also the main constituent of the shells of sea animals, which make their shells from elements acquired from the surrounding waters. Now, the degree of fractionation of the oxygen isotopes as well as the formation of mineral carbonates and of animal shells in sea waters are determined on the basis of the temperature–dependent fractionation of the isotopes of oxygen; the oxygen isotope composition of these materials reflects, therefore, the temperature at the time of their formation. Thus determining the isotope ratio between the stable isotopes of oxygen

$$\frac{\text{oxygen-18}}{\text{oxygen-16}}$$

in sea shells or in limestone provides a most valuable tool for estimating ancient temperatures (Bradley 1999).

Determining the temperature at which ancient sediments were formed entails two successive experimental stages: (1) extracting the carbon dioxide from the sediments and (2) determining the relative amounts of the oxygen isotopes in the extracted carbon dioxide. Treating a sediment with a standard acid, such as hydrochloric or sulfuric acid, dissolves the calcium and magnesium carbonates in the sediments and releases carbon dioxide that is collected and. The relative amounts of the oxygen isotopes in the released carbon dioxide is then measured, usually in a *mass spectrometer* (see Textbox 10). With this measurement at hand, the temperature prevalent at the time of formation of the sediment can be calculated with an accuracy within +1°C and occasionally even within +0.5°C.

The study of oxygen isotopes in water and in sediments also provides information on other paleoclimatological fields, such as wind strength and temperature divergences among different areas of the world. When air moves from warm, equatorial regions toward colder areas near the poles, for example, it loses water rich in oxygen-18 as rain and snow, while the water vapor remaining in the air becomes richer in the lighter isotopes (Grootes et al. 1993; Wilson and Hendy 1971). Determining the relative amounts of oxygen isotopes in rain or snow at particular locations on the surface of the earth makes it possible, therefore, to calculate temperature differences between colder and hotter regions of the world.

Analyzing the oxygen isotopes in ice cores obtained from stratified, frozen sheets in cold regions of the world may provide data on climatic variations in the past, often extending back for many thousands of years. Using hollow drills to cut through layers of sea sediments or the ice of glaciers, for example, provides samples, generally referred to as *sea cores*, in which the oxygen isotopes ratios can then be analyzed; the analysis of each separate layer from a core provides information on the environmental conditions at the time of formation of the sediments and/or the freezing of the water.

6.2. SOIL

Soil is a relatively thin layer of unconsolidated matter on the surface of the earth, in which there is biological activity. The bulk of most soil consists of a mixture of extremely small, loose particles of minerals and organic matter: the mineral particles are derived from the weathering of rocks; the organic matter from the dead remains of living organisms (Rowell 1994; Limbrey 1975). The composition and texture of the soil are altered by human habitation; humans change the natural flora and fauna of entire areas, their activity bringing about chemical and morphological changes of the soil. It is important, therefore, to understand the correlation between the composition of a soil and the ways it is altered by human habitation (Holliday 2004; French and French 2002).

The nature and relative amounts of each one of its components determines the type of soil. The composition and size of the mineral particles (e.g., of quartz, feldspar, mica, clays, salts) determine the texture of the soil. The relative amount of organic matter affects its water-carrying capacity, looseness, and fertility. Water, which makes up the liquid component and, in its vapor form, part of the gases in the soil, is an effective temperature regulator of the soil. It also transports salts and nutrients from the mineral and decaying organic matter particles to the roots of plants and to other organisms. The gases in the soil, which include, in addition to water vapor, mostly nitrogen, oxygen, and carbon dioxide from the air, occupy pores and crevices between the particles and are essential to the life of all living organisms in the soil (see Fig. 47) (Hassett and Banwart 1992).

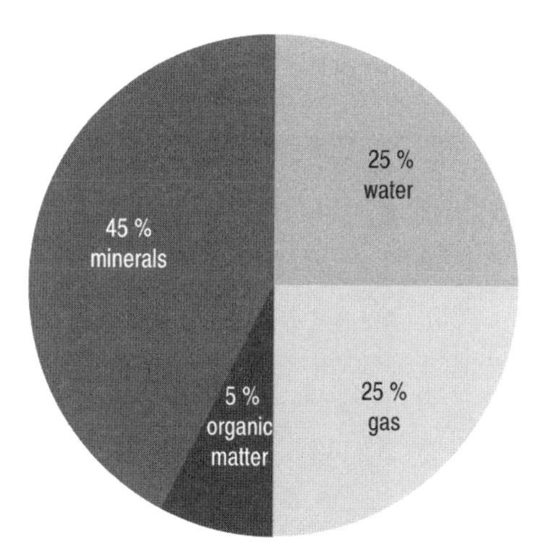

FIGURE 47 The main components of soil. Soil consists of a mixture of solids, liquids, and gases. The solid part includes minerals particles derived from weathered rocks (mostly clay and sand) and organic matter (humus) resulting from the decay of dead plants and animals. Water makes up the liquid and air, the gaseous part. The relative amounts of each one of these components determines the type of any particular soil. The diagram illustrates the composition of an "average soil" from temperate zones. Dessert soil may contain as little as under 1% organic solids, while some wetlands and bogs there is well over 30% organic solids.

Mixtures of such different components give origin to the diverse types of soil, each type having a particular composition, color, and texture. In time, and under the combined effect of mechanical forces – such as those generated by the flow of wind, rainwater and groundwater, and their own density – the particles in the soil become gradually segregated, forming clearly identifiable layered structures, generally known as a *soil horizons*. Each horizon is made up of particles having a definite size range as well as a distinctive composition, physical properties, and biological characteristics. Taken together, the horizons are said to compose a *soil profile*. In most soils, three horizons, known as the *A*, *B*, and *C* are distinguished; horizon *A* is the top layer and *C* is the deepest (see Fig. 48).

The *A horizon*, at the uppermost surface of the soil, is the layer that is plowed during agricultural operations. It contains most of the organic matter and living organisms of the soil and is affected by active weathering and decay processes. Examples of weathering processes active in soils are *leaching* (the dissolution and washing out of soluble substances) and *reprecipitation* (redeposition of the washed-out substances) generally in other soil

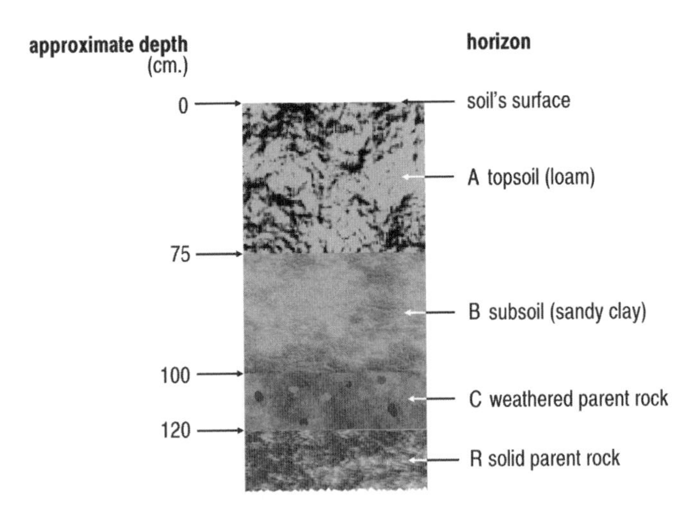

FIGURE 48 A soil profile. Many of the characteristics of soils vary with depth. A convenient way of representing its varying characteristics is by dividing the soil into layers, usually referred to as "horizons," identified by letter symbols. The surface layer, which is known as the *A* horizon, is generally rich in organic matter. Next come the *B* and *C* horizons, each of which may have compositional characteristics and modifications. The deepest soil horizon (*R*) is solid rock. The illustration identifies clearly defined horizons, although in most soils the horizons are not as clear and in some they may be very diffuse.

horizons. The *B horizon* is the layer of intermediate weathering in which small particles, washed down from the *A* horizon, are mixed with reprecipitated minerals initially leached out also from the *A* horizon. The *B* horizon has some distinctive characteristics: it generally is of a color different from that of the *A* horizon, is more intensely red or brown than the *A* horizon, and often includes accumulations of clay and/or organic matter or a combination of these two. In most temperate climates, the *A* and *B* horizons together make up about one meter in depth. The lowermost *C horizon* is composed largely of mineral fragments and rock particles; it is little affected by soil-forming processes and does not share the properties typical of the overlying horizons.

Composition and Properties of Soils

Over 98% of the soil is made up of only eight major chemical elements, listed in Table 51, in order of decreasing abundance. The 90-odd others make up the remaining 2%; many occur in the soil as secondary or minor elements, while a large number of still others are present in only very low, often trace, concentrations. Thus all soils contain main, minor, and trace elements combined into chemical compounds and aggregated into complex particles of varying shape, size, and chemical composition (see Textbox 8).

The size, and particularly the size distribution of the particles (or PSD, as *particle size distribution* is often abbreviated), which is closely related to the texture of the soil, is an important characteristic that determines many of the

TABLE 51 Elemental Composition of Soil

Element	Average abundance (wt%)
Oxygen	47
Silicon	27
Aluminum	7
Carbon	5
Iron	5
Calcium	3
Sodium	2
Magnesium	2
Potassium	1
Other elements[a]	below 1

[a]Totaling 84, these elements include, in descending order of abundance, titanium, hydrogen, phosphorus, nitrogen, barium, and strontium, each one of them in less than one percent.

properties of the soil: it regulates, for example, the flow and the retention of water and the thermal characteristics (*thermal capacity* and *thermal conductivity*) of the soil. Depending on the size of the main constituent particles, it is customary to classify soils as *clayey*, *silty*, and *sandy*, or *loamy*, as listed in Table 52; the latter type is a mixture of the first three types listed.

The pH of Soils. The pH of soils (see Textbox 48) varies widely with location and climate; the local climate of a region usually provides a clue as to the pH of the soil. In high-rainfall regions, for example, soils are generally acid. Alkaline soils, in contrast, typically occur in low-rainfall areas; the actual value of the pH varies over a wide range, from below 5 in very acid

TABLE 52 Soil Particle Size and Soil Types

Soil type	Texture	Particle size (mm)
Clayey	Fine	below 0.002
Silty	Moderately fine	0.002–0.05
Sandy	Coarse	over 0.1
Loamy	Mixture	below 0.002–over 0.1

TEXTBOX 48

ACIDS AND BASES: THE pH OF SOLUTIONS

Most of the water on the surface of the earth occurs as molecules made up of two atoms of hydrogen *bonded* (joined) to one of oxygen, and the formula for water is H_2O (see Textbox 72). A very small percentage of the water molecules are *ionized*, however; that is, they are *dissociated* (separated) into two electrically charged particles, known as *ions*: a positively charged hydrogen *cation* (an atom of hydrogen bearing a positive electric charge, H^+) and a negatively charged *hydroxyl anion* (composed of a hydrogen atom joined to one of oxygen, OH^-):

$$\underset{\text{water molecule}}{H_2O} = \underset{\text{hydrogen cation}}{H^+} + \underset{\text{hydroxyl anion}}{(OH)^-}$$

In very pure water there are very few ionized water molecules, and the positively charged hydrogen cations as well as the negatively charged hydroxyl anions occur in equal numbers: approximately two of each for every billion molecules of water. When substances such as *salts*, *acids*, or *bases* are dissolved in water, however, the balance between the number of hydrogen cations and hydroxyl anions breaks down and the solutions may be either *acid*, *alkaline*, or *neutral*. In *acid solutions* there are more hydrogen cations than hydroxyl anions; in *alkaline solutions* there are more hydroxyl anions than hydrogen cations, and in *neutral solutions*, which are neither acidic nor basic, there are, as in pure water, equal numbers of hydrogen and hydroxyl ions.

Acids

The dissolution of such substances as atmospheric carbon dioxide or sulfur dioxide, the latter an air pollutant, in water, results in the formation, in the solution, of a large number of hydrogen (H^+) cations. Substances that when dissolved in water induce an increase in the number of hydrogen cations in the solution are said to be *acid substances* or simply *acids*. The acid solutions formed by the dissolution of carbon dioxide and sulfur dioxide, which are known as *carbonic acid* and *sulfuric acid*, respectively, differ greatly in their acid properties: water solutions of carbonic acid have a much lower concentration of hydrogen cations than do those of sulfuric acid. *Carbonic acid*, which is the most abundant acid in the earth's environment, is therefore said to be a *weak acid*. *Sulfuric acid*, which is abundant in the polluted atmosphere of modern industrial societies, is a *strong acid*. The solutions of all weak acids, as, for example, carbonic acid, acetic acid (in vinegar), citric acid (in oranges and lemons), and tannic acid (in the bark of many trees, which was widely used for making leather) have a much lower concentration of hydrogen cations than do those of all strong acids such as sulfuric and nitric acids in highly polluted atmospheres.

Alkalies

When substances such as lime, soda, or potash are dissolved in water, they induce the formation, in the solution, of *hydroxyl* (OH^-) anions and a consequent great reduction in the number of hydrogen cations. The solutions of these substances are known as *alkaline* or *basic solutions*, or simply *alkalies* or *bases*. Two common and typical alkalies are calcium hydroxide (quicklime) and sodium hydroxide (caustic soda). The solutions of these two alkalis have, however, highly differing concentrations of hydroxyl ions, and their alkaline properties also, differ widely: other things being equal, the solutions of sodium hydroxide are much stronger alkalies than are those of calcium hydroxide; this means that they have a much higher concentration of hydroxyl anions (or a much lower concentration of hydrogen ions) than do those of calcium hydroxide.

With very few exceptions, naturally occurring acids and alkalies are weak. All acids known in antiquity were of organic origin; some occur in fruits, especially in unripe fruitjuices. Most ancient alkalies were derived from the ash of plants such as barilla, *Salsola soda* and *Salsola kali* (Russian thistle), and kelp.

Properties of Acids and Bases

All *acid solutions* taste sour and are more or less corrosive and chemically quite reactive; they react with most metals, many of which are corroded and dissolved by acids. *Alkaline solutions*, also chemically reactive, are *caustic* (they burn or corrode organic tissues), taste bitter, and feel slippery to the touch. Both acids and bases change the color of *indicators* (substances that change color, hue, or shade depending on whether they are in an acid or basic environment).

Neutral Solutions: Salts

When an acid solution is mixed with an alkaline solution in the appropriate quantities, the two solutions are said to neutralize each other: the hydrogen (H^+) cations of the acid and the hydroxyl (OH^-) anions of the base combine to form water molecules (HOH or H_2O), canceling the (acid or alkaline) properties of the other. The term *neutralization* is used to refer to this reaction because the acid and basic properties of the two solutions are *neutralized* and the solution is *neutral*, neither acid nor basic:

$$\text{Acid solution} + \text{alkaline solution} \xrightarrow[\text{neutralization}]{} \text{water} + \text{salt}$$

Neutral solutions also contain a dissolved *salt*, derived from the neutralization of the acid and the base (a *salt* is an ionic compound formed in a neutralization reaction and is composed of the cation of an alkali and the anion of an acid). When a solution of carbonic acid (formed when atmospheric carbon dioxide dissolves in water), for example, reacts with an alkaline solution of lime, the two solutions neutralize each other and form a salt, calcium carbonate:

$$\underset{\substack{\text{carbonic acid}\\ \text{(acid)}}}{2H_2CO_3} + \underset{\substack{\text{lime}\\ \text{(alkali)}}}{Ca(OH)_2} \rightarrow \underset{\substack{\text{calcium carbonate}\\ \text{(salt)}}}{CaCO_3} + \underset{\text{water}}{H_2O}$$

pH

The acidity and alkalinity of water solutions and, therefore, differences in their acidity or alkalinity, can be quantified and assigned numerical values. One way of doing this is to express the concentration of hydrogen ions in solutions on a numerical scale. Such a scale is provided by the widely accepted *pH scale*, in which the strength or weakness of acid or alkaline solutions is expressed as the negative logarithm of the concentration of hydrogen (H^+) ions in the solution. In mathematical form, the pH is expressed by the following relationship:

$$pH = -\log[\text{concentration } H^+]$$

This relationship provides a relatively simple but exact way of expressing the acidity or alkalinity of solutions: the lowest numerical value of the relationship (for strong acids) is zero, and the highest (for strong alkalies) is 14. The stronger an acid (or the stronger the acidity of its solution), the greater its concentration of hydrogen ions and the lower its pH, and vice versa, the stronger an alkali (or the stronger the alkalinity of its solutions), the lower the concentration of hydrogen ions in the solutions and the higher its pH (see Fig. 49).

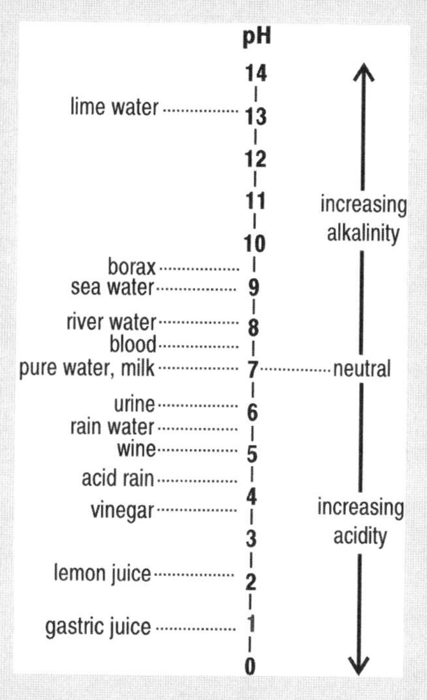

FIGURE 49 The pH of water solutions. The pH of a water solution specifies the numerical value – ranging between 0 and 14 – of the concentration of the ions of hydrogen in the solution. In pure water there are equal amounts of hydrogen and hydroxyl ions and the value of the pH is 7. Pure water has a pH of 7 and is said to be "neutral"; pH values below 7 specify acid solutions – the lower the pH of a solution, the stronger its acidity; pH values above 7 indicate that the solution is alkaline; the higher the pH of a solution, the stronger its alkalinity. A change of one pH unit indicates a ten fold change in the concentration of the ions of hydrogen.

The pH of solutions is generally measured with a *pH meter*, an instrument that in a single, simple operation measures and yields the pH value of any water solution, thus making unnecessary any further measurements or calculations. There are many technically different pH meters, some large, used mainly in laboratories, others portable, easily taken out for field measurements. The pH can, however, also be measured using substances known as *indicators*, which exhibit different colors, shades, or hues at different pH values. *Litmus*, for example, is an indicator that exhibits shades of red in acid solutions, that is, in solutions having pH values between 7 and 1, and shades of blue when in alkaline solutions with pH values between 7 and 14.

TABLE 53 pH of Soils and Natural Waters

Matter	pH
Soil	
Acid	below 4.5– 6.5
Neutral	6.6– 7.3
Alkaline	7.4– 9.0
Water	
Pure	7
Rainwater	5.5– 6
Polluted rainwater	as low as 2.1
Riverwater	7 – 8
Seawater	8 –10

soils to above 10 in strongly alkaline ones. Table 53 lists characteristic pH values of soils.

Archaeological Soils

The terms *archaeological soils, archaeosols, anthroposols,* and *paleosols* are variously used to refer to soils that have been physically and/or chemically altered by human habitation or activity. The soil of a site constitutes an integral part of its archaeological record (Wells 2004). It is a well-known fact, for example, that in areas of intense ancient human habitation the fertility of the soil is higher than that of the surroundings. Dark soils rich in organic matter often define, with considerable precision, areas of past intensive human activity.

Soils altered in the past by human habitation are not difficult to find. Farmers in many regions of North America, for example, noted that the soil of old Indian villages was more productive than adjacent soils. The unique properties of soils from ancient inhabited places have frequently been put to use in many places in the world; farmers in the Ashdod area in Israel and El Phosfat in Egypt, for example, used the soil that they excavated from ancient archaeological sites to fertilize the land they cultivated (Wright 1986).

Alteration of Soil by Human Habitation. Humans inhabiting a site leave buried or partly buried remains of their existence (excreta and dead human remains) and of their activities [ashes, domestic plants, and animal remains (blood, flesh, and bones), food wastes, and byproducts of manufacturing undertakings]. After burial, these remains are affected by weathering and diagenetic processes (see Textboxes 45 and 46) and eventually become part of the soil, altering its composition. Thus, the chemical composition of soils at sites of past human habitation preserves information on changes brought about by human activity. Using modern chemical and physical techniques it is possible to study much of this information (Retallack 2000, 1990; Groenman van Waateringe and Robinson 1988).

The chemical elements that may provide clues for recognizing and locating sites of past human habitation are mainly *carbon*, *nitrogen*, and *phosphorus*. *Carbon* is derived from all forms biological matter; *nitrogen*, mostly from amino acids in plant and animal proteins; and *phosphorus*, from the bones and excretions of humans and animals. The amounts of carbon and nitrogen added to the soil are relatively large (since most discarded wastes are composed of organic matter); these elements are, however, also subject to depletion during decay processes, and much of them is lost: carbon, mainly through biooxidation processes and nitrogen, in a number of ways. Not so phosphorus, which, when deposited in soil, becomes *fixed* and is not subsequently depleted. It has been calculated that a community of 100 people living in an area of one hectare (10,000 m^2) may provide an annual increment to the soil of about 120 kg of phosphorus. Relative to the amounts of this element naturally present in the soil, this is a considerable addition, representing an annual increase of up to one percent of the phosphorus in the uppermost horizon of soil. The relative amount of phosphorus in the soil provides, therefore, an indicator of the extent, duration, and nature of past human habitation. Measuring the relative amount of phosphorus in the soil has, therefore, become a quite widely used tool for surveying archaeological sites, but also to appraise the intensity and/or duration of past human habitation at a site (Bethell and Máté 1989; Rottländer 1983).

Phosphorus Survey of Soils. The practical side of surveying soils for phosphorus is relatively simple; soil samples, usually from a depth varying between 10 and 20 cm below the surface, are collected at a site at regular intervals, according to a predetermined plan. After drying of the samples at a standard temperature, the relative amount of phosphorus in each sample is determined by a suitable analytical method (and usually expressed in terms of milligrams of phosphorus pentoxide per 100 grams of soil). A statistical analysis of the analytical results makes it possible to discriminate between *site samples*, that is, sites having a relatively high phosphorus content, and *nonsite samples*, containing relatively low amounts of the element; often these sites, which can be clearly delineated on the ground, were used in the past for specific activities.

One of the disadvantages of early phosphorus surveys was not long ago, the need to obtain a relatively large number of heavy soil samples, which had to be taken to a chemical laboratory for analysis. In later studies, however, use has been made of portable equipment that makes it possible to analyze, even in the field, very small samples, and statistically appraise the analytical results (Persson 1997).

Current chemical surveys of archaeological sites seem to indicate that small variations in the concentration of other elements, besides phosphorus – such as calcium, magnesium, and potassium, but also strontium, lead, zinc, and some trace elements – may also provide valuable archaeological indicators of ancient human habitation (Entwistle and Abrahams 1997; Linderholm and Lundberg 1994). Alterations to soil caused by ancient human habitation can also be recognized with a number of physical techniques, based on the detection of some optical or other physical characteristics of the surface of the soil; some of these, known by the generic terms *geophysical survey* or *remote sensing techniques*, are listed in Table 54 (Donaghue 2001; Scollar et al. 1990).

TABLE 54 Techniques of Remote Sensing[a]

Technique	Radiation used for detection
Aerial photography	Visible, infrared, and ultraviolet
Earth satellites	Visible and infrared
Thermography	Thermal infrared
Aerial radar	Radar

[a]Remote sensing data are generally gathered with *sensors*, devices for detecting matter or *electromagnetic radiation*, which are not in material contact with the site. Sensors that reveal information relevant to archaeology are usually airborne or spaceborne.

Bogs. A particular type of soil which has received frequent attention in archaeological studies, particularly in northern Europe, is that of *bogs*, waterlogged areas that have a spongy consistency. The solid component of most bogs is *peat*, a very dark, unconsolidated, partially decomposed and carbonized mass of plant remains in which, usually, very low pH values and anaerobic conditions prevail. Some bogs include as little as 2% peat; the remaining 98% is water. In others, the peat may constitute as much as 15% by weight of the bog. The low pH and the anaerobic conditions inhibit the development of spoilage microorganisms. Buried under such conditions the remains of dead organisms may become naturally mummified and remain in a good state of preservation, often lasting for extended periods of time. The prevailing acidity also causes *decalcification* (dissolution of the calcium) as well as removal of most of the inorganic components of the bones; this becomes apparent when human and/or animal mummies with soft, yielding bones are removed from bogs (see Chapter 15) (van der Sanden 1996; Brothwell 1987).

CLAY
POTTERY AND OTHER CERAMIC MATERIALS

Clay is the generic name for very fine grained, unconsolidated earthy minerals that are supple and pliable when wet, become hard and retain their shape when dried, and turn into a stony material, known as *fired clay*, when heated to red heat. It is because of these unique properties that very early in the development of humans, clay became a most important raw material, used to make bricks, pottery, and later, other ceramic materials (Kingery 1986–2000). Clay absorbs and loses water easily, and the water absorbed penetrates the surface of the grains, which swell accordingly and shrink when they lose the water. The grains also *adsorb* gases: their surface takes up and retains gases such as carbon dioxide and water vapor (in the soil, the gases adsorbed on clay particles are used as nutrients by plants).

The meaning of the word clay differs among various fields of interest. In geology, for example, the term is used to refer to very fine grained minerals formed as a result of the weathering of siliceous rocks (see Textbox 45); in chemical mineralogy, a clay is an unconsolidated mineral belonging to the group known as the clay minerals, and in soil science the word clay is used to refer to the inorganic fraction of the soil that is made up of very small particles (below 0.002 mm in diameter) having a very large surface area per unit weight (Murray 1986). Clay is of particular interest in archaeology and in materials science, since, for a very long time, it has been used for making *ceramic materials* such as *pottery* and *fired brick*, which are among the most ancient and most frequently encountered human-made materials in archaeological excavations (Newman 1987; Grimshaw 1980). In these fields clay is defined as an earthy aggregate of mineral particles that when mixed with water exhibits plastic properties, on drying becomes rigid, and when heated to a sufficiently high temperature acquires hardness, strength, and chemical and physical stability (Rice 1987).

Archaeological Chemistry, Second Edition By Zvi Goffer

Chemically, all clays are composed of oxides of silicon, aluminum, and hydrogen – namely, silicon dioxide, aluminum trioxide, and water in a weight proportion that can be expressed by the following general formula (see Fig. 50):

$$Al_2O_3 \cdot SiO_2 \cdot 2H_2O$$

There are a number of different clays, and some of the most common are listed in Table 55. The composition of each clay can be expressed by a formula that differs slightly from the general formula given above. Chemical composition alone, however, is not sufficient for characterizing clays; their crystal structure provides the best way of characterizing any type of clay (see Textbox 21). In many clays, for example, the atoms are grouped in

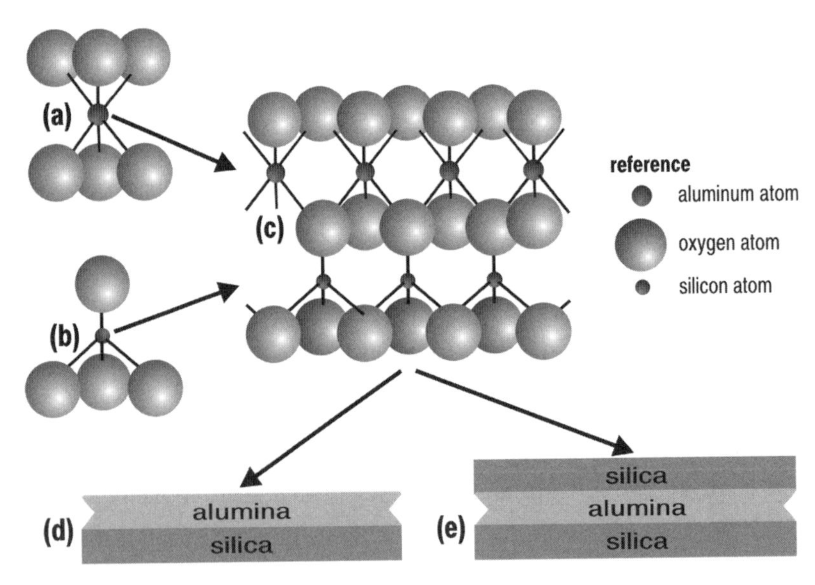

FIGURE 50 The structure of clay. Clays are the end products of the weathering of rocks and minerals; their main components include alumina, composed of aluminum trioxide, with an octahedral structure **(a)**, and silica, composed of silicon dioxide, with a tetrahedral structure **(b)**. Structural studies show that in most clays the silica and alumina are arranged in layers superposed on each other: atoms of oxygen of the alumina and the silica are shared by the superposed layers **(c)**. The distinctive character of each different clay is due to cations of sodium, potassium, calcium, or magnesium occupying positions in and between the sheets. Kaolinite and halloysite, two clays that are suitable for making ceramic materials, have a two-layer (also designated as a 1 : 1) structure, in which one layer of alumina is bound to one of silica **(d)**. Other clays have a three-layer (or 2 : 1) structure, made up of one layer of alumina between two of silica **(e)**.

double-sheetlike arrangements, as shown in Figure 50: one sheet is made up of a network of silicon atoms surrounded, in a three-dimensional tetrahedral arrangement, by atoms of oxygen; the other sheet, made up of aluminum atoms surrounded by iron and/or magnesium, is embedded between two layers of closely packed oxygen atoms (Weaver and Pollard 1975).

More than 20 different types of clay can be actually distinguished. Those most appreciated for making ceramics, for example, kaolinite, are built up of combinations of the basic structural units described above. The particles of most consist of *platelets* (very small, flat sheets) that, when stacked together, form layered arrangements having extensive surface areas, much like the pages of a book. Other common clay particle shapes are fibrous or tubular.

Clays occur naturally either in a relatively pure condition or mixed with other materials and they are therefore classified into one of two large groups: *primary* and *secondary clays*. *Primary clays* are quite pure, uncontaminated by other materials, and have a rather uniform composition. *Secondary clays* are mixtures of clay with other minerals such as quartz, talc, mica, iron oxides, and even organic matter (the latter derived from the decay of dead plants and animals); the particles of most of the contaminating materials are generally of similar size to those of the clay (Kingery et al. 1976).

7.1. PRIMARY CLAY

Primary clay is also known as *residual clay*, indicating that they are either the in situ residue of one type of weathered rock or the transported residue of many types of rocks; most primary clay deposits occur, however, in situ, at the location where the clay particles were formed. The clay is usually quite pure and colorless or white, but very small relative amounts of minerals mixed with the clay, such as quartz and/or iron oxides, may impart to it a yellow, brown, or green color. Primary clay is also characterized by the extreme fineness of its particles, which usually measure below 2 micrometers (0.002 mm) in diameter. The more than 20 different types of primary clay minerals can be distinguished by their chemical composition, which varies widely, and by their physical properties. Primary clays that have been used for making ceramic objects are listed in Table 55.

Kaolin, China clay, terra alba, argille, porcelain clay, and *white bole* are the generic names used to refer to primary clays that include three distinct white minerals – *kaolinite, nacrite*, and *dickite* – all of which share a very similar composition but differ slightly in their structure. Kaolin is rarely found pure, but as a natural mixture with other varieties of clay; together, the various clays make up over 95% of the total weight of the mixture, other earthy

TABLE 55 Primary Clays

Clay	Characteristics
Kaolinite	White platelets; the type of clay most widely used for making ceramics
Halloysite	Similar in composition to kaolin but of different structure
Montmorillonite	White, gray or green; dull microscopic or very small platelets
Illite	Very small platelets, often found in shale and mud stone
Chlorite	Green platelets; color due to varying amounts or iron
Attapulgite (or paligorskite)	Tough, lightweight and fibrous
Sepiolite (or meerschaum)	White or off-white, soft, fibrous earthy masses, easily carved and polished

materials making up the remaining 5%. Kaolin is the main component of most *potter's clay*, the common secondary clay used by potters throughout the world since ancient times (see text below); another ancient use of kaoline is as a white pigment. A particular type of kaolin, distinct from most other types of clay, is *Chinese porcelain stone*, which has been used for nearly 2000 years, mainly in southern China, for making porcelain. Chinese porcelain stone is a mixture of mainly kaolinite, together with *sericite* (a fine-grained white mica), quartz, and other mineral impurities, such as lime and rutile (titanium oxide) (Guo Yanyi 1987).

Montmorillonite, also one of the most common clay minerals, occurs as a soft rock; after it is powdered and mixed with water, it acquires plastic properties. It is the main constituent of *fuller's earth*, used since antiquity for *fulling* (cleansing textile cloth).

Sepiolite, also known as *meerschaum* or *sea foam*, is a white or off-white soft clay of low density. It is composed of hydrated magnesium silicate and occurs in consolidated masses that are easily carved and whose surfaces can be smoothed and endowed with a durable polish. It is for this reason that meerschaum has been used, since antiquity, for making ornaments and decorative carvings. For many centuries meerschaum has been intensively mined in Greece and Turkey.

7.2. SECONDARY CLAY

Secondary clay, also known as *sedimentary clay*, *transported clay*, or *potter's clay*, is a mixture of clay and *nonclay* particles. After it is formed, much clay is transported by such natural agencies as water streams and winds and en route mixes with other mineral particles. When it is finally deposited, at more or less distant places from its site of origin, it does so as secondary clay, generally containing over 50% of nonclay matter such as sand, limestone, iron oxides, and organic matter (the latter is derived from vegetable and animal wastes). The iron oxides impart to most secondary clays a yellow, red, brown, and, occasionally, even a green color; organic matter darkens any clay, whatever its color.

The particles of secondary clay are usually smaller than those of primary clay; the fineness of the particles is the result of the mechanical friction undergone during transportation from their place of origin to where they become settled. The small particle size makes wet secondary clay more plastic and pliable than wet primary clay and therefore highly appreciated by potters for making ceramic objects. Two particular types of secondary clay that have been extensively used for making pottery are *ball clay*, which is rich in organic matter and *red clay*, rich in iron oxides. When extracted from the earth the actual color of these clays is usually yellow, brown, gray, or black, but during firing they turn red; the change in color is due to iron ions in the clay being oxidized and turning a rich red at high temperatures.

7.3. CLAY AND CERAMIC MATERIALS

Three main properties render clay suitable for making ceramic materials: its *plasticity* when wet, its hardness when dry, and the toughness, increased hardness, and stability that it acquires when fired. The addition of water to dry clay produces a clay–water mixture that, within a narrow range of water content, has *plastic properties*: it is deformed, without breaking or cracking, by the application of an external stress, and it retains the acquired shape when the deforming stress is removed. Wet clay mixtures can, therefore, be modeled, molded, or otherwise made to acquire a shape that will be retained after the forming operations. Water-poor mixtures are not plastic, however, and excess water results in mixtures, known as *slips*, that are too fluid to retain a shape, as shown in Table 56.

The plasticity of clay–water mixtures is due principally to two factors: the flat platelet shape of the clay particles and their small size. When clay is wet, the water, which under such conditions is known as *water of plasticity*, envelops each particle, acts as a lubricant between the particles, and allows them, when an external force is applied to the mixture, to readily slide along each other, as shown in Figure 51.

As for the drying properties of clay–water mixtures: when wet clay dries at ambient temperature, the water of plasticity that surrounds the particles evaporates, the particles are drawn closer to each other and, as a consequence, the volume of the clay shrinks. The *drying shrinkage* of clay is defined as the reduction in size that takes place when a mass of wet clay is dried, as illustrated in Table 57.

If wet clay is heated to 100°C (the boiling temperature of water) or above, the water in the mixture evaporates, the clay particles come into close contact with one another, and the mass of clay becomes hard, developing a certain degree of strength; this strength, which is known as *dry strength*, is a measure of the force required to break the mass of clay. Dry strength varies greatly for different types of clay, and its magnitude is relevant when handling dry clay-shaped objects.

TABLE 56 Plasticity of Clay–Water Mixtures

Water (% weight)	Clay–water mixture
below 10	Dry
10–20	Semidry, stiff
12–25	Stiff but plastic
15–30	Plastic
20–40	Oversoft
above 30	Liquid (slip)

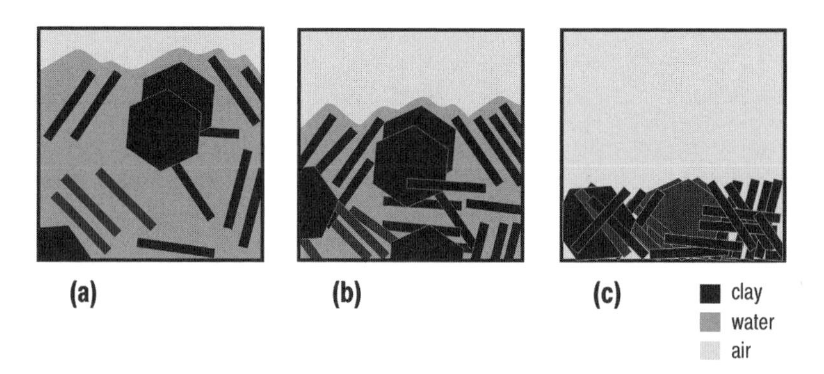

FIGURE 51 Drying clay. The illustration shows a schematic, highly magnified cross section of a mass of wet clay at three different levels of water contents. Initially, all the space around the clay particles is filled with water **(a)**. As the wet mass dries up, water is lost from between the particles, which gradually come nearer to each other **(b)**, and the clay mass shrinks in size. When all the water has evaporated, the particles are in contact with each other and the mass cannot shrink any further **(c)**. From this point onward the bulk volume of the clay mass does not change.

TABLE 57 Linear Shrinkage of Clays on Drying and Strength of Dry Clays

Clay	Linear shrinkage (%)[a]	Dry strength (kg/cm^2)
Kaolin	3–10	0.5–50
Illite	4–11	15 –75
Montmorillonite	2–23	20 –60

[a]Expressed as a percent of the length after drying at 105°C.

TABLE 58 Changes Caused by Heat to Wet Clay

Temperature (°C)	Change
RT[a]– 100	Drying: loss of the water of formation
100– 500	Loss of the water of plasticity
500– 600	Dehydration; loss of chemically combined water and modification of clay structure
600– 900	Breakdown of clay structure and incipient vitrification
900–about 1700	Vitrification and formation of new structures
above 1700	Melting

[a]RT = room temperature.

When clay is *fired* (heated to high temperatures, above about 600°C), it undergoes profound chemical changes and as a result these changes acquires many of the qualities that make *fired clay* uniquely useful. The heat supplied during the firing process provides sufficient energy to dislodge atoms from their positions within the clay and cause them to migrate to more favorable sites; the final result of the firing process is that the clay is converted to a new hard and rigid *ceramic material* that is stable to water, high temperatures, and weathering. Table 58 summarizes the chemical changes that take place when clay if heated.

7.4. CERAMIC MATERIALS

Pottery, one of the earliest human-made ceramic materials, is actually an artificial form of stone, made by combining the four basic elements recognized by the ancient Greeks: earth (clay), water, air, and fire. In fact pottery is made from a circumstantial or deliberately prepared mixture of clay, other solid materials known by the generic name of *fillers*, and *water*. When a wet mixture of clay and fillers is formed into a desired shape, then dried and finally heated to high temperature (above 600°C), it becomes consolidated into pottery. Fired pottery usually has a rough surface and is very hard, stable, and durable, although it fractures easily. The term *pottery* is generally used both as the generic name for artifacts consisting mainly of fired clay and to refer to the lower grades of ceramic materials, distinct from *stoneware* and *porcelain*. It is as the generic name for artifacts made of clay that the term is used throughout this book (Knappett 2005; Orton et al. 1993).

TEXTBOX 49

REFRACTORY AND CERAMIC MATERIALS

Ceramics are nonmetallic materials made mainly from clay and hardened at a high temperature. All ceramic materials are *refractory* – they withstand high temperatures without softening, fusing, changing shape or weight, or undergoing changes in their chemical composition; they are also mechanically strong and resistant to weathering and to attack by chemicals. Most ceramic materials are brittle, however, cracking, fracturing, shattering, and breaking easily. *Pottery*, *stoneware*, and *porcelain* are ceramic materials. Many of their uses rely on one or more of the attributes mentioned above. The materials from which metal-casting molds, crucibles, and the lining of furnaces, kilns, hearths, and domestic ovens are made are also ceramics (Bray 2000). It should be noted, however, that all the ceramic materials of antiquity withstood only temperatures not higher than about 1300°C, the highest temperatures then attainable. At present, only materials that withstand temperatures above 1580°C are considered as refractories. Such a rigorous demarcation thus excludes most ancient ceramics. *Ceramic materials* are, therefore, best defined as refractory to the highest temperatures feasible in the furnaces, kilns, and hearths at a particular period of time during the development of high-temperature-withstanding materials.

Common ancient ceramic materials often found in archaeological excavations, such as fired brick and pottery, were made mostly from a mixture of a secondary clay and fillers. The nature, composition, and properties of clay have been already discussed; the nature of the fillers, the changes undergone by the clay as well as by the fillers during their conversion to ceramics, and the unique properties of ceramic materials, are reviewed in the following pages. Attention is drawn also to studies that provide information on the composition and characteristics of ancient ceramic materials.

Some of these studies contribute to the understanding of ancient pottery-making techniques, others to learning about the provenance of pottery. The craft aspects of potterymaking, fascinating as they may be in themselves, are, however, outside the scope of this book (Rice 1982).

Preceramics

Loess soils are soils composed mainly of very fine particles of clay mixed with other, also fine-particle minerals such as feldspar and mica and little, if any, organic material. Rains on loess soil causes puddles to form on the surface. When the puddles dry, the uppermost layer of the crust of the soil, from which the water evaporates fastest, contracts, forming relatively large, irregular, concave shapes that fissure from each other; as the drying progresses the crusts break away from the soil beneath and from each other, and the broken crusts become compacted as irregular, dishlike concave pieces that may attain an area as large as 100 cm^2 and a thickness of up to 10 mm (see Fig. 52). Firing the fragile crusts, for example in a small open-air fire, results in their consolidation into fired dishlike pieces that can only with difficulty be broken by hand. The physical properties of the fired crusts, such as their hardness, toughness, and porosity, are similar to those of simple pottery.

It is possible, therefore, that early humans may have accidentally made this type of dish when making fire to warm themselves or for cooking. Could such accidentally fired objects have given prehistoric humans the idea of modeling clay by hand and then firing it into pottery? Any such hypothesis

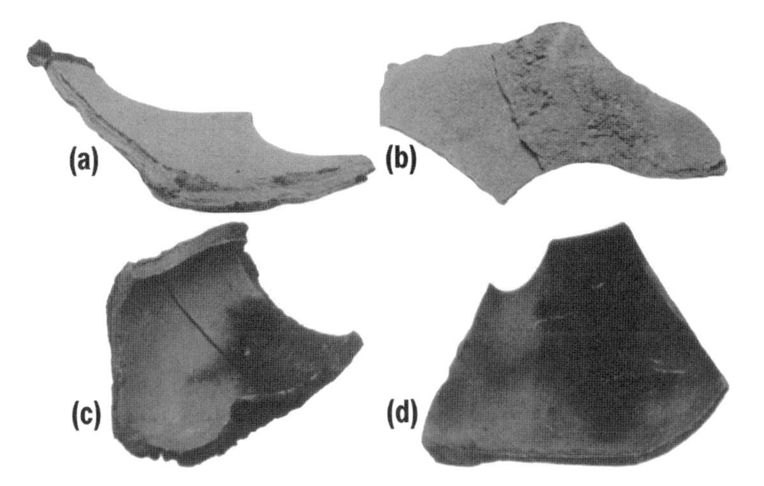

FIGURE 52 Preceramics. Cracked loess crusts after firing in an open fire, found in a ravine: **(a)** and **(c)** inside unfired; **(b)** and **(d)** outside unfired.

is impossible to prove. There is a way, however, to distinguish between such accidentally formed "preceramic" dishes and intentionally made ones. The outer surface of human-made pottery is always relatively smooth, while that of naturally formed soil crusts is normally coarse. If ceramic dishes with coarse outer surfaces were to be found in ancient habitation sites, or if food or other organic residues were to be detected in their inner surface, they would provide evidence of "preceramic pottery" having been used by humans (de Korosy 1975).

Components of Ceramic Materials

Clay is the essential, although not the sole, constituent of a *clay paste* (or *clay body*), as the mixture of raw materials used for making pottery is known. The other essential component of most wet clay mixtures are the *nonclay materials*, which are also known by a variety of other names, such as *fillers, nonplastic fillers, nonplastic materials, inclusions, tempers,* and *additives*; these are added to the clay so as to obtain a mixture with good working and drying properties, which on firing acquires toughness and strength. A vast range of materials, differing in their composition, their effect on pliability, the drying and firing properties of the clay mixture, and the coherence and strength of finished objects made from it, have been used as fillers. A short list includes such varied materials as sand, volcanic ash, crushed rocks, shards and seashells, twigs, straw, and dung (Bronitsky and Hamer 1986).

The method used in antiquity to shape a clay paste into objects can often be established by observing the orientation of the particles of filler within finished pottery: in ring- or coil-shaped vessels, for example, the particles are randomly oriented, whereas in "wheel-thrown pottery" they are oriented in a particular direction, the long axes of the particles lying parallel to the turning plane of the wheel on which the vessels were turned. Some fillers, such as sand and crushed igneous stone, are stable to heat changes and remain unaltered during firing. Others, however, undergo radical changes when heated to high temperatures; organic matter, for example, is oxidized and partly or entirely removed from fired pottery and sedimentary stone or shell (which are composed of calcium and/or magnesium carbonate) are *calcinated* (see Textbox 34).

7.5. MAKING CERAMICS

The generic name used to refer to ceramic objects shaped from a wet mixture of clay and fillers that is then dried and subsequently fired at high temperatures is *pottery*. Making *pottery* involves a number of working stages:

1. Procuring and handling the clay
2. Procuring and comminuting the filler
3. Mixing the clay, fillers, and water to prepare the clay paste
4. Forming and shaping the clay paste into objects
5. Drying the objects
6. Firing; baking and burning

All except stage 4, forming and shaping, which involves mainly stylistic considerations, entail chemical and physical changes to the components as well as to the entire clay paste used for making pottery. All the stages except 4 are, therefore, discussed in some detail in the paragraphs that follow.

Raw clay is excavated from the ground, and it usually requires the removal of foreign, nonclay material before use. This is often followed by grinding, so as to obtain a fine-particle clay, suitable for preparing a smooth and uniform clay paste. The powdered clay is then mixed with water, and the mixture is kneaded into a paste of uniform consistency. *Tempering,* as the addition of fillers to the clay paste is generally known, is done at this stage. Adding some more water or slightly drying the clay paste renders it of the appropriate consistency for shaping objects. Still, the final composition of the clay paste varies, depending on the nature of the local raw materials, the intended use of the finished wares, and the ceramic tradition of the potter. The plastic clay paste is then shaped into objects and, while still wet, the surface of each object is usually smoothed so as to improve its appearance and fill up pores with fine particles of clay (the latter operation makes the objects impervious to liquids) (Bronitsky 1989).

Drying clay objects after shaping is a relatively complicated operation. During drying, *water of plasticity,* the water that makes the clay paste plastic and pliable, evaporates; the particles of clay are then drawn close together and the objects shrink in size (see Fig. 51). If, during the drying process, wet objects are exposed to the heat of the sun, for example, the evaporation of water from the outer surface is faster than the rate of flow of the water from the interior to the surface. This gives rise to *differential shrinkage,* whereby the outer layers of the drying clay paste shrink faster than does the interior. As a result of differential shrinkage, mechanical stress develops between the drier outer layers and the wet interior of the clay paste; if such stress becomes strong, it may distort drying objects and cause them to crack, even shatter. One of the more important effects of adding nonplastic fillers to clay is to obviate differential shrinkage; the coarse particles of the filler occupy space that would otherwise house wet clay particles, thus reducing the amount of water required for the preparation of the clay paste as well as the amount of water evaporating during the drying process.

Slipping, the optional application of a *slip*, a thin mixture of clay and water to the surface of clay objects, is done after drying. The slip further smooths the surface and makes it relatively impervious to liquids. It also makes an excellent substrate for painting and decorating; a pigment added to it changes the color of the surface to which it is applied.

Firing pottery, the process of heating to high temperature a shaped and air-dried clay object, is a critical stage in the process of making pottery. It is during this stage that the clay mixture acquires the hardness, toughness, and durability characteristic of finished ceramic objects. At the initial stage of the firing process, when heated from room temperature to 100°C, the clay paste loses water of plasticity and dries up. As the temperature is gradually increased above 100°C, three successive chemical processes follow: *oxidation*, *dehydration*, and *incipient vitrification*. Between 100°C and about 400°C, provided there is an abundant supply of air, any organic matter in the clay paste, such as straw, twigs, dung, or feathers – an integral part of most ancient clay pastes – is *oxidized*: the carbon in the organic matter combines with oxygen in the air to form carbon dioxide, which is released into the atmosphere. As a consequence of the oxidation, the space that was occupied by organic matter in the clay mixture is left empty and the ceramic mass becomes porous [if the heating takes place under *reducing* conditions, in which there is insufficient oxygen to oxidize the organic matter, however, the latter is *charred* (burned to charcoal) and remains within the pores of the finished pottery as a black residue; thus all pottery fired under reducing conditions is either black or gray]. The temperature and time required for the full oxidation of the organic matter may vary greatly, depending on the actual composition of the clay paste. In general, however, firing under oxidizing conditions is timed so as to allow the oxidation of all organic matter to proceed to completion before the following stage, dehydration of the clay, begins. In this way neither charred organic matter nor carbon particles are trapped within the fired clay.

During the *dehydration* stage (between 450°C and 600°C), hydroxyl (OH^-) ions in the clay are dislodged from their molecules, combine with each other to form water vapor, and are thus removed from the clay structure and released into the atmosphere. It is during this stage, as a consequence of the displacement of the hydroxyl ions, that the chemical composition and the structure of the clay are irreversibly altered and converted to fired clay.

The next stage, *baking*, begins at about 600°C, when small amounts of flux within the clay mixture melt and induce incipient melting of the surrounding clay, which is therefore converted to glass (see Chapter 3). The melting, or *vitrification* process, starts at isolated sites within the clay mixture where there is flux, and it expands and intensifies as the temperature increases. Sufficient vitrification takes place between 600°C and 700°C to cement the clay particles together and give the baked clay mixture, when cooled, a moderate degree of hardness while still remaining highly porous. Above 700°C the clay can be considered fired, and the firing of many types of simple pottery ends at this stage.

The *burning* stage, as the high-temperature firing stage is often referred to, begins above 850°C; as a result of expanding vitrification, the baked clay becomes progressively less porous, more compact, and, after cooling, extremely strong and tough. At still higher temperatures, over 1000°C, the vitrification process expands still further (above about 1300°C, temperatures generally out of reach in ancient times, total melting may occur whereby objects may coalesce into a mass of molten glass).

Following the firing stage, at whatever temperature it is heated, baked or burned pottery is usually kept for some time at the highest temperature reached and only then gradually cooled down. If allowed to cool too quickly, the pottery may crack or fracture, since rapid changes in temperature cause too great a thermal shock for fired clay to stand. A good, practical way of checking whether pottery has only been baked, that is, has been heated below 800°C and therefore undergone only incipient vitrification, or has been burned above 850°C and became extensively vitrified, is to strike two vessels together. If, when stuck, a dull sound is emitted, the pottery has only been baked; the emission of a "good ring" provides evidence that the pottery has been burned. Fired pottery, regardless of whether baked or burned, is stable to most chemicals and is attacked by neither acids nor alkalies; it is also strong, tough, mechanically resistant to impact and abrasion, thermally stable at high temperatures, and resistant to thermal shock.

Pottery Kilns

Pottery kilns, the structures in which clay is heated to change it into a ceramic material, have been made in a wide variety of types: from the earliest open bonfires in which clay objects were fired in prehistoric times, to the efficient structures used today. Most pottery has been fired in *periodic kilns*, in which a thermal cycle is conducted for a number of objects produced in a single operation. An entire thermal cycle involves heating, holding at peak temperature, cooling, and removal of the ceramic objects. First, air-dried objects and fuel (in the past mainly wood, charcoal, or dung) are loaded into the kiln and the fuel is ignited. The temperature of the kiln is then raised at a controlled rate until it reaches a maximum temperature, at which the fired objects are kept for a set length of time. Finally, the kiln and its contents are gradually cooled until they reach room temperature, and the now fired pottery is removed from the kiln (Rice 1997).

The design of periodic kilns varies widely. In most, known as *open-flame kilns*, the heat is used directly, that is, the firebox of the kiln (the place where the fuel is ignited and burned) is usually below the objects being fired. The flames surround the objects and the gases of combustion (mainly water vapor and carbon dioxide) evolved as a consequence of the burning of organic matter, move upward by natural convection (see Fig. 53). The temperature distribution in open-flame kilns is poor; the lower part of the kiln, nearest to the burning fuel, is hotter than the top, and the flow of the combustion gases causes horizontal temperature gradients; this results in irregular firing, some objects reaching much higher temperatures than others.

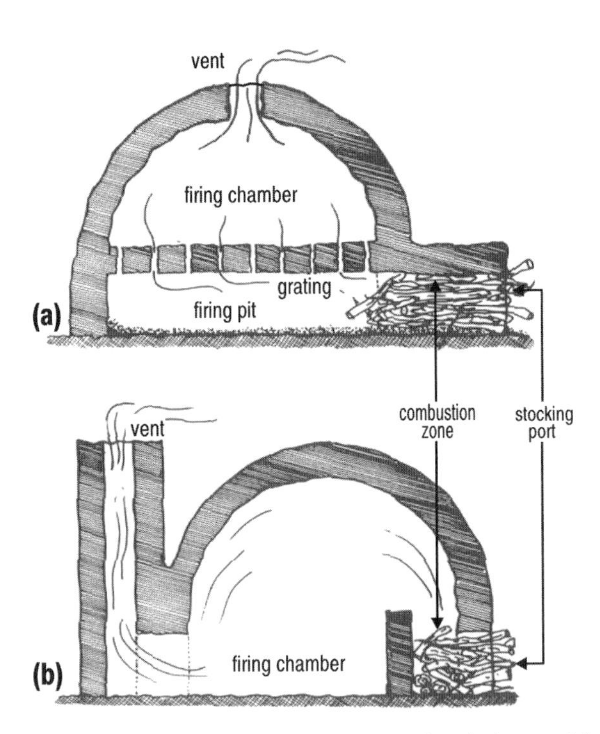

FIGURE 53 Pottery kilns. Kilns, heated enclosures for drying and hardening clay objects, are required to reach very high temperatures; their design is therefore focused mainly on insulation, so as to prevent the loss of heat and the need to add fuel so as to keep high firing temperatures. The illustration shows a cross section through two wooden fired ceramic kilns: an updrought kiln **(a)** and a downdrought kiln **(b)**.

The Color of Fired Pottery

Primary clay, for example kaolin, is colorless, and when such clay is heated to a high temperature it produces white ceramic materials. Most pottery, however, is colored: its color is due to the fact that most of it was, and still is, made not from primary but from secondary clay. Secondary clay contains minerals other than clay, and colored metal ions in them endow the pottery with their color. Iron ions (in iron oxides), for example, tend to make pottery yellow, brown, or red, and manganese ions (in *pyrolusite*, a mineral composed of manganese oxide) make it either dark or black.

The atmosphere in the kiln, as well as the temperature at which pottery is fired, also affects its final color. A draft in the kiln ensures an abundant supply of oxygen and facilitates the oxidation of the metals. When pottery containing iron, for example, is fired under oxidizing conditions, the iron is gradually oxidized and *baked pottery* (fired below 800°C) acquires a buff color; but if it the pottery is *burned* (fired above 850°C), more oxidized iron ions generally cause it to be yellow or red. If the kiln's draft is shut off, however, the atmosphere in the kiln becomes filled with reducing gases (such as hydrocarbons, released by wood in the early stages of the combustion, and carbon monoxide, formed when organic matter is heated in an atmosphere lacking in oxygen). These gases prevent the oxidation of the organic matter additives in the clay paste, which is therefore converted into carbon particles. As a consequence, and depending on the amount of organic matter in the clay paste, pottery fired in a reducing atmosphere is either gray (when there is little organic matter) or totally blackened by the deposition of *soot* (unburned carbon) particles within the fabric of the fired clay. Often, black pottery is made by coating the walls of dry clay objects, before firing, with a layer oil or bitumen. The oil or bitumen initially permeates into the walls of the objects; when these are then fired in a kiln with a reducing atmosphere, the oil or bitumen is charred, forming extremely small particles of *soot* that impart to the surface of fired objects a gray or black finish.

Glazing

The surface of fired pottery is generally porous and readily absorbs moisture. To conceal outside imperfections, decorate the surface, and make the pottery more or less impervious to water and other liquids, many ceramic objects are *glazed*, that is, coated with a thin layer of *glaze* (Tite 2004). The nature of glaze and the glazing process are discussed in Chapter 3.

7.6. COMMON CERAMIC MATERIALS

The physical properties most often used for the characterization of ceramic materials are *porosity, translucency, strength,* and *color. Porosity,* the relative amount of voids within a material, is generally measured as the amount of water that the material absorbs and is expressed as a percentage of the total volume taken up by the voids. The porosity of ceramic materials is determined mainly by the nature of the raw materials used for making them, and the degree of vitrification the objects underwent when fired. The lower the porosity, the higher the quality of a ceramic material, that is, the tougher, stronger, and more resistant it is to mechanical and thermal shock. *Translucency* (the opposite of opacity), the property of materials to transmit light, is related to porosity: the less porous a material, the more translucent it is. Since ceramic materials become less porous during the vitrification stage of firing, their translucence is related to the degree of vitrification; highly vitrified ceramics has no, or little, porosity and is therefore translucent; slightly-vitrified ceramics, on the other hand, is porous and consequently opaque. Using these two parameters, porosity and translucency, ceramic materials can be technically classified within four basic types: *terracotta, earthenware, stoneware,* and *porcelain,* as listed in Table 59; within each basic type there are, however, many variations (Litchfield 1967).

TABLE 59 Ceramics Materials

Material	Firing temperature (°C)	Porosity (%)	Optical characteristics
Terracotta	below 900	greater than 25	Opaque
Earthenware	above 950	up to 10	Opaque
Stoneware	1100–1300	up to 5	Opaque
Porcelain	1300–1450	below 2	Translucent

Terracotta

The simplest and coarsest type of pottery is a lightweight, very porous, and typically red-colored *terracotta* that is fired at temperatures below 850°C. Much of ancient pottery, for example, is of the terracotta type. Excavations in the Near East have revealed that primitive terracotta vessels were being made there more than 8000 years ago; characteristic types of terracotta were developed in China by about 5000 B.C.

Because of the relatively low temperatures at which it is fired, terracotta is only slightly vitrified and consequently is highly porous and totally opaque. The pores occupy over 25% of its total volume, which makes terracotta not only opaque but also relatively soft: it is easily scratched with a hard steel point. The color of most terracotta fired under oxidizing conditions is either red, brown, or yellow; the intensity of the color is dependent on the relative amount of iron oxides in the secondary clay used to make it. If fired under reducing conditions, however, it includes variable amounts of carbon particles and is either gray or black. Most terracotta is not glazed.

A special type of objects made of a very simple and crude type of terracotta are *fired bricks*, used for building and road construction (see Fig. 54 and Chapter 4).The discovery of clay firing was a major technological advance in the development of building materials, since the firing process converts unstable and vulnerable sun-dried mud to hard, mechanically strong, and chemically stable fired clay. Because of the relatively large amounts of iron oxide in the secondary clays used to make brick, fired bricks are red. The distribution of the heat during the brick-firing process is usually uneven, and bricks that are closest to the fire and at higher temperatures are often bright red and sometimes even develop a vitrified surface; those distant from the

FIGURE 54 Burned brick. Sealed Roman burned brick, first–third centuries C.E., Jerusalem, Israel. It seems that the method of making strong and durable bricks by firing blocks of a mixture of clay, various additives, and water was discovered in Mesopotamia; the earliest known burned bricks, dating from about 6000 years ago, were found in Babylonian sites. The technology of burning bricks then probably found its way to the east and west: through Persia and northern India, to China, and through the Middle East to Europe. Sections of the Great Wall of China were built (during the third century B.C.E.) partly of burned brick. The Romans greatly advanced the technology of firing bricks, taking great care in the selection and preparation of the clay mixture used for making the bricks and improving the design of the kilns used for firing them.

fire are usually of a lighter and duller color. There is evidence on the manufacture and use of fired bricks as early as the sixth millennium B.C.E. in Mesopotamia, the Middle East, and Egypt, and its later spread to Rome, India, and China.

Earthenware

When the same ceramic paste that is used for making terracotta is fired at higher temperatures (above 950°C), the material obtained is known as *earthenware* (see Fig. 55). *Earthenware* is more vitrified and therefore less porous and stronger than terracotta, although it is also opaque. Its porosity generally varies within the range 5–10%. Earthenware is often glazed.

Two particular types of archaeological earthenware, among many others, are *majolica* and *faience*. During the Middle Ages the terms *majolica*

FIGURE 55 Earthenware. Large earthenware vessels from the sixth–fourth centuries B.C.E., recovered from under the Mediterranean Sea, at Caesarea Maritima, Israel. Earthenware one of the simplest types of ceramic material, is highly porous and permeable. It is made from clay and a variety of additives fired at about 950°C. Iron oxides impurities in the clay usually make earthenware buff, red, or brown. Most earthenware, like the vessels shown, was not glazed but, if required, sometimes was waterproofed or decorated with a layer of glaze.

and *faience* were used to refer to glazed earthenware made in the Spanish island of Majorca and Faenza in Italy, respectively. At present, however, the names refer to two different varieties of relatively coarse earthenware made in many places, which is generally red or gray, has a porosity of about 15%, and is therefore mostly opaque. To add to the confusion, in archaeology the term *faience* is also associated with a material used mainly in Egypt: *Egyptian faience* (see text below) which is made not from clay but from quartz.

Stoneware

Stoneware is the term commonly used to refer to a very strong ceramic material that is fired at a high temperature (1200°C–1300°C, the highest temperature attainable in most ancient kilns). The high temperature causes quite extensive vitrification and relatively low porosity (generally below 5%), although not sufficiently low to make it translucent. Stoneware is therefore opaque to visible light. Stoneware objects may be of a variety of colors, ranging from gray through cream and brown to red. Its surface may be either glazed or unglazed. Stoneware was probably first made in China during the seventh century C.E. In Europe, distinctive stoneware objects seem to have been made only since the late fourteenth century C.E., initially in Flanders (Belgium) and Germany.

Porcelain

Most if not all ancient terracotta and earthenware, and some stoneware, are made from secondary clay. *Porcelain*, the strongest ceramic material known until the twentieth century has always been made from primary clay, kaolin, also known as *China clay*. The clay paste used for making porcelain consists of a mixture of kaolin and a type of feldspar known in China as *penutse* and in the West as *China stone*; formed porcelain objects are fired at very high temperatures, about 1300°C and higher. Some Chinese porcelain includes also a third mineral such as mica, which varies in different geographic regions and yields characteristic local varieties of porcelain. The extremely high firing temperature results in a high degree of vitrification and consequently a very low porosity (below 2%), which makes porcelain highly translucent. Since the kaolin and feldspar used contain few impurities, porcelain is white. When struck, porcelain produces a prolonged sound (Maneta 1997; Yap and Hua 1994, 1992).

It seems that some types of "protoporcelain" were being made as early as the second millennium B.C.E. The widescale production of porcelain seems to have begun, however, sometime during the seventh century C.E. in China. For many centuries it was produced exclusively by the Chinese, but by the end of the seventh century porcelain was already being exported to the Islamic world, and by the twelfth also to Europe. During the eleventh century porcelain-making spread to Korea, during the sixteenth to Japan, and only during the seventeenth century did the European production of porcelain begin, apparently in Florence, Italy.

7.7. STUDY OF ANCIENT POTTERY

Pottery is very durable, and its use has been widespread for a long time. After breakage, however, pottery fragments (*shards*) totally lack value, and this has resulted in shards being by far the most abundant of all human-made material remains from former cultures. Much archaeological research is therefore based on the study of shards and of ancient pottery reconstructed from retrieved shards. Several systems for classifying ancient pottery, based mostly on stylistic guidelines, are widely used in archaeological studies.

Scientific interest in ancient pottery centers not on stylistic considerations but mainly on its composition, physical properties, and the manufacturing technology. During the nineteenth century isolated shards and specimens of ancient pottery were being analyzed to learn about their composition and properties, but the systematic investigation of pottery was first attempted only by the middle of the twentieth century. Most of the early chemical work was concerned with identifying the main components of pottery (fired clay and tempers). The composition of shards is, however, quite similar throughout the world. Although the relative amounts of the major components of pottery from different regions may differ, these differences are generally of no consequence regarding their origin; often, similar and even greater compositional differences can be detected in pottery from a single region. It was only during the second half of the twentieth century, when modern sensitive analytical techniques were developed, that the systematic study of the composition of ancient pottery came into its own. The availability of rapid and discerning analytical techniques, such as spectroscopy and nuclear activation, have promoted investigations on the nature of the ancient raw materials used for making pottery, pottery manufacturing technologies, the relationship between the composition of pottery and the intended use of the objects made from it, and the provenance of ancient pottery and of the raw material used for making it (Mommsen 2001; Hackens and Schvoerer 1984).

Attic Vase Painting

The technique used in ancient Greece to produce characteristic "black-and-red glaze" decorative patterns on Attic pottery (see Fig. 56) has long intrigued scholars; it was only with the development of scientific techniques, however, that progress was made so as to understand and reproduce the technique. In some ways the term *glaze*, used to refer to the surface layer on Attic pottery, is a misnomer, since a glaze is a thin layer of glass applied to the surface of ceramic objects (see Chapter 3). The outer red-and-black layer of Attic pottery is not made up of a vitreous material and is not, therefore, a glaze; neverthe-

FIGURE 56 Attic lekythos. Attic black figure lekythos (container for storing perfume, oil, or ointment, often placed in graves) said to be from Agrigento in Sicily. Attic pottery is unique and probably the finest type of decorated Greek pottery. It was initially made toward the end of the seventh century B.C.E. and its production peaked between the sixth and fourth centuries B.C.E. There are several types of Attic decorated pottery, including "white ground," "red figure," and "black figure;" the latter type is shown in the illustration. In the *black figure* type, the figures on the surface appear in black against a red background. Sometimes red ochre was applied to the red areas so as to intensify the red. Attic pottery was not glazed. Still, the black-decorated portions are often said to be covered with a black "glaze," although the term *glaze*, in this case, is a misnomer: the black surface areas are not covered with a glaze but were created by the application of a very fine-textured clay slip, much finer than on the rest of the decorated surface, which was then fired in a controlled atmosphere.

less, since the terms *Greek glaze* and *black-and-red glaze* are generally used in the relevant literature, it seems appropriate to retain them here.

Attic glaze is not a coating applied to the surface but an integral part of the clay from which the pottery objects was made. Spectroscopic analysis reveals that there is no chemical element in black-and-red glaze and that it is not a component of the underlying clay mixture; that is, the components of Attic glaze are identical to those of the underlying ceramic material. It seems that to produce the characteristic black-and-red surface coloration, advantage was taken of the different colors exhibited by the ions of iron, which is a component of practically all secondary clays used to make Attic pottery: *oxidized iron* (*ferric*) *ions* are red, but they are black in the *reduced* (*ferrous*) *ion* condition. When pottery is fired in a well-ventilated kiln, under oxidizing conditions, with an abundant supply of oxygen, the iron in the clay is oxidized to the ferric (red) condition and the fired pottery turns red; if the ventilation is restricted during firing, however, and there is practically no oxygen in the kiln's atmosphere, the iron is reduced to black ferrous ion and the pottery is either dark brown or black. Moreover, under reducing conditions the burning fuel produces large amounts of soot which deposites on the surface and within the pores of the pottery being fired, and contributes to intensifying the black.

Were the entire firing process done under oxidizing conditions, both the body and the surface of Attic fired pottery would be red or, under reducing conditions, black. Making black-and-red Attic pottery seems to have required, therefore, a rather elaborate sequence of operations, starting from the selection of the clay and ending with the firing of the vessels in a cycle of oxidizing, reducing, and again oxidizing conditions.

The following is a probable sequence of operations that was successfully tried experimentally (Noll et al. 1975; Hoffmann 1962). First, the carefully selected clay intended for making Attic pottery was separated into two fractions of coarser and finer particles, probably by a process known as *elutriation*; this entails subjecting the clay particles to an upward flow of water so as to create a suspension of the particles in the water. When the upward flow of water is terminated and the suspended particles are allowed to settle, and they do so at different rates: coarse particles settle rather quickly while the finer ones settle much more slowly. The coarse particles were probably discarded, while the slow-settling fraction of fine particles was used for making the vessels. A third fraction of the clay, containing extremely small particles that did not settle during the elutriation process but remained suspended in water, were probably used as the *slip*. This fraction of very fine particle seems also to have then been *peptized* (turned into a *colloid*, a mixture of two or more substances one of which, the *peptized clay* in this case, is thoroughly dispersed through the other, water); for this purpose this fraction was further dispersed in water to which *tannins* (see Textbox 62) were added: the tannins caused the formation of a *colloidal dispersion* of the peptized clay particles in the water, and the slip become colloidal. The properties of the slip were further modified by the addition of an *alkali*, probably derived from *wood ash*. When the pottery was latter fired, the alkali acted as a *flux*, lowering the melting point of the slip particles and altering their *sintering properties* (the *sintering properties* of particles are related to the way in which they become bonded to each other during the firing process (see Textbox 29). The carefully prepared, peptized, and alkalinized slip was then applied only over those areas of the air-dried objects intended to be black, as a very thin layer that was burnished when dry.

The dry, burnished objects were then fired in three successive stages: the first stage was performed under oxidizing conditions, that is, with an abundant supply of oxygen, at about 850 C; this resulted in the oxidation of all the iron to red *ferric ions*, and the body as well as the slip-coated areas turned red. The second firing stage was performed under reducing conditions to develop the black color: fresh, wet wood, rich in resins, was added to the regular fuel while the air vents of the kiln were shut, and water was probably continuously poured on the burning fuel to prevent the access of oxygen. Burning the fuel under such conditions produced hydrogen and carbon monoxide (both reducing gases), which reduced the red ferric ions in the body of the objects as well as in the slip to black ferrous ions; firing under reducing conditions was continued for a short while (probably 5–10 minutes) until the now black slip sintered and became impervious to gases and resistant to oxidation. The result of this second firing stage was, therefore, black porous ceramic objects, certain surface areas of which were covered with an also black and sintered slip, impervious to oxygen and resistant to chemical changes. During the third and last firing stage, the air supply to the kiln was restored, thus reestablishing oxidation conditions, and the temperature was raised, probably to nearly 900°C; areas covered with the sintered slip resisted oxidation and remained black, but the underlying body and those surface areas not covered with the slip were again oxidized and once again became red. The kiln and the now fired objects, having a red body and a black-and-red surface, could be finally cooled in an oxidizing atmosphere. Examination of Attic pottery's glazed surfaces under an electron microscope reveals that the red surface areas are, indeed, rough, while the black areas are smooth; during the last firing stage, oxygen penetrated the rough, porous surface and oxidized the black ferrous ions in the body as well as on the surface of the objects back to the red ferric condition; the burnished, sintered slip layer, impervious to oxidation, remained black.

Coral Red Attic Pottery

Another type of Attic pottery has not black but bright coral red decorations that appear to be painted on. This type of pottery is usually known as *intentional red, sealing-wax* or *coral red Attic pottery*. The coral red hue seems to have been obtained by adding *ocher*, a natural earth pigment (see Chapter 1), to the slip, which was probably prepared using the elutriation and peptization techniques described above. The amount of ocher added varied, depending on the nature of the clay used for preparation of the slip. Analysis of different locations on the surface of coral red Attic pottery seems to indicate that enough ocher was added to the slip to increase its iron content to about 11%, several times more than the underlying pottery. Following a three-stage firing cycle that involved a sequence of oxidation, reduction, and again oxidation conditions, similar to that described for Attic glaze pottery, the color of the slip turned into an intense coral red (Farnsworth and Wisely 1958).

Manganese Black Decoration

The use of a slip made from clay to which *pyrolusite*, a mineral (composed of manganese dioxide) was added, made possible the production of another black-and-red type of pottery, which closely resembles Attic pottery, although it involved a much simpler firing process. Surface areas intended to be black were covered with a pyrolusite rich slip, and the pottery was then fired in only one firing stage, under oxidizing conditions: above 900°C the slip turned intensely black while the remaining surface areas turned red. Because of the simplicity of the firing technology required to produce this type of black decoration, the use of pyrolusite-rich slip seems to have become quite popular in Greece, eventually being used whenever this mineral was available (Noll et al. 1975, 1973).

Islamic Stone Paste

Islamic stone paste, also known as *frit ware* and *quartz frit*, is a characteristic type of pottery that seems to have been first made in Iraq during the ninth century C.E. Between the tenth and twelfth centuries it was made mainly in Egypt, and since then, and until the present time, Islamic paste has been made in most of the Middle East.

Analytical studies, using *scanning electron microscopy* and *electron microprobe* techniques (see Textbox 10), revealed that the mixture used for making this particular type of ceramic material consists mainly of clay, comminuted quartz, or other siliceous mineral, and *glass frit* (see Textbox 20). It seems that while mixing the components together, some type of organic binder, such as a *gum* or *glue*, was added so as to help keep the components of the mixture together while the clay paste was formed into objects. When the pottery was fired, incipient melting of the clay and the glass frit cemented the grains of quartz or siliceous mineral. A glaze, which was fired at a higher temperature than the bulk stone paste, consolidated the finished objects and made them tough and stable (Molera et al. 2001; Mason and Tite 1994).

Egyptian Faience

Egyptian faience (see Fig. 57) is a refractory material that has been used for many millennia in the Middle East for making small decorative objects. Beads, figurines, and rings made of faience were generally, although not exclusively, covered with a blue glaze. Objects made of Egyptian faience first occurred as early as over 5000 years ago and are still made at the present time. Egyptian faience is not made from clay and therefore is neither a type of pottery nor a ceramic material (see Textbox 49). It certainly is, however, a

FIGURE 57 Egyptian faience. Egyptian faience cameo ninth–seventh centuries from Tel Megido, Israel. Egyptian faience is a pseudoceramic material that was used to make beads and small decorative objects. The raw materials of Egyptian faience do not include clay but powdered silica as the main component, and small amounts of soda and lime. The nonsilica components, soda and lime, serve as a flux that, when shaped objects are fired, causes incipient melting of the silica grains; the grains are cemented together and also form an outer glaze layer that surrounds and strengthens the finished objects. The most common color of Egyptian faience is bright blue or green, due to small amounts of copper minerals in the mixture, but other colors include white, hues of yellow, brown, and red.

refractory material that is claimed to have been the "first high-tech refractory". It is worth mentioning here that Egyptian faience should not be confused either with true *faience,* a type of earthenware (see above), or with *Egyptian blue,* an artificial blue pigment also made by the Egyptians, discussed in Chapter 1 (Nicholson and Peltenburg 2000; Friedman et al. 1998).

The raw materials from which the bulk of Egyptian faience is made are small-particle quartz (see Chapter 2) (about 90%), lime (1–3%) (see Chapter 4), and smaller amounts of feldspar, soda, and alumina. See composition listed in Table 60. Making Egyptian faience objects entails shaping or molding a wet mixture of the raw materials, and then drying and firing the mixture at a high temperature (about 900°C). After cooling, the fired bulk is coated with an alkaline glaze in which there is generally some coloring mineral. Copper minerals, for example, endow the glaze with a characteristic blue; iron minerals make it red or brown; and lead antimoniate, yellow. Since the bulk of Egyptian faience includes almost no consolidating components, unglazed Egyptian faience objects are very fragile, as their relative strength is due mainly to the glaze coating. As soon as the very hard glaze is removed, the material crumbles and is easily powdered (Clark and Gibbs 1997; Kaczmarczyk and Hedges 1983). It has been suggested, that in some cases the bulk glazed itself during firing. If so, it is possible that before firing, a flux was applied to the surface so as to induce the self-glazing process.

Experimental refiring of ancient Egyptian faience has shown that the original firing was carried out at temperatures between 890°C–920°C. Under microscopic examination the bulk of Egyptian faience is seen to be made up of grains of quartz partly consolidated by incipient melting. X-rays diffraction analysis (see Textbox 10) provides evidence that different varieties of powdered quartz were used to make it. The glaze is usually either of the sodalime or potassium lime type, both usually referred to by the generic term *alkaline glazes* (Tite and Bimson 1986; Lucas 1962).

TABLE 60 Composition of Egyptian Faience

	Composition (%)						
Component	Silica	Alumina	Lime	Magnesia	Alkali oxides	Iron oxide	Copper oxide
Body material	80–98	15–0.5	0.5–0.3	1.5–0.5	2– 0.4	1–0.1	–
Glaze	80–85	1–0.5	5 –3	1 –0.6	10–15	1–0.5	2–1

Firing Conditions of Ancient Pottery

Assessing the conditions at which ancient pottery was fired has attracted some attention, and a number of physical techniques have been used to estimate the maximum temperature reached by ancient pottery during firing. If shards are fired under controlled conditions, for example, changes in their color or in other properties can often be correlated with temperatures exceeding that at which the pottery from which the shards are derived was originally fired. Other techniques for estimating ancient firing temperatures are listed in Table 61.

The Provenance of Pottery

Given a series of pottery samples, such as shards that have a number of chemical elements in common, a system of sample groups can be defined on the basis of style or of differences in the concentration of some elements in their composition. Shards sharing a common stylistic origin, for example, may be selected, analyzed, and then correlations established between the concentrations of various elements present in them. In this way different types of trace element composition can be established, each one characteristic of a distinct area or period of time. Single samples can then be fitted into one or another sample group with greater or lesser certainty, depending on the extent to which the results of their analysis deviate from the mean of the group. The nature of the groups that can be expected is illustrated in Figure 58, a model that summarizes the predictable differences for a single element.

TABLE 61 Techniques Used for Studying the Firing Conditions of Ancient Pottery

Technique	References
Elastic properties, ultrasonic evaluation	Heimann and Franklin (1981)
Electronic properties: radial electron density	Heimann (1982)
Magnetic properties: remnant magnetization, coercive force	Yelon et al. (1992); Cohey et al. (1979); Barbetti (1976)
Morphology	Rice (1987); Tite and Maniatis (1975)
Mössbauer spectroscopy	Feathers et al. (1998); Bakas et al. (1980)
Electron spin resonance spectroscopy	Warashina et al. (1981)
Colorimetry	Matson (1981)
Thermoluminescence	Sunta and David (1981)
Thermal analysis: differential thermal analysis, dilatometry	Kingery and Frierman (1974)

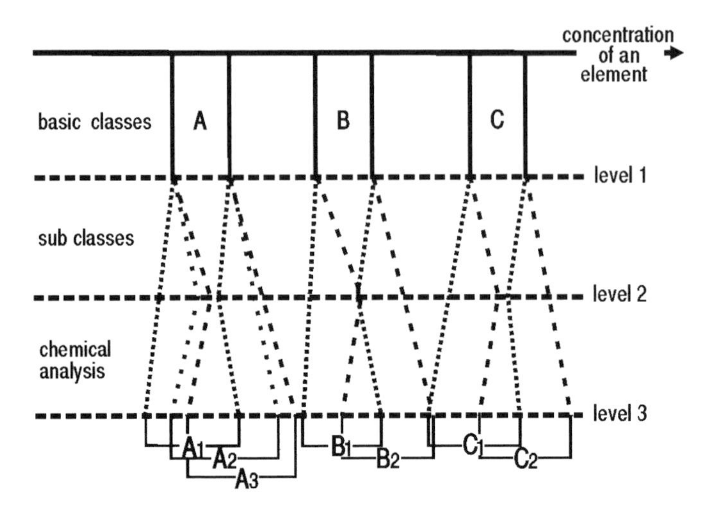

FIGURE 58 The provenance of pottery. The provenance of pottery can often be determined by (1) measuring the abundance of trace and minor elements in pottery fragment and then, (2) comparing their relative abundance with that of elements in raw clay from specific sites or in artifacts made at known pottery workshops. The analytical methods of choice for measuring elemental abundance are neutron activation and X-ray fluorescence. The diagram illustrates the predictable differences of pottery classes when studying a single element, as well as the lack of precision inherent in classifying pottery on the basis of the analysis of a single element.

The model is arranged in three levels: in the first level are depicted the ranges of concentration of an element in different clay beds. The existence of considerable differences in trace element concentration between different clay beds is well established, although at present little is known about the variation of trace element concentration within a single clay bed. The composition of clay from a singular source, such as a particular clay bed, although homogeneous in composition, can differ considerably from the composition of the pottery made from it. This may be due to a number of factors, such as refining of the clay by the potter during preparation of the clay mixture; the addition of tempers, the removal of coarse material, and even the addition of clay from other sources, for example, may all heighten or decrease the concentration of particular constituents of the clay.

As a consequence of these operations, the original concentration ranges of the elements tend to separate into a number of subgroups, as illustrated in the second level in Figure 58. It follows that there is little point in trying to establish a relationship between a given shard and any particular source of clay. Instead, each shard should be assigned to an identified group, with the objective of establishing compositional uniformity within groups.

The combined use of sensitive analytical technique for studying pottery samples, and statistical methods for scrutinizing the analytical results and discriminating between samples of different composition generally provides valuable information for establishing compositional groups and identifying the provenance of the pottery. The requirements from any particular analytical technique to be used for the study of provenance are as follows:

1. The technique should make possible the detection of a large number of elements; the more elements that are determined, the greater the degree of certainty in identifying the source of the pottery.
2. It should be sensitive to detecting elements that may occur in very low, even in trace amounts.
3. It should be technically and economically viable to analyze a large number of samples.

Given these requirements, it emerges that a suitable analytical technique for studying provenance should provide relatively rapid results and preferably be nondestructive, enabling determination of each element, and differentiation among a large number of elements in relatively short periods of time. Techniques that fulfill these conditions for studying the provenance of pottery include several spectroscopic techniques, neutron activation and X-rays fluorescence (see Textbox 10).

Optical emission spectroscopy was used in an early investigation to study the provenance of Greek Minoan and Mycenaean pottery, which was widely used not only in their lands of origin but also in more or less distant areas, from Asia minor and Egypt to Sicily and Italy. The results of these studies made it possible to distinguish between Minoan and Mycenaean pottery and also between other types of Greek pottery, for example, from the Peloponnese and Crete. Moreover, the study also provided information on Mycenaean pottery that was produced in the Peloponnese and then exported to the Greek Islands and to Cyprus (Millett and Cattling 1967; Cattling et al. 1961). The analytical similarities and differences between the pottery groups, as established in this study, were later confirmed, when samples from the same groups were analyzed again, using a different analytical technique, namely, neutron activation analysis (see Textbox 10). Moreover, the neutron activation study provided a useful complement to the emission spectroscopy study, both as a check on the latter and as a means of extending the range of determinable elements.

Several thousand shards were analyzed in a wide-ranging study intended to evaluate the accuracy obtainable when using *neutron activation analysis* to establish the provenance of pottery. After determining the relative concentration of more than 40 elements in each shard, the data were subjected to statistical analysis, disregarding any archaeological information about the samples. This made it possible to divide the samples into "trial compositional groups" and then to ascertain whether single samples fitted into any of the groups (Bieber et al. 1976; Harbottle et al. 1969). Chemical analyses by neutron activation made it possible also to differentiate among pottery groups of different provenance. Obtaining reliable information on the provenance of any specific shard would require, however, a very large amount of data, which would have to be derived from a very large number of analyses (estimations vary between 10,000 and probably more than 100,000 pieces of pottery), too large a number to be analyzed with the analytical equipment and statistical techniques available even today (Adan-Bayewitz et al. 1999; Perlman 1984).

Dating Pottery

Most archaeological pottery has been and still is dated using typological criteria. During the last half-century, however, scientific dating techniques such as radiocarbon (see Textbox 52) and thermoluminescence (see Textbox 24) have reduced the burden that typology once entailed. Even so, a major problem in radiocarbon dating of pottery arises from the contamination of samples during burial, as a result of the absorption of carbon-containing substances from the burial environment. Although this problem may be overcome if the carbon in the pottery samples is purified, the accurate radiocarbon dating of archaeological pottery continues to be an area of ongoing research. Some more recent work suggests that the time of use of pottery can be dated indirectly by the radiocarbon methods: if for example, organic matter such as oil, stored in ceramic vessels was absorbed within the walls of the vessels and preserved there in sufficient amount to be dated by radiocarbon, dating the absorbed matter would provide the time of the latest use of the vessel (Stott et al. 2003).

THE BIOSPHERE
ORGANIC AND BIOLOGICAL SUBSTANCES

The term *biosphere* (from the Greek words *bio*, life, and *sphaira*, globe) was first defined during the 1930s as "the stable and adaptable life sustaining system on the earth's surface" (Vernadskii 1998). The biosphere is that part of the earth in which life is supported and living organisms exist. This includes not only the living organisms themselves but also the wastes of their living activity and their dead remains that have not been fully decomposed. The biosphere extends from the depths of the oceans to the topmost limits of the atmosphere, comprising the outer part of the solid crust of the earth, most of the *hydrosphere*, and the entire *atmosphere*. Altogether, these make up a layer over 20 kms thick; in relation to the size of the planet, the space occupied by the biosphere is just as thin as the skin of a peach.

The existence of the biosphere is made possible by two factors that make life on earth viable: the abundance of water and the ample supply of energy from the sun. Water, which in the earth coexists in three different phases (forms), as water vapor, liquid water, and solid ice, is the main component of living matter, while energy provides the driving force required for all living processes. Thus water and the sun's energy together create the appropriate environment for the existence of the biosphere in the extremely cold as well as in the very hot areas of the planet (Madder 1998; Starr 1997).

Three processes that take place in living organisms – respiration in animals and plants, photosynthesis only in plants, and the precipitation of solids by some aquatic animals – have altered the primeval composition of the outer solid, liquid, and gaseous layers of the earth. Respiration consumes oxygen from the atmosphere and creates carbon dioxide. Photosynthesis, which does the opposite (consumes carbon dioxide and releases oxygen), has

Archaeological Chemistry, Second Edition By Zvi Goffer

transformed the atmosphere of the earth into an oxidizing environment in which there is a relatively high proportion of free oxygen. The precipitation of solids has created much of the sediments in the outer crust of the earth. It can be said, therefore, that the biosphere is that part of the planet's surface which is regulated by a flow of energy, mediated by the process of photosynthesis.

8.1. LIVING ORGANISMS AND CELLS

Living organisms are made up of *cells*, organized aggregations of associate substances arranged in more or less complex structures adapted to fulfill specific living processes (see Textbox 50). Cells are the fundamental units of life. Primitive living organisms, such as bacteria and some algae and fungi, consist of just one cell. Advanced and complex organisms, for example, the animals and the green plants, are made up of millions of cells grouped into organized arrangements that make up the tissues, organs, and organ systems in their bodies.

TEXTBOX 50

THE LIVING CELL

The *cell* is the smallest organized structure of living matter and the basic constituent of all living organisms. To a large extent, the cell is also the building block as well as the factory of all living matter. A cell contains all the components necessary to sustain life: *carbohydrates, lipids, proteins, nucleic acids*, and *minerals*. Depending on its function, a cell may have a variety of shapes and sizes, although most cells are relatively small; their size may vary, however, from less than one micrometer (0.001 millimeter) in diameter in some bacteria up to 75 millimeters in the eggs of ostriches. A human cell is about 100 micrometers (0.1 millimeters) across (Alberts et al. 1998). The diagram in Figure 59 illustrates the major features in an animal cell.

Each cell consists primarily of a *membrane*, which separates it from the environment, preserves its structural integrity, and keeps it apart from other cells or from the surrounding environment. Plant cells, unlike animal cells, also have, in addition to a cell membrane, a *cell wall*, composed of cellulose and lignin. The cell *wall* provides structural strength not only to the vegetable cell itself but to all plant tissues as well. Inside the membrane, the interior of the cell, known as the *protoplasm*, includes two main constituents: the *nucleus* and the *cytoplasm*. The *nucleus*, a dense region at or near the center of the cell, carries the *genetic material*, which encodes heritable information for the maintenance and continuity of the life of the cell. The *cytoplasm*, surrounding the nucleus, is an aqueous solution in which there are a number of suspended inclusions having specialized functions.

The nucleus contains bundles of a fibrous material known as *chromatin*, which is made up of mixed proteins and deoxyribonucleic acid (DNA), the substance that carries the genetic information of the living organism of which the cell is a component. All cells replicate by division. When a cell replicates, DNA in the chromatin of the nucleus passes the genetic information from one generation to the next one. As the cell divides, the chromatin clusters into rodlike structures known as *chromo-*

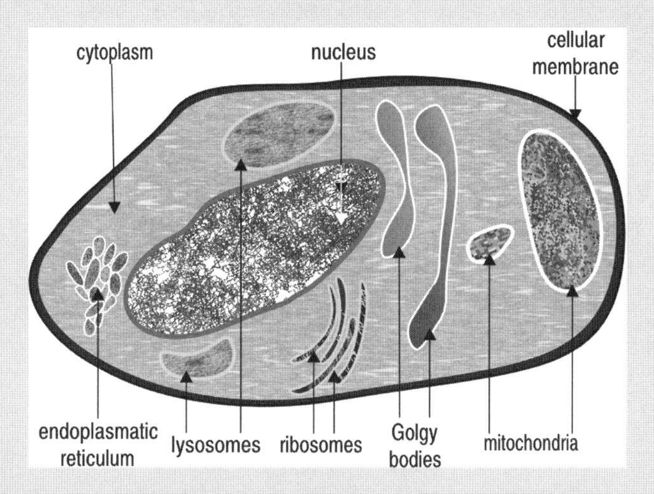

FIGURE 59 The living cell. The cell, the smallest unit of life, has a complex structure, much more elaborate than any machine humans have ever built. Most animal and plant cells range in size between 1 and 100 micrometers and are visible only under high magnification. Even the smallest cells are made up of many billion atoms combined into molecules of mainly proteins, DNA and RNA. Although some organisms are composed of just one cell, most living organisms are "multicellular," composed of many cells; humans, for example, have an estimated 100,000,000,000,000 cells. Groups of cells make up tissues, which make up organs, which make up organ systems, which in turn make up living organisms. All cells share a similar basic architecture that includes three essential components: an outer membrane, a central nucleus and, in between, the cytoplasm, which contains various organelles, structures having distinctive functions. Vegetable cells have a rigid membrane; animal cells lack a rigid membrane, which allows the animals to develop a greater diversity of cell types, tissues, organs, and organisms. The illustration provides a schematic view of an animal cell.

somes, which are made up of linear segments of DNA. All cells in any given living organism contain the same number of chromosomes; in different plant and animal species, however, the number of chromosomes varies (see Table 62).

The cytoplasm of plants and animals is a gelatinous fluid that includes discrete membrane-surrounded structures, collectively known as *organelles*, which perform specialized functions in the life of a cell. The *lysosomes*, *ribosomes*, *mitochondria*, and the *endoplasmatic reticulum* and the *Golgi apparatus*, for example, are all organelles. The function of the *lysosomes* is to break down substances taken up by the cell. Proteins and other substances required by the cell are synthesized in the *ribosomes* and *endoplasmatic reticulum*. In the *mitochondria*, sugars are broken down to keep the cell supplied with energy. A difference between plant and animal cells is that the plant cell also contains a *vacuole* (a fluid containing cavity), which controls pressure in the cell, and *chloroplasts*, in which *photosynthesis*, the biosynthesis of sugars, takes place.

TABLE 62 Number of Chromosomes in the Cells of Living Organisms

Living organism		
Common name	Genus and species	Chromosomes
	Animals	
Cat	*Felix maniculata*	38
Cattle	*Bos taurus*	60
Dog	*Canis familiaris*	78
House fly	*Musca domestica*	12
Human	*Homo sapiens*	46
Mosquito	*Culex pipiens*	6
Rat	*Rattus norvegicus*	42
	Vegetables	
Clover (white)	*Trifolium repens*	32
Corn (maize)	*Zea mays*	20
Oak	*Quercus robur*	24
Onion	*Allium cepa*	16
Pine tree (yellow)	*Pinus ponderosa*	24
Potato	*Solanum tuberosum*	48
Tomato	*Lycopersicon esculentum*	12

Heredity and Cell Structure

One of the most essential features of the cells, and hence of living organisms, is their ability to reproduce by duplicating themselves: all living cells can divide into two identical daughter cells. Prior to the division of a cell, the chromatin in the nucleus of most living organisms clusters into *chromosomes* (self replicating threadlike structures), and when the cell divides, each chromosome is also divided into two new, identical chromosomes, one for each new cell. Under normal conditions the chromosomes of the body of a living organism always occur in pairs; the cells of the human body, for example, contain 46 chromosomes, arranged into 23 pairs. Those of other organisms, vegetable as well as animal, have different but characteristic numbers of paired chromosomes. The sex cells of plants and animals differ from their body cells in that they have only half as many chromosomes as do body cells. In a human sex cell, for example, there are 23 unpaired chromosomes; each chromosome in a sex cell is one of a potential pair in the body cells. Table 62 lists the number of chromosomes of body cells as well as sex cells in several plants and animals.

The chromosomes, only a few micrometers long, are made up of a large number of *genes*, which are sequences of *deoxyribonucleic acid* (DNA), the substance that carries, encoded within its structure, the genetic characteristics of living organisms (see Chapter 12). The genes in each chromosome are arranged one next to the other, much like beads on a necklace; they are also paired, in a way similar to the chromosomes in the nuclei of the body cells. In other words, each chromosome pair has paired genes. Since the genes carry the genetic characteristics of living organisms, there are two genes, located opposite each other on a pair of chromosomes, depicting each characteristic of the organism. During fertilization, a male sex cell joins a female sex cell and, as a result, the body cells of the new organism have the combined number of chromosomes and genes of the sex cells of the mother and father.

8.2. BIOLOGICAL MATTER: ORGANIC AND BIOINORGANIC SUBSTANCES

Living processes involve a vast number of *biological* or *biogenic* substances, the substances that make up all living organisms and those that they create. Chemical and biological studies have elucidated the composition and structure of a great number of these substances, the ways they are created in living organisms, their function in the organisms, and the manner in which their characteristics are passed from one generation to the next (Simmonds 1992). Most of the biological matter that makes up the cell consists of *organic substances*, an immense variety of forms of matter, all composed of only a few light elements (see Textbox 51).

TEXTBOX 51

ORGANIC SUBSTANCES

Organic substances are composed mainly of carbon atoms linked either to a few other elements or to other atoms of carbon by carbon–carbon bonds. Most *living matter* consists of organic substances. The term *organic substances* related, during the nineteenth century, not to the nature or composition of the substances, but to their source: organic substances were then obtained only from living organisms – plants or animals – whereas inorganic substances were obtained from mineral sources; since the midtwentieth century, however, many organic substances are made by humans in chemical laboratories and industrial plants.

Most of the bodies of all living organisms (about 85% of the human body for example) consist of *organic substances* that are also those involved in practically all biological processes. The main constituent of the molecules of organic substances are, however, relatively few: mainly *carbon* combined to *hydrogen*, but also to *oxygen*, *nitrogen*, and a few others, such as *sulfur*, *phosphorus*, and even a few *metals*. (It should be noted, however, that a small number of inorganic substances that comprise carbon as a major constituent of their molecules are not considered organic substances; these include mainly, mineral carbonates abundant on the earth's surface, such as limestone and marble, which are composed of calcium carbonate.)

Carbon, the main constituent of the organic substances, is almost unique among the chemical elements in that it has the property of *catenation* (chaining): atoms of carbon *bond* (chemically join) to other atoms of carbon, forming a wide range of chained structured substances; in some of these, a few bonded atoms of carbon form relatively short and simple chained molecules, while in others, many carbon atoms build up large and complicated molecules shaped as long chains, branched chains, rings, chains linked to rings, or rings linked to rings. In all cases, however, the carbon atoms are also bonded to atoms of hydrogen, oxygen, nitrogen, phosphorus or sulfur. The composition of such substances in writing is, by universal convention, expressed using *chemical formulas*.

The formulas of organic substances may be written in two main forms: as *molecular* or as *structural* formulas. *Molecular formulas* provide the lowest level of information about the composition of a substance, including only the symbols of the elements and the number of atoms of each element in a molecule of that substance. A molecule of glucose (a carbohydrate), for example, is composed of 6 atoms of carbon, 12 of hydrogen, and 6 of oxygen. Thus the molecular formula of glucose includes only the symbols for carbon (C), hydrogen (H), and oxygen (O) and the number of atoms of each element, 6, 12, and 6, respectively, so that its molecular formula is $C_6H_{12}O_6$. In a similar way, the formula of acetic acid, a component of vinegar, whose molecule is composed of two atoms of carbon, four of hydrogen, and two of oxygen, is written as $C_2H_4O_2$. Although molecular formulas provide accurate information on the elements composing a substance and on their relative weights in a molecule of the substance, they do not reveal any details on the internal arrangement of the atoms in the molecule.

Structural formulas also provide information on the way the atoms are arranged and bonded to one another within a molecule. The structural formula of substances not only specifies the type of atoms and how many atoms of each type there are in the molecule of a compound; it also provides an outline of the structure of the molecule, pinpointing exactly where each atom is located. Each element in a structural formula is represented by its symbol, and the bonds between atoms are indicated by lines connecting the symbols (see Fig. 60). Thus, structural formulas not only provide information on the type and number of atoms in a molecule of a substance but also depict the internal structure of the molecule of the substance.

The molecules of two organic compounds are sometimes composed of the same type and number of atoms, but arranged in different ways. The molecular formula of each one of such compounds, which are known as *isomers* (for example, isoleucine and alloisoleucine, shown in Fig. 73), is therefore identical to that of the other; only the *structural formulas* of the two isomers show the differences between their molecules (see Textbox 63).

name	molecular formula	structural formula
1 - alcohol (chemical name: ethanol)	C_2H_5O	$CH_3—CH_2—OH$
2 - sugar (chemical name: sucrose)	$C_{12}(H_2O)_{11}$	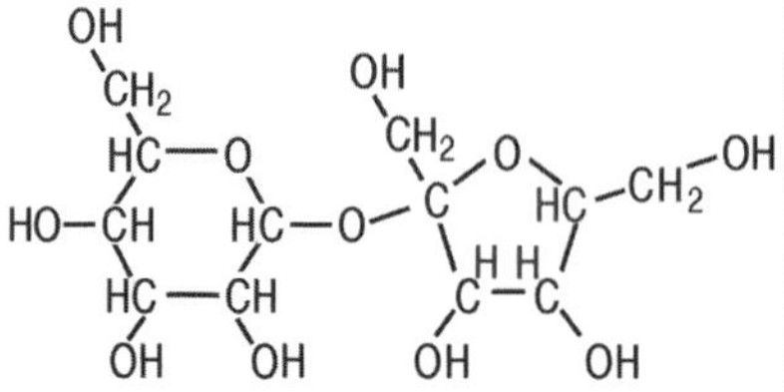
3 - glycine (an amino acid)	$C_2H_5NO_2$	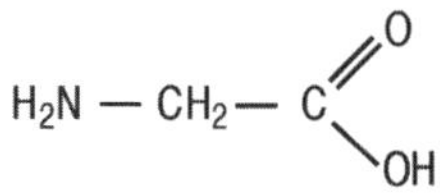 $H_2N — CH_2 — C(=O)OH$
4 - kermes (a red dye) (chemical name: kermessic acid)	$C_{16}H_{10}O_8$	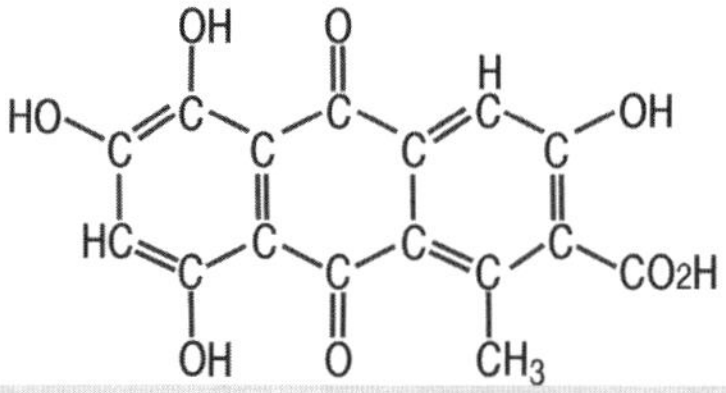

FIGURE 60 Formulas of organic substances. The molecular formula of an organic substance conveys information about the nature of the component elements (expressed by symbols) and the number of atoms of each element that make up a molecule of the substance; if greater than one, the number of atoms of each element is indicated by a subscript. The structural formula provides a two-dimensional representation of the arrangement of the atoms in the molecule, showing how they are attached to one another and the type of bonds involved.

Organic Substances

The *organic substances* that constitute most of biological matter, particularly the matter that makes up the structure and performs the functions of living organisms, belong to one of four fundamental chemical groups: *carbohydrates* (or sugars), *lipids* (or fats), *proteins*, and *nucleic acids* (see Table 63). One of the main characteristics of these substances is that their molecules are relatively large, much larger than those of most inorganic substances, some organic

TABLE 63 Main Constituents of Living Matter

Constituent	Composition	Function	Occurrence (in living organisms)	Relative amount in the human body (%)
Carbohydrates	Sugars, starch, cellulose	Structural components of plant cells; easily released energy storage in plants and animals	Sugars in fruits; starch and cellulose in plants; glycogen in animals	below 1
Lipids	Fats and oils	Energy storage (reserve fuel)	Vegetable oils, animal fats, cholesterol, hormones	1–30
Proteins	*Biopolymers* composed of *amino acid* units	Structural component of animal cells, but also of vegetables	Blood, muscles, skin, hair, enzymes	about 17
Nucleic acids	Complex biopolymers	Storage and transfer of genetic information and makeup of proteins	Nuclei and cytoplasm of living cells	about 2
Bioinorganic materials	Carbonated hydoxyapatite in mammals; silica or calcium oxalate in weeds	Provide structural support to the body	Bones, teeth, and shell in animals; phytolithes in plants	about 5
Water	H_2O	Solvent for most body chemicals	All cells	50–70
Elements	Potassium, sulfur, sodium, chlorine, and magnesium	Physiological functions	All cells	below 2.5
Trace elements	Vary widely	Essential for physiological processes	Cells	below 0.01

molecules comprising many thousands of atoms. Such molecules, which are usually referred to as *macromolecules* or *biopolymers*, are produced by the living organisms from relatively small molecules, known as *monomers*, joined together in a repeating fashion. Well-known examples of natural biopolymers are the *carbohydrates cellulose*, the most abundant substance in the biosphere and *starch*; also most of the *proteins* and *nucleic acids*, all of which are an integral part of, and fulfill specific functions in all living organisms.

Since time immemorial, humans learned to utilize organic substances and materials consisting of mixtures of organic substances derived from plants or animals for particular tasks: mostly for food, but also for making body coverage and adornment, building dwellings, making tools and weapons, and countless others functions. The materials listed below, all organic materials, are examples often found in archaeological excavations:

Amber	Fat	Horn	Paper	Skin
Antler	Flax	Incense	Papyrus	Sugar
Charcoal	Food	Ivory	Parchment	Vellum
Cork	Glue	Lacquer	Perfume	Wood
Cotton	Gum	Linen	Resin	Wool
Dyes	Hair	Oil	Silk	Wax

The nature and properties of the organic substances, the ways in which humans have used them, and the information that may be derived from the study of their composition and characteristics in archaeological contexts are discussed in the chapters that follow.

Bioinorganic Substances

Organic substances are not, however, the only substances created by living processes. Some biological processes, quite widespread in nature, result in the formation of inorganic substances that are thus known as *bioinorganic substances* or *biominerals*. Biominerals of archaeological interest are listed in Table 64 (see also Chapter 15). Some bioinorganic processes proceed in ways that may be of use for solving archaeological problems. The exoskeletons (shells) of some invertebrate animals, such as snails, for example, increase in size during the growth process by accretion, that is, by the addition of new material at the edge of earlier formed skeletal structures. The rate of growth of the accreting material is affected by variations in environmental conditions, such as temperature and rainfall; this gives rise to the

TABLE 64 Biominerals

Occurrence and function	Main mineral form	Composition
Bone and teeth (endoskeletons of vertebrates)	Carbonated hydroxyapatite	Carbonated calcium phosphate
Shell, coral, pearl (exoskeletons of molluscs) and encasements of eggs	Limestone and/or aragonite	Calcium carbonate with variable amounts of magnesium carbonate
Phytolytes and valves of diatoms	Amorphous silica or calcium oxalate	Silicon dioxide or calcium oxalate

formation of lines, known as *growth lines* at the border where new material was added to earlier formed skeletal material, and these lines in the shells of such animals can be used for dating archaeological events (Lowenstam and Weiner 1989; Clark 1974).

8.3. ANCIENT ORGANIC MATERIALS

In most archaeological excavations, organic materials, particularly those of biogenic origin, are much scarcer than inorganic ones, such as stone, pottery, or glass. This is because organic substances are highly susceptible to decay processes that lead to their decomposition and eventual total breakdown in much shorter periods of time than inorganic materials. Following the death of living organisms, most of their dead bodies is either consumed as food by microorganisms, insects, and some animals, or are rapidly and efficiently decomposed by natural processes. The hair, nails, and teeth of a mammal's corpse, for example, become detachable just a few weeks after death. Some time later, after a month or so, most tissues become liquified as a result of *autolysis*, the spontaneous destruction of dead tissues, which is brought about by the chemical action of enzymes in the corpse. Together with microbial activity, the enzymes break down most of the components of the soft tissues and, after a year or so after death, only the bones and teeth, the parts of the body most resistant to decay, are practically all that is left of the corpse (see Chapter 16). In addition to these natural processes, bacteria and fungi may cause the putrefaction of biological substances, particularly of the proteins, into simpler substances, a process that is usually accompanied by the emergence of foul-smelling gases such as the *mercaptans*.

Thus, unless kept under special environmental conditions, most organic substances that make up the body of vegetable and animal organisms break

TABLE 65 Decay of Biological Substances (under Temperate Environmental Conditions)

Substance	Source	Rate of Decay
Carbohydrates		
Cellulose	Plants	Fast
Starch	Plants, some algae, bacteria	Fast
Sugar	All organisms	Fast
Fats (lipids)		
Oils	Plants	Very fast
Fats	Animals	Very fast
Waxes	Mostly plants but also some animals	Fast
Proteins		
Albumin	Animals	Fast
Enzymes	All organisms	Fast
Collagen	Animals	Slow
Keratin	Animals	Slow
Nucleic acids		
DNA	All organisms	Fast
RNA	All organisms	Fast
Others		
Lignin	Plants (mostly in wood)	Slow
Resin	Plants (secretions from wounds)	Slow
Galls	Plants (tissues grown in wounds)	Slow

down quite readily. The extent to which they are preserved varies, however. Some biological substances, in particular those of vegetable origin such as lignin and resins derived from wood, are quite stable even under generally hostile environmental conditions. Sugars, proteins, and DNA, on the other hand, are preserved only under very specific conditions. Table 65 empirically grades the rates of decay of biological substances under regular environmental conditions.

8.4. DATING ORGANIC MATERIALS

Organic materials can be dated by the radiocarbon dating technique, the best-known and most widely used dating technique in archaeological research. Discovered and first developed during the mid-twentieth century by a group of scientists led by W. F. Libby, the technique of radiocarbon dating was one of the most significant scientific discoveries of the twentieth century. Its development provided a means for dating finds independently of stratigraphic or typological relationships and made possible a worldwide chronology, thus transforming archaeological investigation (Bar Yosef 2000;

Taylor et al. 1994). After over half a century of further developments and use, there are many significant uses of radiocarbon dating not only in archaeology but also in a variety of different fields, including environmental studies, ecology, geology, climatology, hydrology, meteorology, and oceanography. Radiocarbon dating provides the most consistent technique for dating materials and events that occurred during the last 50,000 years on the surface of the earth. At the end of the twentieth century there were, around the world, well over 100 laboratories that determined radiocarbon dates on a vast range of materials (Lowe 1997; Bowman 1990).

Radiocarbon Dating

Radiocarbon dating is a *radiometric* technique based on measuring the relative amount of radiocarbon (*carbon-14*), a *radioactive isotope* of carbon in matter containing carbon as a component. The method, which is most useful for dating organic materials, provides reliable absolute dates not only of organic and biological materials but also of carbonate sediments and other inorganic carbon-containing materials.

Carbon, and therefore also radiocarbon, is a natural component of the earth's atmosphere, where it occurs combined with oxygen in the form of carbon dioxide. All living organisms acquire radiocarbon, as well as all the other isotopes of carbon, in the same way: plants, acquire it in the form of carbon dioxide, through the process of photosynthesis; animals acquire it when feeding from plants. When vegetable or animal organisms die, their intake of carbon, and therefore also of radiocarbon, ceases. Since radiocarbon undergoes radioactive decay, the dead remains begin to lose radiocarbon. This loss, which is regulated only by the *half-life* of the isotope, results in a continuous reduction in the total number of atoms of radiocarbon, as well as in the relative amount of atoms of radiocarbon to those of stable carbon-12 in the dead tissues. Therefore, the age of dead remains, or of any carbon-containing matter, can be determined by measuring the relative amount of radiocarbon they contain.

TEXTBOX 52

RADIOCARBON DATING

Carbon, a common element in the outer crust of the earth, and the main component of all biological and organic substances, occurs in three isotopic forms: carbon-12 or C-12 for short (whose chemical symbol is ^{12}C), carbon-13 or C-13 (^{13}C), and carbon-14 or C-14 (^{14}C) (see Fig. 8 and Table 66).

Two of these isotopes, carbon-12, the most abundant, and carbon-13 are stable. Carbon-14, on the other hand, is an unstable *radioactive* isotope, also known as *radiocarbon*, which decays by the *beta decay process*: a *beta particle* is emitted from the decaying atomic nucleus and the carbon-14 atom is transformed into an isotope of another element, nitrogen-14, N-14 for short (chemical symbol ^{14}N), the most common isotope of nitrogen:

$$\underset{\text{radiocarbon}}{\text{C-14}} \xrightarrow{\text{beta decay}} \underset{\text{nitrogen}}{\text{N-14}} + \underset{\text{beta particle}}{\beta}$$

The decay of radiocarbon (see Fig. 61) into nitrogen-14 proceeds at a constant rate, and its *half-life* is 5730 ± 40 years (see Textbox 14). This means that in any material containing carbon, some radiocarbon atoms disintegrate before 5730 years have elapsed and others later; after 5730 ± 40 years have elapsed, however, only half of the original atoms of the carbon-14

TABLE 66 Natural Isotopes of Carbon

Isotope	Natural Abundance (% total carbon)	Remarks
Carbon-12	98.9	Stable
Carbon-13	1.1	Stable
Carbon-14 (radiocarbon)	0.000000000002	Radioactive; half-life 5730 ± 40 years

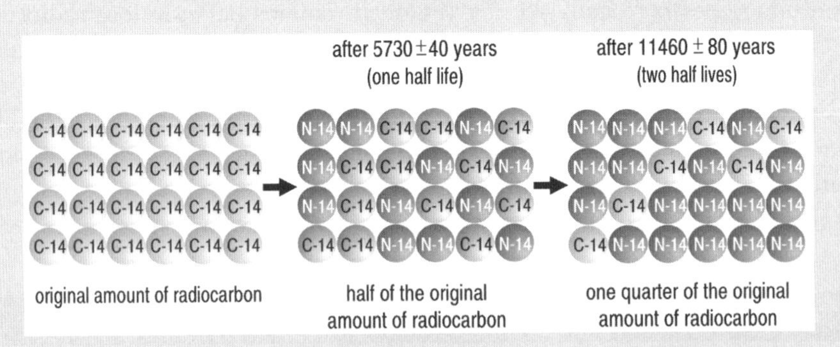

FIGURE 61 The decay of radiocarbon. Radiocarbon is a radioactive isotope whose half-life is 5730 ± 40 years. This means that half of the original amount of radiocarbon in any carbon-containing sample will have disintegrated after 5730 years. Half of the remaining radiocarbon will have disintegrated after 11,400 years, and so forth. After about 50,000 years the amount of radiocarbon remaining in any sample is so small that older remains cannot be dated reliably.

isotope remain in the sample, the other half having decayed to nitrogen-14. The rate of decay does not change, regardless of the nature of the material of which radiocarbon is a component (whether it is part of a living organism or the dead remains of a plant or animal) or its state (whether it is combined as a solid, liquid, or gas).

The source of all the radiocarbon on earth is in the top layers of the *atmosphere*, where it is created as the result of a series of events. Initially, *cosmic rays* (highly charged and rapidly traveling particles from outer space) enter the atmosphere, interact there with atoms and other particles, and create free *neutrons*. Soon after being created, some of the neutrons collide with and penetrate the *nuclei* of *nitrogen-14* atoms, the gas that makes up four-fifths of the atmosphere. The penetration of the neutrons (a process known as *neutron absorption*) into the nuclei of the nitrogen-14 atoms results in the *transmutation* of the nitrogen atoms into atoms of radiocarbon, while a *proton* (an atom of hydrogen bearing an electrical positive charge) is released from each reacting nucleus of nitrogen:

$$\underset{\text{nitrogen-14}}{\text{N-14}} + \text{neutron} \rightarrow \underset{\text{radiocarbon}}{\text{C-14}} + \underset{\text{proton}}{H^+}$$

The rate of radiocarbon formation in the upper atmosphere depends on a number of factors, which include the intensity of the incoming cosmic radiation, the activity of the sun, and the *magnetic field* of the earth (the latter affects the way cosmic rays travel). It can be safely stated, however, that radiocarbon is formed at a steady rate that averages just about 2.4 atoms of radiocarbon per second for every square centimeter of the earth's atmosphere outer surface.

Soon after their creation, the newly formed radiocarbon atoms react with atmospheric oxygen to form *radioactive carbon dioxide*, molecules of carbon dioxide in which the carbon atom is radiocarbon:

$$\underset{\text{radiocarbon}}{\text{C-14}} + \underset{\text{air oxygen}}{O_2} \rightarrow \underset{\text{radioactive carbon dioxide}}{(\text{C-14})O_2}$$

The carbon dioxide molecules including a radiocarbon atom are chemically undistinguishable from those of ordinary carbon dioxide, with which it mixes, and eventually, carbon dioxide, including a radiocarbon atom, is homogeneously distributed throughout the earth's atmosphere and hydrosphere. Thus there is a state of constant production, distribution, and decay of radiocarbon, which results in the relative amount of radiocarbon in the atmosphere and hydrosphere remaining constant. In this homogeneously distributed condition, radiocarbon enters the *carbon cycle* – as the circulation of carbon through the atmosphere, hydrosphere, and biosphere is usually known (see Fig. 62).

During the process of *photosynthesis*, atmospheric carbon is taken up (as carbon dioxide) from the atmosphere by plants, and becomes a basic component of all vegetable cells and tissues (see Textbox 53). Since its chemical properties are identical to those of ordinary carbon, radiocarbon also undergoes the photosynthesis process. Because the carbon isotopes are *fractionated* (their relative abundance is changed) during the process, however, radiocarbon becomes a part of the cells and tissues of plants in a lower isotopic proportion than in the atmosphere. The radiocarbon assimilated by the plants is, in turn, consumed by land and sea animals and by humans. Radiocarbon becomes, therefore, uniformly distributed not only in the atmosphere and hydrosphere but also throughout plants, animals, and the whole of the biosphere, forming part of what is known as a *carbon exchange reservoir*, the entire inventory of carbon in all its isotopic forms and chemical combinations. This includes the three isotopes carbon-12, carbon-13, and carbon-14 in a wide variety of chemical combinations, which circulate freely between the different components of the reservoir. The time it takes for the isotopes to circulate and mix within the reservoir varies greatly: only a few years, for example, are required for a carbon isotope to circulate between the stratosphere and the troposphere, but much longer periods of time are needed for it to do so between other components (the hydrosphere and plants and animal tissues) of the reservoir. Whatever its length however, the time of circulation is much shorter than the half-life of radiocarbon (5730 $\pm$ 40 years).

The mean relative amount of the carbon isotopes on the earth's surface, that is, the ratio carbon-12 : carbon-l3 : carbon-14, which is known as the *isotopic ratio of carbon*, is known to be constant. There must therefore be a sufficient amount of radiocarbon in the planet to ensure that the rate of decay of the isotope balances its rate of formation. The conclusion to be drawn from this is that radiocarbon in the earth is in a state known as *secular equilibrium*, which means that for every radiocarbon atom created by cosmic radiation in the upper layers of the atmosphere, there is one decaying atom in the carbon reservoir of the earth (the total amount of radiocarbon in the earth, incidentally, has been calculated to be slightly above 80 tons). The equilibrium between the formation and the depletion of radiocarbon is also maintained in living organisms, which exchange carbon with their surroundings. The specific activity of living matter – that is, the intensity of radioactivity due to radiocarbon per unit mass (gram) – averages about 14 disintegrations per minute, equal to that in the atmosphere.

The death of a living organism, however, stops the intake (into that organism) of radiocarbon, and its dead remains are said to have been with-

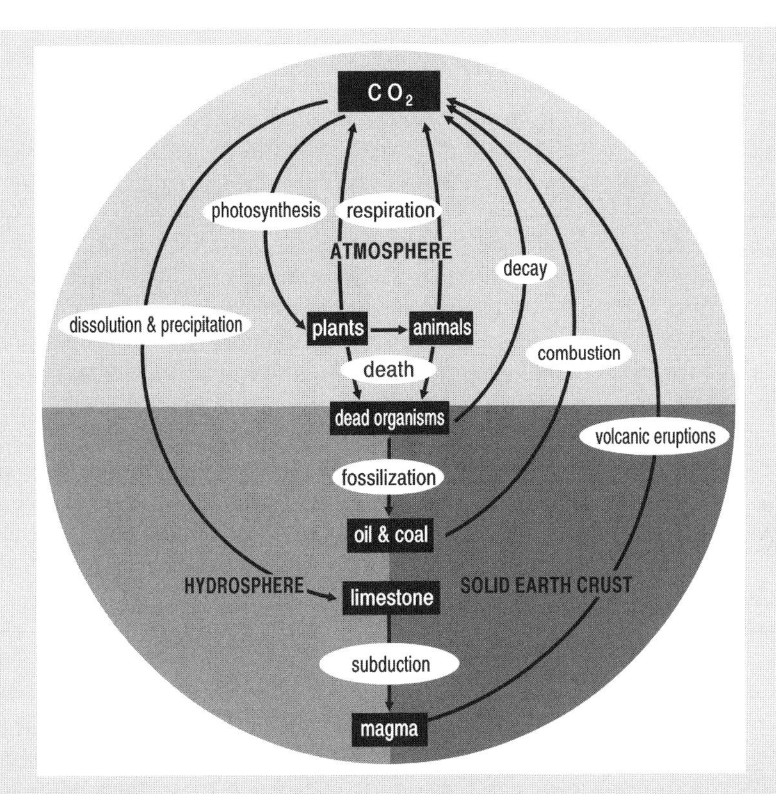

FIGURE 62 The carbon cycle. Carbon, the element that is the building block of most living matter, cycles through the earth: it circulates between the chemical components of its atmosphere, hydrosphere (rivers, seas, and oceans), outer solid crust, and biosphere. In the atmosphere, for example, carbon occurs combined with oxygen as carbon dioxide, which dissolves in the seas and oceans, creating carbonic acid. Carbonic acid, a weak acid, reacts slowly but continuously with ions of calcium and magnesium also dissolved in the water; the products of these reactions, calcium carbonate and magnesium carbonate, are insoluble in water and precipitate out, forming minerals such as limestone and dolomite. Parts of these minerals are, in turn, dissolved by rainwater and carried again to the seas and oceans, while other parts are drawn below the earth's surface into liquid magma. Eventually, however, during volcanic eruptions, the carbon from these minerals is returned to the atmosphere as carbon dioxide. Carbon enters the biosphere through the processes of photosynthesis in plants, which converts atmospheric carbon dioxide into sugars, which sustain practically all forms of life on earth. From the plants, which serve as food for herbivorous and omnivorous animals, carbon circulates into the animal world. The respiration of plants and animals enables the metabolism of the sugars and the production of the energy required for all living processes while also returning carbon, in the form of carbon dioxide, into the atmosphere. Thus photosynthesis provides the foundation of the biological fixation of carbon in living matter, while respiration returns the biologically fixed carbon back to the atmosphere.

drawn from the carbon exchange reservoir. As a consequence of the withdrawal, the amount of radiocarbon in dead plant and animal matter does not remain constant, as in the carbon exchange reservoir: since radiocarbon undergoes radioactive decay, its relative amount in the dead remains of organisms decreases gradually and continuously. The rate of decrease is independent of the nature of the organism of which the radiocarbon is a component, its physical condition, and the nature and conditions of the surroundings of the dead remains: it is determined only by the half-life of the isotope, 5730 ± 40 years. This means that the remains of any type of vegetable or animal matter 5730 years old contain only half as much radiocarbon as contemporary living matter, one-quarter after 11,460 years, only one-eighth after 17,190 years, and so on (see Table 67). Thus, although there are some differences in isotopic abundance between different parts of the exchange reservoir, the *mean ratio* between the isotopes (the proportion carbon-12 : carbon-13 : radiocarbon) in the biosphere is kept constant throughout, at approximately 1 : 0.01 : 0.000000000001 (10^{-12}). In other words, for every atom of carbon-12 in the biosphere there are only 0.01 atoms of the isotope carbon-13 and as little as 10^{-12} atoms of radiocarbon. Such a minute concentration of radiocarbon is practically impossible to detect, let alone to assess by most chemical and physical techniques. It can, however, be detected and accurately measured because of its radioactive properties. Measuring either (1) the amount of beta radiation emitted by the decaying radiocarbon or (2) the actual number of radiocarbon atoms provides a means to determine the total amount of radiocarbon in any sample.

TABLE 67 Decay of Radiocarbon in Dead Remains

Time since death of an organism (years)	Number of half-lives	Number of atoms[a]	
		Carbon-14	Carbon-12
0	0	$1N$	$1N$
5,730	1	$\frac{1}{2}N$	$1N$
11,460	2	$\frac{1}{4}N$	$1N$
17,190	3	$\frac{1}{8}N$	$1N$
22,920	4	$\frac{1}{16}N$	$1N$
28,650	5	$\frac{1}{32}N$	$1N$
34,380	6	$\frac{1}{64}N$	$1N$
40,110	7	$\frac{1}{128}N$	$1N$
45,840	8	$\frac{1}{256}N$	$1N$
51,570	9	$\frac{1}{512}N$	$1N$
57,300	10	$\frac{1}{1,026}N$	$1N$

[a] N = number of atoms at the time of death.

From the amount or radiocarbon in archaeological remains, it is relatively easy to calculate the time of death of a living organism, that is, to determine the time elapsed since its dead remains became separated from the carbon exchange reservoir. It is important, particularly in archaeology, to remember that radiocarbon dates provide only the time of death of organisms. Materials such as wood, for example, which were used in the past for building or for making tools, might have been derived from trees that had been dead for a long time before their wood was used. It should also be noted that the time of removal of a tree from the radiocarbon exchange reservoir, that is, the time of death of its wood, may be valid only for the outer part of the trunk or branches. Often, the inner part of the trunk and branches of a tree are dead, while the outer layers are still growing. The time of death of the inner core may be much earlier than that of the outer wood – often even before the latter was only being formed. In any case, the radiocarbon method provides only the maximum possible age of a material, revealing solely when living matter died, not when it was put to use (Hedges 2000).

The list below provides a rather limited but nevertheless impressive inventory of materials that have been dated with the technique:

Antler
Avian eggshells
Blood
Bone
Cement
Charcoal
Coprolites
Coral
DNA
Fats
Feathers
Ferrous alloys
Fibers
Glue
Gum
Hair
Hide
Horn
Incense
Ivory
Leather
Mortar
Oils
Paints
Paper
Papyrus
Parchment
Peat
Plaster
Pollen
Pottery
Resins
RNA
Seeds
Shell
Skin
Slags
Soil
Teeth
Textile fabrics
Varnish
Vellum
Waxes
Wood

Methodology of Radiocarbon Dating. The natural concentration of radiocarbon in materials on earth is extremely low; add to this the fact that the beta radiation emitted by radiocarbon is very weak, and the conclusion is that the measurement of natural levels of radiocarbon is a rather difficult task. Indeed, very elaborate physical and chemical procedures are required to obtain accurate radiocarbon measurements and dates.

Two basically different technical procedures are at present in use to measure the amount of radiocarbon in matter:

1. An older and long-established technique, *radiocarbon decay counting*, also known as the "conventional" method of radiocarbon dating, is based on detecting and counting the amount of beta radiation emitted in unit time by radiocarbon atoms in a sample of known weight.
2. A more recently developed technique, known as the *accelerator mass spectrometry* (AMS) radiocarbon dating technique, based on counting, in a *mass spectrometer*, the relative amount of radiocarbon to stable carbon isotopes in a sample (see Textbox 10).

The conventional radiocarbon decay counting technique generally provides reliable results, but it has some limitations; the following are worth mentioning:

- A relatively large sample (several grams but often more) is required to obtain reliable data.
- A long time (usually several hours) is required for counting the beta radiation emitted by the disintegrating radiocarbon.
- Since the counting of radioactive disintegration events is subject to statistical considerations, the dates obtained with this method are inherently subject to some degree of uncertainty.

The AMS technique, on the other hand, requires only very small samples (usually below one gram in weight), and the time necessary to measure the amount of radiocarbon in each sample is only a few minutes, much less than that required for counting disintegration events in the conventional technique. AMS demands, however, a large initial financial investment for building a dating facility, which greatly increases the cost of measurements as compared with those derived by the radiocarbon decay counting technique.

Although the size of a sample required for radiocarbon dating is determined by the actual method used (see text above), it is also dependent on the actual composition of the material to be dated; Table 68 lists the amounts of various materials usually required for either of the radiocarbon dating techniques. The importance, particularly in archaeology, of using small samples (weighing only a few milligrams, all that is required for AMS) cannot be overemphasized: removing only small samples makes it possible to sample valuable and irreplaceable objects (fractional remains of parchment or textiles for example) that might otherwise be destroyed to provide a sample of sufficient size to be assayed by the radiocarbon decay counting

TABLE 68 Typical Sample Size for Radiocarbon Dating

	Dating technique and sample size (g)			
	Radioactivity counting		AMS	
Material	Optimum	Minimum	Optimum	Minimum
Bone	200–2000	100	0.30– 0.050	0.002
Bone (charred)	100– 500	30	0.03– 0.050	0.002
Charcoal	5– 12	1	0.05	0.001
Dung	10– 30	5–10	0.02– 0.050	0.005–0.010
Plant tissue (nonwoody)	20– 50	7–10	0.02– 0.050	0.005–0.010
Peat	10– 30	2–10	0.03– 0.010	0.010–0.015
Textile fibers	10– 20	1	0.03– 0.100	0.500
Shell and avian shell	10– 100	5	0.03– 0.001	0.005
Skin	10– 40	5	0.02– 0.050	0.010
Soil	200–2000	200	2 –10	1
Wood (dry)	10– 40	5	0.02– 0.050	0.005

technique. Moreover, the small sample size required for AMS allows one to carefully select appropriate samples, ignoring, for example, parts that may be contaminated with modern carbon (Harris 1987). Still, despite the advantages of the AMS technique, the relatively low cost of operating a radiocarbon decay counting facility causes this technique to continue to be used, particularly when large samples are available.

Dating with Radiocarbon. The important information held in a sample to be dated by radiocarbon is its present radiocarbon concentration; comparing this concentration to that of radiocarbon in the atmosphere, which is considered to be constant (however, see discussion below), yields the conventional radiocarbon date of the sample. All that is required to establish the age of a sample, therefore, is to determine the present-day relative amount of radiocarbon in the sample. Once this has been determined by either the conventional radiocarbon decay counting or by the AMS method (see Fig. 63), a number of internationally established conventions and assumptions are used to calculate the age of a material or object:

1. Radiocarbon ages are reported in years "before present" (B.P.), which, by convention, refers to "before the year 1950 C.E."
2. The *half-life of radiocarbon* used to calculate radiocarbon dates is 5568 years, a value known to be about 3% in error with respect to the actual half-life of radiocarbon, 5730 years. This is done to avoid confusion

with dates measured during the early days of the radiocarbon method, the fifth decade of the twentieth century. The discrepancy is taken into account and corrected, together with others caused by natural variations in radiocarbon.

3. The relative amount of radiocarbon in the earth is assumed to have been the same during geologic times.
4. The relative concentration of radiocarbon in the samples dated is corrected for isotope fractionation, based on the ratio C-13/C-12 (see text below).

Practical considerations limit the use of the radiocarbon dating technique to a range of 200–50,000 years; samples less than 200 years old cannot be called

Radiocarbon laboratory **AMS radiocarbon instrumentation**

FIGURE 63 Dating with radiocarbon. Radiocarbon dating is the most widely used method for determining the age of archaeological materials containing carbon. Accurate radiocarbon dating requires great scientific expertise and advanced equipment. Elaborate chemical instrumentation **(a)** is required, for example, when using the radiocarbon decay counting technique, just to isolate the carbon in the material before determining the relative amount of the radiocarbon isotope. Acceleration mass spectrometry (AMS) involves the use of large devices, as illustrated by the tank **(b1)**; in the large tank, the atoms of carbon in the sample are first ionized (converted into ion) by removing a negatively charged electron, and then accelerated to precise velocities and separated according to their mass. From the measured amount of carbon 14 relative to that of the other stable isotopes of the element, the age of a sample can be determined. All the AMS operations are accurately controlled in a complex, computerized control panel **(b2)**.

anything but "modern"; by contrast, in samples older than 50,000 years, the relative amount of radiocarbon is insufficient to measure using the conventional radiocarbon dating method.

Errors Inherent to the Radiocarbon Dating Method. The decay of radiocarbon is radioactive, involving discrete nuclear disintegrations taking place at random; dates derived from the measurement of radiocarbon levels are therefore subject to statistical errors intrinsic to the measurement, which cannot be ignored. It is because of these errors that radiocarbon dates are expressed as a time range, in the form

Age ± [error] years B.P.

The error expresses the *standard deviation* of the actual age. The *standard deviation* is an index of variance used in statistics to characterize the dispersion of measured values (see Fig. 64). This implies that there is a 68% probability, that is, a likelihood of 2 to 1, that the real age is within the indicated

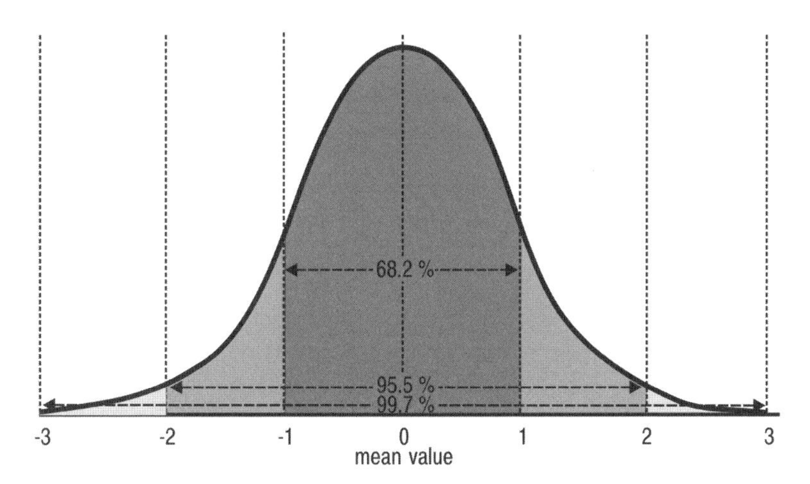

FIGURE 64 The standard deviation. The standard deviation is a statistical value used to describe the variability in experimental measurements. Its theoretical basis is complex and need not trouble most users, but it can be conceived of as a measure of the average amount by which observations deviate from a mean value. In the graph, one standard deviation away from the mean, in either direction on the horizontal axis (the darkest area in the graph), accounts for 68.2% of the values measured. Two standard deviations away from the mean (the darkest and middle shaded areas together) account for roughly 95.5% of the values. Three standard deviations (the darkest, medium shaded, and lightly shaded areas) account for about 99.7%, representing practically all of the values.

limits; 95% probable, or a likelihood of 20 to 1, that it lies within twice the given limits, and a likelihood of 300 to 1 that the date is somewhere within those limits. The measured half-life of radiocarbon, for example, is stated to be 5730 years ± 40 years. This means that there is a likelihood of 2 to 1 that it is somewhere in the range 5690–5770 years, a 20 to 1 likelihood that it is within the range 5650–5810 years and a 300 to 1 likelihood that it is within the range 5610–5850 years.

Uncertainties, Limitations, and Complications in Radiocarbon Dating. Discrepancies between a measured radiocarbon age and the otherwise verified age of certain specimens are sometime found. These discrepancies are due to deviations from the basic assumptions on which the radiocarbon method rests, which are essentially, as follows:

1. The half-life of radiocarbon is immutable.
2. There is constant formation of radiocarbon in the upper layers of the atmosphere.
3. Radiocarbon is uniformly distributed in the carbon exchange reservoir.

There are no doubts about the stability and lack of fluctuation of the rate of decay of radiocarbon, which is constant and immutable; the value of 5730 years ± 40 years for the half-life is the most accurately measured at the present time. The internationally accepted value used for dating is, however, 5568 years, which is based on earlier, obsolete measurements. The decision to retain an obsolete value as the accepted half-life of carbon-14 is the outcome of a consensus of opinion among scientists engaged in radiocarbon dating. Changing the old value for the new and more accurate one would have required the revision of many thousands of previously published radiocarbon dates. It was also recognized that discrepancies between the radiocarbon chronology and other chronologies arising from the old value would not have been corrected by the change.

Experimental evidence indicates that for a very long time the formation of radiocarbon has proceeded at a relative constant rate; comparing the age of ancient wood samples determined by *dendrochronology* (see Chapter 10) with those determined by radiocarbon-dating the same sample has shown, however, that during previous ages there have been some changes in the rate at which radiocarbon is formed. It is also well known that this rate changes in a cyclic fashion, as two cycles are apparently in operation: a long-term cycle with a period lasting about 10,000 years, probably due to geomagnetic oscillations in the earth, and a short-range period, only a few years long, caused by the electromagnetic activity of the sun. Both cycles, combined, seem to be the cause of systematic errors of several hundred years that are generally allowed for when dating with radiocarbon.

A mass of evidence seems to confirm that the mixing rate of radiocarbon in the atmosphere is rapid, and that with respect to its radiocarbon content the atmosphere can be considered as a homogeneous entirety. The *contamination* of samples with matter from an extraneous source can nevertheless invalidate this assumption. Two types of contamination can be differentiated: *physicochemical contamination* and *mechanical intrusion*. There are two forms of *physicochemical contamination*. One is due to the dilution of the concentration of radiocarbon in the atmosphere by very old carbon, practically depleted of radiocarbon, released by the combustion of fossil fuel, such as coal and oil. The other is by the contamination with radiocarbon produced by nuclear bomb tests during the 1950s and later in the twentieth century. The uncertainties introduced by these forms of contamination complicate the interpretation of data obtained by the radiocarbon dating method and restrict its accuracy and the effective time range of dating.

Mechanical intrusion is the penetration of the matter to be dated by carbon of a different age from that of the sample itself; if not taken into account, mechanical intrusion, too, leads to erroneous ages. The penetration of rootlets from growing plants into buried specimens, the infiltration of windblown organic matter, and the accidental insertion of fibers from brushes or other instruments used to clean a sample are examples of likely *modern carbon intrusions* into prospective samples, which lead to assigning to a sample later dates than the true ones; *old carbon intrusions*, such as those caused by the penetration of carbonate minerals from groundwater, or of petrol or oil from excavating tools, on the other hand, are conducive to assigning earlier dates than the true ones.

Fortunately, however, the deviations from "true dates" caused by all these factors are usually small for dates falling within the last three millennia. *Correction curves* are used to correct dates that fall between several millennia B.C.E. and the present day, so that the dates determined with radiocarbon are concordant with historical dates (Pearson et al. 1989; Suess 1965).

Radiocarbon Dates. Historical dates are usually expressed in calendar years, but dates determined by the radiocarbon method are expressed as *radiocarbon years*. This is done on the assumption that during the past the relative concentration of radiocarbon in the atmosphere has been constant;

since it is known that it fluctuates, however, radiocarbon ages, measured in years before present (B.P.), are only close approximations of ages expressed in calendar years. Radiocarbon dates are therefore calibrated to give calendar dates, and *standard calibration* curves are available (Ramsey 2005; Stuiver et al. 2005). The *calibration* is based on comparing radiocarbon dates with dates derived from *tree rings*, ice cores, deep-ocean sediment cores, lake sediment varves, and speleothems. Trees, for example, hold a record of past levels of radiocarbon concentration in the atmosphere (see Chapter 10). Wood in general and *tree rings* in particular, are made by the trees using carbon derived from atmospheric carbon dioxide through the process of photosynthesis (see Textbox 53). During this process, the carbon, in all its isotopic forms and in the concentration (in the atmosphere) characteristic at the time of the formation of the tree rings, is incorporated into and preserved in the cells of the rings. Thus tree ring (*dendrochronologic*) measurements provide a tool for correcting errors introduced into radiocarbon measurements by the incorrect assumption that the level of radiocarbon concentration in the atmosphere has been constant (Schweingruber 1988; Hillam 1987). The calibration curves thus obtained, such as those shown in Figure 65, can vary significantly, so that comparing uncalibrated radiocarbon dates, by plotting them in a graph, for example, is likely to give misleading results.

The many possible errors, as well as the difficulties inherent to the radiocarbon dating method itself, serve to emphasize the need for a close collaboration between archaeologists and natural scientists when dating archaeological samples by the radiocarbon method.

Archaeological Applications of Radiocarbon Dating. Regardless of the relatively minor drawbacks mentioned, the radiocarbon dating method has become a standard technique of major importance in archaeology. The experience gained in the years since the technique was first introduced indicates that the method is reliable in simultaneity; in other words, contemporary samples taken from any place in the world will give the same date for any time in the past. Absolute dates, on the other hand, may occasionally be subject to errors large enough to invalidate them. This is due to the variability in time of such natural phenomena as the intensity of cosmic radiation, or to ecological contamination, as, for example, the fossil fuel effect mentioned above. The proper use of radiocarbon dates provides chronological solutions to many archaeological problems, affording a chronology worldwide in scope. Experience has shown that radiocarbon dates are generally reliable, but it is advisable to beware of the unqualified acceptance of radiocarbon ages derived from measurements on single archaeological specimens.

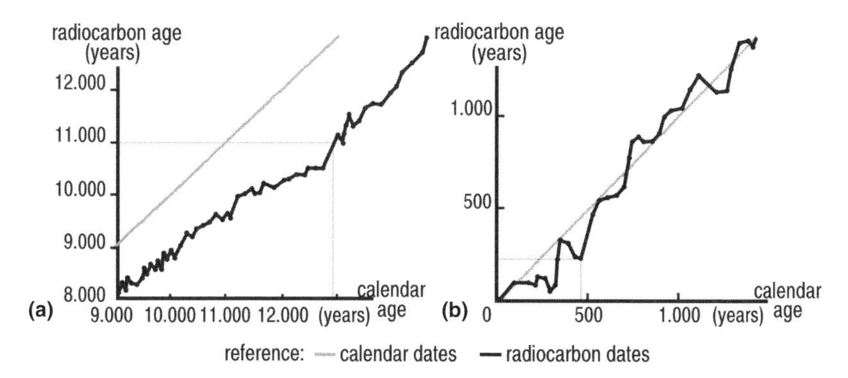

FIGURE 65 Radiocarbon calibration. Radiocarbon dates are always reported in terms of years before present (B.P.). The actual values of these dates are calculated on the assumption that the relative concentration of radiocarbon in the atmosphere has always been constant. The concentration of radiocarbon in the atmosphere however, has not always been the same; significant variations have taken place in the past. Consequently, there is no fixed relationship between radiocarbon dates and calendar dates; radiocarbon dates have to be corrected or "calibrated" into calendar dates. The conversion process, which is known as *radiocarbon calibration*, is not straightforward: it is based on comparing radiocarbon dates with dates derived with other dating methods. The most reliable and relevant of these is tree ring counting (dendrochronology) that involves measuring the relative amounts of radiocarbon in tree rings, calculating their radiocarbon dates, and then comparing these dates with the dendrochronological date of the rings. Calibrated calendar dates are generally followed by the suffix "Cal B.P." or simply B.P. The two graphs show the relationship between radiocarbon dates and calendar dates. Had the concentration of radiocarbon in the atmosphere been constant over time, the relationship between the dates would have been linear and the radiocarbon and dendrochronological dates would be on a straight line. As can be seen in the two graphs, there are significant mismatches: in graph **(a)** are shown dates between 9.000 and 13.000 years BP and in **(b)** dates between 0 and 1.500 years BP. A radiocarbon age of 11.000 years bp, as in **(a)** for example, corresponds to a calibrated (or calendar) age of about 13.000 bp and a radiocarbon age of about 200 years bp in **(b)** corresponds to a calibrated age of between 300 and under 500 years bp (Hughen et al. 1998, 2004).

The amount of information gathered over the decades since the radiocarbon dating method was developed is very extensive, so that even the briefest of reviews on the subject is out of the question here. The interested reader is referred, therefore, either to *Radiocarbon*, the periodical dedicated to the regular publication of new radiocarbon data, issued by the University of Arizona, or to one of the excellent reviews on the subject such as those by Hedges 2001 and by Lowe 1997.

CARBOHYDRATES
WOOD, GUMS, AND RESINS

The *carbohydrates* are the chemically simplest and most abundant natural biological substances; their molecules are composed of only three elements: carbon, hydrogen, and oxygen, in a ratio that approximates 1–2–1. *Sucrose* (ordinary table sugar), *starch* (the substance that stores energy in plants), and *cellulose* (the main component of vegetable tissues and the most abundant biological substance), for example, are all carbohydrates. Common foods such as flour, cornmeal, and honey are composed primarily of carbohydrates, as are the textile fibers cotton and linen, the natural gums, and most of the resins used since ancient times. The carbohydrates are created by green plants through the process of *photosynthesis*, one of the most fundamental biological process in nature. In the plants, the carbohydrates serve either for building structural materials or to store reserve energy acquired from the sun. Living organisms other than green plants cannot synthesize carbohydrates; they derive those that they require from the plants (see Textbox 53).

TEXTBOX 53
CARBOHYDRATES

The *carbohydrates* are substances made by green plants from carbon dioxide that they extract from the air and from water absorbed from the soil. The term *carbohydrate* is derived from the chemical formulas of the compounds, which can be written so as to express their composition in terms of atoms of carbon combined with molecules of water; the formula for sucrose (table sugar), for example, is $C_{12}H_{22}O_{11}$, but it can also be written as $C_{12}(H_2O)_{11}$, indicating that the molecule consists of 11 molecules of water

Archaeological Chemistry, Second Edition By Zvi Goffer

and 12 atoms of carbon, so it is a *hydrated carbon* or *carbohydrate*. The process by which green plants make carbohydrates, known as *photosynthesis*, beyond doubt the most important biological process on earth, requires the energy provided by sunlight: in its simplest form, the photosynthesis process involves the production of relatively simple carbohydrates, such as glucose, from carbon dioxide and water, as represented by the following equation:

$$6CO_2 + 6H_2O \xrightarrow{\text{sunlight energy}} \underset{\substack{\text{or } C_6(H_2O)_6 \text{ glucose} \\ \text{(a carbohydrate)}}}{C_6H_{12}O_6} + 6O_2 \uparrow$$

Among the many carbohydrates created by the photosynthesis process are *cellulose*, the main structural material of plants and the most abundant organic material on earth, and *starch*, which stores the plants' excess energy. The process is also crucial for animals, since it supplies them with carbohydrates, their principal source of energy. Because the oxygen released during the photosynthetic reaction constitutes the main source of oxygen in the earth's atmosphere, the photosynthesis process also sustains the oxidizing atmosphere in which animals breathe, burning carbohydrates and generating the energy they require for their living processes (Hall et al. 1994). Although the above equation suggests that photosynthesis is a simple process, the actual course that it takes in plants is rather involved: it varies from occasionally quite simple to very complex, and often evolves through a number of different paths and stages that may produce a variety of intermediate compounds. Of great interest in archaeological studies is that some of the intermediate compounds formed during the photosynthesis process in different plants provide ways to study the ancient diets of animals as well as of humans (see Textbox 56).

In addition to making some simple carbohydrate molecules, known as the *monosaccharides* (one sugar, in Greek), plants can combine two molecules of monosaccharides together to form *disaccharides* (two sugars), such as sucrose (table sugar) and maltose (the sugar in germinating grains), and many monosaccharides into extremely large macromolecules known as *polysaccharides* (many sugars, in Greek). Starch and cellulose are examples of polysaccharides made by plants: starch to store energy and cellulose to make up the principal structural material of plants. (See Table 69).

Cellulose, the most abundant biological substance on earth, is the basic and major component of most plant tissues; it is the substance that makes up the walls of vegetable cells and therefore the framework of vegetable fibers. The molecules of cellulose are long, linear polysaccharides, made up of many (several hundreds and even over 1000) aligned glucose *monomers* bonded to each other, as shown in Figure 66. The cellulose contents of some natural fibers are listed in Table 70 (Zeronian and Nevell 1985).

Animals are unable to combine carbon dioxide and water to make carbohydrates as do plants, and they rely on their feeding on plants for their

TABLE 69 Natural Carbohydrates

Carbohydrates	Source
Monosaccharides[a]	
Glucose	The bloodstream
Fructose (also known as fruit sugar)	Sweet fruits and fruitjuices
Galactose (also known as milk sugar)	Milk
Disaccharides[b]	
Sucrose, composed of glucose and fructose	Sweet fruits
Lactose, composed of glucose and galactose	Milk
Maltose, composed of two glucose molecules	Malt (steeped and germinated barley or other grain)
Complex saccharides[c]	
Starch	Plant tubers and seeds (e.g., potato and wheat)
Cellulose	Many plant tissues (e.g., wood, cotton, linen)

[a] Molecules consist of six joined carbon atoms and six water molecules.
[b] Their molecules are made up of two joined monosaccharides.
[c] Also known as *polysaccharides* or *polymeric carbohydrates*; their molecules are *biopolymers* made up of many monosaccharides joined together.

TABLE 70 Cellulose Content of Vegetable Matter

Vegetable matter	Cellulose (%)
Fibers	
Abaca	77–85
Cotton	95–99
Linen	75–85
Jute	70–80
Ramie	80–90
Sisal	65–73
Wood	40–50
Wood bark	20–30

regular supply of carbohydrates. Herbivorous animals get carbohydrates directly, when they eat and digest plants; carnivorous animals acquire carbohydrates indirectly, from the meat of other animals, and omnivorous animals, including humans, acquire them from both plants and animals. Although animals are unable to produce carbohydrates, most animals can combine molecules of monosaccharides into larger polysaccharide molecules, such as those of *glycogen*, which serves as a reserve store of energy in their bodies.

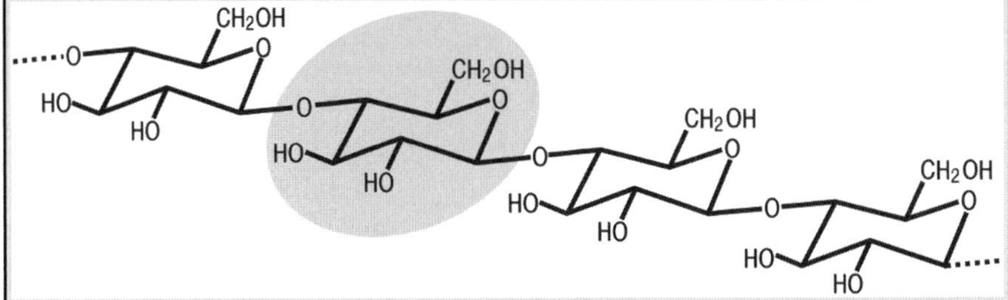

FIGURE 66 Cellulose. Cellulose, a strong yet flexible material, is the major component of all plant tissues, constituting about half of a plant's dry weight. The cell walls of wood, for example, are made up of a mesh of cellulose fibers surrounded by and interlinked with supporting materials, namely, hemicellulose and lignin. The molecules of cellulose are long, unbranched polymers made up of many molecules of glucose (a carbohydrate) linked together. The figure illustrates just four of as many as 1500 glucose units that may make up a cellulose polymer. Because of the length of its molecules, cellulose makes excellent natural fibers as, for example, those of cotton and linen.

Plants synthesize two different types of carbohydrates, generally known as *simple* and *complex carbohydrates*. All the simple carbohydrates, which are also known by the generic name sugars, are sweet-tasting crystalline solids, soluble in water; their molecules basically consist of either one or two combined molecules of carbohydrates: those of the *monosaccharides* – of which glucose (as in ripe grapes and in honey) and fructose (fruit sugar) are examples – include only a single carbohydrate molecule. The molecules of the *disaccharides* include two monosaccharide molecules joined together; the molecules of sucrose (table sugar), for example, a disaccharide synthesized by sugarcane, sugarbeet, and many ripe fruits, are made up of one molecule of glucose combined with one of fructose. Plants also join together simple carbohydrate molecules into long linear or branching polymeric chains (generally referred to as *complex sugars* or *polysaccharides*); starch and cellulose in plants, and glycogen and chitin in animals are examples of polysaccharides. Their molecules are extremely large and serve to provide mechanical support and protection for the living organisms. *Starch*, for example, which stores energy in plants and is hoarded in such organs as the stems (e.g., of onions and potatoes) and/or the roots (e.g., of cassava and sweet potato) of the plants, is made up of many units of glucose joined end to end to each other. Since starch stores energy and humans consume vegetables as their food, plant tissues that hoard starch provide the main sources of carbohydrates in the human diet. *Cellulose*, another polymeric carbohydrate, is the main constituent of most vegetable tissues (see Table 70). Its molecules are even larger and more complex than those of starch; they are made up of thousands of glucose units bonded to each other, forming long polymeric fibers (see Fig. 66) The fibers of such plants as *cotton*, *flax*, and *agave*, for example, which have long been used for making yarn, textile fabrics, cordage, and paper (see Chapter 13), consist of long cellulose molecules. Also *wood*, the most abundant biological material on the surface of the earth, is composed mainly of consolidated cellulose fibers.

9.1. WOOD

Much wood has been used for the construction of dwellings and public buildings, vehicles and vessels and for functional and decorative objects for everyday life and/or ceremonial rituals. Until the eighteenth century wood was also the main source of fuel for heating and illumination and, since the second half of the nineteenth century, one of the major raw materials for making paper. It is of interest, therefore, to discuss the composition and properties of wood in general as well as the characteristics of wood from different types of trees and shrubs that were used in the past (Hackens et al. 1988).

The Nature of Wood

Wood is the fibrous matter inside the bark that makes up the supporting material of the trunks, branches, stems, and roots of trees and shrubs (see Textbox 54). Wood fibers start out as the walls of living cells and, when the cells die, they continue to provide the structural matter that makes wood firm and mechanically strong. Most of the cells of wood are elongated directionally along the axis of growth of the plants and aligned along the length of the trunks, branches, stems, and roots of trees and shrubs. This orientation accounts for the fibrous nature of wood (see Textbox 66) and for its anisotropy, the fact that its properties are dependent on direction, as compared to *isotropic* materials, whose properties are the same regardless of direction.

The Composition and Physical Properties of Wood

Wood is a *composite material* that is made, up basically of a mixture of three main constituents, *cellulose*, *hemicellulose*, and *lignin* (see Textbox 54), all of them biopolymers synthesized by the plants, which differ from one another in composition and structure (see Textbox 58). The physical properties of any type of wood are determined by the nature of the tree in which the wood grows, as well as on the environmental conditions in which the tree grows. Some of the properties, such as the *density* of wood from different types of trees, are extremely variable, as can be appreciated from the values listed in Table 71. No distinctions as to the nature of a wood, whether it is a hardwood or a softwood, for example, can be drawn from the value of its specific gravity.

Greenwood, from living trees and shrubs, contains much water. When wood dries in air, much of the water evaporates and the wood shrinks until it reaches a certain degree of dryness. In the resulting *air-dried wood*, also known as *seasoned wood*, *timber*, or *lumber*, the walls of the wood cells remain saturated with water but the cavities of the cells become relatively dry and the wood attains dimensional stability. The relative amount of water in seasoned wood depends on the nature of the tree from which the wood is derived and the relative humidity and temperature of the surrounding environment. When air-dried wood absorbs or releases moisture (under varying environmental conditions), it undergoes dimensional changes: it increases in size when it absorbs water and shrinks when the water is released. Because of its anisotropy, however, the dimensional variations are of different magnitude across and along the wood's grain, causing wood that is exposed to extremely humid or dry conditions to become warped. To minimize these changes and make it dimensionally stable, most wood used for construction and building is seasoned through exposure to the surrounding environment (Desch and Dinwoodie 1996; Rowell and Barbour 1990).

TEXTBOX 54

WOOD: HARDWOODS AND SOFTWOODS

Wood, the material that makes up the solid interior of trees, consists of a mixture of three major components: two carbohydrates, *cellulose* and *hemicellulose*, and *lignin*, which is not a carbohydrate. *Cellulose*, the main constituent of wood, is a tough fibrous material that constitutes the walls of the wooden cells; *hemicellulose* provides the matrix on which cellulose builds the structure of wood and lignin, a kind of cementing substance, is ubiquitously distributed in all mature wood tissue. Lignin permeates cell walls as well as the intercellular regions, consolidating the structure built of cellulose and hemicellulose into the strong, rigid, and stable composite material that is wood (see Textbox 32). The exact nature of lignin, a chemically complex substance, has not yet been fully elucidated (Glasser et al. 2000).

Most wood can be classified into two broad categories, *hardwood* and *softwood*, terms that distinguish between the nature of the trees rather than between the properties in wood. The term *hardwood*, for example, refers to the wood of deciduous, broad-leaved trees such as birch, oak, and poplar, from the temperate and tropical regions of the world. *Softwood*, on the other hand, is used to refer to the wood of coniferous trees, needle-leaved, evergreen trees such as fir, hemlock, and pine, which grow mainly in temperate and cold regions (see Table 71). The terms *hardwood* and

TABLE 71 Density of Common Types of Wood

Wood type	Density (g/cm^3)
Hardwoods	
Balsa (*Ochroma pyramidale*)	0.14–0.17
Beech (*Fagus sylvatica*)	0.65–0.90
Maple (*Acer* spp.)	0.54–0.76
Oak (*Quercus* spp.)	0.60–0.90
Teak (*Tectona grandis*)	0.68–0.75
Softwoods	
Birch (*Betula* spp.)	0.55–0.65
Fir (*Abies* spp.)	0.32–0.50
Larch (*Larix* spp.)	0.48–0.59
Pine (*Pinus* spp.)	0.43–0.53
Redwood (*Sequoia sempervirens*)	0.35–0.42

softwood do not refer to the actual hardness of the wood, since the wood of some hardwoods, such as those of poplar and aspen, are actually soft, while several softwoods, for example, that of yellow pine, are hard.

The two types of wood differ, however, in their nature and structure. The main structural characteristic of the hardwoods (which are botanically known as *angiosperms*, plants that flower to pollinate for seed reproduction) is that in their trunks or branches, the volume of wood taken up by dead cells, varies greatly, although it makes up an average of about 50% of the total volume. In softwoods (from the botanical group *gymnosperms*, which do not have flowers but use cones for seed reproduction) the dead cells are much more elongated and fibrous than in hardwoods, and the volume taken up by dead cells may represent over 90% of the total volume of the wood.

The trunk and branches of trees grow thicker by producing new layers of wood. A transverse section of the trunk or branches of many trees, for example, is a succession of ring-shaped layers that lie more or less concentrically around the pith (a strand of spongy tissue) at their center (see Fig. 67). Each *wood ring* consists of two layers of tissue, one denser than the other. The rings, which are clearly discernible in many types of wood, are not equidistant from each other, nor are they of uniform thickness: they occur in varying patterns of alternated narrow and wide layers of characteristic shape and color. These features are due to the wood growing under changing temperature, humidity, and solar radiation conditions. The growth of the rings is generally vigorous in the spring and gradually declines toward the winter; the variation in the thickness, color, and density of the layers of each ring reflect variations in the dominant environmental conditions while the wood was growing. The more moisture that is available, for example, the wider the annual rings. The variation in ring width is, in a way, also characteristic of particular tree species. In trees from temperate zones, each ring consists of tissue grown during one year, and the rings are therefore known as *annual growth rings*. Whereas in many softwoods the growth rings are clearly defined and visible, in most hardwoods they are less defined because of tubular vessels (extending longitudinally through a trunk or stem), which cause the rings to appear diffused and not well delimited.

Dendrochronology and Dendroclimatology

The study of the nature of wood growth rings revealed, during the first half of the twentieth century, that there is a direct relationship between the

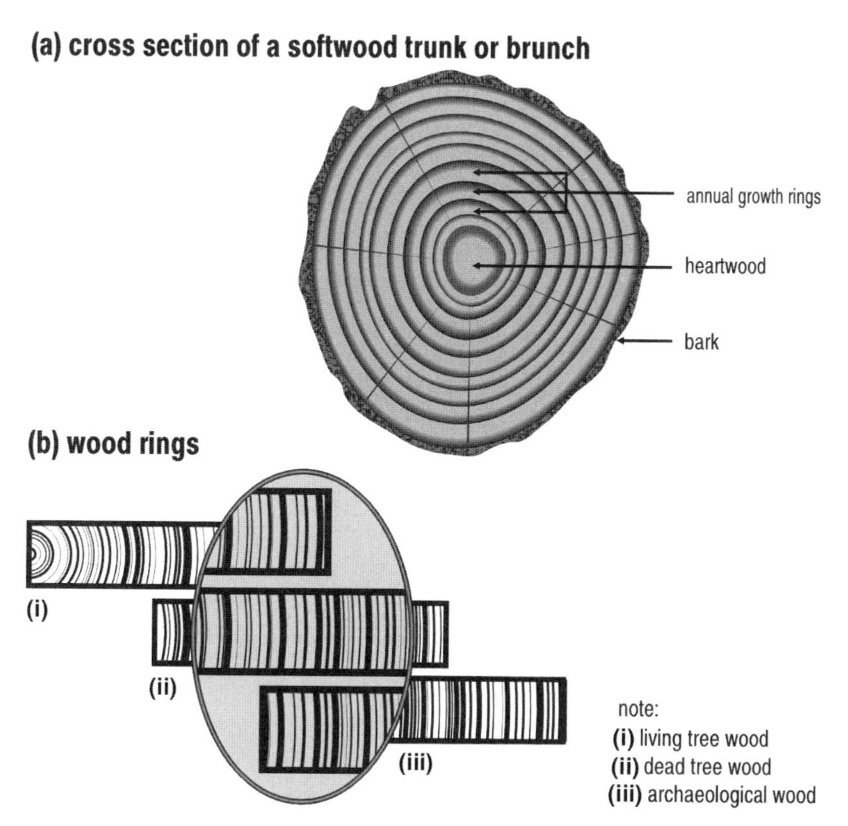

FIGURE 67 **Wood.** Wood is the tough fibrous substance under the bark of the trunk, roots, and brunches of trees, which provides the trees with mechanical strength and serves for the conduction of water and storage of the trees' food reserves. A cross section through a tree trunk, root, or branch reveals a succession of roughly concentric rings **(a)**. The center of a trunk, root, or branch, known as the "heartwood," is the oldest part of the tree. In temperate climates a new ring, known as a "growth ring," is formed each year. Each growth ring is composed of two parts: a light and relatively thick inner portion (which grew during the high-rain season, when the tree received plenty of water) and a surrounding narrower and darker border (which grew during the drier season). Thus the wood growth rings in the trunks and branches of trees preserve a continuous record of annual environmental events and, in particular, of climate. Studying the nature of the rings, as seen on cross sections of trunks or branches, provides information on past environmental changes. Dendrochronology, for example, a method of absolute dating, is based on counting and analyzing wood growth rings. Matching the patterns of rings from different trees that grew under similar environmental conditions makes it possible to determine the exact year in which each ring was formed. The procedure is known as "cross dating." **(b)** the magnified, matching pattern of growth rings from the cross sections of three different trees enables cross dating the wood of the trees.

environmental conditions under which trees grow and the nature of the rings. This finding led to the development of two new disciplines of much relevance to archaeological studies: *dendrochronology* and *dendroclimatology*. *Dendrochronology* is a method of dating wood based on analyzing and counting annual growth rings (Dean 1997; Hillam 1987); *dendroclimatology* is the study of the relationship between the nature and thickness of the rings and the environmental conditions in the region in which a tree grew (Baillie 1995; Fritts 1976). In seasonal climates, the fluctuations in temperature and humidity are annual, and so are the growth rings formed: thick rings are associated with rainy years and thin ones with dry years. In northern hemisphere climates, temperature is the dominant factor regulating the growth and thickness of rings; tree rings formed during cold years are therefore thinner while those formed in warmer years are thicker (see Fig. 68).

The seasonal variations in climatic conditions in any particular region are never exactly the same. Since the thickness of tree rings is related to the environmental condition in the region where a tree grows, the pattern of rings displayed by trees from any particular region is unique, differing from that of trees from other regions. The pattern closely matches, however, that

FIGURE 68 Wooden bowl. A wooden bowl made of boxwood, first century C.E., from Qumran, Israel; the wood used to make the bowl seems to have been imported from Turkey. The excellent preservation of the bowl is due to extremely hot and dry environmental conditions in the region. Three conditions are necessary for wood to decay: (1) a favorable temperature (0–32°C), (2) moisture in excess of the fiber saturation point (above 25–30%) and (3) an adequate supply of oxygen. If any one of these is eliminated wood remains well preserved for long periods of time.

of rings in trees from other regions in which similar climatic conditions prevail. This makes it possible to identify, by comparison, similar ring patterns in trees from different regions: a similar, but not identical, tree ring pattern in wood from different regions indicates that the rings were formed during the same time period and under similar climatic conditions.

Not all wood can, however, be dated using growth rings. Only trees that grow in temperate areas produce clearly defined rings that can be used for dating. In regions outside temperate areas there are no clearly distinct annual seasons, and the trees growing in these regions often lack a well-differentiated ring pattern. These trees may even have more than one period of growth during the year, which invalidates the basic assumptions of the dendrochronologic method.

Dead Wood

Most wood is durable and, under favorable environmental conditions, can last for extremely long periods of time. This is why in most areas of the world wood remains from all periods are often found in archaeological sites. Like all materials, however, wood is susceptible to *weathering* and *biodegradation*.

Weathering is brought about by the combination of physical and chemical changes caused by variations in the environmental conditions; heat, light (particularly ultraviolet radiation), humidity, and the abrasion caused by the mechanical forces of wind and flowing water, for example, first affect the outer surface of wood, which is initially discolored. Under continued exposure the surface eventually becomes rough, and at a still later stage cracks develop in the bulk of the wood, which becomes increasingly friable and disjointed (Feist and Hon 1984). As it is a biological material, wood is generally also affected by *biodegradative processes* brought about by bacteria, fungi, plants, and animals. Microorganisms, bacteria, fungi, and rodents, for example, use wood as food and cause the decay of much wood. Sometimes, biodegradative processes (generally known as *delignification*) are initiated by microorganisms that dissolve lignin (the noncellulose component of wood) and leave the cellulose fibers exposed to further biological attack and to their gradual conversion into simpler substances. A significant loss of strength takes place during the early stages of delignification that may eventually lead to the total disappearance of the wood (Florian 1990; Kent Kirk and Cowling 1984). Occasionally a reverse process to the one just described takes place, when microorganisms at the surface consume and digest the outer wood's cellulose, leaving soft and mechanically weak lignin.

Decayed wood is eventually converted to *humus*, a dark and amorphous form of organic matter in the soil, or, after very long periods of time (running

TABLE 72 Factors Affecting Wood Decay

Decay-causing factor	Outcome
Weathering	
Heat	
Slight	Darkening
Intense	Conversion to charcoal, burning, and total destruction
Light (visible and ultraviolet)	Color changes and chemical degradation
Mechanical (wind and flowing water)	Surface roughening and comminution
Chemical factors	Staining, discoloration, and loss of strength
Biodeterioration	
Bacteria, fungi, and algae	Discoloration and biochemical decay
Insects, birds, and mammals	Boring, pecking, and cutting

up, at ordinary temperatures, to millions of years), and compaction by external pressures, to *coal*. Some wood may, however, be preserved from total decay by *fossilization* processes, whereby the vegetable tissues are gradually replaced by minerals derived from the environmental surroundings (see Table 72 and Chapter 16).

Burned Wood

When wood burns in air, the carbon and hydrogen in the wood combine with oxygen in the air and are converted to carbon dioxide and water. When cellulose burns, for example, the following reaction takes place:

$$\underset{\text{cellulose}}{(C_6H_{10}O_5)_n} + \underset{\substack{\text{oxygen} \\ \text{(from the air)}}}{6nO_2} \rightarrow 6nCO_2\uparrow + 5nH_2O\uparrow$$

The burning process leaves very little solid remains: only *ash*, made up of inorganic salts that rarely make up more than a few percent of the total mass of wood. When wood burns with a restricted supply of air, however, and there is insufficient oxygen to combine with all the carbon in the wood, the remains are made up of *charcoal*, a very porous and impure form of carbon. Charcoal is extremely stable; it does not decay, nor is it altered by most microorganisms, and it may be preserved for very long periods of time; charcoal often also preserves the morphology of the burned wood. Because of its stability, charcoal residues are often found in archaeological sites where wood was either used as fuel or otherwise burned.

Archaeological Wood

Most ancient wooden objects recovered in archaeological excavations are usually in a decayed, weak, and friable condition that requires stabilization before the objects can be safely handled and studied. Stabilization of wood and decayed wooden objects, generally includes the use of *consolidants*, liquid solutions of a resin that impregnates and fills gaps in the wood and on drying solidifies, strengthening its fragile, deteriorated structure (Thompson 1991; Rowell and Barbour 1990).

Of particular interest in archaeological studies is *waterlogged wood*, wood that has been submerged in water or buried in very wet soil for long periods of time. When submerged in water, wood undergoes chemical and biological processes that, at an advanced stage, leave it with very little solids, containing more water than the total mass of dry matter (Jensen and Gregory 2006). Waterlogged wood is, therefore, very soft and mechanically weak, retaining its shape only as long as it remains wet. If removed from the site where it was submerged or buried and then exposed to the atmosphere, the water soon evaporates and, as the wood dries, it undergoes severe dimensional changes. This introduces internal stresses that further reduce the strength of the wood, which develops cracks, cause flaking and may even result in the total breakdown of the wood. The preservation of waterlogged wood removed from a wet environment requires, therefore, specialized treatment beginning even before removal from the submersion or burial site (Jordan 2001; Grattan and McCawley 1981).

9.2. GUMS

Gums are complex carbohydrates exuded from plants, or produced by the decomposition of vegetable matter, that have been used since remote times as *adhesives*, *sizes*, and *binders* (see Table 73). Most gums are tasteless and odorless solids that either dissolve or swell in water to form adhesive, viscous mucilages. When the water evaporates from a mucilage, the

TABLE 73 Natural Gums

Gum	Source
Gum arabic	*Acacia segalencis* and other acacia trees, native to eastern Africa
Gum kadaya	*Sterculia urens* trees from India
Locust bean	Seeds of carob (*Ceratonia siliqua*) trees
Tamarind	Seeds of tamarind (*Tamarindus indica*) trees
Tragacanth	*Astragalus gummifer* shrubs from southern Europe and the Middle East

remaining dry gum acquires considerable adhesive and consolidating power (Parry 1918). Most gums are derived from plants that grow in dry regions of the world, although only a few of those available have been of importance to humans.

TEXTBOX 55

ADHESIVES; SIZES AND BINDERS

The *adhesive cements*, or simply *adhesives*, are viscous liquid preparations used mainly for joining together different objects or parts of objects into coherent units, for *sizing* (sealing) porous surfaces, and as *binders* in the preparation of paints (see Textbox 18). Until the end of the nineteenth century the only adhesives known to humans were of natural origin, derived either from vegetable secretions or from animal fluids and tissues. Few modern adhesives are now derived from natural substances; most are artificial, human-made (synthetic) substances developed since the twentieth century.

Gums, Glues, and Resins

The substances mostly used as adhesives since antiquity were the *gums* and *glues*. *Gums* are mucilaginous saps or exudations of plants that include carbohydrates in their composition. Most gums are soluble in water, or swell and then become suspended in water. *Gum arabic*, *gum tragacanth*, and *ghatti* (see Table 73) are examples of gums that have been widely used as adhesives, binders, and sizes. Other substances of vegetable origin that were also used in adhesive preparations include a few natural *resins* such as *rosin*, from coniferous trees, and *rubber*, from ficus trees.

Glues, also mucilaginous preparations, are derived from processed animal hides, bones, and fish offal. They are composed mainly of proteins derived from the hydrolysis of collagen extracted from animal tissues (see Textbox 61). Glue has been made for many centuries by a relatively simple, basic process in which bone, skin, or fish offal is first washed to remove dirt (skins are also soaked in lime water for some time to remove hair, in which case they are washed again). The clean bone, skin, or fish offal is then heated in water (to temperatures above 50°C) for some hours, to extract the glue (which dissolves in the water), and the glue solution is then boiled to evaporate the water. After all the water has evaporated, the glue remains as a very viscous brown liquid that solidifies on cooling. Heating the animal tissues in water is often repeated several times; the temperature is slightly increased each time so as to extract all the glue. Other frequently used types of glue include *blood*, *casein* (derived from milk), and *egg albumen* (see Table 74).

TABLE 74 Natural Adhesives (Classified by Origin)

Organic			Inorganic
Vegetable	Animal	Mineral	
Gum	Glue	Asphalt	Mud
Latex	Casein	Bitumen	Lime cement
Starch	Blood	–	Gypsum cement

Adhesive Preparations

Most adhesive preparations consist of at least two main components: an *adhesive* and a *solvent*. The *adhesive* is the binding substance that, when dry, attaches and joins together adjacent surfaces, or sizes (covers and seals) porous surfaces. Water was, for many millennia, practically the only solvent used in adhesive preparations. Adhesives soluble in water form solutions and, if insoluble, become suspended in the water, usually in such concentration as to render the preparation viscous and tacky. As well as the adhesive and the solvent, a variety of *additives* have often been added to adhesive preparations so as to improve their spreading properties or modify other qualities. Additives, which increase the viscosity of the preparations, are known as *fillers*, while those that increase their volume are referred to as *diluents* (Eagland 1988; Masschelein-Kleiner 1985).

9.3. RESINS

The *resins* are substances derived mostly from vegetable sources; some have been used as binders in the preparation of paints and varnishes, others as incense burned in ritual ceremonies, and a few, such as amber, have been used on their own, as semiprecious stones. It should be noted, however, that since the midtwentieth century the term *resin* has acquired a new meaning: that of a synthetic pliable "plastic" material that can be shaped, mostly when hot. Synthetic resins are used mainly for packaging and for making textile fibers and automobile parts. In the discussion that follows the term *resin* is used to refer only to resins of natural origin (Serpico and White 2000a; Parry 1918).

The chemical composition of the natural *resins* is very diverse, and their molecular structure may be highly complex; most resins of vegetable origin, however, are chemically related to the carbohydrates. Dry vegetable resins

TABLE 75 Natural Resins

Resin	Source	Uses
Vegetable		
Amber	Extinct *Pinus succinifera* trees	Gemstone; varnish
Dragon's blood	Fruits of *Daemonorop draco* palms	Varnish
Dammar	Trees of family *Dipterocapacaea*	Torch fuel; varnish
Frankincense	Trees of genus *Boswellia*	Incense
Kauri	*Agathis australis* pine	Varnish
Labdanum	*Cistus landaniferus* trees	Incense
Lacquer	*Melanorrhoea usitata* trees	Varnish
	Rhus succedaneae trees	Varnish
	Rhus vernicifera trees	Varnish
Mastic	*Pistacia lentiscus* evergreens	Varnish
Rosin	Coniferous trees	Adhesive; varnish
Sandarac	*Callitis quadrivalis* trees	Varnish
Animal		
Shellac	Secretion of *Karria lacca* insects	Varnish
Mineral		
Gilsonite	A type of bitumen	Adhesive; lacquer

are usually hard, brittle, noncrystalline translucent solids. When struck, they break with a conchoidal fracture, and if ignited, they burn with a sooty flame. Only a few resins are soluble in water, the vast majority are not, but are soluble in organic solvents such as alcohol. This makes most resins suitable for use as *sealers*, protective coatings on porous materials such as wood, skin, and textiles. Among the most widely used resins of vegetable origin are *amber*, *dammar*, *copal*, *dragon's blood*, *mastic*, and *rosin*, all of which are either exudates from living, dead, or even fossilized plants, or extracts from the stumps or heartwood of living trees. Table 75 lists natural resins that have been used since ancient times.

Amber. *Amber* is a relatively hard, transparent, or translucent fossil resin that was exuded during the Oligocene geologic period (over 25 million years ago) from now-extinct coniferous trees of the species *Pinites succinifer*. It generally occurs in irregular rod- or drop-shaped accumulations whose color may vary from shades of yellow, through brown, to orange, and occasionally red. Its main constituents are a mixture of organic substances that include carbohydrates and succinic acid (Fraquet 1987). Amber is easily shaped, either by carving and polishing its surface or by heating, since it softens (at about 150°C) and can then be molded. Because it is easy to shape, amber has been used, since early prehistoric times, for making ornamental and decorative objects such as beads, amulets, and boxes. Moreover, as it dissolves in some vegetable oils, amber has also been used as a binder in the preparation of varnishes. One of the oldest and best known sources of amber is the Baltic Sea coast of Germany, but there are also rich deposits off the coasts of Sicily and England in Europe, in Myanmar (Burma) in Asia and in the Dominican Republic and Mexico in the Western Hemisphere. The provenance of archaeological amber has been extensively studied using a variety of physical methods (Lambert et al. 1996; Beck 1970).

Rosin. *Rosin* resin, also known as *colophony*, is collected from one of over 100 different types of trees that grow throughout Europe, Asia, North America, and New Zealand. The term *colophony* seems to derive from the ancient city of Colophon, in Lydia, which produced a high-quality resin. Rosin is drawn directly from living trees in a tapping process that entails inducing outflow of the resin from the trees. Rosin has been used as an *adhesive* and *size* and in the preparation of *paints*.

Mastic. *Mastic* is the resin obtained from the small mastic tree *Pistacia lentiscus*, of the sumac family, found chiefly in Mediterranean countries. When the bark of the tree is injured, the resin exudes as drops. Mastic is transparent and pale yellow to green in color. The main ancient uses of mastic were as an *adhesive*, for making *varnish*, as a *medicine*, and for *flavoring*.

Dragon's Blood. *Dragon's blood*, a red resin described by Dioscorides, the Greek botanist from the first century C.E. and by other early writers, was derived from a number of different plants. A main source of the resin seems to have been *Dracaena cinnabari*, a tree of the agave family, from which it is exuded as garnet colored drops when the trunk or branches of the tree are injured. The early Greeks and Romans believed dragon's blood to have medicinal properties. Its main use in the ancient past, however, was as a coloring material and, since the end of the eighteenth century Italian crafters used it as a varnish for violins.

Shellac. *Shellac* is outstanding in that it is the only resin of animal origin. Shellac is a yellow, orange, or brown solid derived from *lac*, the secretion of the scale-like *Kerria lacca* insects that inhabit trees in areas of India and Thailand. To obtain the resin, twigs encrusted with the insect's secretion are cut down from the trees, the incrustation is removed from the twigs, coarsely crushed, washed with water, and filtered: a red dye, known as the *Lac dye*, is dissolved by the water and can be subsequently obtained from the filtrate; shellac, which is insoluble in water, remains as a solid residue (Bose et al. 1963; Parry 1935). Although insoluble in water, shellac is soluble in alcohol, and the solutions, at various concentrations, have been used as adhesives and sealers for wood, as lacquers on wood and other solids, and as components of cosmetic preparations.

Lacquers

Lacquers are resinous substances obtained from certain trees and used as varnish, which have been and still are used in the East, particularly in China and Japan, for making "lacquerware": wooden objects covered with a highly regarded, durable varnish layer having a fine luster. Resins used for preparing lacquer are derived from the sap of various trees but mainly from *Rhus vernicifera* (generally known as the "varnish tree"), *Rhus succedaneae*, *Melanorrhea usitata*, and from the Dipterocarpacaea and Burseraecea families. It seems that the use of the resins from these trees for making lacquer began over 4000 years ago, when they were used to coat furniture and personal items such as combs and earrings. The resin derived from trees from the Dipterocarpacaea and Burseraecea families, known as *Damar resin*, was also used in many Pacific Islands and in the Malay Peninsula as a fuel for lighting torches.

Incense

Resins such as *myrrh*, *frankincense*, and *labdanum* achieved prominence early in antiquity for their fragrant smell and were used for making perfumes and medicines as well as for burning as *incense* during religious services and ritual ceremonies (Morris 1984). Some well-known incense resins are listed in Table 76.

TABLE 76 Incense Resins

Incense	Source	References
Frankincense	Trees of genus *Boswellia*	Watt and Sellar (1996); Martinetz et al. (1988); Groom (1981)
Galbanum	*Ferula galbanifera* plants	Moldenke and Moldenke (1952)
Myrrh	Trees of genus *Commiphora*	Watt and Sellar (1996); Martinetz et al. (1988); Groom (1981)
Storax	Probably from *Styrax officinalis* plants	Moldenke and Moldenke (1952)

9.4. CARBOHYDRATES, ISOTOPES, AND THE STUDY OF ANCIENT DIETS

Animals, including humans, create their body tissues from the food they consume; all the substances that constitute their bodies are derived from food. Food is, therefore, essential for their survival, and food and the search for adequate supplies of food have determined, and still determine, the life-patterns of ancient as well as of modern people. The study of past human feeding habits and their diets is, therefore, central to the understanding of ancient societies. A large number of studies have shown that there is strong correlation between the type food consumed and the relative amounts of the stable isotopes of some elements in the body or in its dead remains. This seems to confirm an axiom generally recognized in science, that "You are what you eat."

Studying the relative amounts of the isotopes of such elements as carbon and nitrogen in the tissues of human remains and in the plants that were consumed as food in specific areas makes it possible to reconstruct the diet of the humans living in that area. Carbon and nitrogen are common constituents of the earth's atmosphere as well as of plant and animal tissues. Each of these two elements has a light and a heavy stable isotope. The light stable isotopes, carbon-12 and nitrogen-14, make up most of the bulk of these elements on earth. The heavy, stable isotopes, carbon-13 and nitrogen-15 respectively, occur naturally to an extent of less than one percent of the total amount of each one of the two elements. Carbon also has a unstable isotope, carbon-14 (radiocarbon), which makes up only two parts in 1,000,000,000,000 (a million million) of the total amount of carbon (see Table 66). Because of specific chemical processes, especially *isotopic fractionation* during living processes, the weight ratios between the isotopes of each of these elements is altered in distinctive ways in different plants and animals and it is also altered along the food chain: some tissues become enriched in the heavy isotopes while others are depleted of them as the elements pass from plants to herbivore, to carnivore and/or to omnivore animals. Consequently, the stable isotopes of oxygen and nitrogen occur in plant and animal tissues in different and characteristic weight ratios. Much can be learned about what people ate in the past by analyzing the isotopic composition of their remaining tissues such as bone or hair (see Textbox 56) (Katzenberg 2000; Burton 1996).

TEXTBOX 56

STABLE CARBON ISOTOPES AND ANCIENT DIETS

The outcome of the photosynthesis processes is ultimately similar in all green plants: carbon dioxide from the atmosphere is taken up by the plants, where it reacts with water to form carbohydrates and oxygen; the carbohydrates are assimilated by the plants while the oxygen is released to the atmosphere (see Textbox 53). Extensive studies have shown that the conversion of carbon dioxide and water into carbohydrates in different plants may follow, however, one of three different photosynthetic pathways, which are usually referred to as the *C3*, *C4*, and *CAM paths*. Each type of plant follows just one of these three pathways.

More than 95% of the plants of the world follow the *C3 path*, whereby the carbon dioxide is initially incorporated, by the plants, into intermediate compounds made up of three atoms of carbon – which is the reason these plants are known as *C3 plants*. Only about 1% of all plants, including maize, millet, sorghum, and sugarcane, follow the *C4 path*, incorporating carbon dioxide into intermediate compounds made up of four atoms of carbon – which is the reason these plants are known as *C4 plants*. The remaining 4% of plants, mostly succulents (cacti), follow the third, *CAM*, *path*, which has no bearing on the discussion that follows.

The relative weight of the two stable (nonradioactive) isotopes of carbon, *carbon-12*, and *carbon-13* (see Table 66), has been constant for quite a long time. Thus the molecules of carbon dioxide in the atmosphere may include atoms of carbon-12 or carbon-13 as one of their components. The intermediate C3 and C4 compounds synthesized by the plants from atmospheric carbon dioxide also include, therefore, atoms of both carbon-12 and carbon-13. The two isotopes, as well as the compounds they are part of, react (chemically) in the same way. Since they are of slightly different size and weight, however, whenever there is a chemical reaction involving carbon atoms, compounds containing the lighter carbon-12 react faster than do those containing carbon-13. The difference in *reaction rate* results in a process known as *isotopic fractionation*, whereby the relative abundance of the two isotopes in the products of a reaction is different from that in the reactants. Thus, since C3 and C4 plants follow different chemical pathways during the photosynthesis process, the two types of plants *fractionate* the isotopes of carbon in entirely different ways, and each isotope, carbon-12 and carbon-13, occurs in a different and characteristic relative amount in the tissues of C3 or C4 plants. Moreover, the *weight ratio* carbon-13 : carbon-12 (known as the *isotopic signature*) in which the two isotopes occur in C3 and C4 plants is well known. Determining the *carbon isotopic signature* in any particular type of plant from any region of the world thus makes it possible to establish whether the plant is of type C3 or C4 (Stanford 1993; Van der Merwe 1982).

When plants are consumed as food by herbivorous animals, the isotopic signatures in the plants are passed on to the consumers. Therefore, provided the isotopic signatures of C3 and C4 plants are known, determining the isotopie signatures in the tissues of herbivorous animals enables one to determine the relative amounts of C3 and C4 plants that the animals consumed as food, and to reconstruct their diets. Moreover, since carnivorous and omnivorous animals, including humans, feed on herbivorous animals as well as on plants, determining the isotopic signatures of the isotopes of carbon in tissues of ancient animals and humans makes it possible to elucidate the components of their diets.

Many isotopic studies provide perspectives on particular aspects of ancient diets. In the American continent, for example, maize, a *C4* plant, was a dominant food that played a fundamental role in the development of prehistoric societies, and its consumption had important health significance. A study of isotopes in human bone from eastern North America revealed when maize was adopted as a food and when it became a major component of human diets. Isotopic analyses of carbon from human tissues disclosed that, before the ninth century C.E., C4 plants had little or no place in human diets. Soon afterward, however, C4 plants became part of the diet, a change that grew more pronounced with the passing of time. By the thirteenth century C.E., C4 plants (maize) seem to have become a primary human food, representing as much as 70% of the diet of the inhabitants of the North American woodlands, for example. A similar change detected also in human remains from Panama and Venezuela (although the change occurred at different times than in North America) points to the introduction, also into these regions, of the cultivation of C4 plants (Ambrose 1987; Larsen 1997). In the Old World, maize become part of the human diet only after discovery of the Americas. Other C4 plants, identified in human remains from before that time, have been studied, including sorghum in Nubia (White and Schwarcz 1989) and millet in central Europe (Murray and Schoeninger 1988) and northern China (Schwarcz and Schoeninger 1991).

Also the stable isotopes of nitrogen, like those of carbon, are fractionated when nitrogen is incorporated into the plants. Determining the isotopic ratios between nitrogen-15 and nitrogen-14 in animal remains also reveals information on the diets of ancient animals and humans (White 1999; de Niro 1987).

LIPIDS
OILS, FATS, AND WAXES; SOAP

Lipids, from the Greek word *lipos*, for animal fat or vegetable oil, is the generic chemical term for *oils*, *fats*, and *waxes* (see Table 77). In the body of living organisms the lipids serve as highly concentrated, long-term storage substances of the excess energy acquired or created by the body. If and when the organism requires it, the energy stored in the lipids can be released back to the body, yielding more than twice the amount of energy supplied by equal weights of carbohydrates or proteins. In animals, fat under the skin also provides a thermal insulating layer that prevents heat from escaping from their bodies and protects them from extreme external temperatures.

10.1. OILS, FATS, AND WAXES

All the lipids are viscous when liquid and greasy (lubricious) to the touch. Some, as for example tallow and suet, are solid and hard at ambient temperature, others, such as lard and butter, are solid and soft, while still others are liquid, as are olive and linseed oils. Lipids that are solid at room temperature are generally of animal origin and are traditionally known as *fats*; those that are liquid at ambient temperature are derived mostly from plants and are known as *oils*. The properties of a wide range of lipids, natural fats, oils, and waxes have been known to humans for a long time and have been put to multifarious uses; some are consumed as foodstuffs, others are burned as fuel for lighting, or used as lubricants (for turning wheels and moving heavy objects), to protect – insulate and waterproof – or to decorate the surface of a variety of solids (Evershed et al. 2001; Serpico and White 2000a).

Archaeological Chemistry, Second Edition By Zvi Goffer

TABLE 77 Lipids: Oils, Fats, and Waxes

Lipids	Examples	Uses
Fats	Butter	Food
	Tallow	Making candles
Oils		
Drying oils	Linseed oil	Making paint
Nondrying oils	Olive oil, soybean oil	Food
Waxes	Beeswax	Making molds and candles
	Lanolin	Cosmetics

TEXTBOX 57

THE LIPIDS

The *lipids* are a large group of organic compounds including *oils*, *fats*, and *waxes*. Akin to the carbohydrates, the molecules of the lipids are composed of only three elements: carbon, hydrogen, and oxygen. They differ from the carbohydrates, however, in their chemical properties and in that they are insoluble in water and lubricious to the touch. If lipids are exposed to the environment, particularly to oxygen in the air, for extended periods of time, they are oxidized. Some oils (known as the *drying oils*), for example, solidify when oxidized. Most other oils, as well as the fats, do not solidify when oxidized but become *rancid*, gradually developing an unpleasant taste and exuding an objectionable stale odor (Gunstone 2004).

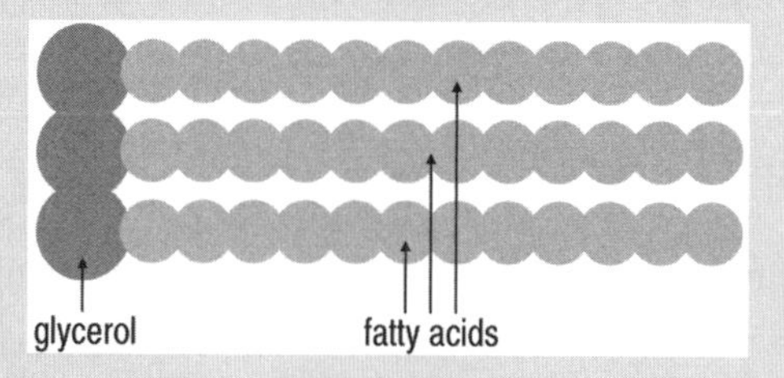

FIGURE 69 The triglycerides. "Triglycerides" is a generic name for the oils and fats, which are also known as "lipids." The molecules of all triglycerides consist of three molecules of fatty acids bonded to one of glycerol (an alcohol also known as "glycerine"). The molecules of the fatty acids are made up of long chains of atoms of carbon linked to each other; the chain length generally varies from about 16 to 22 carbon atoms. All the chains of the triglycerides may be of the same length (as illustrated), only two the same, or all of different lengths (see Table 78).

The molecules of all lipids can be described as being derived from the reaction between two different types of molecules: one of *glycerol* (an alcohol with a hydroxyl group on each of its three carbons) and three of *organic acids*, commonly known as *fatty acids* (see Fig. 69). Each fatty acid is made up of a chain of between 12 and 22 carbon atoms with hydrogen atoms attached to each carbon in the chain. Thus another generic term for the lipids, *triglycerides*, is derived from the composition of their molecules: three molecules of fatty acids attached to one of glycerine. At room temperature, triglycerides may be liquid or solid; those that are liquid are called *oils*, and the solid ones at room temperatures are called *fats*. Triglycerides also differ in the composition and structure of their component fatty acids (see Table 78).

TABLE 78 Main Fatty Acids in Common Oils and Fats

Lipid	Fatty acid components	Number of carbon atoms in fatty acid chain
Fats		
Animal fats	Palmitoleic and stearic acids	16 and 18
Butterfat	Butyric, caproic, and vaccenic acids	4, 8, and 18
Oils		
Coconut oil	Caprylic and lauric acids	8 and 12
Corn oil	Linoleic acid	18
Olive oil	Oleic acid	18
Palm oil	Palmitic and myristic acids	16 and 14

Oils

Most oils, are derived from the seeds, nuts, fruits, and other organs of plants. To express their oil, the relevant plant organs are usually crushed and pressed. Oils are conveniently classified into two main types: *drying* and *nondrying oils*. The *drying oils* solidify, they are said to "dry" on exposure to the atmosphere. Actually, they do not dry but *polymerize*, that is, they combine with oxygen and water in the air to create relatively hard and tough polymeric films (see Textbox 58). It is because of the property of polymerizing into solid, cohesive, and resistant films that drying oils have been used primarily as *binders* for the preparation of paints and varnishes and as *sealers*, to seal the pores and protect the outer surface of porous solids such as wood, skin, and leather. *Linseed oil*, probably the best known drying oil, seems to have been known since as early as the second century B.C.E. Linseed oil–based paints and varnishes were quite widely used in northern Europe since the thirteenth century C.E. In southern Europe, where oil painting was introduced at later times, *walnut oil* was initially preferred. Apparently only during the fifteenth century C.E. did linseed oil begin to be used throughout Europe for the preparation of paints and other surface coatings (Mills and White 1996; Eastlake 1960).

TEXTBOX 58

POLYMERS

Polymers are substances whose molecules are very large, formed by the combination of many small and simpler molecules usually referred to as *monomers*. The chemical reaction by which single and relatively small monomers react with each other to form polymers is known as *polymerization* (Young and Lovell 1991). Polymers may be of natural origin or, since the twentieth century, synthesized by humans. Natural polymers, usually referred to as *biopolymers*, are made by living organisms. Common examples of biopolymers are *cellulose*, a carbohydrate made only by plants (see Textbox 53); *collagen*, a protein made solely by animals (see Textbox 61), and the nucleic acid DNA, which is made by both plants and animals (see Textbox 64).

Some biopolymers consist of relatively few, linearly aligned monomers: some biopolymeric carbohydrates, for example, among the smallest natural biopolymers, are formed from the combination of as few as five molecules of glucose. Others involve extremely large numbers of monomers joined to each other in either linear or complex tridimensional arrangements. These include biopolymers such as cellulose and collagen, that constitute much of the biological matter in the body of living organisms.

Proteins, the main constituents of the animals body, are *polypeptides*, biopolymers consisting of many amino acid molecules (the monomers) combined together (see Chapter 11); collagen, for example, the main component of animal skin, is a complex protein consisting of many molecules of amino acids combined together into *polypeptide chains* (see Fig. 71). *Polysaccharides*, the essential constituents of plants, also consist of many monosaccharide molecules combined together. Cellulose, the most abundant biological material on earth, which makes up most of the structural matter of plants, for example, is a biopolymer that consists of many thousands of molecules of glucose (the monomer) interconnected in a linear arrangement (see Fig. 66). The nucleic acids, which occur in animals and plants, are also *polynucleotides* made up of *nucleotides*, themselves biopolymers made up of three different monomers: *sugar*, *nitrogen bases*, and *phosphate groups* (see Fig. 74). Many natural fibers, such as cotton, wool, and silk, consist of linear biopolymers. Cotton, composed of cellulose, for example, consists of polymerized carbohydrate monomers; wool and silk are proteins composed of polymerized amino acids.

Since the early twentieth century, a vast and ever-increasing range of artificial polymers have been synthesized. Nylon, polythene, styrofoam, and plexiglass are just a few of the myriad artificial polymers that have greatly enlarged the range of materials available to humankind, serving as structural materials, plastics, rubbers, surface finishes and textile fibers (Nigam and Prasad 1992).

The *nondrying oils*, which are comestible, do not polymerize and therefore, do not sdidify when exposed to the atmosphere. On exposure to the air for long periods of time, however, they combine with oxygen and water and become *rancid*, develop an offensive odor and acquiring a bad taste. *Olive oil* is a good example of a nondrying, comestible oil. Pressing the pulp of the fruit of the olive tree (*Olea europea*) expresses the yellow-green viscous oil that has been used since antiquity as food. In the Mediterranean Sea area, to which olive trees are indigenous, olive oil was used not only as a foodstuff but also as a cleansing agent, a fuel for illumination, and an ingredient of ointments employed for cosmetic and ritual purposes (Frankel et al. 1993). Other nondrying oils widely used as foodstuffs are listed in Table 79a.

Fats

The animal *fats* are greasy solids, insoluble in water, that melt at a low temperature, generally below 100°C (see Table 79b). They are obtained either by *rendering* (the process of separating fat from fatty tissues, such as muscle or skin, by melting) or by *churning* (vigorously agitating milk). *Lard*, the fat along the back and underneath the skin of the hog, *suet*, the hard fat around the kidneys and loins of cattle and sheep, and *tallow*, the overall fat of cattle and sheep, are made by rendering fatty tissues; *butter*, a mixture of water and fat is obtained by churning milk (Baer and Indictor 1973; Jamieson 1932).

TABLE 79a Natural Oils

Oil	Source	Habitat	Uses
Nondrying Oils			
Olive	Olive tree (*Olea europea*) fruits	Mediterranean sea area	Food, cosmetic
Soybean	Soy plant (*Glycine max*) beans	China	Food
Palm oil	Palmtree (*Elaeis guinnesis*) fruits	China, India, West Guinea	Food
Drying Oils			
Linseed	Flax plant (*Linum usitatissimum*) seeds	Europe, Asia, America	Binder in paints
Tung	Tung plant (*Aleurites fordii*) nuts	East Asia	Varnish; binder in paints

TABLE 79b Natural Fats

Fat	Melting temperature (°C)	Source	Uses
Tallow	42–43	Adipose tissues of cattle, sheep, and other animals	Making candles, dressing leather
Suet	45–50	Surrounds the kidneys and loin of cattle, sheep, and other animals	Food
Lard	38–40	The abdomen of pigs	Food, cosmetics
Butter	32–35	Milk	Food
Lanolin	38–44	Wool	Cosmetics

Waxes

The natural *waxes* are hard and often brittle substances at room temperature, which on warming acquire and retain a degree of plasticity, until melting at higher temperatures. Although they do not feel greasy to the touch as do the fats, the waxes are lipids and therefore chemically related to the oils and fats. Some waxes made by living organisms, vegetables as well as animals, are secreted from the surface of the leaves of plants, the exoskeleton of insects, the feathers of birds, or the furs of mammals. Some insects, such as bees, for example, produce wax that they use as a material to build their hives. Most waxes are *hydrophobic* (they repel water), and therefore make excellent water-proofing materials and are used to protect the external surface of many materials. Many natural waxes are extensively distributed on the surface of the

TABLE 80 Natural Waxes

Wax	Melting range (°C)	Source	Uses
Vegetable			
Candelilla	67–71	Stems of plant *Pedilantus pavonis*	Varnish
Carnauba	80–86	Leaves of carnauba palm (*Copernicia cerifera*)	Varnish
Animal			
Beeswax	60–65	Common bee (*Apis mellifica*)	Cosmetic, painting, sealer
Indian beeswax (or Ghedda wax)	65–68	Indian bees (*Apis indica* or *Apis peroni*)	Cosmetic, painting, sealer
Japanese beeswax	64–67	Japanese bees	Cosmetic, painting, sealer
Ambergris	53–83	Intestines of sperm whale	Perfume, cosmetic
Spermaceti	43–47	Head cavity of sperm whale	Making candles, leather and textiles dressing
Chinese insect wax (or pe-la)	79–83	Scale insect (*Coccus ceriferus*)	Cosmetic, sealer
Mineral			
Bitumen	varies	Bitumen	Cement, sealer
Ozokerite	70–80	Ozokerite	Making candles, leather dressing

earth; different waxes are indigenous to different climates. Generally, the waxes are classified, by origin, as vegetable or animal waxes, as shown in Table 80. It should be mentioned that some materials commonly referred to as *mineral waxes* are not lipids but solid derivatives of petroleum; they are not chemically related to the waxes, which are all derived from living organisms.

Beeswax. The most widely used type of wax practically throughout the world has been *beeswax*; in matter of fact, when the word *wax is* used without further designation, it is generally accepted that it refers to beeswax. Since remote times humans have known the properties of beeswax and have used it for numerous and varied applications. The ancient Egyptians, for example, used beeswax over 6000 years ago to preserve mummies, embedding in wax the wrappings that encased embalmed corpses. They also coated and sealed with wax coffins in which mummies were kept (Warth 1956). Wax was also used for making writing tablets that could be rubbed down or remelted and then reused, as a component of many cosmetic preparations and at times as fuel for candles and illuminating lamps (Evershed et al. 1997). From early times beeswax was used by artists and artisans for creating encaustic pictures (see Textbox 17) and for modeling three-dimensional objects, as in the lost wax metal casting process (see Textbox 38). Encaustic painting was probably first developed by the ancient Greeks; Egyptian portraits on sarcophagi from the Fayum period, dating from the second century C.E., are at present still remarkably well preserved, providing evidence of both the long-term stability of the wax and the method of painting (Walker and Bierbrier 1997; Doxiadis 1995).

10.2. ANCIENT OILS, FATS, AND WAXES

Lipids are ubiquitous in ancient remains, and structural differences between lipids provide a tool for relating their structural characteristics to particular animals or plants. This has given rise to the use of lipids as *biomarkers* (from *biological markers*), a term used to refer to compounds used to correlate processes in biological systems. Lipid biomarker studies are possible because the composition and structures of lipids can be rapidly and reliably identified using modern instrumental techniques such as gas chromatography (GC), mass spectrometry (MS), and combined gas chromatography – mass spectrometry (GC-MS). In archaeological studies, lipids have been used as biomarkers to investigate *paleodiets*, the way artifacts were used, and even some ritual activities. (Evershed 1993). Other biomarker studies have been devoted, for example, to identify fecal matter in archaeological soils and sediments, to study the history of dyestuffs used in ancient textiles, and to analyze nucleic acids in pre-Hispanic populations (Peters et al. 2005).

10.3. SOAP

Dirt is attracted to and held on the skin and clothing mostly by fats and oils that are insoluble in water. *Soap*, an artificial, human-made (synthetic) substance, cleanses by making fats and oils dispersible in water; soap acts as an *emulsifier*, a substance that disperses solids into liquids in which they are usually immiscible. When used as a body cleanser, for example, soap combines with oils and fats together with dirt particles on the skin, emulsifying and dispersing them in water in such a way that they can be disposed of with the water.

Soap is formed when an alkaline solution that include salts of sodium or potassium reacts with a fat or an oil, in a chemical reaction known as *saponification*. Soap has been made since antiquity by the saponification of animal fat or vegetable oil with some alkaline substance such as wood ash, soda, or potash. One of the earliest formal accounts of making soap was written on Sumerian clay tablets from the midthird millennium B.C.E. and the writings disclose the making of soap from cassia oil, ash, and water. Clay cylinders containing a soaplike substance found in Babylon probably provide evidence that soapmaking may have being practiced there at even earlier times, during the early third millennium B.C.E. Inscriptions on the cylinders explain that the contents of the cylinders were made by boiling fat mixed with ash, a method of making soap that is practiced to this day. Pliny, the first-century C.E. Roman author of an encyclopedic natural history, described the process of making soap from goat's tallow and wood ash.

It is quite probable that in antiquity soap was used only for cleansing cloth and garments and for the treatment of disease, but not necessarily for personal cleanliness. Galen, a second-century C.E. Greek physician, for example, recommended bathing with soap for some skin conditions. The use of soap for personal cleanliness apparently became relatively widespread only during the late Roman Empire; as late as the eighteenth century in many places of the world however bathing was considered an oddity, not the norm.

PROTEINS
SKIN, LEATHER, AND GLUE

Proteins, from the Greek word *proteios*, meaning holding the first place, are the essential structural and functional components of most animal tissues. In the average human being, the proteins constitute over 15% of the total mass of the body: they make up the skin, muscles, and ligaments and even provide the basic structural components of bones. Proteins in the blood transport around the body substances necessary for its preservation and upkeep. *Enzymes*, the substances that stimulate biochemical reactions in the body (organisms could not function without enzymes), are also proteins; so are the *hormones* and nerve receptors that send and receive messages within the body, and the *antibodies* that fight off disease. Proteins even play a part in the ability of animals to see, since the lens of the eye is also a protein. After animals, including human beings, die, the proteins in their bodies are consumed as food either by microorganisms or by other animals, or they decay. Under particular environmental conditions, however, some proteins, particularly those in the skin and bones of animals (again including humans), may be preserved for very long periods of time. If discovered after long periods of time, the study of such proteins may reveal valuable information not only on their nature but also on the nature and characteristics of the organisms of which the proteins were the basic constituents (Tokarski et al. 2003; Gerneay et al. 2001).

The *proteins* are organic compounds composed of carbon, hydrogen, oxygen, nitrogen, and often also sulfur, the most characteristic element being nitrogen. Chemically, the proteins constitute a large group of complex substances made up from associations of *amino acids* (see Textbox 59). Plants synthesize amino acids from inorganic substances that they derive from elements and inorganic compounds acquired from the air and the soil. Animals cannot synthesize amino acids from inorganic substances, as do plants,

Archaeological Chemistry, Second Edition By Zvi Goffer

TEXTBOX 59

AMINO ACIDS, POLYPEPTIDES, AND PROTEINS

Amino Acids

Amino acids are the *monomers* (molecular units) that make up the (polymeric) *proteins*. They are organic substances whose molecules consist of at least four elements: carbon, hydrogen, oxygen, and nitrogen. Three of these, carbon, hydrogen, and oxygen, also occur in the carbohydrates and the lipids. The fourth element, nitrogen, does not occur in either of the latter; it is characteristic only of the composition of the amino acids and proteins. Another common feature that causes amino acids to be classified as a class of compounds is that, within their molecules, these elements occur as well-defined and characteristic groups, as shown in Figure 70a. The common name of these compounds, *amino acids*, is derived from the combined names of the two atomic groups, the *amino* and the *acid groups*, shown in the figure, that are characteristic of their constitution. There are several hundred naturally occurring amino acids, some of which have more than one amino or acid groups. Only 20, however, listed in Table 81, make up virtually all proteins in mammals, including humans. The possible combinations of the different amino acids are almost infinite, and some proteins include thousands of combined amino acids.

Plants synthesize all the amino acids they require. They do so using as raw material carbohydrates, which they make during photosynthesis, and nitrogen, derived from nitrate ions absorbed from the soil. Animals cannot synthesize all the amino acids required for their regular living, health, and growth. Those they cannot synthesize, known as the *essential amino acids*, are acquired from plants and/or animals they consume as food. Human beings, for example, acquire nine essential amino acids from their diet.

Polypeptides and Proteins

In living cells, amino acids are linked into polymeric chains known as *polypeptides*, typically containing between 10 and 100 amino acids [the

TABLE 81 Amino Acids in Proteins

Alanine	Glutamic acid	Leucine[a]	Serine
Arginine	Glutamine	Lysine[a]	Threonine[a]
Asparagine	Glycine	Methionine[a]	Tryptophan[a]
Aspartic Acid	Histidine[a]	Phenylalanine[a]	Tyrosine
Cysteine	Isoleucine[a]	Proline	Valine[a]

[a] The essential amino acids that cannot be synthesized by humans.

links (or *bonds*) between the amino acids within the polypeptides are referred to as *peptide bonds*] (see Fig. 70). The large number of possible combinations of the 20 different amino acids in mammals (see Table 81) gives rise to an extremely large number of polypeptides.

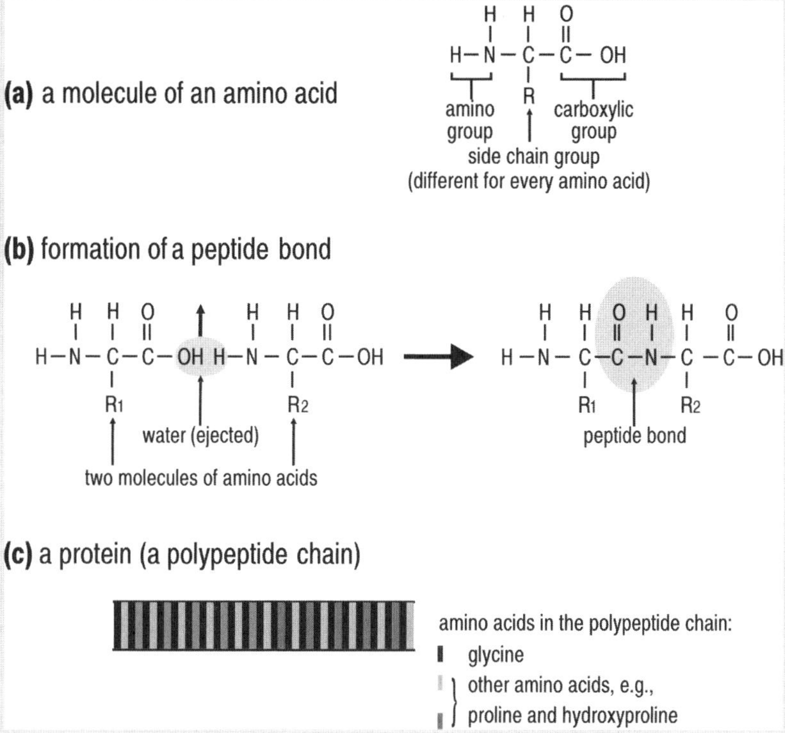

(a) a molecule of an amino acid

H H O
H–N–C–C–OH
R
amino group
carboxylic group
side chain group (different for every amino acid)

(b) formation of a peptide bond

H–N–C(R_1)H–C(=O)–OH + H–N–C(R_2)H–C(=O)–OH → H–N–C(R_1)H–C(=O)–N(H)–C(R_2)H–C(=O)–OH

water (ejected)
two molecules of amino acids
peptide bond

(c) a protein (a polypeptide chain)

amino acids in the polypeptide chain:
glycine
other amino acids, e.g., proline and hydroxyproline

FIGURE 70 Amino acids, peptides, and proteins. The amino acids are the building blocks of the proteins, the substances that make up the bulk of animal cell matter. The molecules of all amino acids contain three atomic groups **(a)**: an amino group, a carboxylic group, and a side (R) group. The side group is what distinguishes each amino acid from any other. Molecules of amino acids combine to each other, creating "peptide bonds" **(b)**, which are formed when an hydroxyl ion from a carboxylic group in one amino acid and an hydrogen atom from the amino group of another amino acid combine to form water; the water is ejected while the CO group in one amino acid becomes attached to the amino group in another, forming "peptides": dipeptides, tripeptides, and polypeptides). The proteins are extremely large polypeptides composed of sequences of hundreds or thousands of amino acids **(c)**. The shape, properties, and hence the function of a protein depend on the type, sequence, and number of amino acids from which it is made up. The sequence of amino acids in a protein, which is generally referred to as the "primary structure" of the protein, is determined by the nature of the DNA, that is, by the genetic code of the living organism in which the protein is made.

The term *polypeptide* is often used synonymously with *protein*. A *protein*, however, may consist of one or more specific polypeptide chains. The dividing line between the use of the terms *polypeptides* and *proteins* is somewhat undefined, although polypeptides are less likely to constitute the structural element of living bodies. The type of amino acids and the order in which they are linked together within a polypeptide chain determine the properties of particular proteins. A change in just one amino acid, out of a total of the several hundreds or even thousands that make up many a protein, and even a change in the position of an amino acid within the chain, results in proteins having different properties and therefore being functionally different.

Some proteins may be very large, indeed. *Titin*, for example, a muscle protein and the largest known protein, includes in its molecule over 26,000 combined amino acid units.

although they synthesize a very few from their food intake. Almost all of the amino acids that animals require but cannot synthesize are taken up as food from plants or from other animals, which, in turn, previously acquired them from plants. Also, humans, like other animals, are unable to synthesize all the amino acids they require for their regular living; nor can they assimilate proteins in their food, because during digestion, proteins are broken down into their component amino acids. They can, however, assimilate the amino acids they ingest and use them to build all the proteins necessary for their subsistence.

TEXTBOX 60

THE PROTEINS

The *proteins* are complex biological substances that make up the structural elements of the body of animals and fulfill many body functions (see Textbox 59). Each protein has different and unique functions. Their uniqueness depends on the number and order of amino acids within their polymeric chains. Proteins are required for the structure, function, and regulation of the cells, tissues, and organs of living organisms. Some proteins make up the structural elements of the body, such as specific organs, muscles, skin, blood, or part of the bones. Others perform specific body functions; examples are some *hormones*, known as *peptide hormones*, which control the activity of cells or organs, and the *antibodies*, which protect the body from foreign particles and organisms. The information for making specific proteins is determined by the nature of the DNA of a living organism. The proteins can be conveniently classified by either their function or their shape. In Table 82 are classified some common proteins by their function in living organisms.

TABLE 82 Some Common Proteins

Protein	Natural function	Human uses
Structural Proteins		
Collagen, myosin	Make up muscles, skin, and part of the animal skeleton	Food
Keratin	Makes up hair, wool, nails, and feathers	Making yarn
Fibroin	Makes up silk	Making yarn
Motorial Proteins		
Myosin, actin	Contract and change shape	Food
Storage Proteins		
Seed storage	Store nutrients until germination	Food
Hemoglobin	Stores oxygen	Adhesive, food
Egg albumin	Stores nutrients	Adhesive, food
Milk casein	Stores nutrients	Adhesive
Regulatory Proteins		
Hormones	Regulate metabolic processes	
Enzymes	Regulate catalytic activity	
Defensive Proteins		
Snake venom	Provide body protection	

When classified by shape, two main different types of proteins can be distinguished: *fibrous* and *globular proteins*. *Myosin*, the protein that makes up the muscles, *keratin*, which makes up the nails and hairs, and *collagen*, in the bones, tendons, and skin of mammals, are all *fibrous proteins*. They are insoluble in water and are mostly tangled around each other and stretched out (see Fig. 71). In the *globular proteins* the polypeptide chains are folded into the shape of balls – as, for example, in the *hemoglobin* of blood, which transports oxygen to the various parts of the body, or the *albumin* of eggs. A moderate increase in temperature can provide enough energy not only to unfold the polypeptide chains but also to totally break down the shape of the proteins. Proteins that lose their shape because of an increase in temperature are said to be *denaturated*. Soft- and hard-cooked eggs and cooked meat are examples of denaturated proteins.

Enzymes

The *enzymes* are complex proteins that *catalyze* reactions (alter the rate at which chemical reactions proceed) in living organisms, but are not themselves altered. The synthesis of the enzymes, as that of all the proteins in the living body, is regulated by DNA when *proteins* are synthesized. Enzymes are active, for example, in the formation of biopolymers in the body of living organisms, as when amino acids polymerize into proteins. The synthesis of proteins, which without enzymes might take extremely long periods of time to complete, takes very short times, a matter of seconds or even a fraction of a second when catalyzed by enzymes. Enzymes also accelerate the synthesis of DNA and RNA. Enzymes known by the generic name *polymerases* are used in the *polymerase chain reaction* (PCR) for enlarging fragments of DNA (see Textbox 64).

Most enzymes are very specific in their activity, and each chemical reaction in a living organism requires a specific enzyme. Their specificity arises from what is known as an *active site*, a location in the enzyme's molecule that has a shape matching that of a part of the molecule with which it reacts. The activity of the enzymes is affected by such factors as temperature and pH, each enzyme functioning best within a specific range of temperatures and pH. Outside this range the enzymes are structurally altered and their activity is either impaired or terminated.

The proteins are complex substances, much more complex than the carbohydrates or lipids, the other main components of the body and the sources of energy of living organisms. When carbohydrates or lipids are in short supply, living organisms can replace them by proteins that are then chemically altered and made to release energy. There are, however, no substances that can take the place of proteins in the building of new cells, the maintenance of existing ones, or the replacement of dead cells in the body (Creighton 1983). One of the most essential proteins in animal tissues, particularly vertebrates, is *collagen*, the main component of skin and a major component of bone, cartilage, and connective tissues (see Textbox 61).

TEXTBOX 61

COLLAGEN

Collagen is the most abundant animal protein in the body of animals, where it makes up as much as one-quarter of all the proteins. It is a *fibrous protein* that provides structure to and protects and supports soft tissues; it also connects tissues to the skeleton. Collagen forms, for example, most of the resilient layers that make up the skin and the filaments that support the internal organs. Interwoven with bioinorganic components, collagen also makes up the bones and teeth of vertebrate animals (see Chapter 15).

Collagen is a rather complex protein composed of three long polypeptide chains wound together to form a tight triple helix, as shown in Figure 71. Each chain is over 1400 amino acid units long, arranged in repeated sequences of three, every third amino acid of which is *glycine*; many of the remaining positions in the chains are filled by *proline* and *hydroxyproline* (see Table 81) (Woodhead-Galloway 1980; Faraday Society 1953).

Different tissues in the body have different types of collagen (there are 27 in total) classified as *collagen types* I, II, and so on. *Type I* collagen, for example, the most abundant collagen of the human body, makes up the tendons and the organic part of bone; *type II* collagen makes up articular cartilage; *type IV* makes up the eye lens; *type VII* and *type XI* colla-

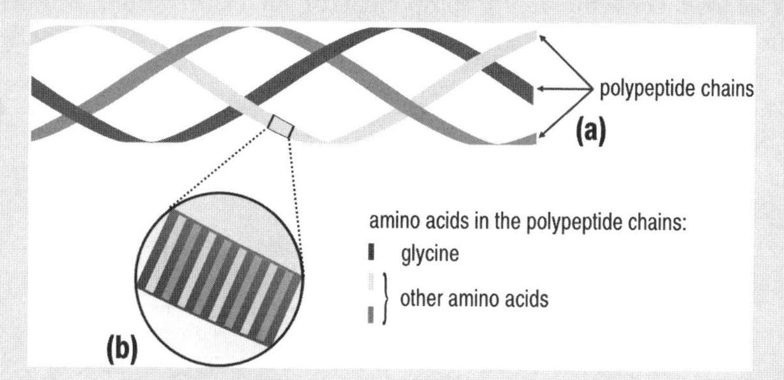

FIGURE 71 Collagen. Collagen, the major structural protein in the body of mammals, is the main component of the skin and tendons and a major component of the bones. The rather complex molecules of collagen are composed of three polypeptide chains wound together forming a tight triple helix. The illustration shows only a small segment of collagen chains **(a)**, which are over 1400 amino acids long; every third amino acid in the chain is glycine, as shown in the magnified section of one of the chains **(b)**.

gen, associated with type II, also makes up cartilage. The differences between the different types resides in the way the amino acid molecules are linked to each other within the polypeptide chains that make up the basic structure of collagen.

When collagen is heated, it is *denaturated*, losing its structure: the polypeptide chains separate and unwind, and, as the collagen cools down, it soaks up surrounding water and forms *gelatin*.

11.1. ANIMAL SKIN

The *skin* is is the outermost protective covering layer of the body of animals; in humans, the skin represents under 15% of their body weight; and, in adults, ranges in area within one and two two square meters. It is a tough but flexible membrane that surrounds the body and fulfills many physiological functions: it bounds the size and shape of animals, protects their bodies, regulates their temperature, and prevents the loss of water and the invasion of foreign microorganisms and viruses. All these functions are possible because of the particular composition and structure of the skin, which is quite similar for all vertebrates: it is made up of intertwined bundles of fibers of collagen—the main component of all skin—water, and lesser amounts of fats, nucleic acids, and coloring matter (see Table 83) (Woodhead-Galloway 1980).

The skin of mammals, including humans, consists of three layers (see Fig. 72):

TABLE 83 Approximate Composition of Animal (Mammal) Skin

Component	Percent	
Protein	32	
Collagen		29
Keratin		2
Albumin		1
Fat	2	
DNA	below 1	
Pigments	about 0.5	
Mineral salts	about 0.5	
Water	65	

1. A thin outer, sometimes hairy layer, known as the *epidermis* or *cuticle*
2. A middle layer, the *dermis, corium,* or *cutis,* which contains the hair follicles and the sebaceous and sweat glands
3. An inner layer, known as the *hypodermis* or *subcutis,* which is made up of fat and, in those animal that can wriggle the skin, muscle

Collagen fibers are long and thick in the middle dermis but become increasingly fine toward the outer epidermis as well as toward the inner hypodermis. Apart from this structural characteristic, which is common to all animals, the skin of each species has a different and unique morphology that significantly affects its properties (Calnan and Haines 1991).

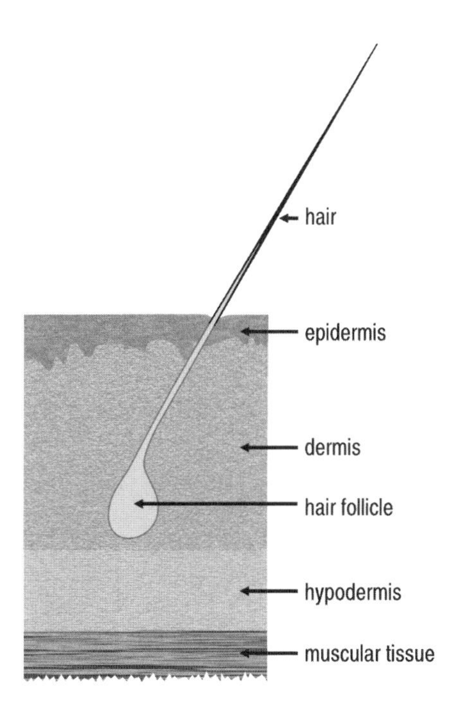

FIGURE 72 Cross section of a mammal's skin. The skin forms a protective barrier around the body of animals, preserving their shape and maintaining a constant body temperature. The skin of mammals is made up of three layers. The outer, tough and protective layer, the epidermis, gives the skin its color and shields the body from radiation of the sun. The second layer, the dermis, located under the epidermis, contains hair follicles, sweat and oil glands, and nerve endings. The innermost layer, the hypodermis, is composed mainly of fat (in some animals also muscle). The average skin of adult humans is about 2 mm thick, has an area of about 2 m^2 and weighs about 2.5 kg.

The natural weave of the fibers of collagen bundles not only varies through the thickness of the skin but also depends on the location in the body of the animal. Skin that covers the back of an animal, for example, has more compactly entwined bundles of fibers and a denser growth of hair than that of the belly area. This results in the belly area of the skin being generally thinner, weaker, more pliant, and having a greater tendency to wrinkle. The least compact structures and the ones having the sparsest hair growth are the axillae, where the fiber bundles are poorly intertwined.

Skin and Hide

In common language the terms *skin* and *hide* are generally used synonymously to refer to the skin of animals. Technically, though, the two terms have different meanings; *skin* is used to refer to the natural covering of small mammals such as sheep, goats, and pigs, as well as birds, fish, and reptiles, while *hide* is generally restricted to the skin of large mammals, such as cattle and horses. The skins of sheep, goats, calves, deer, and pigs, and the hides of cattle, buffalo, and horses have been most widely used for building shelters and making artifacts and clothing. Skin freshly removed from animals carcasses is generally referred to as *raw* or *green skin*. Unless raw skin is *cured*, that is, preserved in an appropriate way, it putrefies and decays soon after removal from the carcass.

Curing may involve *salting* or *brine pickling*, two techniques that have been used since early antiquity practically throughout the world. *Salting* entails spreading common salt (sodium chloride) over stretched and often stacked skins that are then left to cure for several weeks. In *brine pickling*, a relatively faster process, the skins are immersed in *brine*, a concentrated solution of common salt in water, and left to soak for a few days to allow the brine to thoroughly penetrate the skin thickness; from time to time the solution is agitated, and finally the skins are removed from the solution and left to dry. Other methods for curing skins are *drying* in the sun, practiced in hot and dry regions, and *freezing*, in cold, mostly northern areas of the world, which results in the skins becoming stiff and unyielding.

After skins have been cured by any one of the initial preservation processes mentioned above and made relatively resistant to decay, any inner fat or fleshy tissue is removed with sharp instruments. When so required, the outer hair is also removed, generally using one or more of a number of hair removal (*dehairing*) processes, such as *sweating*, *liming*, or *scudding*, which leave the *dermis*, or *corium*, the most useful part of the skin, almost unaltered. *Sweating* is a controlled putrefactive process that loosens the epidermis without damaging it. In some areas of the world sweating was carried out by burying the skins in the ground for a few days; in others the skins were spread with fish roe or placed in urine for a few days. *Liming*, an alternative to sweating, was carried out by treating the skins with lime, which breaks down the hair. *Scudding* consists of loosening and scraping with a blunt knife the outer skin layer, the epidermis, including the hair or wool, as well as the fleshy hypodermis (Kuhn 1986; Reed 1972).

Animal skin that has undergone one of the dehairing processes is hard and stiff. Making the skin soft and supple has been done, for thousands of years, by one of a number of processes that entail bacterial activity and controlled fermentation, such as *bating*, *puering*, and *drenching*. For *bating*, the skins are immersed and fermented in an infusion of pigeon and/or hen dung. For *puering*, a faster process than bating, dog dung is used, while for *drenching*, they are immersed in a bath of barley, bran, or the husks of cereals in water. Following any one of these processes the skins become pliable and smooth; after they are washed and dried, they are also flexible and ready for use (Wood 1912).

Fur

Fur is the name given to the skin of some mammals processed so as to preserve the hair and thus provide a high degree of thermal insulation. Since prehistoric times furs have been appreciated for the warmth they provide and have, therefore, been used for making clothing and footwear. Furs have also been appreciated for their *beauty* and for a long time, the furs derived from some animals were associated with status and luxury. Not all hairy skins or hides, however, provide "true" furs; only those that have a unique combination of long hairs, known as *guard hairs*, that shed moisture, and short soft hairs, known as the *underfur*, that act as the thermal insulating layer are considered true furs. The skins and hides of most domestic animals (lambs and cows, for example) either lack guard hair or lack underfur. "True" furs are derived from animals from cold or temperate climates, for example, raccoons, beavers, ermines, minks, and sables.

The process for making fur is known as *fur dressing*. After the skinning of an animal, the hairy skin intended to be made into fur is first cured and cleansed of any remains of fat and fleshy tissue. It is then processed, as are most skins, by one of the skin processing methods mentioned above (although one that does not remove the hair) so as to make a fur that is stable, resistant to decay, and also soft and supple. Only very infrequently were furs *tanned*, that is, converted to leather. Sometimes, the guard hairs of some furs were plucked out, or the underfur sheared close to the skin. At times, furs were also *glazed*, that is, made lustrous and glossy by applying a vegetable gum suspended in water, which, after drying, formed a thin shiny layer on the fur's hairs.

Parchment and Vellum

Parchment and *vellum* are animal skins that, probably since the third century B.C.E., have been treated so as to make them suitable for writing. Parchment has generally been made from the skins of sheep but also from those of goats and asses. Fresh skins intended for making parchment or vellum are first cured, washed, and dehaired. They are then soaked for a few days in slaked lime to soften the hair, which is then scraped off with a dull knife. Following the dehairing process, the skins are stretched, often on wooden frames; there they are left to dry and, while still under tension, are again scraped with a sharp knife, this time to remove any surface imperfections. The surface is then further flattened by rubbing it with large smooth stones; sometimes a mild abrasive powder is used as an aid for smoothing. This treatment usually leaves the inner, fleshy side of the skin, with a smoother surface than the outer hairy side.

Making parchment occasionally also entails splitting the skin into two thin layers whose surfaces are then subjected to smoothing. Vellum, whose texture is much finer than that of parchment, was made from unsplit skins. Generally, the younger the animal from which the skin is derived, the finer the parchment or vellum (Reed 1975). Both parchment and vellum are tough but flexible materials that, under favorable environmental conditions, withstand mechanical stress. They are highly *hygroscopic*, that is, they easily absorb water, their moisture content being in equilibrium with the relative humidity of the surroundings. When subjected to high humidity or dipped in cold water for a relatively long time, parchment and vellum swell and eventually decay, forming a mucilaginous fluid that consists of a solution of glue in water (see text below) (Ryder 1964).

11.2. LEATHER

Leather is the material made from animal skin by the process of *tanning*, which entails chemically altering the composition of the skin so as to make it durable and resistant to decay. Leather is therefore not a protein but a protein derivative. Although the tanning process alters the composition of skin, leather retains the fibrous structure and utilitarian functionality that make skin suitable for multifarious applications. Shelter, clothing, and decorative objects made from leather are, unlike skin or hide, stable to physical, chemical, and biological decay under dry or wet conditions (O'Flaherty et al. 1965;

Gustafson 1956). The essential differences between skin and leather are in their stability in damp and humid environments: when damp or humid for extended periods of time, even at ambient temperatures, skin is gradually hydrolyzed into a viscous treacle; leather remains practically unaltered even after prolonged immersion in boiling water (Stambolov 1969).

Tanning

Making leather, or *tanning*, as the technology for making leather is known, is very ancient, although it is difficult to substantiate chronologically at what time the tanning of skins was first practiced. Many remains of ancient clothing and utensils made from leather have been found in archaeological sites throughout the world, and the 4200-year-old Iceman, popularly known as Ötzi, discovered in the Italian Alps, was clothed in leather. Tanning seems to have been practiced in the Mediterranean Sea region before the second millennium B.C.E.. The Egyptians apparently made leather as early as during the second millennium B.C.E.; the inhabitants of the American continent, long before the arrival of Europeans.

The term for leathermaking technology, *tanning*, is derived from just one specific process out of several for converting skin into leather: the only one in which vegetable *tannins* are used (see Textbox 62). Other tanning processes (in which tannins are not used) include *mineral tanning, oil tanning*, also known as *chamoising*, and *smoke tanning*. The common characteristic of all these basically very different processes is that they all convert animal skin into leather, a stable material that, while retaining the original structure of the collagen fibers in the skin, is stronger as well as being resistant to decay, and often more supple than unprocessed skin (Bienkiewicz 1983). Before skin – whether fresh, but most often cured – is tanned by any process, it is soaked and washed to dissolve and remove any excess salt added for preservation purposes, and bring it into a condition in which it can readily interact with a tanning agent. After removal of the salt, the skin is also dehaired and trimmed, and only then is it ready for one of the different tanning processes.

Vegetable Tanning. *Vegetable tanning*, one of the oldest known tanning processes, has been practiced for at least 4000 years. Clean, dehaired skins are immersed in a solution (usually at a low concentration) of vegetable tannins (see Textbox 62). After a relatively long period of time (several weeks and often even months) during which the collagen in the skin reacts with the tannins, the immersed skins become tanned, that is, converted to leather;

the leather is then removed from the tanning solution, washed, and dried. Following the tanning process, vegetable-tanned leather is often impregnated with natural oils or greases so as to make it pliable and impervious to water.

TEXTBOX 62

TANNINS

The *tannins* are acid substances found in the bark, wood, roots and stems of many plants, and especially in the bark of Oak (*Quercus*) trees and in grape vines; all the tannins are soluble in water and have a bitter taste and *astringent* qualities, tightening the pores and drawing out liquids from animal skin. The term *tannin* (derived from the Celtic word for oak), is not related to any particular substance but reflects a technology: that of making *leather.* To *tan* is to cause changes to *animal skin* so as to make it resistant to weathering and decomposition, while leaving it strong and flexible. Tanned skin is known as *leather*. Only one of several tanning processes practiced since antiquity, however, entails treating skins with *extracts* or *infusions* of tannins (tannin solutions prepared by steeping tannin containing vegetable tissues in water).

The tannins are synthesized by plants and are abundantly distributed in many different forms of plant life. Common sources of tannins include for example, the bark, leaves, fruit and roots of many plants; most tannins, however, have been and are still derived from the bark of a few trees and shrubs, such as oak, chestnut, hemlock, mangrove, quebracho, and wattle, from which they are generally extracted with water.

The chemical nature of the tannins varies between different plants and parts of plants. Although the composition and structure of many tannins is not yet well known, they are believed to be *glucosides*, substances derived from *glucose*, a carbohydrate. A common characteristic of all the tannins is that although the composition of each one is different from that of others, all tannins have at least one *phenolic group* in their molecules [a *phenolic group* is an arrangement of atoms of carbon in a cyclic molecule to which a hydroxyl group (OH^-) is attached].

All the tannins readily react with proteins, forming insoluble, stable compounds; when they react with *collagen*, the main constituent of animal skin, they form leather, a material that is resistant to *hydrolysis*, *oxidation*, and biological attack and therefore stable to weathering and resistant to decomposition. Since tannins from different plant sources have different chemical compositions, each tannin used for tanning skin produces a leather having slightly different properties and color. Tannins that have been used since antiquity and their botanical sources are listed in Table 84 (Howes 1953).

TABLE 84 Tannins

Trees from Which Tannins Have Been Extracted
Birch (*Betula* spp.)
Eucalyptus (*Eucalyptus* spp.)
Myrtle (*Mirtus* spp.)
Oak (*Quercus* spp.)
Pine (*Pinus* spp.)
Quebracho (*Scinopsis balansae*)
Wattle (*Acacia* spp.)
Willow (*Salix caprea*)
Plant Parts Rich in Tannins
Bark
Fruits
Fruit pods
Galls
Wood

In addition to their use for making leather, tannins have also been used as *mordants* for *dyeing* (see Chapter 14), for clarifying turbid solutions such as wine, and as components of some types of *ink*. The best sources of tannins for making ink are the *gall nuts* (round growth swellings caused by the egg-laying activity of the gall wasp) of oak trees.

Oil Tanning. *Oil tanning*, also known as *chamoising*, entails incorporating oils, fats, or fatty substances (such as eggyolk or brain matter) into animal skin (some Eskimos, for example, still tan leather by rubbing it with a macerated mixture of animal brain, liver tissue, and water). For this purpose, clean and dehaired skin is rubbed with an oil that penetrates and impregnates the skin. The oiled skin is then exposed to the atmosphere so that the impregnating oil becomes oxidized (reacts with atmospheric oxygen). At the end of the process, after some days or weeks, depending on the prevailing temperature (the higher the temperature, the shorter the process), the *chamois leather*, as oil tanned leather is known, is washed and dried. Oil tanned leather is very flexible and readily absorbs and expresses water. The oil tanning process often leaves some collagen fibers unaltered, which, in temperate and hot climates, constitutes a source of decay of the leather. If and when such fibers are exposed to a humid or wet environment for some length of time, they are hydrolyzed (react with the water) and begin to decay. The water in such environments also becomes a source of decay processes: it promotes bacterial activity and results in most oil-tanned leather having a relatively short life.

Mineral Tanning. Mineral tanning, a process often also called *tawing*, entails immersing skins in a solution of specific minerals known as *alums* (composed of mixed sulfates of aluminum and other metals; alunite, also known as *alum stone*, for example, an alum composed of mixed sulfates of aluminum and potassium was widely used for mineral tanning in antiquity). Skins to be mineral tanned are drenched in a solution of alum for a rather long period of time (lasting from several days to several weeks, depending on the thickness of the skins and the ambient temperature), until the collagen fibers in the skin become thoroughly impregnated and react with the alum. The alum-soaked skin is then taken out of the solution and the excess alum is removed by repeated washing in water. The freshly prepared leather is finally dried, either by exposure to the air or by *staking* (a process that in addition to exposure to air, also includes mechanical stretching and smoothing).

Mineral tanning was probably first practiced in ancient Mesopotamia and then spread to Egypt, the Middle East, and the Mediterranean Sea area (Levey 1958). Mineral-tanned leather is soft to handle, has a velvety texture, and is almost white, a color practically impossible to achieve by other tanning processes. It is, however, very sensitive to humidity and water; under wet conditions the alum in the leather is *hydrolyzed* (decomposed by water), forming sulfuric acid, a very strong acid that attacks the leather and causes its rapid decay. Mineral-tanned leather that has been humid or wet for a more or less extended period of time loses some of its characteristic properties, such as softness, pliability, and strength, and becomes hard, horny, and brittle.

Smoke Tanning. In the *smoke-tanning* process, clean skins are exposed to the smoke of burning wood, which results in the reaction of the collagen fibers in the skins with some of the compounds in the smoke, and the conversion of the skins to tanned leather. Although the process can be practiced exclusively of others, smoke tanning was often associated with initial oil tanning: the skins were first rubbed with oil or animal fat and only then were they smoked; the result was a combined oil–smoke tanned leather that was, and still is, made by some people in central Asia, the Arctic region, and the Americas.

Aftertanning Processes; Finishing

When freshly tanned leather is dried, it is often stretched on a frame so as to prevent shrinkage during the drying process. Overdry leather becomes rigid and unbending, but it can be made flexible again by wetting. The water serves as a lubricant between the leather fibers, and the introduction of water allows the fibers to slide along one another, restoring the flexibility and suppleness of the leather.

11.3. GLUE

Glue is an organic *adhesive* and *sizing* substance derived from animal tissues. The product of the hydrolysis of proteins, glue has been made and used since early antiquity (see Textbox 55). The ancient Egyptians, for example, used glue for making furniture; since their time, and probably since much earlier, the use of glue, throughout the world continued uninterruptedly until the twentieth century. At that time, synthetic glues were developed and animal glues became relegated to only specialized tasks.

Glue is an impure form of gelatin, a jellylike protein formed when collagen or other animal proteins are *hydrolyzed*, that is, chemically altered by reacting with water. Most glue was made from the skins, bones, or offal of animals but also from casein (milk protein) and albumen (blood protein). Boiling skin, bone, and/or animal offal in a slightly acidified water solution for several hours *denaturates* (changes the structure) and hydrolyzes the collagen fibers and dissolves the products of the hydrolysis. Straining the resulting solution, then evaporating the water, and finally cooling down the remaining dark viscous liquid produces hard, solid, and usually brown glue.

The detailed handling of the raw material, before boiling, varied with the type of tissue that was used for making the glue, as skin or offal, for example, required specific methods of preparation. When the glue was prepared from bones, the bones had to be first cleansed of other tissues, washed with water and crushed before boiling; when derived from skins or hides, however, these were first washed and soaked in a slaked lime solution to remove hair and proteins other than collagen, before boiling. The remainder of the process, boiling the raw material in water and then evaporating the water, were quite similar for all different types of raw materials.

Glue is a brittle, hard solid at room temperature that, depending on the nature of the tissue from which it is derived, melts at about 120°C into a very viscous liquid. Also, when dissolved in water, glue forms viscous solutions. Both hot molten glue and solutions of glue in water are used as adhesives;

only the water solutions, however, are suitable for *sizing*, that is, for coating, filling, or stiffening such varied porous materials as skin, cloth, wood, and paper (Alexander 1923; Smith 1923).

11.4. DATING ANCIENT PROTEINS – AMINO ACID RACEMIZATION DATING

Many substances exhibit the property of *isomerism*: they occur in two or more molecular forms that have the same composition but differ from each other in structure and in their properties. One type of isomerism, known as *optical isomerism*, is exhibited by molecules that have the same constituent atoms but are arranged in different spatial distributions, where one of the optical isomers is a mirror image of the other (see Textbox 63).

TEXTBOX 63

OPTICAL ISOMERS, CHIRALITY, AND RACEMIZATION

Chemical *structural formulas* depict, in two-dimensional display media such as a sheet of paper or a computer screen, the composition and structure of molecules that generally have three-dimensional structures (see Textbox 51). In most molecules containing carbon, for example, the carbon atoms are generally bonded (linked) to four other atoms or molecular groups. If the four atoms or molecular groups are identical, the molecules formed are symmetric and the four bonds are arranged in space in a tetrahedral shape: the carbon atom is located at the center of a tetrahedron and the other atoms or chemical groups at its corners (see Figure 73 (a)). If they are different from each other, however, they may be arranged in two different forms, as illustrated in Figures 73 (b) and (c). In both these forms, the molecules are not symmetric and the central carbon atom in each molecule is thus said to be *asymmetric* or to have *chirality* (the term *chirality* relates to the structural characteristic of molecules that makes it impossible to superimpose some molecules on their mirror image). The two possible molecules have the same components but differ in their spatial arrangement, and are known as *stereoisomers* or *optical isomers* (the term *stereoisomer* refers to two or more molecules with the same type and number of atoms, yet different geometric and spatial arrangement). Each stereoisomer is the mirror image of the other; to distinguish between the two in writing, the prefixes D- and L- are, by international agreement, arbitrarily assigned. Many biological substances – as, for example, the amino

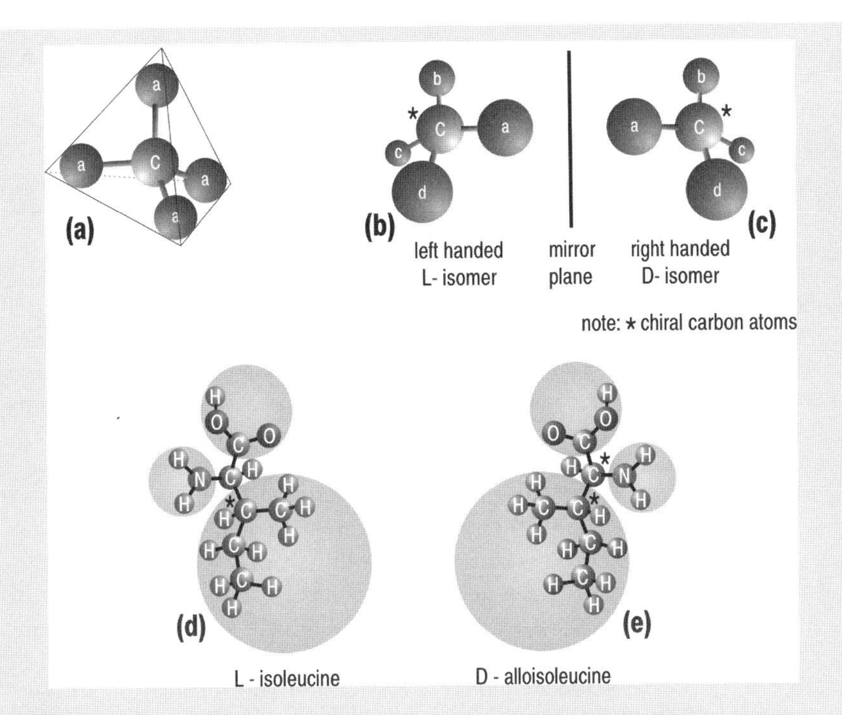

FIGURE 73 Optical isomers and amino acid racemization dating. In molecules in which an atom of carbon is linked to four identical atoms or chemical groups, the four atoms or chemical groups are tetrahedrally distributed around the central atom of carbon, forming a symmetric, triangular pyramid **(a)**. If the linked atoms or chemical groups are different, however, the molecule is *asymmetric* and the four atoms or chemical groups are arranged in two different spatial distributions around the carbon **(b,c)**; the two molecules are said to be *optical isomers* or *stereoisomers*. Each one of the two stereoisomers is known by a slightly different name: one is known as the "*L-isomer*" and the other, as the "*D-isomer*." The amino acids L-isoleucine **(d)** and D-alloisoleucine **(e)** for example, are optical isomers. The racemization of L-isoleucine into D-alloisoleucine is used to estimate the age of dead remains by the "amino acid racemization dating" method. The * next to carbon atoms indicates that the carbon is "asymmetric".

acids – include in their composition at least one *chiral* or *asymmetric atom* of carbon and therefore exhibit stereoisomerism. For reasons that have as yet not been explained, biological substances in living organisms occur only as the L-isomers; thus all amino acids in living organisms occur exclusively as L-isomers.

The L-isomer of any compound can transform into the D-isomer, and vice versa. A mixture of equal amounts of the L- and D-isomers of a compound is said to be a *racemic mixture*, and the process whereby an isomer, whether L- or D-, is transformed into a racemic mixture is known as *racemization*. Amino acids, which in living organisms occur in only the L-isomeric form, racemize spontaneously after organisms die and are converted to racemic mixtures.

Unless an amino acid in a dead organism decomposes or is consumed by other organisms, therefore, it racemizes and is converted, after some longer or shorter period of time, into a racemic mixture. The *half-life* of the racemization process of many amino acids (see Textbox 14) has been measured and is well known. Thus, in the remains of a living organism, measuring the relative amounts of each of the L- and D-isomers of an amino acid that has a well known half-life, makes it possible to determine the time of death of the organism of which the amino acid was a constituent (Bada 1990; Weiner et al. 1980). *L-Isoleucine*, for example, is one of the amino acids that occur in proteins, but its stereoisomer, *D-alloisoleucine*, does not occur at all in living organisms. When a living organism dies, L-isoleucine in the dead remains immediately begins to racemize into D-alloisoleucine, the process continuing until half of the original L-isomer is converted into the D form and a racemic mixture is formed (Figs. 73 (d) and (e)). At a temperature of 20°C, the half-life of this reaction has been measured to be about 110,000 years, a time long enough to make the racemization of L-isoleucine suitable for dating the remains of ancient organisms (Bada 1972).

Some of the amino acids, the building blocks of the proteins, for example, exhibit optical isomerism: their molecules occur in one of two isomeric forms, generally known as the D and the L forms, which have the same composition but differ in structure. For reasons not yet fully understood, the amino acids in living organisms occur almost exclusively in the L form (in the past it was assumed that all amino acids in natural proteins occur in the L form, but it is now known that there are a few exceptions to this assumption). From the moment of death, however, a *racemization* process begins, whereby part of the L-amino acids are converted into D-amino acids, which do not naturally occur in living organisms: the relative amount of the L-amino acid decreases, while that of the D form increases from zero, at the moment of death, until there are equal amounts of the L and D forms in the dead remains. The racemization process is not instantaneous but gradual, taking time to complete. Some amino acids racemize at very slow rates, many take thousands of years until completion, while others do so at much faster rates. Thus, measuring the extent of racemization of particular amino acids in the remains of dead animals provides a tool for dating such remains as skin and bone, and the technique for doing so is known as *amino acid racemization dating*, or *AAR dating*, for short.

There is, however, a drawback to this method of dating. The *half-life* (see Textbox 14) of the racemization process is greatly affected by temperature: it is shorter at higher temperatures and slows down as the temperature decreases. An uncertainty of ±2°C in the temperature history of dead remains can lead to an error of about ±50% in the age determined. For the racemization of amino acids to be applicable for dating, therefore, it is crucial that the temperature history of the environment where the amino acids (that is, skin or bone) have been deposited should be well known.

Because of the limitations imposed by temperature variations, ages determined with the amino acid racemization dating method may be rather uncertain. Nevertheless, although the accuracy of the method is nowhere near that provided by radiocarbon dating, it is of use, for example, in estimating the age of remains too old to be dated by the radiocarbon method. Among the places where temperatures are expected to undergo little variation over long periods of time are the ocean depths and the inside of caves, where paleotemperature studies indicate essentially constant temperatures over long periods of time (see Textbox 47). Moreover, if temperature changes have occurred in a cave during the past, it is still possible to determine the actual variation, from the *oxygen isotope ratios* in cave deposits. Also, because of the strong effect of temperature on the half-life of the racemization of amino acids, determining the extent to which an amino acid of known racemization half-life is racemized may be of use for estimating ancient temperatures. Provided that the age of ancient animal remains at a site is known, the ancient temperature at the site can be estimated from the extent of racemization of an amino acid in the remains (Manley et al. 2000).

THE NUCLEIC ACIDS
HUMAN TRAITS; GENETICS AND EVOLUTION

The *nucleic acids* are substances produced by living cells to preserve and carry hereditary traits of living organisms from one generation to the next; they store and transmit the genetic information that regulates all the cell's activities. There are two main classes of nucleic acids: *deoxyribonucleic acids* (DNA) and *ribonucleic acids* (RNA). DNA carries genetic hereditary information and controls all cellular functions in most living organisms; RNA transfers the genetic information encoded in DNA to the site in the cell where proteins are synthesized and there translates this information into proteins.

12.1. DNA AND RNA

Two different types of DNA occur in distinct locations in the living cell: *nuclear DNA* in the cell's nucleus and *mitochondrial DNA* (mt-DNA) in the mitochondria (see Textbox 50). RNA occurs in several types: *messenger RNA*, *transfer RNA*, *ribosomal RNA*, and others, all in the cytoplasm of the cells (see Table 85) (Bloomfield et al. 2000).

The chemical composition of *nuclear DNA*, also called *genomic DNA*, is unique to each organism and different from that of all others. This implies that nuclear DNA carries, embedded in its structure, the genetic code of every living organism (together with some proteins, nuclear DNA makes up the genes and chromosomes that constitute the heredity units in the nuclei of the cells). When cells divide, DNA preserves and transfers, in its chemical composition and structure, the characteristics of each living organism from one generation to another; RNA carries this information from the DNA in the chromosomes to the cytoplasm, so that specific proteins, characteristic to every organism, needed for life processes can be built by the cells.

TABLE 85 Nucleic Acids

Abbreviation	Full name	Characteristic function
DNA	Deoxyribonucleic acid	Carries genetic instructions for making living organisms
n-DNA	Nuclear deoxyribonucleic acid	DNA enclosed within the chromosomes, in the nucleus of the cell; inherited from both parents
mt-DNA	Mitochondrial deoxyribonucleic acid	DNA in cell mitochondria; inherited only from the mother
RNA	Ribonucleic acid	Transfers genetic information from DNA to proteins synthesized by the cell
m-RNA	Messenger ribonucleic acid	Transfers genetic information from nuclear DNA to the protein synthesized in cell ribosomes
r-RNA	Ribosomal ribonucleic acid	A stable, structure component of ribosomes
t-RNA	Transfer ribonucleic acid	Picks up specific amino acids and transfers them to the appropriate location in m-RNA during protein synthesis

TEXTBOX 64

THE NUCLEIC ACIDS: DNA AND RNA

The *nucleic* acids, *deoxyribonucleic acid* (DNA) and *ribonucleic acid* (RNA), which carry embedded in their complex molecules the genetic information that characterizes every organism, are found in virtually all living cells. Their molecules are very large and complex biopolymers made up basically of monomeric units known as *nucleotides*. Thus DNA and RNA are said to be *polynucleotides*. The nucleotides are made up of three bonded (linked) components: a *sugar*, a *nitrogenous base*, and one or more *phosphate groups*:

- The nature of the *sugar* varies, depending on whether the polynucleotide is RNA or DNA; in RNA the sugar is *ribose* ($C_5H_{10}O_5$) while in DNA it is *deoxyribose* ($C_5H_{10}O_4$). The prefix *deoxy-*, often found in relation to nucleic acids, means "without oxygen"; thus *deoxyribose* denotes ribose lacking (some) oxygen.
- Four different nitrogenous bases may occur in each nucleotide: *adenine* (A), *cytosine* (C), *guanine* (G), and *thymine* (T); in DNA, *uracyl* (U), substitutes for the thymine in RNA.
- The *phosphate groups* are electrically charged atomic groups composed of phosphorus and oxygen atoms (the formula of the phosphate group is PO_4^{3-}). The large amount of phosphorus in the nucleic acids, it should be mentioned, is another unusual characteristic of the nucleic acids among biological materials.

In each nucleotide, the three components are joined to each other and arranged in the repeating sequence: sugar, phosphate group, nitrogenous base:

```
—sugar—phosphate group—
              |
            N base
```

In the molecules of the polynucleotides, those of DNA and RNA, the nucleotides are linked to each other, forming long, continuous strands in the pattern

```
—phosphate—sugar—phosphate—sugar—phosphate—sugar—
    |                |                 |
  N base           N base            N base
```

Since the sugar in DNA is deoxyribose, the polynucleotide that makes up DNA can be written as

```
—phosphate—deoxyribose—phosphate—deoxyribose—
    |                      |
  N base                 N base
```

and for RNA, where the sugar is ribose, as:

```
—phosphate—ribose—phosphate—bose—
    |                 |
  N base            N base
```

The molecules of RNA and of DNA are therefore of similar overall chemical composition, although there is a very important structural difference between them – the RNA molecules consist of single strands of polynucleotides, while those of DNA are double-stranded (see list of strand components in Table 86). Each strand is complementary, although inverted in relation to the other, in what is known as an *antiparallel arrangement* that adopts a helical configuration, as illustrated in Figure 74. The two complementary strands are also linked to each other by connecting segments, made up of paired nitrogenous bases that create a ladder-type structure, as shown in the figure. Each of the paired bases within the helical arrangement is joined to another in the complementary strand in a very specific and unique way; for instance, adenine joins only with thymine and cytosine joins only with guanine.

In different molecules of DNA the four nitrogenous bases (adenine, cytosine, guanine, and thymine) appear in different order, and the genetic code (the genetic information carried by DNA) is determined by the order of the bases within the molecules. Therefore, it can be stated that the genetic code is made up of four letters, A, C, G, and T, each letter representing a base: (A) adenine; (C) cytosine, (G) guanine, and (T) thymine, respectively. In the DNA molecules the four bases are arranged in groups of three, and the code may thus be AAA, ATG, CCG, TAC, and so on. A sequence of bases in a DNA molecule might appear as follows:

TABLE 86 Components of Nucleic Acid Strands

n bases		Sugar		Phosphate
In DNA	In RNA	In DNA	In RNA	In DNA and RNA
Adenine (A) Cytosine (C) Guanine (G) Thymine (T)	Adenine (A) Cytosine (C) Guanine (G) Uracil (U)	Deoxyribose	Ribose	Phosphate groups

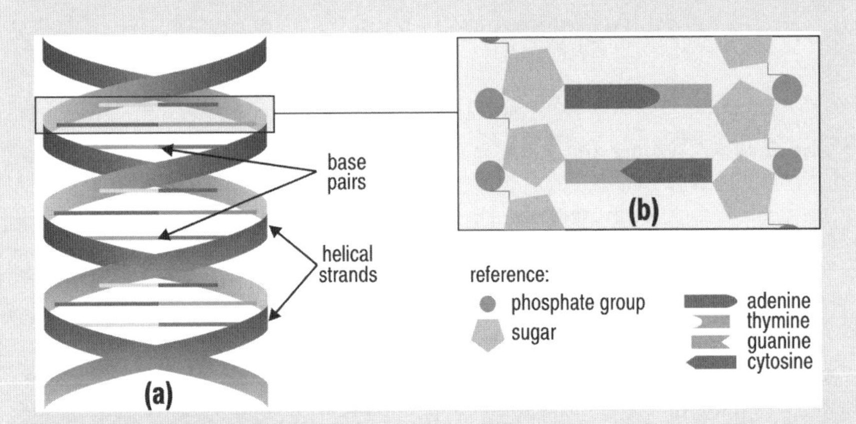

FIGURE 74 DNA. The molecules of deoxyribonucleic acid (DNA) are very large, polymeric, and double-stranded, the two strands spiral around each other in the shape of a double helix **(a)**. The long helical strands are made up of sugar and phosphate groups. Within the DNA molecule the strands are arranged in what is known as an "antiparallel, complementary sequence," meaning that the chemical groups within each stand are the same as in the other but run in opposite directions: the head of one strand is always opposite the tail of the other strand. Moreover, the two strands are held together by bridging nucleotides, arranged in pairs and usually referred to as "base pairs" (bp). In the nucleotides, adenine bonds only to thymine and cytosine only to guanine **(b)**. The number of base pairs is often used as a measure of the length of a DNA segment; thus,"a DNA segment 700 bp long."

—GAG TGA GGC CTC CTC TTC—

This order might provide the genetic code for the formation of, say, a protein in the skin collagen of a human being. At the same location in the DNA molecule of another human the order of the bases may be only slightly different, such as

—GAG TGA GGC TTC CTC CTC—

Because the order of the bases in the two sequences is not identical, the new order provides the code for the formation of a different protein; different nitrogenous base sequences provide different genetic codes, each particular sequence supplying instructions for a different and particular characteristic. One molecule of human nuclear DNA contains about 3,000,000,000 nitrogenous bases, enough to form well over 1,500,000,000 coded pieces of genetic information. If each coded piece corresponded to a letter of the alphabet, it would take approximately 2500 volumes, each roughly the size of this book, to genetically describe a human being!!!

Nucleic Acids and Heredity

When a living cell divides, it produces two new cells, both cells having identical characteristics. During the process of cell division, the DNA in the cells is said to *replicate*; the process of replication, which is quite complex, may be visualized as taking place in two stages: during the first stage the connecting segments that give the DNA molecule its ladderlike structure open up along their middle, creating two "half-ladders." Following the opening of the connecting segments, the bases (A, C, G, and T) as well as the phosphate groups and the sugar join the open "half-ladders" of DNA (it should be remembered that A joins only with T, and C joins only with G). Thus, the joining of the nitrogenous bases, sugar, and phosphate groups of each of the two half-ladders completes the formation of two new DNA molecules, each an exact copy of the original one. Most of the protein production in the cell takes place not in the nucleus, where DNA is located, but in the cytoplasm. The transfer of genetic information from nuclear DNA, where most of the genetic information is stored, to the cytoplasm where proteins are made, is carried by one particular type of RNA known as *messenger RNA*. *Messenger RNA* copies the genetic code of the nuclear DNA and flows from the nucleus to the cytoplasm; there, in the *ribosomes*, it is said to *transcribe* the code into a sequence of amino acids and polypeptides characteristic of each protein:

DNA	→	RNA	→	protein
(genetic code)		(messenger)		a sequence
a sequence of bases		a sequence of bases		of amino acids

In most organisms, the basic unit of nuclear DNA is a double-stranded molecule derived from those of both its ancestors, the father and the mother. The information it carries provides clues, therefore, not only to the genetic constitution of the organism but also on the characteristics it shares with its ancestors.

The molecules of *mitochondrial DNA* are separate and different from those of nuclear DNA. In contrast to nuclear DNA, which is derived from the DNA of both ancestors of the organism, mitochondrial DNA is derived exclusively from the mother's eggs. Sperm has virtually no cytoplasm and therefore no mitochondrial DNA. All the mitochondrial DNA of an organism is thus passed along from females, and its composition and structure reveal maternal links within a community or population. In each cell there are a large number of molecules of mitochondrial DNA, many more than that of nuclear DNA, and, because of its relative abundance, mitochondrial DNA provides a powerful and effective tool for the genetic study of ancient organisms. The mitochondrial DNA of the Ice Maiden found frozen in the Andes mountains of Peru, an Inca child who died about five centuries ago, for example, provided an opportunity to establish her genetic origin; a comparison of her mitochondrial DNA with that of other ancient inhabitants of the American continent indicates that her mitochondial DNA matches that of one of the first humans on the American continent (Powledge and Rose 1996).

DNA After Death

After matter is separated from living organisms (as wastes discharged from animals, for example), or when organisms die, the nucleic acids in the severed parts or in the dead remains may either vanish (consumed as food by other organisms) or decay and disintegrate. As they decay they break down and are initially split into small fragments, which eventually disintegrate and evanesce. Occasionally, under exceptional environmental conditions, some remains of nucleic acids may be preserved for long periods of time, usually in very small, almost unmeasurable amounts. If eventually found and identified, such remains provide the means by which much can be learned about the traits and relationships of ancient organisms (Keyser-Tracqui and Ludes 2005; Gugerli et al. 2005). Because each cell has only one nucleus but a relatively large number of mitochondria (several hundred in the cytoplasm of a cell), there is much more mitochondrial DNA than nuclear DNA in the remains of ancient organisms. Often, therefore, fragments of mitochondrial DNA are the only remains of DNA left in ancient remains. Whatever their origin, nucleic acids in ancient remains are generally in an advanced state of degradation, broken down into small fragments. To study them it is necessary, therefore, to replicate the fragments so as to obtain manageable samples whose composition and structure can be determined (Geigl 2002).

12.2. THE POLYMERASE CHAIN REACTION (PCR)

A scientific and technical breakthrough at the end of the twentieth century was the discovery of the *polymerase chain reaction* (*PCR*), a technique that makes it possible to replicate minimal amounts of nucleic acid fragments (Hummel 2002; McPherson et al. 2000). Sometimes referred to as "molecular photocopying," PCR is a technique for generating a practically unlimited number copies of nucleic acids fragments from a few and even from single, usually short and often damaged fragments (see Textbox 65). By using enzymes known as *polymerases*, which copy genetic material in living organisms, PCR replicates specific fragments of nucleic acids extracted from tissue remains derived from ancient animals, plants, or microorganisms, regardless of their age (Cano 2000).

TEXTBOX 65

THE POLYMERASE CHAIN REACTION (PCR)

The *polymerase chain reaction* (PCR) is a laboratory procedure for generating multiple copies of short fragments of DNA. It is a cyclic reaction repeated again and again, simulating the way nature creates DNA. The technique is generally performed in a single vessel in which a few segments of DNA are mixed with nucleotides, generally referred to as *primers*, and an enzyme known as *polymerase* (see Textbox 60). Heating the mixture to a temperature just below 100°C causes the separation between the paired fragment strands of the original DNA; cooling it afterward, to about 50–70°C, results in the primers [which serve as *templates* (patterns or molds) for the formation of the product] reacting with the now separated original strands to form new strands. These have the same composition and structure as the original ones.

After the complementary strands are separated from each other, the primers attach themselves to the ends of the strands; one primer becomes attached to one *template* strand, the other, to the complementary strand. Next, the enzyme causes bases to be added at the end of the primers, extending the formation of the complementary strands initiated by the primers. Because each template produces a new double-stranded fragment, the original sequence of bases in the sample of the DNA is doubled in quantity during each cycle, and the products of each cycle are two new, double-stranded fragments of DNA, both identical to the original fragment (see Fig. 75). The net result of each cycle is the doubling of the number of DNA fragments.

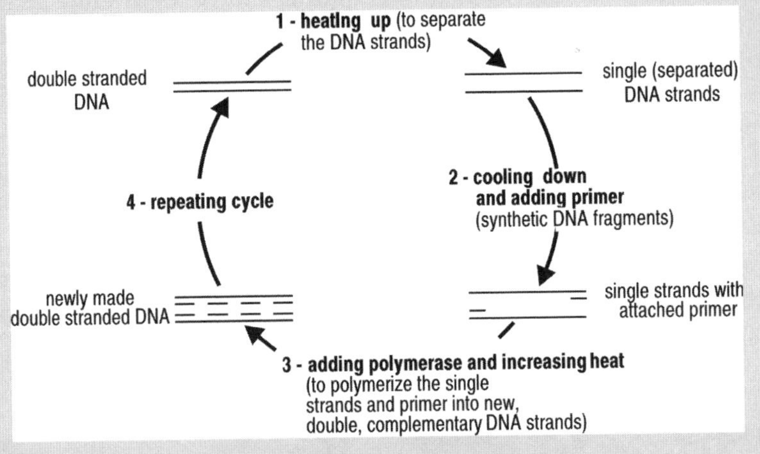

FIGURE 75 The polymerase chain reaction (PCR). The PCR reaction, which replicates (the term used is actually "amplifies" or "multiplies") single fragments of DNA, is widely used for preparing samples of DNA large enough to be studied. In practical terms, the PCR procedure involves repeatedly heating and cooling a fragment (or fragments) of DNA through a cycle of well-established temperatures and, at at some stage of the procedure, adding to the DNA two reactants: one known as "polymerase" and the other, as a "primer." A double-stranded DNA fragment (or fragments) is initially separated by heat into strands **(1)**; adding the primer and reducing the temperature results in the primer becoming attached to the now separated strands **(2)**; when the polymerase is added and the temperature is increased again, the primer triggers the formation of new double stranded DNA fragment **(3)**.

At the end of the first cycle, which takes only minutes or an even shorter time, the number of DNA fragment is doubled. The cycle can be repeated indefinitely; after 30 cycles, for example, which are completed in only a few hours, there will be about a billion copies of the original DNA fragment, sufficient for studying its nature and characteristics (see Fig. 76).

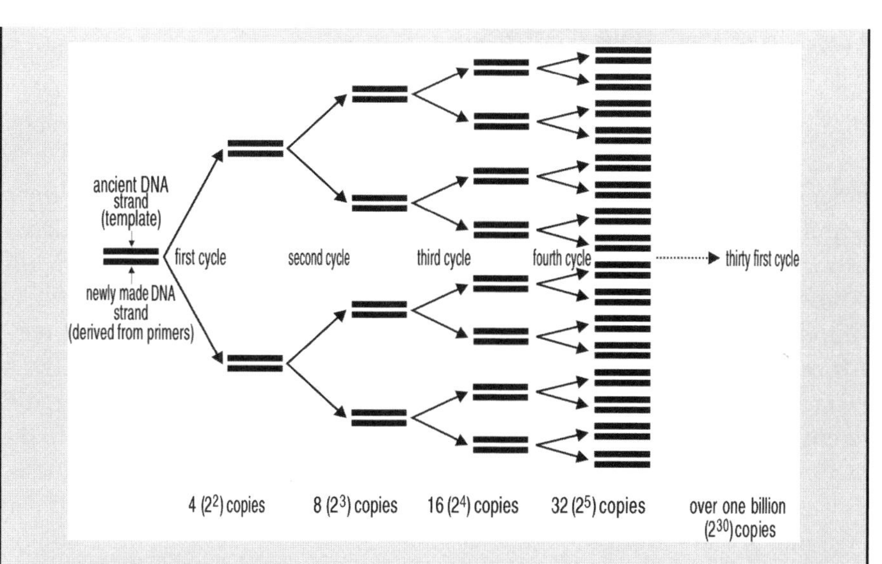

FIGURE 76 PCR and the amplification of DNA. During the polymerase chain reaction there is an exponential increase in the number of copies of the segments of DNA (the increase is proportional to the number of strands at any given moment). Thus, if there is only a single strand at the beginning of the reaction, there will be two after one cycle, four after two cycles, eight after three cycles, sixteen after four cycles, and so on. After thirty cycles, for example, which only take a few hours to perform, there will be well over one billion copies of the original DNA strand.

Collecting samples of ancient nucleic acids is a delicate operation that requires what are basically surgical procedures. It is advantageous, whenever possible, that the samples be collected at excavation sites and precautions taken to ensure that they do not become contaminated with other, particularly more recent, nucleic acids. At high temperatures and humidity, nucleic acids decay quickly. Well-preserved ancient nucleic acids can, therefore, be expected in sites where low temperatures and a dry environment prevail, as, for example, in cold, desert areas of the world. Once collected, the samples need to be isolated from any other remaining materials until they can be amplified by PCR and their chemical composition and structure can then be studied.

12.3. ANCIENT DNA STUDIES

Since the development of PCR, ancient nucleic acids, particularly DNA studies, are providing clues for solving quite a wide range of anthropological and archaeological issues important for understanding the evolution of modern humans. The relationships, or lack thereof, between ancient human populations, for example, has been elucidated through the study of the DNA extracted from ancient bone. Also, the occurrence of different forms of DNA (known as *phenotypes*) in different human populations is understood to support the hypothesis of an initial divergence between African and non-African human beings (Mountain et al. 1993). And mitochondrial DNA extracted from bone of a Neanderthal skeleton was shown to be quite different from the mitochondrial DNA of modern-day humans, attesting to the theory that Neanderthals did not pass on their DNA to modern populations. This is probably because Neanderthals were a separate branch of the human evolutionary line from the one that gave rise to *Homo sapiens*, the common ancestor of all modern human beings (Krings et al. 1997). Other archaeological studies to which the investigation of ancient nucleic acids may provide useful information include those on human migrations, paleodiets, and paleodiseases (Pääbo 1993; Brown and Brown 1992). A study of ancient DNA in bones, for example, confirmed the presence in the DNA of sequences of bacteria that cause leprosy and other infectious diseases (Taylor et al. 1996; Raffi et al. 1994). Another study revealed DNA from tuberculosis pathogens in Peruvian mummies dating from over 600 years ago, thus probably putting an end to a long-continuing discussion as to whether the first European explorers of the American continent brought the disease with them (Morell 1994).

The large and ever-increasing number of archaeological questions that have been elucidated by the study of ancient nucleic acids have given rise to the development of a relatively new and rapidly expanding field of study known as *molecular archaeology* (Evershed 1993; Hedges and Sykes 1992). Studies in this field, particularly since the large international scientific activity involved in the Human Genome Project during the 1990s, have improved and expanded techniques for extracting, replicating, identifying, and sequencing nucleic acids. Interdisciplinary work is revealing information not only on human ancestry and genetic development during the last few million years, but also on such diverse fields as paleopathology, ancient medical practice, food, and many other fields of human concern. As technology improves and human knowledge in molecular archaeology increases, the collection and study of ancient nucleic acids is opening new areas of study inconceivable in the past (Pääbo et al. 2004; Poinar 2002).

FIBERS
YARN, TEXTILES, AND CORDAGE; WRITING MATERIALS

Ancient human beings, like all the other anthropoids, had a large amount of body hair, which provided them with a natural layer of thermal insulation. During evolution, however, most of the hair was lost and replaced by coverage (clothing) that was made, at first, probably from the furs of animals, and in time from *textiles*. Both these types of materials, animal skins as well as textiles, are *fibrous materials*: in animal skin (the outer covering of the body of animals) collagen fibers are naturally interwoven with one another (see Chapter 11); in human-made textiles, fibers derived from either vegetable or animal sources are first separated from other tissue matter, spun into yarn, and then woven into cloth or knitted into garments (Joseph et al. 1992). Also threads for sewing and cordage for binding, trussing, and rigging are made from spun fibers. Ancient *fibers*, *yarn*, *textiles*, *threads*, and *cordage* are the main subjects discussed in the present chapter; animal skins are discussed in Chapter 11 (Chawla 1998; Grant 1954).

13.1. FIBERS, TEXTILES, AND CORDAGE

Most natural *fibers* are derived from vegetable and/or animal sources (see Textbox 66). Those derived from vegetables consist of the stems of plants, as for example linen and sisal, or of the husks of some fruits or seeds, such as cotton (this type of fiber is known as *bast fiber*). *Animal fibers* occur mainly as threadlike strands that grow on the skin and form a coat on a mammal, as for example the hair of camels and the wool of sheep. An exception is silk, which is the solidified secretion of the silkworm insect (Gohl and Vilensky 1987; Adovasio 1970). Thin and soft fibers, such as those of linen, cotton, and wool, used for making textile cloth, are generally known as *fine fibers*, while thicker, rougher, and irregular fibers, such as those of jute, sisal, and hemp, more appropriate for making cordage (string, rope, and twine), are usually referred to as *coarse fibers* (Weber-Partenheim 1971). Countless archaeological remains provide evidence on the use of fibers derived from plant and animal tissues since prehistoric times: fragments of linen fabrics from before the sixth millennium and somewhat later cotton textiles, and the remains of ancient string and cord-made hemp and sisal, for example, bear witness to the of use of fibers in ancient times (Wilson 1979; Gilbert 1954).

TEXTBOX 66

FIBERS AND FIBROUS MATERIALS

Fibers are forms of matter having a high ratio of length to width and thickness, a characteristic that makes them flexible and strong. There are natural fibers of vegetable, animal, and inorganic origin (see Table 87). A number of natural fibers have been used since primeval times, for making *yarns* and, from them, either *textiles* for coverage and shelter or *cordage* for sewing, tying, and binding. All the natural fibers derived from living organisms, whether of vegetable or animal origin, are biopolymers: substances having long molecules with a chainlike structure. Inorganic fibers are derived from fibrous minerals.

The main constituent of all *fibers of vegetable origin* (see Table 88) is, almost exclusively, *cellulose*, a polymeric carbohydrate (see Textbox 53). Vegetable fibers are resistant to alkalies and to most organic acids but are destroyed by strong mineral acids.

Animal fibers consist of *polymeric proteins*. Depending on the type and arrangement of the amino acid monomers that make up the protein, the animal fibers are classified into two major compositional groups: *hair* and *silk* fibers (see Table 89). *Hair fibers*, grown by fur-bearing mammals such as sheep (whose hair fibers are known as *wool*), camels, and goats, are composed mainly of *keratin*, the tough protein that is also the main component of nails, scales, horns, and hooves. Differences in the relative amounts as well as sequence of the amino acids within the keratin molecules determine the characteristics of every type of hair fiber. *Silk*, which is composed mainly of *fibroin*, is extruded as long *filaments* by insects such as the silkworm. The molecules of keratin are more complex than those of fibroin: the amino acid components of the two proteins are somewhat different, and the chains they form in keratin are slightly longer than those in fibroin (Gohl and Vilensky 1987; Asquith 1977). Animal fibers are resistant to most organic acids and to some mineral acids but are damaged by alkalies and bleaches.

Mineral fibers useful to humans occur naturally as *asbestos*, the generic name for a group of fibrous silicate minerals that are strong, highly flexible, and chemically and thermally stable; *chrysotile*, the most abundant asbestos mineral, is composed of hydrated silicate of magnesium. The main use of asbestos was, for many centuries, for making fireproof objects. In the midtwentieth century, however, it was found that asbestos causes a fatal lung condition and, consequently, its mining and use have been banned in most countries.

TABLE 87 Fibers and Filaments Used in Antiquity

Source	Variety	Nature	Composition
Vegetable	Cotton, linen	Fine fibers	Cellulose, a polymeric carbohydrate
	Jute, sisal	Coarse fibers	
Animal	Hair, wool	Staple fibers	Keratin, a protein
	Silk	Filaments	Fibroin, a protein
Mineral	Asbestos	Fibers	Fibrous mineral silicates
	Gold, silver	Filaments	Precious metals (silver or gold)

The first fibers used by humans were probably those that occur naturally as tissues or excretions of either vegetables or animals (see Table 87). At much later times, after metals had been discovered, humans also learned to manufacture – from some of the ductile metals, mainly gold, silver, and their alloys – thin filaments (not fibers, however), which have since been used to decorate textile fabrics. It was only during the twentieth century, after synthetic plastics were discovered, that it became possible to make artificial human made fibers. The great majority of the natural fibers, such as cotton and wool, occur as *staple fibers*, short fibers whose length is measured in centimeters. Silk is different from all other natural fibers in that it occurs as extremely long and continuous *filaments* several hundred meters long.

Making Textiles and Cordage. After they are separated from their vegetable or animal sources, most fibers generally undergo a form of mechanical processing known as *spinning*, which by twisting the fibers around each other, locks many fibers together to form long, continuous strands known as *yarn*; the greater the number of twists that the fibers undergo during spinning, the stronger the yarn. Most textile fabrics are made either by *weaving*, that is, interlacing yarn in two directions, or *knitting*, interlocking loops of a single thread of yarn. To make *cordage* (which includes cable, rope, and string), several strands of yarn are spun around each other.

Another very ancient process for making textiles, and one that does not require either spinning, weaving, or knitting, is *felting*; *felted textiles* are made of interlocking fibers so as to produce a *nonwoven* type of material. When wool fibers, for example, are subjected to moisture, heat, and pressure, they shrink and curl, interlocking with each other to form a tight, intertwined mesh, a textile known as *felt* (Gordon 1980; Burkett 1979). *Felting* (making felt) has been practiced since antiquity and is still practiced today. Nomadic people used felt to build tents, the Romans used it for making vests, and in many regions of the world felt is still used for making blankets, boots, and hats (Shams 1987; Whewell 1972).

Vegetable Fibers

Basically, all the vegetable fibers have a common chemical nature: their main component is cellulose (see Chapter 10); notwithstanding this major determinant of their characteristics, fibers from different parts of the plant, stem, leaf, seed, or fruit are generally put to different uses. Vegetable fibers are therefore usually classified according to the part of the plant from which they are derived. In Table 88 the most widely used vegetable fibers, out of several hundred fibers that have been recognized in plants, are listed and classified by origin (Friesen 1995).

TABLE 88 Vegetable Fibers

Fiber type	Fiber	Source
Fruit fibers	Coir	Coconut (*Cocos nucifera*)
Leaf Fibers	Abaca	*Musa textilis* plants
	Henequen	*Agave fourcroydes* plants
	Palm	*Arecaceae* plants
	Sisal	*Agave sisalana* plants
	Yucca	*Agavacea* plants
Seed fibers	Cotton	Plants of genus *Gossipyum*
	Kapok (Java cotton)	*Ceiba pentandra* trees
	Milkweed	*Asclepias* trees
Stem fibers	Esparto	Esparto (*Stipa tenacissima*) grass
	Hemp	Hemp (*Cannabis sativa*) plants
	Jute	Two species of *Corchorus* plants
	Linen	Flax (*Linum usitatissimum*) plants
	Kenaf (Manila hemp)	Hibiscus (*Hibiscus cannabinus*) plants
	Ramie (China grass)	*Bohemia nivea* grass

Cotton. The most widely used natural fiber in the world, *cotton,* is derived from the seed hair of plants of the botanical genus *Gossypium;* cotton varieties from different parts of the world belong to different species of this genus. Typical mature cotton fibers are round; their length varies for different cotton species from less than 1 to 5 cm. Most types of cotton are white, although naturally colored varieties are known. Unique brown fibers, for example, were derived from *Gossypium barbadense* plants grown along the northern coast of Peru as far back as 5000 years ago; light brown and mauve varieties have also been found in other places (Vreeland 1999).

Cotton ignites easily and burns quickly. Insects and fungi damage and may even degrade it. Exposure to sunlight for prolonged periods of time discolors (yellows) and degrades the fibers. Discolored cotton fabrics may, however, be *bleached;* that is, their undesired color may be removed safely with appropriate *bleaches* (agents that cause whitening or decolorizing of fibers, particularly of cotton, by exposure to sunlight or to chemicals) (Wendell 1995; Brown and Ware 1958).

Bleaching has indeed been practiced since antiquity. Until the nineteenth century chemical bleaching entailed one of two different processes: either a wet process, based on the use of an alkaline solution, or a dry one, known as *gassing*. When bleaching was done with an alkaline solution (such as stale urine in water), the excess alkali remaining in the bleached yarns or textiles was neutralized with some natural acid, such as sour milk. For *gassing*, the yarns or textiles were exposed to sulfur dioxide, a gas formed when sulfur, a pale yellow chemical element that ignites easily, is burned in open air. Gassing with sulfur dioxide was practiced by the ancient Greeks, as early as the fourth century B.C.E. (Easton 1971).

Linen. Derived from the stems of the flax plant, *Linum usitatissimum, linen,* also known as *flax,* was probably the first vegetable fiber to have been used by humans. The fibers make up the food-conducting tissue of the plantstems. They extend from the roots to the full size of the plant, which may grow nowadays to nearly one meter high, but probably much less in ancient times. Separating the linen fibers from the associated woody tissue was done, since early antiquity, by *retting,* a bacterial process that softens and decomposes woody tissue and separates the linen fiber bundles into single strands.

Two slightly different retting processes seem to have been practiced: *land* (or *dew*) *retting* and *water retting*. In *land retting,* soil fungi soften and degrade the woody tissue that holds the linen fibers together; in water retting the woody tissue is degraded by bacteria. After retting for a more or less extended period of time, lasting for several days (water retting is much faster than land retting), the woody tissue is broken down and softened by the activity of the microorganisms. The partly decayed stems are then beaten and shaken (the operation is known as *scutching*) to remove the decayed woody tissue, the detached linen fibers are heckled (to separate long and short fibers) and are finally dried. The length of linen fibers is variable and may have reached, in antiquity, lengths of up to 30 cm. Linen is resistant to fungi, most insects, and sunlight, although it is discolored and degraded after extended exposure to light. Like cotton, linen can also be bleached.

Animal Fibers

The *hair fibers* derived from furry mammals are mainly made up (over 80%) of the structural protein *keratin*. The distinction between *wool* and *hair* is not compositional, but related to size: *wool fibers* are generally fine and short, whereas those of hair are usually thicker and longer. The molecule of *keratin* consists essentially of a combination of amino acids: about 18 amino acids make up the keratin molecule (see Textbox 67). The nature of the amino acids, their relative amounts, and their sequence and arrangement within the molecule of keratin vary from one animal species to another but are characteristic of any variety of wool or hair (Asquith 1977) (see Table 89).

TABLE 89 Animal Fibers

Component	Fiber	Source
Keratin	Alpaca	Alpaca (*Llama glama*)
	Cashmere	Cashmere goat (*Capra hircus laniger*)
	Llama	Llama (*Llama glama*)
	Mohair (angora wool)	Angora goat (*Capra hircus*)
	Vicuña	Vicuña (*Llama vicugna*)
	Wool	Sheep of animal genus *ovis*
Fibroin	Silk	Silkworm (*Bombix mori*)
	Tussah (wild silk)	Silkworms other than *Bombix mori*

Wool. *Wool* was procured in the past as either *sheared* or *pulled* wool: it was *sheared* from living sheep and generally *pulled* from the hide of dead ones. Sheared wool is said to be of better quality than pulled wool. The length of the fibers, which is determined by the breed of sheep, their growth conditions, and the length of time between shearings, ranges from 3 cm to as much as 35 cm, although it was probably less in antiquity. The fineness and color of the fibers vary with the breed of the sheep. Wool fibers are naturally protected by a layer of *lanolin*, a wax. Since lanolin attracts dirt and foreign matter, it is generally removed before making wool yarn or felt (Lipson 1953). Prolonged exposure of wool to sunlight induces the *photochemical* *degradation* (a form of breakdown related to the chemical effects of ultraviolet radiation) of the keratin that makes up the fibers and, eventually, the total destruction of the wool. Wool is resistant to bacteria and some fungi. If buried in the soil, however, wool is eventually destroyed by rot-producing bacteria. Some insects, beetles, and larvae (e.g., of moths) use wool as food and, therefore, destroy wool yarns and textiles. Wool burns easily, albeit slowly and without a flame; the burning process is accompanied by slight sputtering.

Silk. *Silk*, the only natural fiber that comes in filament form, has been and still is one of the most appreciated and valued textile fibers. Silk filaments are secreted by the larvae of several types of silk moths to make their cocoons. Most silk is derived, however, from the larvae of the *Bombyx mori* moth, which has been widely cultivated in China for over 5000 years. Fragments of silk fabric dated to the late fourth millennium B.C.E. were found at Qianshanyang, in the province of Zhejiang, in China. There are, however, even earlier indications of the use of silk; silk remains were found together with an eleventh-century B.C.E. mummy in Egypt, probably also providing evidence of ancient trading routes between the Far and Middle East.

The larvae of the *Bombyx* moth secrete not one but two parallel filaments of silk bound together into a single strand by sericin, a wax also secreted by the larvae. After they have secreted the strand of silk, the larvae also wind it into *cocoons*, protective cases in which the larvae envelop themselves before passing into the chrysalis state, between larva and fully developed moth. Processing the silk filaments for making yarn involves three working stages: (1) unwinding the strand from the cocoons; (2) separating the twin filaments, which are very thin and slender, to be woven into textiles; and (3) spinning (twisting) several filaments together so as to obtain a substantial yarn that can be either woven into a textile fabric or made into cord (Varron 1938).

The silk filaments are made up mostly of the fibrous protein *fibroin*, which is characterized by reiterations of amino acid units. The properties of the silk are determined by its amino acid composition, the complexity of the repetitive units, and the arrangement of these units into higher-order arrays within the fibroin molecules. *Sericin*, the natural wax that covers and joins the twin silk filaments secreted by the silk larvae, stiffens the filaments but also provides some protection during processing. To remove the sericin, the silk filaments or the yarn made from them are boiled in water to which some detergent, which combines with the wax and facilitates its removal, is added. In early ancient times all the detergents were of natural origin, mostly derived from plants; more modern detergents, such as soap, are human-made (see Chapter 10).

TABLE 90 Wild Silk

Type	Silkworm	Host plant	Habitat
Muga silk	*Antheraea assamensis*	*Lauraceae* and *Magnoliaceae* trees	India
Tassar silk	*Antheraea paphia*	*Combretaceae* and *Dipterocapaceae*	India
Tussah silk	*Antheraea pernyi*	Oak (*Quercus*) trees	China
Mexican silk	*Eucheira socialis*	Madrone trees	Mexico

The larvae of *Bombyx mori*, the cultivated moth from which most silk has long been and still is made, feed on leaves of mulberry trees. In addition to cultivated silk, small quantities of "wild silk," also known as *nonmulberry silk*, have been derived in many parts of the world from the cocoons of moths other than *Bombyx mori*. Table 90 lists wild silks and the insect species that produce them (Peigler 1993; Jolly et al. 1979).

Nowadays, most silk filaments are several hundred meters long and some may even exceed one kilometer in length. In antiquity they were probably much shorter. Silk is soft but strong and very absorbent. Silk textiles combine thermal insulation with softness, high luster, translucence, and lavish appearance. Silk ignites and continues to burn as long as near a flame, but it is self-extinguishing, the end product of burning is a crisp, brittle black ash. Silk is more susceptible than wool to photochemical decay: ultraviolet radiation in particular causes the relatively fast, total destruction of silk. It is also generally resistant to attack by fungi and most types of bacteria, although some bacteria may cause silk to decay and even to be totally destroyed.

Inorganic Fibers

Asbestos. *Asbestos* is the only natural fiber of inorganic origin that has been used since antiquity to make yarns and fabrics. Asbestos occurs in nature in various forms of fibrous silicate mineral, for example *chrysotile*, the most abundant asbestos mineral, which is composed of hydrated silicate of magnesium. The length of its fibers varies from a few millimeters up to 5 cm. The fire resistant properties of asbestos were well known to the ancient Egyptians, who used it for making burial clothes. Its name is derived from the Greek, *a* meaning not and *sbestos* quenchable; the ancient Greeks used asbestos for making fire-resistant lamp wicks. In Rome, where it was known by the Latin name *amianthus*, asbestos was used to make cremation clothes and also lamp wicks. The Japanese made fire resistant clothes from it

(Zussmann 1972). For many centuries, small cloths woven from asbestos fibers were a luxury item used for handling hot domestic and industrial articles.

Metal Threads. *Metal threads*, not really metal fibers but metal filaments, are human-made. In antiquity, metal threads were made for ornamental or decorative purposes from precious, ductile metals or alloys, particularly silver and gold and their alloys. Such threads were either applied with adhesives to finished fabrics or wound around ordinary textile yarn cores; the metal-covered yarn was then either woven into textile fabrics or embroidered on the textile fabric (Járó and Tóth 1991; Lee-Whitmann and Skelton 1984).

13.2. ARCHAEOLOGICAL FIBERS AND FIBROUS MATERIALS

The identification of ancient fibers is an important part of the study of archaeological textiles and cordage. The most important preliminary information usually required, regarding archaeological fibers, is their qualitative identification, that is, determining their origin and nature, whether they are of vegetable or animal origin. A relatively easy test used for this purpose is the *burning test*, which generally provides reliable information: igniting fibers and observing how they burn and the products of the burning process usually provides dependable evidence on their nature. Typical results of fiber burning tests are listed Table 91.

For blends of vegetable and animal fibers, however, the burning test may be indecisive. Moreover, dyes and mordants on yarns or fabrics may affect

TABLE 91 Burning Characteristics of Fibers and Filaments

	Burning process				
Nature of fiber	Near flame	In flame	After removal from flame	Odor	Residue
Vegetable	Ignites; does not shrink	Burns rapidly	Continues burning and afterglows	Similar to burning paper	Feathery, light to dark gray
Animal	Curls away from flame	Burns slowly	Self-extinguishes	Similar to burning hair	Small black bead
Inorganic (metal or mineral)	Does not burn	Glows	Does not burn	None	Unaltered

the results of the test. Particular consideration needs to be given, therefore, to the examination of mixed or colored yarns and textiles and the evidence derived from burning tests should be supported by the results provided by other techniques. Scanning electron microscopy (SEM), for example, makes it possible to differentiate between vegetable and animal fibers. Atomic absorption spectroscopy provides information on the elemental composition of the fibers. Infrared spectroscopy reveals details not only on the nature of fibers but also on their internal structure. Chromatographic techniques provide means for separating, identifying, and quantifying the actual remains of archaeological fibers as well as the products of their decay (Kouznetsov et al. 1996; Ryder and Gabra-Sanders 1985).

13.3. WRITING MATERIALS

The use of specific materials as media for writing has followed a long and interesting trail that probably started when images were scratched on stone or wood surfaces. Materials that seem to have been almost specifically intended for writing purposes include wood bark, textile fabrics, *papyrus*, *parchment*, *vellum*, and various forms of *paper*. The last four in this list, papyrus, parchment, vellum, and paper, have been widely used for writing during the last five millennia. *Parchment* and *vellum*, made from animal skin, are discussed in Chapter 11. *Papyrus* and *paper*, both of vegetable origin, are briefly discussed in the following paragraphs.

Papyrus

Papyrus is one of the earliest materials made in the form of flats sheets specifically for writing. It was being made as early as the third millennium B.C.E. in ancient Egypt, from the stalks of *Cypera papyrus*, the papyrus plant (El Ashry et al. 2003). To prepare papyrus sheets, the stalks were first cut off the plant and then sliced lengthwise into thin, even strips. The slices were then arranged into overlapping, usually crisscrossed, layers, then hammered or beaten and finally pressed together until dry. This sequence of operations caused the slices, which adhered to each other because of a natural gum they contained, to bond together into single, flexible, and cohesive sheets (Leach and Tait 2000; Parkinson 1995).

In ancient Egypt papyrus sheets were often joined into long bands, known as *scrolls*, which could then be rolled on a wooden rod. These scrolls were known collectively as *volumes*, a name derived from the Latin word *voluere*, to roll. By the beginning of the Christian era, *parchment* and *vellum* (see Chapter 11) began to be used for writing. Sheets of these materials were bound into books, or *codices*, as the earliest books are known, which began to replace papyrus scrolls for writing texts. Papyrus, it was found, was too fragile a material for folding, stacking, and binding together, and it was gradually and almost entirely superseded by parchment and vellum. After the tenth century C.E. parchment and vellum began, in turn, to be superseded by paper.

Paper

Paper, made mostly for writing in the past, consists almost exclusively of cellulose fibers derived from straw or wood. Paper was first made in China, apparently at the beginning of the first century C.E., from fibers of the hemp plant. Making paper in relatively large quantities, however, began only about 150 years later, also in China. By the middle of the eighth century, when the Arabs conquered central Asia, papermaking began to spread throughout the Arab world, gradually displacing papyrus, parchment, and vellum as a writing material. In Europe, paper did not become widespread until the late fourteenth century (Hunter 1978).

The basic principles of *papermaking* (not so the technology) have remained virtually unchanged since the process was first invented. To make paper, the raw material – cotton or linen rags, straw, or nowadays mainly wood – are first ground, mixed with water, and macerated so as to break them down into separate cellulose fibers (*maceration* is a natural biochemical process that softens and separates tissues, such as those that make up wood and straw). After the maceration process is completed, the detached cellulose fibers are separated from the water by pouring the suspension of fibers onto a screen. This leaves on the screen a matted layer of intertwined fibers that, after being subjected to pressure and finally dried, make up a sheet of paper (Goedvriend 1988).

Rice paper is the widely used misnomer for two entirely different materials also made in the form of thin sheets: Chinese kung-shu, which is not paper (see text below) and *washi*. Also known as *Japanese rice paper*, washi is paper made from the cellulose fibers derived from the bark and branches of mulberry trees (*Broussonetia kajinoki*) (Inaba and Sugisita 1988; Barrett 1988).

Nonpaper Flat Materials

Kung-shu, a material akin to paper, is made by drying thin sheets cut spirally from the inner pith of *Tetrapanax papyriferum*, the kung-shu tree, which grows in the mountains of Taiwan. In China kung-shu has been used as a substrate for writing and painting (Perdue and Kraebel 1961).

Amate (or amatl) is another paperlike material that was derived by the Maya inhabitants of ancient Mexico from the bark of *Ficus* (amatl) trees. The Mayas used amate, which they called *"huum,"* for writing codices, a tradition that was carried on by the Aztecs, who also improved on the Maya methods of making amate: they burnished the surface of amate with flat stones so as to close the pores and provide a smooth, nonblotting writing surface. Their burnishing technique was similar to that of European Renaissance papermakers, who also burnished paper by rubbing it with smooth stones (Wiedemann 1995; von Hagen 1977).

Palm leaf was the main material used for writing in southwest Asia before the introduction of paper. It was made by drying and processing the leaves of various palm trees in manners that varied in different areas; the most commonly used leaves were those of *Corypha umbraculiphera*, the fan palm, and of *Borasus flabellifer*, the palmyra palm.

Tapa, also known as *bark cloth*, another paperlike material used for writing as well as for decoration, is still made from the inner bark of various plants in tropical regions of the world, including Asia, some of the Pacific Islands, Africa, and South America. Although there are differences in the methods of making tapa in each region, the process is basically similar everywhere. In some regions the inner bark is first dampened and then soaked in water for lengths of time varying from several hours to several weeks; in others the bark is cooked for several hours. In the southern Pacific Ocean islands, for example, tapa is made even today by beating the damp bark of mulberry trees (*Broussonetia papyrifera*) into thin sheets (Bell 1983). Because of biological destructive factors, mainly consumption by bacteria, fungi, and insects, however, very little ancient tapa has survived.

DYES AND DYEING

14

Dyes, often also called *coloring matters* or *colorants,* are intensely colored, soluble organic substances used to impart color to fibrous materials (see Textbox 66). Not all colored and soluble organic substances, however, are dyes; only those whose molecules have a considerable structural complexity are useful for imparting color to other materials. Moreover, a substance is considered a dye only if the color it imparts is fairly *permanent,* that is, resistant to fading and disappearance.

The use of dyes seems to have began over 10 millennia ago, when dyes of vegetable origin were apparently applied to the skin simply for amusement, for ritual purposes, or to identify or differentiate status or social group. All the dyes used in the past and up to the nineteenth century, when artificial dyes were first synthesized, were of natural origin; most were extracted from plants, some from animals (Verhecken 2005). Common dyes well known since antiquity are listed in Table 92 (Kirby 1988; Celoria 1971).

14.1. STAINS AND STAINING

Sometimes, inorganic substances such as, for example, some mineral *pigments,* have been used to color fibrous materials, a process generally referred to as *staining.* Pigments are not soluble in water, the almost universal solvent used for dyeing, nor do they react with the fibrous substrate. They generally become attached to the substrate by a binder, and the words *stain* and *staining* imply that the pigment (the coloring matter) simply becomes attached to the substrate. Table 93 lists mineral stains that, dispersed in water, have been used for staining fibrous materials.

Archaeological Chemistry, Second Edition By Zvi Goffer

TABLE 92 Ancient Dyes

Color	Dye (common name)	Source
Vegetable Source		
Black	Carob	*Caesalpina brevifolia* trees
	Sticky alder	*Aldus glutinosa* trees
	Walnut	Shells of *Juglans nigra* nuts
Brown	Chestnut	Fruit of *Castanea sativa* trees
Blue	Indigo	Indigo (*Indigofera*) plants
		Woad (*Isatis tinctoria*) plants
Green	Myrtle	*Myrtus communis* shrub
	Yarrow	*Achillea milleforium*
Orange	Henna	Various species of *Lawsonia* shrubs
Purple	Archil (or argol)	*Lecanora tartara* and other lichens
Red	Alkanet	Roots of *Anchusa tinctoria* plants
	Annato	Fruits of *Bixa orellana* shrubs
	Brazil wood	Various trees of the *Caesalpine* species
	Madder	Roots of *Rubiacea* plants
	Sandalwood	Wood of *Pterocarpus santalinium* trees
Violet	Ficus	*Ficus tinctoria* trees
Yellow	Berberry root	*Berberis vulgaris* bush
	Fustic	*Chlorophora tinctoria* trees
	Quercitron	*Quercus discolor* and *Quercus tinctoria*
	Young fustic	*Rhus cotinus* trees
	Gambier	Resin exuded by *Uncaria gambir* shrubs
	Safflower	Bastard saffron (*Carthamus tinctoria*) plants
	Saffron	Saffron (*Crocus sativus*) plants
	Turmeric	*Curcuma* plants
	Weld	*Reseda luteola* plants
Animal Source		
Light blue	Tekhelet	Several varieties of *Janthina* molluscs
Purple	Tyrian purple	*Murex* molluscs
Red	Cochineal	*Coccus cacti* insects
	Kermes	Female *Kermococcus vermilia* insects
	Lac dye	Lac resin, secreted by *Kerria lacca* insects

TABLE 93 Mineral Stains

Color	Stain
Black	Charcoal, pyrolusite
Brown	Clay, ocher, umber
Red	Burned clay, cinnabar, antimony red
Yellow	Ocher, orpiment

14.2. THE DYEING PROCESS

Most fibrous materials readily absorb water. If textile fabrics such as cotton or wool, are soaked in water in which a dye is dissolved, part of the dye leaves the solution, reacts with the fibers that make up a fabric, and imparts its color to the fibers. The textile fabric is then said to be *dyed*. The strength of *fixation*, as the strength of the bond between a dye and a fiber is usually termed, is determined by the chemical affinity between them, that is, by the nature of the chemical bond formed between the dye and the fiber. Dyes that bond strongly to fibers generally produce *fast colors*, resistant to fading by washing or exposure to light. Such dyes are generally known as *direct dyes*. Others, which bond only weakly to fibers, usually fade when the dyed fibers are washed or exposed to light; such dyes are known as *fugitive dyes* (see text below) (Padfield and Landi 1966).

14.3. MORDANTS

Many dyes that have no chemical affinity to fibrous substrates can be attached to such substrates by intermediary (go-between) substances known as *mordants*. These are either inorganic or organic substances that react chemically with the fibers as well as with the dyes and thus link the dyes to the fibers. Mordants are traditionally classified into two main classes, *acid* and *metallic mordants*. The *acid mordants* are organic substances that contain *tannins* (see Textbox 64) as for example, gall nuts and sumac. The *metallic mordants* are inorganic substances, mostly mineral oxides and salts that include metal atoms in their composition. Table 94 lists mordants of both these types, which have been used since antiquity.

The mordants may be applied to the substrate either before, during, or after application of the dye. *Premordants*, as the mordants applied before the dye are known, seem to have been the most commonly used in antiquity. Some mordants not only are instrumental in attaching the dye to the fibers but also alter the shade and even the hue of some dyes; a single dye often provides a range of hues when used with different mordants. The color or hue of textiles dyed with madder, weld, and logwood, for example, are determined by the chemical nature of the mordant used; Table 95 lists hues obtained when dyeing with the same dye but with different mordants.

14.4. THE NATURE OF DYES

Direct and Mordant Dyes. The ancient dyes (see Fig. 77) may be classified into three main groups: *direct dyes*, *mordant dyes*, and *vat dyes* (see

TABLE 94 Mordants Used in Antiquity

Mordant	Source	Composition
Acid Mordants (of Vegetable Origin)		
Gull nuts	Vegetable excrescences (resembling nuts), on vegetable tissues such as of oak trees. Dried leaves, twigs and fruits of shrubs	Tannins make up 50–70% of the weight of gull nuts mixture of tannins and organic acids
Sumac	Trees from *Rhus* order of plants	
Metal Mordants		
Alum	Mineral	Sulfate of aluminum and other metals (e.g., potassium alum)
Borax	Mineral	Hydrated sodium borate
Natron	Mineral	Natural mixture sodium carbonate and sodium bicarbonate
Iron vitriol	Mineral or synthetic	Hydrated iron sulfate

TABLE 95 Dyes, Mordants, Hues

Mordant	Color
Hues of Textiles Dyed with Madder but with Different Mordants	
Aluminum	Red or pink
Iron	Purple-black
Aluminum + iron	Brown
Calcium	Blue-red
Tin	Orange-yellow
Hues of Textiles Dyed with Weld but with Different Mordants	
Aluminum	Lemon yellow
Iron	Olive
Tin	Green-yellow

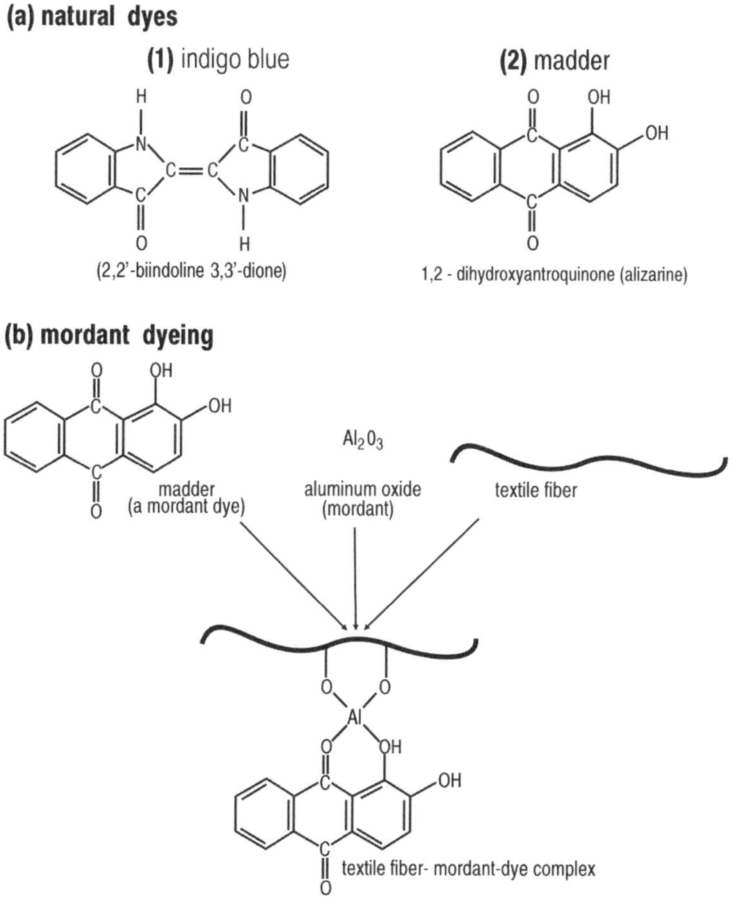

(a) natural dyes

(1) indigo blue

(2,2'-biindoline 3,3'-dione)

(2) madder

1,2 - dihydroxyantroquinone (alizarine)

(b) mordant dyeing

madder
(a mordant dye)

Al_2O_3
aluminum oxide
(mordant)

textile fiber

textile fiber- mordant-dye complex

FIGURE 77 Dyes and dyeing. Since earliest times people have used natural dyes to color textiles, hides, feathers, and even their own skin. Indigo, a blue dye **(a1)**, for example, seems to have been used in Europe since neolithic times. The earliest written record of the use of natural dyes seems to come from China, dating from as early as the middle of the third millennium B.C.E. Chemical analysis of red fabrics in the tomb of King Tutankhamun in Egypt (fifteenth century B.C.E.) showed the presence of alizarin, the coloring matter in madder **(a2)**, a dye of vegetable origin. "Direct dyes" have affinity for fibrous materials and are directly attached to them, while "mordant dyes" require a mineral such as alumina **(b)**, known as a "mordant," to "fix" the dye on dyed fibers. The mordants (whose name is derived from the French word for "to bite") are substances, mostly mineral salts or metal oxides, that form chemical bridges between dyes and dyed fibers. Madder requires a mordant for dyeing, whereas indigo does not.

text below and classification based on usage in Table 96). The *direct dyes,* also known as *substantive dyes,* are those that become directly attached to the fibers to which they are applied. Wool and silk, for example, are generally dyed directly by substantive dyes which have strong chemical affinity for the proteins that constitute most of these animal fibers (see Chapter 13). Direct dyeing of vegetable fibers such as cotton and linen is often impossible, however, because of the low affinity between the dyes and cellulose, the main constituent of all vegetable fibers, and dyeing them generally requires the use of mordants. Also, mordant dyes, such as saffron, kermes, and madder, seldom become attached on their own to most fibers, but attach only through a mordant. Alum and natron, two minerals employed since very early times, were among the most widely used ancient mordants. Dyeing with madder, for example, seems to have entailed a sequence of mordanting and dyeing operations often involving as many as seven different stages and requiring several days for completion.

TABLE 96 Dyes, Classification Based on Usage

Dye type	Application Method	Main substrates
Direct	Slightly *alkaline* dye bath	Cotton
Mordant	In conjunction with a *mordant*	Wool, silk, linen
Vat	First solubilization (by chemical *reduction*) and conversion of the dye into a colorless *leuko* intermediate; after impregnation of the substrate with the *leuko* intermediate, restitution (by chemical *oxidation*) of the dye's original color	Cotton, wool

Vat Dyes. *Vat dyes* are insoluble in water. *Indigo,* for example, an ancient blue dye, is probably the best-known example of an ancient vat dye; others include *woad* and *Tyrian purple.* Since the process of dyeing requires that the dye be in solution, dyeing with a vat dye (or *vat dyeing,* as the process is known) is possible only after the vat dye has been made soluble by a relatively long and somewhat complicated chemical procedure. The terms *vat dye* and *vat dyeing* are probably derived from the large tanks or "vats", in which the process was carried out in ancient times.

The *vat dyeing* process involves three successive processing stages: (1) solubilizing the dye, (2) impregnating the substrate to be dyed with the now-soluble dye, and (3) ensuring that the impregnated substrate attains the desired color. Solubilizing a vat dye in water is possible only by chemically altering it by chemical reduction. The reduction process, however, also results in the loss of dye's color (the reduced and colorless vat dye is generally known as a *leuko compound*, from the Greek word *leukos*, meaning white). Leuko compounds are, therefore, reduced and soluble forms of vat dyes. Oxidizing the leuko compound regenerates the color of the dye. In the second, impregnation stage of the process, the substrate is immersed in the colorless solution, whereby the reduced and dissolved leuko compound becomes attached to the substrate. In the third and final stage, the color of the dye is restored by exposing the dye-impregnated substrate to the atmosphere: as atmospheric oxygen oxidizes the dye, its color is regenerated.

In practical terms, when vat-dyeing yarns or textiles, the insoluble dye is first mixed with and suspended in water under alkaline and reducing conditions (i.e., in the absence of oxygen). The alkaline suspension of the dye in water is then subjected to a controlled *anaerobic* putrefactive process that converts the insoluble dye into the chemically reduced, soluble and colorless intermediate leuko compound. The yarn or textile is next immersed in the colorless solution and becomes impregnated with the leuko compound. When the impregnated fibrous material is removed from the solution and exposed to air, the leuko compound is oxidized by oxygen in the air, recovers its original color, and becomes firmly attached to the substrate. A difficulty inherent to the vat dyeing process is that, until the color of the dyed fibers is finally developed, it is impossible to foretell what the end color will be; the actual depth of color is not related to the concentration of the dye in the dye bath but depends on the number of impregnations undergone by the fibers in the solution of the leuko compound. The second and last stages of the dyeing process, involving the impregnation and restitution of the dye's color, may be repeated a number of times, if required, to intensify the color of the dyed material (Clark et al. 1993; Fox 1948). Vat dyes are among the fastest dyes known and continue to be used today, whenever color stability is required.

14.5. ANCIENT DYES

Only natural dyes were known until the nineteenth century. By trial and error and probably also by chance, humans learned to extract and use a large variety of dyes of vegetable and animal origin. Dyes were extracted from the roots, trunk bark, and branches of trees, the stems, leaves, flowers, and fruits of plants, the bodies of insects and mollusks, and the eggs of insects. All the dyes obtained from natural sources are rather impure, and hence the accurate reproducibility of colors was almost impossible during antiquity. Still, many of the dyes and dyeing techniques used in antiquity were highly developed and remained in use until the discovery of the synthetic dyes in the middle of the nineteenth century (Colombo 1995; Robinson 1969).

Blue Dyes

A single blue dye, *indigo*, seems to have been predominant in the ancient world since it first became known to humans, over 4000 years ago. Some historians believe indigo to be the oldest known dye. Indigo is a vat dye that may be derived from either of two different types of plants: *indigo* or *woad*. Indigo plants belong to the genus *Indigofera*, which grows naturally and was also cultivated in many areas of the world, as for example, Egypt in Africa, India in Asia, and Brazil and Peru in South America. The woad plant, *Isatis tinctoria*, is native to southeastern Europe, from where it spread quickly throughout Europe in prehistoric times (Hurry 1930). Although the dye in both types of plants is the same, namely, indigo, in the *Indigofera* (indigo) plants the dye is about 30 times more concentrated than in *Isatis* (woad) plants. The dye is composed of a mixture of two different organic substances: *Indigo blue*, the major component (see Fig. 77), and small amounts of *indigo red*, chemically known as *indirubin*. Trace amounts of indirubin are probably present in all samples of the natural blue dye. Some indigo plant varieties, such as some from Java, are particularly rich in indirubin, containing as much as 6% indigo red. Indigo blue has been traditionally extracted from the leaves and to some extent from the midribs of the plant. After the leaves and midribs have been collected, they are soaked in water; the mixture of water and plant tissue is then pounded into a paste that is left to ferment for a few days. Following the fermentation process, the liquor is filtered out of the solid residue and exposed to air; oxygen in the air oxidizes the dye, which precipitates out of solution as a powdery dark blue solid that is then separated by filtration from the liquid and finally dried.

Since it is a vat dye, indigo cannot be used for dyeing without prior chemical modification to make it soluble. Thus, during the dyeing process indigo is first chemically reduced to its colorless intermediate, *leuko indigo*, which is soluble in water. Yarn and/or textiles dipped in the colorless solution absorb the reduced leuko indigo, which, when exposed to air, is oxidized and yields the characteristic indigo blue color (Seefelder 1982).

Ancient people in such widely separated regions as India, Egypt, and Peru independently discovered the preparation of the blue dyestuff from different indigo plants indigenous to the regions they inhabited. The dye has been identified in woolen fabrics from upper Egypt dating from 200 to 700 C.E., in cloth from the Judean Desert in Israel, and in Inca textiles from Peru. The ancient Celts tattooed their bodies with designs colored with indigo that they obtained from the woad plant (Balfour-Paul 2000; Saltzmann et al. 1963).

Green Dyes

It seems that no green dyes were known in the ancient world. Most ancient green-colored textiles appear to have been obtained using mixtures of dyes having complementary colors, such as yellow and blue. Examination of green textiles from Egypt, for example, revealed that the color was obtained by mixing indigo blue with a yellow dye whose chemical nature was not determined (Pfister 1935).

Purple Dyes

Two dyes were extensively used in antiquity for dyeing purple: *archil* and *Tyrian purple. Archil,* also known as *argol* or *lichen purple,* is derived, as its name suggests, from a variety of lichens. The best-known variety, *Lecanora tartara,* grows on rocks in many Mediterranean islands. The preparation of the dye entails a rather involved sequence of operations beginning with maceration of the lichens, followed by oxidation of the macerated product, and then fermentation of the oxidized solution, during which a powdery residue is formed. The residue is *archil,* which is finally dried. The coloring matter in the dye is the organic compound *orcein.*

Tyrian purple, also known as *royal purple* and *purple of the ancients,* was undoubtedly, one of the most renowned and highly valued ancient dyes, which had a high standing as a symbol of wealth and distinction. Cloth dyed with Tyrian purple was very costly, and it seems that only royalty and priests could afford it. The production of Tyrian purple flourished, as the name suggests, in the ancient town of Tyre in southern Lebanon, on the shore of the eastern Mediterranean Sea. Quite accurate information about the importance of purple in the ancient world has long been available, but only during the twentieth century did its composition, the way it was produced, and its true appearance become known. Although it is often claimed that ancient purple possessed fastness and brilliance of color, purple-dyed fabrics actually had a reddish shade tending toward violet. The color that seems to have dazzled the ancients actually fell short of the beauty of shades and tints produced by modern dyes; scientific research on Tyrian purple laid to rest a very long accepted but unproved tenet.

Tyrian purple was derived from the "purple snail," the common name for what, in reality, are several species of mollusks of the genus *Murex*. Each one of the mollusk species yielded a slightly different variety of purple. In Tyre, where the most prized purple dye was produced, *Murex brandaris* snails were those most abundant and generally used, while in Sidon, not far to the north of Tyre, an amethyst purple variety of the dye was obtained from

Murex trunculus snails. The dye is one of the components of a colorless secretion produced by the *Murex* mollusks. When the secretion is exposed to oxygen in the air and to solar radiation, it acquires color, turning first yellow then green and finally purple. In the course of an investigation on the production of Tyrian purple, only 1.4g of the dye were obtained from about 12,000 mollusks! The coloring matter in Tyrian purple was identified in the early twentieth century as the organic compound 6,6″-dibromoindigo (Cooksey 2001; McGovern and Michel 1991, 1985). Because of the heavy demand for purple and its high price, many substitutes, designed to either dilute the dye or imitate its color, were used in the ancient world. Thus mixtures of indigo and other dyes that yield shades of purple were often used as substitutes. At the time of the decline of the Roman Empire the manufacture and use of Tyrian purple seems to have subsided, being gradually replaced by dyes easier to produce, such as archil, kermes, and madder. It has been suggested that a purple dye similar in composition to Tyrian purple was produced in the South American continent from the mollusk *Purpura patula Pansa* (Doumet 1980; Saltzman et al. 1963).

Red Dyes

Quite a variety of red dyes of both vegetable and animal origin were used in antiquity, although only a few of them ever attained practical importance. Among these were *madder*, of vegetable origin, probably the most widely used and in its heyday also the most important red dye, and *kermes* and *cochineal*, both derived from the bodies of insects.

Madder, also known as *Turkey red*, is a scarlet dye extracted from perennial herbaceous plants of the order *Rubiacea*, of which there are about 35 species (Chenciner 2001; Farnsworth 1951). A well-known plant from this order is *Rubia tinctorum*, found naturally in Palestine and Egypt, abundant in Asia and Europe, and extensively cultivated in the ancient world, was widely used for production of the dye since remote antiquity. The use of madder for dyeing seems to have originated in the Middle East: it was identified in many textiles found in Egyptian tombs and in woolen fabrics from the Judean Desert in Palestine. It was also used by the ancient Persians, Greeks, and Romans. Madder from other varieties of *Rubiacea* plants were used by the Incas in ancient Peru (Schaefer 1941; Fieser 1930).

The dye is concentrated mostly in the roots of the plants, older roots being richer in the dye than young ones. After removal from the soil the roots are washed with water, dried, and chopped. The dye is extracted from the plant tissue with hot water, in which it is soluble, but precipitates as a powder when the solution cools down. The powder is then separated from the liquor and finally dried. The main coloring material in madder, as well as in all the red dyes obtained from most varieties of *Rubiacea* plants, is *alizarin*, an organic compound (see Fig. 77). A minor component of the dye extracted from *Rubia tinctorum*, but the main one in the roots of *Rubiacea* plants from South America is *purpurin*, also an organic compound that resembles alizarin (Fester 1953).

Dyeing with madder yielded brilliant, permanent reds. When the dye was used in conjunction with different mordants, however, it provided a variety of hues, as listed in Table 95. Madder was also used for the preparation of red lakes (see Textbox 20), as for example, that applied to colored Greek terracotta statuettes (Abrahams and Edelstein 1967; Farnsworth 1951). For many generations and up to the present time madder has also enjoyed popularity in the Middle East for its medicinal properties. Madder has been identified in human bones buried in Qumran at the Dead Sea in Israel. People living at Qumran apparently believed that madder had medicinal and even magic properties, and drank a concoction of madder as a beverage (Steckoll et al. 1971).

Henna is another ancient red dye of vegetable origin, that was not widely used for dyeing textiles, but mainly for staining human hair, skin (mostly the hands), and nails. Henna is derived from the small shrub *Lawsonia inermis* and from other species of *Lawsonia* plants that are native to the Middle East, northern Africa, Persia (Iran), and India. The dye is derived from the leaves of the plant. To extract the dye, *Lawsonia* roots are soaked in water and ground into a paste. The coloring matter in henna is the organic compound *lawsone*, which is soluble in water and henna is, therefore, obtained as a powder, after filtering the solution and evaporating to dryness. Henna yields bright orange hues when applied directly to fibers or textiles, but yellow or brown when applied in conjunction with different mordants (Cox 1938).

Three red dyes of animal origin were known in the ancient world: *kermes*, *cochineal*, and *lac*; all three are derived from insects. *Kermes*, which is scarlet-red, is probably the earliest one on which there are records. Derived from the insect *Kermococcus vermilia Planch* (formerly known as *Coccus ilicis*), which lives on oak trees of the species *Quercus cocciferaea*; it is widely distributed in Armenia, the Middle East, northern Africa, and Spain. Kermes seems to have been the most important red dye known to the ancient Babylonians and was also used by the Egyptians, Greeks, Hebrews, and Phoenicians. To prepare the dye, the female insects are collected prior to hatching eggs and killed either by exposure to the vapors of vinegar or by immersion in a solution of vinegar in water. Crushing the bodies of the dead insects and then adding to them a vinegar–water solution results in the dissolution of the dye; when the water evaporates, kermes precipitates as a powdery residue. The coloring matter in kermes is the organic compound *kermesic acid* (Schweppe and Runge 1986; Dimroth and Scheurer 1913).

Cochineal, also derived from the bodies of insects, was known for many centuries to the pre-Columbian inhabitants of the Americas. Its native habitat seems to have been Mexico, although there is evidence that it was also used for dyeing in Peru during the Inca period (Dahlgren 1990). Cochineal was apparently unknown outside the American continent until the beginning of the sixteenth century, when, following the Spanish conquest of Mexico, it was brought to Europe. It has been suggested, however, that a red dye similar to cochineal may have been derived from a variety of insects native to the Ararat valley in Turkey (Verhecken and Wouters 1988–1989). In ancient Mexico cochineal was extracted from the dried bodies of *Coccus cacti*, insects that live on the cactus plant *Nopalea cochenillifera*. To obtain the dye, the insects were first collected and then killed by one of three methods: immersing them in scalding water, wrapping them in a bag and exposing the bag to the heat of the sun, or placing the bag in a hot oven. The dye, which is soluble, was then extracted from the dead insects by immersing them in water. After filtering out the exhausted solids from the solution, the water was evaporated and the dye was obtained as a red powder. The coloring matter in cochineal is the organic compound *carminic acid* (Schweppe and Runge 1986).

Lac is derived from *lac resin*, the hardened secretion of the lac insect, the only known resin of animal origin. The lac insect, *Kerria lacca*, formerly known as *Laccifer lacca*, is a natural parasite of a variety of trees in large areas of southern Asia. Three different products are derived from lac resin: *lac dye, lac wax, and shellac*. To obtain the lac resin, twigs encrusted with the secretion of the insects are cut down from the trees, then the incrustation is separated from the twigs, washed with water, and filtered. The wax and shellac, which are insoluble in water, remain as a solid residue of the filtration, while the soluble red dye (lac) is obtained as a powder when the water from the filtered solution is evaporated. The coloring matter in lac dye is an organic compound known as *laccaic acid*.

Yellow Dyes

A number of yellow dyes were known in antiquity; *weld* and *saffron* seem to have been the most widely used, but *barberry root, turmeric, Persian berries,* and *safflower* have also been identified in ancient fibers. *Weld*, probably the oldest European-known yellow dye, is derived from the herbaceous plant *Roseda luteola*, which is indigenous to central Europe. The dye is distributed throughout the entire plant, although it is concentrated in the upper branches and seeds. To extract it, the leaves and stems of dried plants are washed with boiling water, in which the dye is soluble. After the solution cools, the dye precipitates as a yellow powder. Dyeing with weld yields the purest and fastest shades of yellow; used in conjunction with different mordants, however, it provides a variety of hues (see Table 95) (Mell 1932).

Saffron, a rare and costly dye in antiquity as well as in modern times, has been valued not only for dyeing textiles but also as a coloring and a fragrance in food. It is derived from the dried stigmas of the saffron plant, *Croccus sativus*, a species of iris that grows naturally in central Asia as well as in the Spanish peninsula in Europe. The coloring matter in saffron is the organic compound *crocin*. Because of its rarity and high cost, saffron was and still is often replaced or adulterated with the more common safflower and/or turmeric.

Safflower, also known as *bastard saffron*, is a yellow dye that has been used for well over three millennia, having been identified in fabrics from the Egyptian twelfth dynasty. It is derived from the safflower plant, *carthamus tinctoria*, native to southern Asia and the Middle East. The coloring matter in the plants is a mixture of two components: one is yellow, known as *safflower yellow B*; the other, *carthamin*, is red. Safflower yellow B dissolves in water when fresh safflower flowers are washed with acidulated water. Evaporating the water from the filtered solution leaves the dye as a residue in the form of a powder. Following removal of the yellow component, the red constituent of safflower, carthamin, can be extracted from the flowers by washing them with hot water. In the East, carthamin was widely used in the past, mainly for making cosmetic preparations.

Turmeric, also known as *curcuma*, is an easily fading yellow dye that was used in Mesopotamia many centuries B.C.E. and later became popular in ancient Rome. It is derived from the turmeric plant, *Curcuma longa*, and other varieties of *Curcuma* indigenous to China and Southwest Asia. The dye is extracted with hot water from the shredded rhizomes of the plant and then dried into a yellow powder. The coloring matter in turmeric is the organic compound *curcumin*.

Barberry root is a yellow–red dye that has been used since prehistoric times; it is extracted with hot water from the stems, bark, and roots of *Berberis vulgaris*, a bush that grows indigenously in Europe as well as in North America. The coloring matter in the dye is the organic compound *berberine*. Silk and wool can be dyed directly with barberry root, yielding a yellow color; however, for dyeing cotton, a mordant is required to attach the dye to the substrate fibers.

Persian berries is the common name of a yellow-greenish dye, well known in antiquity. It is extracted from the seed-bearing fruits of a variety of species of *Rhamnus* plants that grow indigenously in the Middle East and southern

Europe. To extract the dye, the juice of unripe berries is pressed out and then dried. The coloring matter in the dye is a mixture of three organic compounds: *rhamnazin*, *rhamnetin*, and *quercetin*.

14.6. IDENTIFICATION AND CHARACTERIZATION OF ANCIENT DYES AND MORDANTS

Only occasionally is a quantity of preserved dye discovered in an archaeological excavation. Most of the ancient dyes are identified when still attached to, or after being removed (by extraction) from dyed yarns or textile fabrics. The extraction usually entails using chemical techniques for detaching the dye from the substrate and is generally based on differences in the solubility of the dye and the substrate in liquid organic solvents such as ether, acetone, or benzene. When dissolved in suitable solvents, dyes form solutions of characteristic color that can be well defined by their characteristic optical properties. Following extraction, therefore, dyes are identified by examining the spectral absorption of their solutions, using either visible, infrared, or ultraviolet spectroscopy (Taylor 1990; Hofenk de Graff and Roelofs 1978).

The main disadvantage of most techniques that require, the removal of a dye from its substrate, before identification, is that they are destructive. Since archaeological dyed fibers are often irreplaceable, destroying some of them for analytical purposes may not be possible, and the use of nondestructive analytical techniques for identification of the fibers and dyes is usually preferred. Techniques such as *Fourier transform infrared* (FTIR) and *Raman spectroscopy* overcome this disadvantage, allowing the characterization of dyes attached to a substrate. Moreover, using nondestructive techniques also make it possible to identify extremely small amounts of dyes, much below those required when extraction of a sample is necessary before characterization is made possible (Gillard et al. 1994; Guineau 1989).

Other techniques that have also been widely used to identify and characterize ancient dyes are those included under the generic name *chromatographic techniques*. *Thin-layer chromatography* in particular has been found most sensitive for detecting and identifying extremely small dye samples (Walton and Taylor 1991; Masschelein-Kleiner and Heylen 1968). A combination of chromatographic and spectroscopic techniques has often been used to separate, identify, and characterize ancient dyes. In a study combining these two techniques, for example, a large variety of dyes used on textiles dating to between the fourth century B.C.E. and the seventh century C.E. were identified (Koren 1996).

BIOINORGANIC MATERIALS
BONE, IVORY AND SHELL; PHYTOLITHS

15

Bone is, after stone and pottery, probably the most frequently recovered material in archaeological excavations. Its composition and structure provide much information on the animals of which the bones were originally part, serving as indicators of their anatomical and physiological characteristics, dietary habits, age, pathological affections and, since the 1990s, of their genetic traits. Much scientific endeavor is therefore devoted to deriving anatomical, physiological, chronological, pathological, and genetically related evidence from the composition and structure of ancient bones (Mays 1998; Larsen 1997).

Bone and shell are hard and resilient *composite materials* (see Textbox 32) that form the skeleton of the body of animals. *Bone* makes up most of the skeletons of all vertebrate and some invertebrate marine animals; *shell* is the main component of the exoskeletons of many marine invertebrates (mollusks, echinoderms, and crustaceans) as well as the encasement of avian eggs. Both these materials consist of a structural framework made up of essential organic substances interwoven with *bioinorganic materials*, also known as *biominerals*; see Table 64. The *biominerals* are somehow intriguing substances that challenge the perception (based on the historical, but erroneous, definition of inorganic and organic substances) that biological materials are made up exclusively of *organic* substances created by living organisms, while inorganic substances belong solely to the inanimate world. Although the composition of the biominerals is inorganic, they are biological materials, created by living organisms. Thus the biominerals link the apparently lifeless inorganic compounds to the living world.

Bone, shell, and coral are not, however, the only biominerals created by living organisms. The kidney and liver of animals, for example, often synthesize biominerals in the form of pathological stones (known as *calculi*) of varied composition (mostly of calcium oxalate, calcium phosphate, or

Archaeological Chemistry, Second Edition By Zvi Goffer

magnesium ammonium phosphate). Birds make *eggshells* from calcite (a mineral form of calcium carbonate) together with aragonite, another mineral form of calcium carbonate. Some microorganisms, for example, the diatoms (microscopic algae) produce hard symmetric exoskeletons, mainly from silica, and many plants synthesize extremely small biomineral structures, known as *phytoliths*, made up of silica or calcium oxalate. Grasses, too, contain large numbers of phytoliths, which are one of the main causes of the erosion and wear of the teeth of animals and humans who feed on grass (Piperno 1988). Altogether, over 60 different bioinorganic substances are known to be synthesized by living organisms, including metal carbonates (aragonite and calcite), citrates, oxalates, phosphates (dahlite and apatite), silicates (opal), sulfates (gypsum), sulfides (pyrite and gelena), and many others (Wilkins and Wilkins 1997; Lippard and Berg 1994).

15.1. HARD ANIMAL TISSUES

Bone

Bone, the compact and firm material that makes up the bulk of the skeleton of vertebrate animals, combines two extremely useful properties, light weight and strength; bone is, for example, relatively stronger, although a great deal lighter, than concrete. Regardless of differences in the composition of bone according to species and sex, significant quantitative facts are general and well known. The typical, approximate composition of fresh bone is given in Table 97 (Price 1989; Chaplin 1971).

The bones and teeth of humans and other vertebrate animals, for example, consist mainly of a composite material made up of an organic substance, *collagen*, and a biomineral, *calcium carbonate phosphate* (see Textboxes 32 and 61). The latter, which makes up about two-thirds of the total dry weight of bone, is composed of calcium phosphate containing between 4–6% calcite (composed of calcium carbonate) as well as small amounts of sodium, magnesium, fluorine, and other trace elements. The formula $Ca_4(HPO_4)(PO_4)_5(CO_3)(OH)$ approximately represents its composition; its crystal structure is akin to that

TABLE 97 Fresh Bone Composition

Bone component	Relative amount (%)	
Inorganic component		
Carbonated hydroxyapatite	70	
Organic components	30	
Collagen		28
Fats and carbohydrates		2

of the mineral hydroxyapatite (composed of calcium hydroxy phosphate) (Katzenberg and Harrison 1997). The complex composite material created by the intimate blend between collagen and carbonated calcium phosphate maximizes the ratio between the strength and the weight of bone and optimizes its load-bearing qualities. Thus, although the whole skeleton makes up only about 15% of the total weight of the human body, the bones of the skeleton provide practically all of the mechanical support that the body requires (Sandford 1993; Woodhead and Galloway 1980).

Visual examination of fresh bone reveals two major components: an outer layer of hard compact matter that makes up the casing of the bone, and inside, blending with the casing, cancellous, spongy tissue having a fine, loose structure. In long bones, such as those of the arms and legs, the spongy tissue makes up the internal part of the bone, including the marrow, as illustrated in Figure 78. The thickness of the hard, dense casing varies in

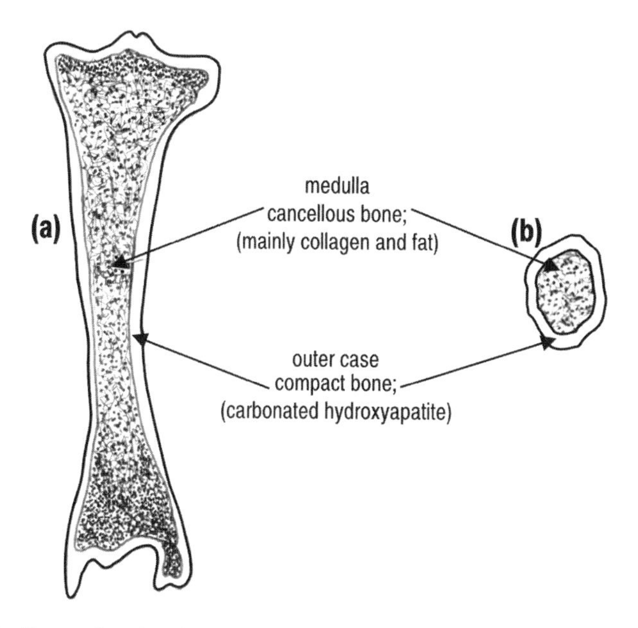

FIGURE 78 Bone; longitudinal and transversal cross sections. Bone, the dense, rigid tissue that composes the skeletons of mammals, is made up of a matrix of soft organic matter impregnated with a hard bioinorganic substance. The organic matrix consists mainly of collagen fibers mixed with lesser amounts of fats. The impregnating bioinorganic component is carbonated hydroxyapatite, which imparts rigidity to bone. Macroscopically, two varieties of bone tissue can be distinguished: compact bone and cancellous (spongy) bone; compact bone, made up primarily of carbonated hydroxyapatite forms the exterior mass of the long bones and the surface layers of others. Cancellous bone, a soft, pulpy tissue that consists mainly of fat and collagen, fills up the interior of short and flat bones and also the ends of long bones. The illustration shows longitudinal **(a)** and transversal **(b)** cross sections of a long bone.

FIGURE 79 Bone carving. A seventh-century B.C.E. decorative carving on bone that was inlayed in wood, Tel Malhata, Israel. Bone has been crafted into practical and decorative objects since the dawn of time. Bone carvings hold great detail, and the surface polish that can be achieved is high. Many bone-made objects have survived, partly because it was widely used, but also because buried bone is generally well preserved in many types of soil.

different parts of any one bone as well as according to the type of bone. Microscopic examination of the bone tissue reveals that the collagen fibers are organized in bundles, around which are packed *lamella* (small plates) of carbonated hydroxyapatite.

Teeth

Teeth, the hard conical structures embedded in their jaws that vertebrate animals use to chew food, consist of two layers of compact matter surrounding a core of soft, living tissue. The inner layer is composed of *dentine,* also known as *ivory,* whose composition is similar, but not identical, to that of bone; it contains less collagen and is harder than bone. The thin outer layer of the teeth, the teeth's *enamel,* includes even less collagen and other organic matter than dentine and is the hardest substance produced by animals (Hilson 1986a; Kurtén 1986b; 1982).

Ivory

Ivory, or *dentine,* the main constituent of the teeth of mammals, is a relatively hard, cream–white material that can be carved or mechanically formed, and its surface can be polished to a high shine (O'Connor et al. 1987; Wills 1968). Of particular interest is the ivory that makes up the tusks (large incisor teeth) of large mammals such as elephants, hippopotami, whales, narwhals, and

FIGURE 80 Ivory. Ivory amulet, chalcolithic period, Beer Safadi, Israel. Ivory has been extensively used for carving decorative objects and ornaments. The ivory of the tusks of elephants and ancient mammoths has long been valued for its close-grained texture, hardness, pleasing color, and the smoothness of its surfaces. As early as palaeolithic times humans engraved representations of animals on tusks. Ivory carving was practiced in ancient Europe and in Asia. In China and Japan, for example, ivory was used for making small statues and ornaments of great precision and for inlay work.

walrus, which have been used for making utilitarian and, especially, ornamental objects (see Fig. 80). The only source of "true ivory," however, is the tusks of elephants and, in ancient times, also of mammoths. Tusks have a cavity at their base, filled, in living animals, with a mixture of proteins and fatty material. Unlike most animal teeth, tusks consist entirely of dentine (ivory), with no outer layer of enamel. Most elephant ivory originates in the equatorial regions of Asia and Africa. Mammoth ivory, still found in Siberia, grew in spiral form to lengths of up to 4 m, and each tusk weighed as much as 100 kg (Kühn 1986; McGregor 1985). A material derived from the beak of a bird, the helmeted hornbill (*Rhinoplax vigil*) though structurally it is not ivory, was widely used as ivory in China for many centuries (Kane 1981).

Horn

Horn is the hard organic material that makes up the horns and hooves of many mammals, among them cattle, sheep, goats, and antelopes, and the carapace of tortoises and armadillos. Horn consists mainly of successive layers of *keratin*, a fibrous, complex protein that is also a component of the skin, hair, nails, and claws of mammals, and the feathers of many birds. In horn, claws, and hooves, the keratin surrounds bone protrusions from the skeleton of the animals. In marine turtles, a thick outer layer of keratin on the turtle's carapace serves as the animal's protective armor. The protein chains that make up keratin have a high proportion of the amino acid *cysteine*, which forms various types of bonds between parallel chains of proteins. Depending on the actual type of bond formed between the chains, the keratinous material may be either soft or hard: in soft keratin, as for example in the continuously renewed outer layers of the epidermis of mammals, birds, and reptiles, the protein chains are randomly oriented; in hard keratin, such as that which makes up horn and tortoise shell, the protein chains are oriented along a major axis, a structural arrangement that makes them strong and resilient. Because of its chemical composition, keratin (a biopolymer) has *thermoplastic* properties and can be easily and repeatedly softened by heating: keratin is hard and rigid at room temperature but soft and pliable at boiling water temperature (100°C) and in direct heat. Heating breaks down some of the bonds between the protein chains of keratin, so that when heated, horn and tortoiseshell become soft and can be shaped or impressed with a new form; when cooled under pressure, new bonds between the chains are re-created and the material acquires its new shape or form. At relatively higher temperatures (above about 200°C), horn and tortoiseshell melt, making it possible to join pieces together (Kühn 1986).

Antler

Antler is the name of the bony material that makes up a deciduous pair of protrusions shed every year known as "antlers" on the heads of animals of the deer family, for example, reindeer, elk, and fallow deer. Antler has a composition similar to that of horn. Like bone, antler is made up of a hard and compact outer layer surrounding a core of spongy tissue. Since it is regularly shed from the body of the animal, it differs morphologically from horn, which is not shed (O'Connor et al. 1987).

Shell

Shell is the term generally used to refer to the hard exoskeleton of some animals, such as the mollusks, as well as to the protective covering of the eggs of birds and some amphibians and reptiles. The shell of many mollusks, for example, consists of three layers: the innermost is a shiny layer of *nacre*, also known as *mother of pearl*, which includes a mixture of the mineral aragonite, composed of calcium carbonate interlaced with *conchiolin*, a protein closely related to keratin (conchiolin is synthesized by the mollusks to bond and hold together the crystals of calcium carbonate that make up nacre). The central layer consists mainly of calcite, while the outer layer, known as the *periostracum*, which consists mostly of conchiolin, is very resistant to chemical breakdown and to decay. In oysters and in some other mollusks, aragonite and conchioline are also synthesized by the organism to build up a layer of mother of pearl around and isolate foreign bodies inside the animal's shell, to create pearls, which have been used in jewelry for over 6000 years. About 86% of most shell is calcite, a very soft, fine-grained mineral composed of calcium carbonate. Large accumulations of shells created by early forms of life were deposited on the surface of the earth, mainly during the period of time between about 135 and 90 million years ago, known as the *Cretaceous period*, a name derived from *creta*, the Latin word for chalk (calcite). These deposits gave rise to many geographic features such as marine reefs, islands, and many mountain ranges (Mann and Walsh 1996; Lippard and Berg 1994).

15.2. ARCHAEOLOGICAL BONE

When bone is disposed of, it can be fresh, cooked, or putrescent. Fresh or putrescent bone usually contains a relatively high proportion of accompanying organic matter, such as flesh, fat, and ligaments; therefore, for quite some time after disposal bone retains much of its original resilience, toughness, and strength. If bone is cooked, it loses some or even most of its organic matter, although it may retain part of the natural fats (Roberts et al. 2002); depending on the extent of boiling, it may become porous and therefore somewhat crumbly and brittle, but not as brittle as roasted bone (roasted bone loses most of its organic matter and is extremely brittle).

Whatever its previous history before disposal and deposition at a site, the alteration of bone buried or exposed to the elements is determined, mainly, by the combined effect of the physical, chemical, and biological conditions at the site where it is deposited; these include the seasonal characteristics, average temperature, relative humidity, amount and flow motion of water, pH value, extent of aeration, and the nature of the microorganism population (Millard 2001; White and Hannus 1983).

Diagenesis of Buried Bone

Buried bone undergoes two different but parallel alteration processes, generally referred to as *diagenetic processes*, during which both its organic and inorganic constituents are altered (see Textbox 46). Organic matter is generally altered and lost quite rapidly, particularly the fats, which are either consumed by microorganisms or undergo decay processes (Trueman and Martill 2002). Traces or small amounts of collagen and DNA may, however, remain and be preserved in bone for extremely long periods of time; if uncovered at a later date, the remaining DNA provides an important source of information on the organism of which it was originally a part; see Chapter 12 (Collins et al. 2002; Colson et al. 1997).

The mineral constituents of buried bone may be selectively dissolved by groundwater, foreign substances from the surrounding environment may be deposited within its framework, and combined dissolution and deposition of matter may even occur simultaneously, drastically altering the composition of the bone. In acid soils, for example, some of the bone's minerals are dissolved away, the rate of their dissolution depending on the acidity and the amount and the mode of flow of the groundwater. On the other hand, chemically basic soil, that is, soil having a pH above 7, as for example calcareous soil, provides, advantageous conditions for the preservation of the mineral components of the bone and for the deposition, within its original structure, of additional elements introduced by groundwater. Moreover, in acid as well as in calcareous soil, roots of plants penetrating buried bone may be acid and therefore active in the localized dissolution of bone components.

Despite these generalizations, since the factors affecting diagenetic processes are very variable and extremely complex, it is impossible to predict their outcome. Under particular conditions the diagenesis of bone, teeth, or ivory interred for extremely long periods of time may result, in an indirect way, in their actual preservation, while under slightly different environmental conditions they will be entirely altered chemically or totally dissolved (Nielsen-Marsh 1997; Millard 1993). Moreover, taphonomic processes during burial (see Textbox 69) may result in the *fossilization* of bones and even of entire organisms. During the *fossilization* process, bone and other tissues are converted to lithic matter in which the original composition of the tissues is wholly altered but their structure and texture are preserved, sometimes in microscopic detail. The composition of the fossils is ultimately determined by the local environmental conditions in which they are formed (Lambert and Grupe 1993; Donovan 1991). The entire process of fossilization can, therefore, be defined only in general terms, as whatever happens to bone or other dead tissues as a consequence of diagenetic changes at the place of burial (see Textbox 46).

Dating Bone

Most bone is, at present, dated by radiocarbon dating the carbon-containing components of the bone (see Textbox 52). Relatively large samples are usually required: several grams of bone carbon for the conventional radiocarbon dating technique, and much smaller ones, only a few milligrams, for the accelerator mass spectrometry (AMS) technique (Taylor and Aitken 1997; Gillespie et al. 1984). If the organic matter of ancient bone has been almost entirely lost, or if the bone is older than 50,000 years (the maximum age datable with the radiocarbon method), other dating techniques, discussed in the paragraphs that follow, need to be applied.

Uranium Dating. Buried bone provides a favorable substrate for the deposition or the depletion, by groundwater, of minor and trace elements; depending on the environmental conditions at a burial place, bone may become enriched or depleted of elements, and these may serve as time indicators. Studying the buildup or the depletion of particular elements in archaeological bone may, therefore, provide means for determining its age. One such element is *uranium*, a heavy element. Trace amounts of uranium enter the body of living organisms via the normal intake of food and water, in which uranium is an impurity. Part of this uranium is incorporated into the blood, circulates through the body, and in the bones is exchanged with calcium and becomes attached to them. In this form all living animals acquire small amounts of uranium.

Natural uranium consists of a mixture of several isotopes, mainly *uranium-235* and *uranium-238*, but also others, most of which are radioactive (see Textbox 16); the concentration of uranium in bone can, therefore, be easily determined by measuring the amount of radioactive radiation that the radioactive uranium isotopes in bone emit. If uranium had been neither added nor removed from the bone after the death and burial of an organism, the absolute age of the bone could be determined by measuring the decrease in the radioactivity it emits. Most uranium in buried ancient bones is, however, acquired after burial. When an animal dies and its remains become buried at a site where the groundwater contains dissolved uranium, all or part of the uranium in the groundwater is deposited within the bone and the total amount of uranium in the buried bone gradually increases; the longer the bone remains in contact with uranium-containing groundwater, the more uranium the buried bone will contain (Millard 1993; Oakley 1963). The rate of increase depends on the environmental conditions at the burial site, such as the concentration of uranium in the groundwater, the volume of groundwater flowing through the bone site, and the permeability of the surrounding soil. Analytical studies show that, although the total amount of uranium in buried bone from different sites varies, there is correlation between the uranium content and the age of the bone; different bones of the same age, from a single site, have similar levels of radioactivity – the older the bone the larger the amount of uranium. Thus, measuring the relative amount of uranium in archaeological bone by radiometric assay, for example, provides a way for determining the approximate age of the bones and for differentiating between older and more modern bones (Howell et al. 1972; Szabo et al. 1969).

Uranium Series Dating. Reliable dating of bone is possible when using the technique known as uranium series dating (see Textbox 16). The technique, which is also based on measuring relative amounts of uranium, makes possible dating very old bones, beyond the range that can be dated with radiocarbon, that is, over 40,000 years and up to 500,000 years old (Schwarcz 1997; Ivanovich and Harmon 1992).

Amino Acid Racemization Dating. Some of the amino acids that make up the molecules of collagen have *asymmetric* carbon atoms, exhibit optical *isomerism*, and are of use for dating ancient bone (see Textbox 63). The *racemization* of *L-isoleucine*, for example, an amino acid that occurs in bone collagen, is slow even in geochronologic terms; its half-life, at 20°C, is in excess of 100,000 years. It is of use, therefore, to date bone too old to be dated by the radiocarbon dating method. In general, however, dates determined using the racemization of amino acids reaction may have an uncertainty as large as ±20–30%, nowhere near the accuracy obtained with radiocarbon (see Chapter 15).

Fluorine Dating. Probably the oldest scientific dating technique is *fluorine dating*, which, although seldom used today, is discussed here because of its historic interest. Fluorine dating of bone centers on an irreversible process whereby the inorganic component of buried bone is slowly and gradually, transformed into a more stable compound (see Textbox 67) (Carnot 1893; Middleton 1845).

TEXTBOX 67

THE FLUORIDATION OF BONE

The bone of most living animals generally includes only small amounts of *fluorine*. In living bone, for example, the concentration of fluorine is generally below 0.05% of the total amount of *carbonated hydroxyapatite* (composed of calcium hydroxyphosphate and calcium carbonate), the bioinorganic component of bone (see Table 98). When groundwater in which there are dissolved fluoride (F^-) ions wets bone buried in the ground, the fluoride ions in the water react, slowly though steadily, with the bone,

TABLE 98 Fluorine in Bone and Fluorapatite

Sample	Maximum age (years)	Fluorine content (%)	Fluorine: phosphate ratio
Recent bone	11,000	below 0.3	0.06
Pleistocene age bone	2,000,000	1.5	0.39
Tertiary age bone	30,000,000	2.5	0.65
Mesozoic age bone	130,000,000	3.5	0.91
Fluorapatite	–	3.8	1

which gradually becomes fluoridized; the fluoride ions replace the hydroxyl (OH^-) ions in the carbonated hydroxyapatite of the bone and progressively convert it to *carbonated fluorapatite*, a much less soluble and more stable mineral:

$$\underset{\substack{\text{carbonated hydoxyapatite}\\ \text{(in buried bone)}}}{Ca_5(PO_4)_3(CO_3)_3} + \underset{\substack{\text{fluoride ions}\\ \text{(in water)}}}{F^-} \rightarrow \underset{\substack{\text{carbonated fluorapatite}\\ \text{(in buried bone)}}}{Ca_5(PO_4)_3(CO_3)_3F + OH^-}$$

Once fluoride ions react with bone, they are not easily dissolved out or exchanged by other elements. If bone is buried for long periods of time, the relative amount of fluorine in the bone gradually increases as a function of time; the "fluoridation" process continues until the maximum amount of fluorine (necessary to convert all the hydroxyapatite to fluorapatite) is reached. The total concentration of fluor in carbonated fluorapatite can reach levels as high as above 3%. There is ample room, therefore, for an increase in the relative amount of fluorine in buried bone. Determining the relative amount of fluorine in buried bone may thus serve as a tool for dating bone.

Fluoride ions may be relatively abundant in groundwater at one location and practically absent in that at another site; hence the rate of fluoridation of the bone (the rate of increase in the relative amount of fluorine in the bone) varies from site to site. For instance, bones buried for a short time at a site in which the groundwater is rich in fluoride may acquire much more fluorine than bones buried for a very long time at a place where there is little fluoride in the groundwater. Therefore, fluorine analysis does not provide a tool for estimating the absolute age of buried bone, but only for dating bones at the same site, comparative to each other. The relative amount of fluoride in buried bone at a particular site thus provides a clue as to the length of time the bone has been buried.

In the first half of the twentieth century much bone was dated using this method, and dates derived with it greatly contributed, for example, to unfolding the nature of the Piltdown bones (see Chapter 18). Experience has shown that bones buried in sand or gravel are suitable for dating with fluorine, while those buried in volcanic soils rarely yield useful dates; dates derived from fluorine analysis results should, therefore, be used with caution (Schurr 1989; Oakley and Hoskins 1950).

Summarizing, it can be said that the accumulation, depletion, or alteration of some compounds or elements in buried bone take place at definable functions of time. Determining the relative amounts of such compounds or elements in bone provides, therefore, information about their age. The information thus obtained should, however, be interpreted with caution and any inherent errors taken into consideration.

15.3. BONE, STABLE ISOTOPES, AND ANCIENT DIETS

The isotopic composition of carbon in the proteins, which are one of the main components of animal tissues, reflects the nature of the food resources in the diet of the animals. Determining the relative abundance of the *stable isotopes of carbon* in the proteins of bones or hair, for example, can facilitate understanding of the effects that different types of plants or animal food resources had on ancient diets (Katzenberg 2000; Burton 1996).

The weight ratio between the stable isotopes of carbon in the tissues of plants and animals is passed along the food chain, from the plants to the animals. In the tissues of animals or human beings, the weight ratio of those isotopes reflects the mixture of isotopes in the food that they consumed. A study of the weight ratios between the isotopes of carbon-13 and carbon-12 in animal bone protein, for example, makes it possible to deduce the composition of the ancient diet that the animal consumed. From such studies it was deduced, for example, that the ancient Maya inhabitants of Cuello, Belize, used dog meat as a significant component of their diet (Tykot et al. 1996; de Niro and Epstein 1978). Studies of *stable carbon isotopes* in the bone of deer from ancient Central America have been used to examine the use of ancient land (Emery et al. 2000), while similar studies on African elephant ivory help distinguish between ivory originating from different regions and seem to provide a tool to establish the provenance of the ivory (van der Merwe et al. 1990).

Nitrogen Stable Isotopes and Diet

Nitrogen is one of the main components of the amino acids; the weight ratio between its stable isotopes, *nitrogen-15* and *nitrogen-14* reflects the nature of the diet of animals and humans. Studying this weight ratio, nitrogen-15/nitrogen-14, thus provides a tool to determine the source of proteins in the body of an animal consumed by humans in the past (Schwarcz and Schoeninger 1991). It has been shown, for example, that the relative amount of the heavier isotope (nitrogen-15) increases as nitrogen passes along the food chain, from plants to animals; also, that there is relatively more nitrogen-15 in herbivore animals than in plants, as herbivores have relatively less nitrogen-15 than do omnivores, which, in turn, have less nitrogen-15 than do carnivores.

Strontium Isotopes

The stable isotopes of *strontium, strontium-86*, and *strontium-87* provide a tool for studying the provenance of ancient animals (including humans) and their mobility. Strontium is taken up by organisms in their food and water intake, and the relative amounts of its isotopes seem to be unaltered by living processes: the relative amounts of strontium isotopes in the soil, plants, and animals are related to those of the underlying geologic formation and local hydrology of a particular geographic region. Determining the relative amounts of strontium-86 and strontium-87 in the tissues of animals provides, therefore, a basis for the reconstruction of their place of residence at the time when the tissues were formed.

It is known that different geographic regions have distinct strontium isotope compositions; the relative amounts between the isotopes measured in bone and teeth can, therefore, be used to infer the geographic region that an animal or human inhabited. The strontium isotope composition of the remains of an animal or human that ingested strontium in one region will thus differ from that of the remains of animals (or humans) that ingested strontium in another region. Moreover, the relative amount between these isotopes in different types of bone represents the average amount of strontium that was ingested by the animal while the bone was formed. Because different types of bone grow and exchange strontium at different stages during the lifetime of an animal, determining the relative amounts between the isotopes of strontium in different types of bone can be used to infer residence or changes in geographic location at different stages during the life of an animal.

Strontium incorporated into the teeth, for example, becomes immobilized after formation of the teeth. The relative amounts of the isotopes of strontium in the teeth is, therefore, related to the average strontium isotope composition that an animal ingested in food or water while growing. Thus teeth retain a kind of record of both early exposure to the isotopes of strontium in the water and food ingested during their growing, early life. In the bones, however, strontium undergoes continuous biological processing and, as a consequence, the relative amounts of strontium isotopes represent the average composition over, say, the last 10 years of an animal's life. Determining the average amounts of the isotopes of strontium is best done by analyzing bone fragments from control groups, such as animals that have the same feeding habits as the animal being studied. Analyses of the isotopes of strontium are now widely used for studying ancient diets, the reconstruction of habitats, and the investigation of human migrations (Montgomery et al. 2003; Price et al. 2002).

SOME ANCIENT REMAINS
MUMMIES, FOSSILS, AND COPROLITES

The remains of most dead organisms are either consumed as food by other living organisms, or decay and vanish relatively fast. Decay processes begin just a few hours after death; within a relatively short period of time (only a few weeks or months) after death, the carcass of an animal, for example, is reduced to a bare skeleton. Under extremely unusual circumstances, however, the remains of dead organisms are neither consumed by other organisms nor do they decay. In exceptionally dry places and in anaerobic environments, for example, they undergo slow compositional changes that may ultimately result in the preservation, sometimes down to microscopic detail, of the body and its morphology. Natural and human-induced processes that result in the preservation of dead organisms are discussed in the following pages.

16.1. MUMMIES AND MUMMIFICATION

One of the natural forms of preservation of dead remains is mummification; a *mummy* is simply the body of a dead animal whose tissues have been diagenetically altered but whose morphology and structure have been preserved. Although the term *mummification* is generally used in relation to the purposeful preservation of dead humans and animals, it actually has a wider meaning – it also refers to the natural preservation of dead corpses (see Fig. 81).

Dead bodies can be naturally mummified under a variety of environmental conditions, such as (1) at extremely low temperatures, about and below the freezing point of water, in extremely cold regions of the world; (2) in very dry and hot environments, as in desert areas and in some caves and rock shelters; or (3) under anaerobic (oxygen-free) conditions, as in bogs (see Chapter 8).

Archaeological Chemistry, Second Edition By Zvi Goffer

The chemical reactivity of materials increases as the temperature increases and diminishes as it is lowered. Freezing conditions, at which the reactivity of most biological matter is very low, provide an appropriate environment for the preservation of biological matter and, therefore, for the natural mummification of corpses. Frozen remains of dead animals, for example, are generally preserved for extremely long periods of time, some occasionally being uncovered after many thousands (or even millions) of years, still in a good state of preservation. Although quite rare, such frozen remains often include the practically unaltered skin, hair, soft inner tissues, and occasionally even the last meal before death of the animal. Striking examples of well-preserved frozen mummies are those of Inca children discovered in a relatively good state of preservation in the heights of the Andes

FIGURE 81 Naturally mummified corpse. The naturally mummified corpse of a child, northern Chile. Natural mummification occurs under favorable soils and climate conditions, particularly in cold or arid regions and in peat bogs. Their tattoos, clothing, and accompanying tools provide clues as to their everyday lives. With the modern advances in archaeological sciences, the study of mummified remains has grown increasingly sophisticated: the analysis of blood and DNA from mummies provide genetic and medical information; the contents of the stomach reveals data on local ecology and subsistence, and trace analysis of specific tissues can disclose exposure to particular, sometimes toxic, elements.

mountains of northern Argentina and Chile and southern Peru (Reinhard 1996), the Ice Maiden from the Altai mountains in Siberia (Polosmak 1994), and the Iceman found in the Tyrolean Alps, at the border between Italy and Austria (Kutschera and Rom 2000; Barfield 1994).

The chemical reactivity of most biological matter is also greatly diminished in the absence of water. Under temperate environmental conditions, therefore, an important factor for the preservation of biological matter and hence, for the mummification of dead animals or plants, is the absence of water from their dead remains. The hot and dry conditions prevalent in desert regions, such as the Egyptian sands, for example, provide a suitable environment for the desiccation of corpses that became mummified soon after burial. Thus were mummified the bodies of dead Egyptians since the fourth millennium B.C.E., wrapped just in clothing or matting; their dehydrated muscles and other soft tissues of the body shrank almost to nothing, encased within the tightly stretched dry skin (Fleming 1980).

Anaerobic conditions (that include little or no oxygen) as exist, for example, in *bogs*, provide an environment appropriate for the preservation of biological matter. *Bogs* are spongy wetlands in which the soil consists mostly of *peat* (a dark, fibrous accumulation of partly decomposed vegetable matter with little associated mineral sediment), and in the stagnant water humic acid and tannins derived from the dead vegetable matter are dissolved (see Textbox 62). The scarcity or total absence of oxygen in bogs inhibits the decay and disintegration of animal remains; the tannins in the acid water tan the skin and other animal tissues, converting the skin into leather, and the acid water dissolves most of the inorganic components of the bones, which become spongy and soft. Corpses buried in bog environments are generally naturally preserved. Since the bones are partly dissolved away, the corpse is usually flaccid; its skin, however, is generally well preserved, as is also the hair and the contents of the digestive system (Brothwell 1987).

Not all corpses recovered from bogs are naturally preserved in exactly the same way. In some cases, all muscle and skin have decayed, leaving only the skeleton. Sometimes the opposite occurs, and all the bones in a bog's corpse are dissolved away, leaving only soft tissue and tanned skin. Some of these discrepancies are difficult to explain. Peat bog conditions seem to be detrimental to the preservation of DNA. Still, several thousands of more or less well preserved bog corpses, dating from about 100 B.C.E. to 500 C.E., have been found during the last two centuries in northern Europe, mainly in Scandinavia. Some, such as the Tollund and Grabaulle men from Denmark and the Lindow man from England, have been thoroughly studied and described (Glob 1972; Turner and Scaife 1995).

Embalming

The terms *embalming* and *mummification* are often employed as synonyms to refer to the deliberate preservation of corpses so that they keep, as much as possible, their lifelike appearance. To *embalm* or *mummify* a dead body is to preserve it by artificial, chemical means. *Dehydratation*, the removal of water, for example, provides suitable conditions for the preservation of organic mater in general and of corpses in particular; many ancient corpses have been mummified by dehydration. In some ancient societies, after the corpse was dry it was impregnated or filled with aromatic substances, usually known as *balms*, such as molten resin, pitch, or tar, preventing it from becoming unsightly.

A great deal of what was known until the twentieth century about ancient embalming and mummification processes was based on the writings of early historians, such as Herodotus, who carefully recorded Egyptian embalming procedures during his travels in the fifth century B.C.E. (Herodotus 1958). Much more about the methods of mummification, the mummies themselves, and the culture at the time they were mummified can be learned, however, from the actual examination and analysis of the mummies (David 2000; Andrews 1998). Although quite a large number of mummies were studied during the nineteenth and early twentieth centuries, it was only during the second half of the twentieth century that it became possible to study mummies without harming or destroying them. Nondestructive methods of examination, particularly *imaging techniques*, based on the use of penetrating radiation such as radiography, tomography, and MRI (see Textbox 68), now provide information on the present condition of mummies with almost no need even to unwrap them (Tchapla et al. 2004; Aufderheide 2003).

TEXTBOX 68

IMAGING

Imaging is the process of creating visual representations of some measurable property of objects, living organisms, or phenomena. Throughout the greater part of human development, images were almost solely of an artistic nature, mostly flat drawings and paintings and three-dimensional carvings and castings reproducing visual impressions or recollections of people, animals or plants, objects, or places. During the nineteenth century technology extended the range of imaging techniques so as to include photography, microscopy (which was used mainly in biology), and telescopy (in astronomy). The discovery of X-rays in the late nineteenth century, and of radioactivity during the early twentieth century, drastically changed and vastly expanded the development of imaging techniques: new tools became available to see and record internal details of objects and living bodies that had been invisible before. The development of television and digital technologies, during the second half of the twentieth century, further increased the range of imaging targets; many modern imaging systems (e.g., spectroscopy, radiography, and magnetic resonance) are based on the interaction of energy with matter to produce images that are then digitally processed (see Table 99).

Present-day imaging systems generally include a source of radiation that interacts with the object to be imaged, a detector of the radiation after it has interacted with the object, and a device for processing and

TABLE 99 Imaging Techniques Frequently Used in Archaeological Studies

Technique	Radiation used	Applicable to	References
Magnetic resonance (MRI)	Microwave radiation	Soft tissues	Hanel et al. (1998); Weishaupt and Köchli (2003)
Microscopy	Visible light, ultraviolet, ionizing radiation[a]	All	Delly (1988); Pluta (1989)
Photography	Visible light, infrared, and ultraviolet	All	Jackson (1982)
Radar	Radar	Buried or distant objects	Toland (2000)
Radiography	Ionizing radiation[a]	Stone, metal, bone	Lifshin (1999)
Computed tomography (CT)	Ionizing radiation[a]	Stone, metal, bone	Bushong (2000); Baruchel et al. (2000)
Thermal	Infrared radiation	Distant objects	Donaghue (2001)
Ultrasound	Ultrasound	Complex structures	Sellers and Chamberlain (1998)

[a]X-rays, beta- or gamma rays, or neutron beams.

displaying the acquired image. In archaeological studies, imaging techniques may reveal three-dimensional images of otherwise unaccessible parts of human or animal remains, of ancient objects or structures, and even of unexcavated sites (Hornak 2002). Imaging science is a relatively new and rapidly evolving technology that explores new imaging techniques and strives to gain a better understanding of the images obtained by such techniques (Doucet 2003; Saxby 2001).

Such methods also make it possible to learn about the state of health of the dead before death, the diseases from which they suffered, their age at the time of death, the method used for their mummification, and even the cultural environment in which they lived and were mummified (Cockburn et al. 1998; Harris and Wente 1980). The conception and development of the polymerase chain reaction (PCR) at the end of the twentieth century made it also possible to study the genetic characteristics of the mummies and of the populations to which they belonged (see Textbox 65).

Embalming was practiced in Egypt as far back as the early fourth millennium B.C.E., although the basic requirements of the embalming process became well understood only much later; it was only about the middle of the sixteenth century B.C.E., that embalming became proficient. The actual process of embalming varied according to the prominence or wealth of the deceased person. The first stage of a typical Egyptian embalming process was *dehydration* of the corpse; this was usually brought about by one of two methods: either (1) exposing the corpse for a period of time (lasting several weeks) to the heat of the sun on the desert sand, or (2) treating it with *natron,* a mixture of salts that absorbs water and also serves as a mild sterilizer that furthers the preservation process. Natron is a natural mixture of salts that occur in the beds of dry lakes in the Egyptian desert. Its major components are soda (sodium carbonate) and sodium bicarbonate together with minor amounts of common salt (sodium chloride) and sodium sulfate. Using solid natron provided the most convenient procedure to dehydrate and preserve corpses. Sometimes, however, corpses were soaked in a solution of natron, in water. The latter approach required relatively large containers as, for example, the one used, to dehydrate the corpse of the mother of King Cheops (her corpse, still soaked in a solution of natron, was discovered after many millennia inside a large stone vessel). After dehydration, corpses in the process of mummification underwent some treatment, such as removal of the brain and bowels. The cavities left after removal of these organs were then filled with preservatives, mostly fragrant oils, spices, and resins (balms); pitch and/or tar were often also smeared over the skin. Smearing the skin with tar gave the mummies a black appearance that is now popularly associated with Egyptian mummies. In the last stage of the preservation process the mummies were thoroughly wrapped in linen cloth whose exterior was impregnated with a gum. Simpler and cheaper mummification processes involved a less thorough dehydration and the sparing use of balm, gum, and wrappings; when such mummies are discovered and unwrapped they are not well preserved: the skin is very brittle, and the limbs and ears easily fall off.

Millions of corpses, of humans as well as of pet animals, were embalmed and mummified in the period between 4000 B.C.E. and the present time, not only in Egypt but in other parts of the world as well. Ancient Ethiopian tribes mummified their dead in a manner similar to that of the Egyptians. So did the indigenous inhabitants of the Canary Islands, about 900 B.C.E. The ancient Persians and the people of Mesopotamia, on the other hand, preserved corpses by placing them in jars filled with honey or wax to prevent air, and therefore oxygen and bacteria, from accessing the corpses, thus preventing their decay.

16.2. FOSSILS AND FOSSILIZATION

Fossils, the mineralized casts or impressions of ancient plants and animals that can be found on or below the surface of the earth, may be formed under many and varied conditions. Sometimes, plants or the bodies of animals become embedded in volcanic ash or in quicksand and leave exact fossilized reproductions of both their external and internal structures. The engulfment of living bodies, or their dead remains, in a resin or asphalt, which prevents the access of oxygen and inhibit decay, for example, may result in their fossilization. Thus were fossilized insects suddenly engulfed by flowing amber that hardened around them. Fossilization processes also occur when flowing groundwater containing dissolved minerals from the surrounding soil penetrates and impregnates the remains of living organisms; eventually, under slightly changed environmental conditions, the dissolved minerals gradually precipitate, replacing tissues and preserving the shape of the dead organisms without destroying, or even obscuring, their internal structure. Such processes change the overall composition of the remains, turning them from biological into inorganic, lithic matter, while preserving their original structure. The variety of minerals in petrified plants and animals often creates strikingly vivid colors. Although the original biological matter of the dead remains is removed, the natural cast often creates a precise mineral reproduction of plant or animal remains (Donovan 1991; Fortey 1991).

The course taken by any particular fossilization process is, therefore, determined by the physical and chemical factors prevalent in the environment of the dead remains. The physical factors include temperature, degree of aeration, and rate of flow of groundwater. The nature of minerals and rocks, and of the groundwater at the site of burial, are the most important chemical factors. Reconstructing and explaining the processes undergone by dead remains, from the time of death to when they are fully fossilized, is the concern of *taphonomy*, the study of the processes taking place when dead remains pass from the biosphere to the lithosphere (see Textbox 69).

TEXTBOX 69

TAPHONOMY

Taphonomy is the branch of science that studies the processes of *decay* and *fossilization* of the remains of dead organisms. The term *taphonomy* (from the Greek *taphos*, burial and *nomos*, law) was introduced during the first half of the twentieth century to describe the study of the transition of the dead remains of organisms from the biosphere to the lithosphere. Archaeologically related taphonomic studies began much later, near the end of the century (O'Connor 2005).

Taphonomic studies examine the processes taking place in the remains of dead organisms during their gradual conversion from biological substances into the inorganic, lithic materials. After organisms die, their remains undergo compositional degradation, passing through two successive and distinct taphonomic stages: *necrolysis* and *diagenesis* (Plotnick 1993; Shipman 1981). *Necrolysis* is the decomposition and dissolution of the remains of living matter. Most dead remains vanish during this stage, living no trace. Soon after death, parts or the whole of dead remains decompose or vanish altogether as a result of the combined activity of aerobic and anaerobic bacteria and the scavenging of the remaining soft tissues by insects, reptiles, birds, and mammals. Often, only the bones remain, which are attacked by other organisms. Under exceptional environmental conditions, however, some dead remains withstand and outlast the necrolysis stage in a somehow recognizable form. If and when they do so, they enter the diagenetic stage, when they are gradually transformed into rock. *Diagenesis* involves changes resulting from the physical and chemical processes taking place when buried remains interact with the surrounding environment (see Textbox 46). These processes, which are brought about by water percolating through and by the pressure of sediments deposited on top of the remains, ultimately result in either the partial or entire dissolution of the remains or in the formation of new minerals within their framework and their consequent gradual transformation into fossils (Martin 1999).

Fossilized primates, especially fossilized humans, are, unfortunately, particularly rare, and this partly explains why the evolutionary history of human beings is incomplete and continuously being revised. Still, the few fossilized human remains that have been preserved provide the main foundations for modern human evolution theories (Reader 1981).

16.3. ANIMAL EXCRETIONS

Whatever animals eat is processed by their bodies. However, only a few components of the food intake – those required for the building, sustaining, and functioning of the body – are selectively extracted and retained. The remaining bulk is more or less regularly disposed of as excreta, including gases, liquids, and solids. The gases are released into the atmosphere; the liquid (urine) is usually absorbed by the soil, and the solids (feces or dung) in a variety of compositions, shapes, textures, and colors, are soon recycled by biological activity (most are used as food by microorganisms and insects) and disappear from the surface of the earth shortly after their excretion. Under suitable environmental conditions some solid excreta may, however, be preserved for extremely long periods of time. If unearthed, for example during archaeological excavations, ancient feces constitute a substantial source of archaeological information (Sobolik 2000; Bryant 1974).

Human feces are normally composed mainly of water, which, although commonly about 80%, may range from less than 50% to over 90% of their weight. Most of the remains consist of insoluble fibrous matter, fats (up to about 10% of the total weight), the residues of dead cells from the digestive system, inorganic compounds (such as calcium phosphate and nitrogen compounds), and bacteria. Humans recognized early in their development that feces (especially the calcium phosphate and the nitrogen compounds in them) are beneficial for the growth of plants. Since time immemorial they have therefore used their own feces and those of domesticated animals as *manure*, or *fertilizers*, to improve the quality of agricultural soils and the productivity of cultivated plants (see Chapter 6). Furthermore, much of the organic matter in feces is combustible, and in many parts of the world dry feces – such as those of cows in India and of camels in many desert regions – have been burned as fuel in fires used for domestic, crafts, and manufacturing purposes.

Coprolites

The feces of domesticated animals are often used as fertilizers and fuel, but the bulk of feces are recycled soon after excretion (Putnam 1983). Under adventitious environmental conditions, however, feces may endure natural recycling processes, be preserved, and become what is known as *coprolites*, hard, brittle concretions often found in archaeological excavations, particularly in desert climate areas or in caves. Their name, coprolites, means "dung stone" (*kopros* is dung and *lithikos* is stone in Greek); they are dry feces whose original composition has often been preserved practically unaltered, having undergone only slight changes (Bryant and Williams-Dean 1975). The preservation of feces and their eventual conversion to coprolites is related to the nature of the organic matter they originally contained: the higher the relative amount of organic matter they incorporate, the faster they decay. But their water content and particularly the environmental conditions at the site where the feces were excreted and became preserved are the main determining factors for their preservation. The coprolites of carnivorous animals, for example, are more likely to be preserved than are those of herbivores, since carnivores derive a relatively high amount of minerals from the bone of the prey they consume as food. When these minerals are excreted they provide better initial preservation conditions for the feces. Appropriate environments for the preservation of coprolites exist in sites where dehydration (loss of water) is rapid and anaerobic (lacking in oxygen) conditions prevail. Dehydration is rapid in desert areas, dry caves, and floodplains associated with rivers (in the latter, feces deposited on dry parts of a plain are somewhat dehydrated before being rapidly buried by river floods). Anaerobic conditions prevail at the bottom of some ponds and swamps, particularly in bogs.

The study of coprolites, particularly of their composition, throws light on the *paleodiets* (the feeding behavior) of ancient animals as well as humans and on the diseases that affected them. Coprolites composed only of plant material, for example, are indicative of a herbivorous diet; bone remains in the feces denote carnivorous behavior, while remains of both plant and animal matter point to an omnivorous diet. Coprolite studies also provide information about the habitats and environment of animals and human beings, their health, and internal parasites, as well as about the presence of humans in areas lacking other human remains (Fry 1985; Callen 1969). The study of ancient DNA and the development of the PCR reaction during the 1990s, opened new horizons in the study of coprolites. Analysis of the DNA found in a coprolite from Gypsum Cave, Nevada, from the Pleistocene period, for example, indicates that it was derived from a now-extinct sloth, a slow-moving herbivorous mammal that became extinct 11,000 years ago. In the coprolite was also identified DNA characteristic of plants that formed part of the diet of the sloth (Poinar et al. 1998). Studies of DNA in coprolites are also of use for the identification of the animal species that produced the coprolites and to determine the nature of the parasites in the animals (Panagiotakopulu 1999). Such studies also make it possible to learn and to understand the evolution and phylogeny (the evolutionary history and genetic lineage) of animal species, particularly, of humans (Takahata and Satta 1997; Horai 1995).

THE ENVIRONMENT AND DECAY OF ARCHAEOLOGICAL MATERIALS

The *environment* is everything that makes up a particular surrounding and affects matter and life. On the earth it includes the atmosphere that surrounds the planet, the water that covers most of its surface, the soil that lies on top of its dry solid crust, and the living organisms, plants, and animals that are its inhabitants. Still, the earth's environment varies and there are a great variety of environmental conditions: in desert regions, for example, the environment is dry and in temperate areas it is wet; in equatorial regions it is hot and in the poles it is cold. Each type of environment imposes distinctive conditions on, and interacts differently with different forms of matter and with the living organisms it surrounds. The environment is, therefore, everything that surrounds and plays any role in the existence of any form of matter or living organism: from the air, water, and soil that surround them the forces to which they are subjected (as, for example, those exerted by the wind, flowing water, and gravitation), and the radiation to which they may be exposed (such as heat, light, and radioactive and cosmic radiation).

Nonliving, physical and chemical environmental factors, such as the composition of the atmosphere and of the soil, the humidity, pH, temperature, and sunlight, make up the *abiotic* (inanimate) environment. Living organisms, plants, animals, and microorganisms derived from living organisms, such as the biological matter or the refuse they create, make up the *biotic* (of biological origin) environment, which includes plants and animals, their wastes, and their remains. The interaction between the abiotic and the biotic components make up the total environment of inanimate matter and living organisms.

Archaeological Chemistry, Second Edition By Zvi Goffer

Only in relatively recent times, during the 1950s, began the scientific study of the ways that matter and living organisms are affected by, and how human beings alter, the environment. The study of the interaction of archaeological remains with the environment is a particularly complex subject that requires the use of a wide range of disciplines, including physics, chemistry, geology, meteorology, and microbiology, among many others. This chapter discuses the main environmental factors that affect archaeological remains – namely, the atmosphere, water, and soil – and the ways they bear on the existence, durability, and decay of the remains.

17.1. AIR AND THE ATMOSPHERE

The atmosphere is the layer of colorless and odorless gases commonly known as *air,* which surrounds the earth and sustains life on the planet. Three regions of the atmosphere are generally differentiated: the *troposphere, stratosphere,* and *ionosphere* (see Fig. 82). The lowest region is the *troposphere,* which extends from the surface of the earth to heights varying between 10 and 18 km above sea level, although it is thinner at the poles and much thicker at the equator. The troposphere is the only atmospheric region where there is water in a variety of states (vapor, clouds, rain, and snow), where weather occurs, and where there is life. The middle layer, the *stratosphere,* extends between the troposphere and up to about 50 km above sea level. It is a very dry region that contains about 1000 times less water vapor than the troposphere and the atmospheric region where most ozone is produced. The *ionosphere,* the uppermost layer of the atmosphere, extends above the stratosphere up to about 400 km above sea level. Its name is derived from the fact that energetic radiation from outer space, such as solar and cosmic radiation, causes ionization of most of the gases that constitute the ionosphere; most atoms and molecules of the different gases there are electrically charged with either a positive or a negative charge.

Because of the weight of the column of gases above any point on the earth's surface, the atmosphere exerts pressure at every point on the earth's surface. This pressure, which is variously known as *atmospheric, barometric,* or *air pressure,* varies according to the altitude above sea level and the weather conditions (see Textbox 70).

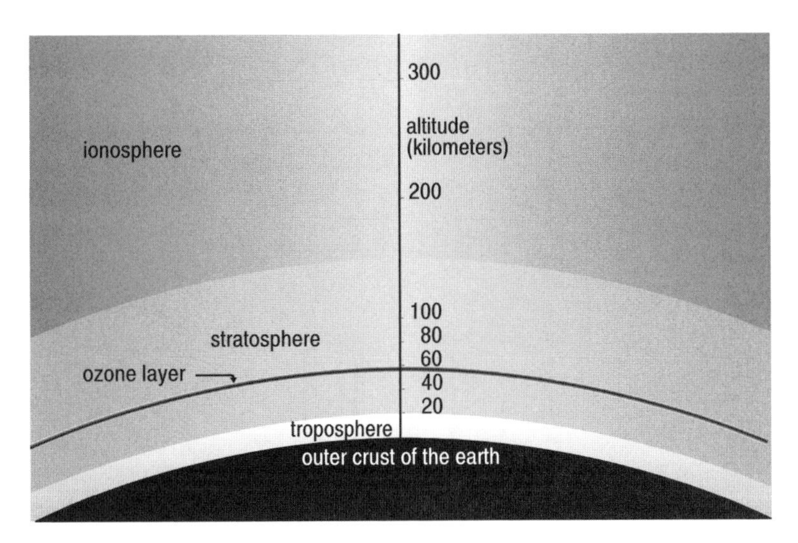

FIGURE 82 The earth's atmosphere. The atmosphere, the layer of air that surrounds the earth, is held in place by the pull of gravitational attraction. Almost all (99%) of the atmosphere is contained within a layer approximately 50 km thick, but life in it occupies only a thin layer making up about 10% of its total thickness. The atmosphere protects living organisms from the vacuum and the energetic radiation from outer space. It also recycles water within the lowest 20 km over the surface of the earth, causing the circulation of water to water vapor, to rain, and thus creating the weather. This includes such phenomena as clouds, wind, fog, rain, snow, and storms. Weather cycles, together with variations in temperature and the circulation of air (wind), cause the erosion and weathering of the surface of the earth and of objects on that surface.

TEXTBOX 70

ATMOSPHERIC PRESSURE

The gravitation of the earth attracts the gaseous components of the air, which exert a force, known as *atmospheric pressure*, on the surface of the planet. The pressure on any particular place on the earth's surface depends on the amount of air above the place. It follows that the atmospheric pressure decreases at high altitudes, increases at low altitudes and below sea level, and is also affected by changes in weather. Measuring the atmospheric pressure is usually done with physical instruments known as *barometers* (see Fig. 83).

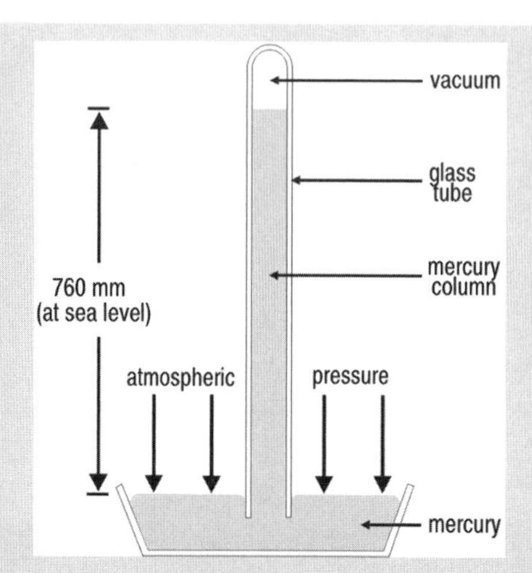

FIGURE 83 Barometer. A most basic and widely used instrument for measuring atmospheric pressure is the centuries-old "mercury barometer." This device measures the atmospheric pressure as that pressure that exactly balances, at any specific temperature, the pressure exerted by the height of a column of mercury measured in millimeters: the higher the mercury level, the greater the air pressure. At sea level and at a temperature of 25°C, for example, the height of a mercury column is 760 mm and the pressure exerted is known as one "standard atmosphere." Pressure values higher than 760 mm of mercury are characteristic of places below sea level, or at sea level where, as a result of particular weather conditions, there is an increase in the atmospheric pressure. Values below 760 mm are typical of places at altitudes higher than sea level, or at sea level but affected by special weather conditions that reduce the atmospheric pressure.

Composition of the Atmosphere

The *air* that envelops the earth is made up of a mixture of gases (see Fig. 84). In the lowest layer, the *troposphere*, the air is made up almost entirely of three elements, nitrogen, oxygen, and argon, although others, some of them air pollutants (see text below) are present in small but significant amounts. The exact composition of the air varies slightly, however, depending on the altitude above sea level, the location, and the time of the year.

Nitrogen. *Nitrogen*, a colorless, odorless, and tasteless gaseous element, is the main component of the atmosphere, which makes up about 78% of its volume; since it is also an important constituent of living organisms, nitrogen circulates between the atmosphere and living organisms. Plants and animals cannot, however, use nitrogen directly from the atmosphere. Plants obtain it from nitrates (compounds in which nitrogen is combined with oxygen and a metal), which are created mainly by bacteria in the soil but also by lightning, during storms. The plants then convert the nitrates to amino acids and proteins; herbivorous animals acquire nitrogen when eating plants, carnivorous animals, when eating herbivorous animals, and omnivorous animals, including human beings, acquire it either from plants or from herbivorous or carnivorous animals.

Oxygen. *Oxygen*, also a colorless, odorless, and tasteless gaseous element, constitutes about 21% by volume of the atmosphere, where it exists as molecular oxygen. Each molecule of oxygen is composed of two atoms of the element. Oxygen is also the most abundant element in the crust of the earth, making up almost 50% of the total mass of the solid planet, where it occurs not as molecular oxygen as in the atmosphere but combined with many elements such as hydrogen, with which it forms water, and the metals

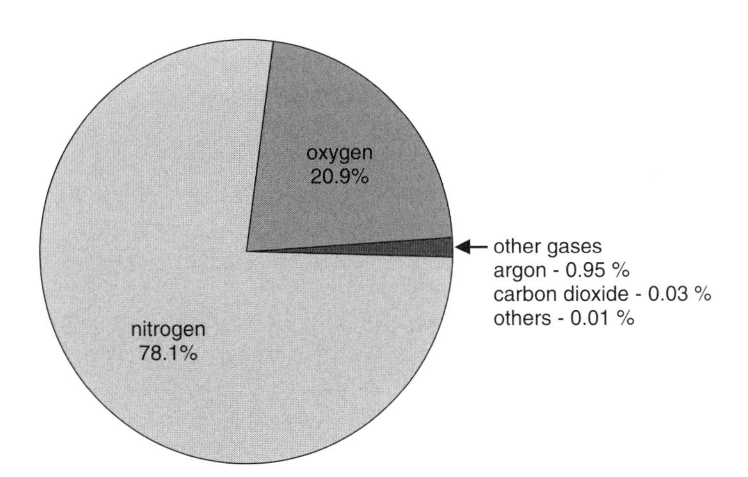

FIGURE 84 The composition of air. Air is a mixture of gases. Some of these gases, such as oxygen and nitrogen, are elements; others are compounds as, for example, carbon dioxide. The composition of dry air is relatively uniform around the world. The graph depicts the composition of "standard dry air," which is composed mainly of three gases: nitrogen (about 78%), oxygen (about 21%) and argon (less than 1%). Local additions to the composition of air vary with site location and with time. Additional components may result from geologic or biologic processes or from home, agricultural, or industrial human activities. Circumstantial components quite common in atmospheric air include methane and sulfur dioxide, which can be present in the parts per million range, but also ammonia, nitrous oxide and nitrogen dioxide.

with which it forms solid metal oxides. Oxygen in the air is the main cause of corrosion in metals, the process whereby oxygen combines with most metals, which thus lose their metallic properties. Rust, composed of iron oxide, for example, is formed when atmospheric oxygen combines with iron (see Textbox 42). Oxygen also combines with nonmetallic elements – such as carbon to form carbon dioxide and sulfur with which it forms sulfur dioxide; carbon dioxide and sulfur dioxide are air pollutants and are discussed below. Oxygen is also essential in the respiration of plants and animals, which use it to break down food and provide energy for their bodies while releasing water and carbon dioxide. In the daytime, under the effect of solar radiation, however, green plants release oxygen during the process of *photosynthesis* (see Textbox 53). Atmospheric oxygen consists of a mixture of three stable isotopes: oxygen-16 (99.76% of the total amount of oxygen), oxygen-17 (0.20%), and oxygen-18 (0.04%). In minerals in which oxygen is combined with metals or nonmetallic elements there are variations in the relative amounts of these isotopes. Such variations are significant in archaeological studies, since they make it possible to study palaeotemperatures and paleoclimates (Bradley 1999; Aitken and Stokes 1997) as well as the provenance of some natural materials (Giuliani et al. 2000; Craig and Craig 1972).

Argon. Argon, a gas that makes up just under 1% by volume of the atmosphere, is an inert element and a noble gas, that does not combine with other elements.

Carbon Dioxide. Carbon dioxide, also a colorless and odorless gas, makes up about 0.03% of dry air. Carbon dioxide is introduced into the atmosphere by several natural processes: it is released from volcanoes, from burning organic matter, and from living animals as a byproduct of the respiration process. It is for this latter reason that carbon dioxide plays a vital role in the *carbon cycle* (see Fig. 62), which makes possible one of the more important scientific tools in archaeology, radiocarbon dating (see Textbox 52).

Water. Water generally occurs in air in low or relatively low concentrations, mostly in the form of *atmospheric moisture*. Its importance cannot, however, be overemphasized, since atmospheric moisture, unique to the surface of the earth, is a determining factor in the *water cycle* (see below) and in living and other processes. Moisture is, therefore, one of the most important and probably the most relevant atmospheric components for the majority of living processes.

TEXTBOX 71

HUMIDITY

Humidity is a general term used to refer to water vapor in the air. The actual amount of water vapor that the air can hold increases with the temperature: the higher the temperature, the greater the amount of water vapor that the air can hold. Various measures are used to express humidity: *absolute humidity* and *relative humidity* are the two most widely used. *Absolute humidity* is the amount (by weight) of water vapor in the air at a particular temperature and pressure and it is usually measured in grams per cubic meter.

For most practical purposes, however, humidity is expressed as the *relative humidity* (RH), the percentage of moisture in the air at a given temperature in relation to the amount of moisture that the air could hold at that temperature before condensing as dew. Since the latter amount is dependent on the temperature, the relative humidity is a function of both moisture content and temperature. A value of 50% RH, for example, means that the air holds half the water vapor it can hold at the prevailing temperature. At 100% RH, moisture condenses and falls as rain. The relative humidity is measured with instruments known as a *hygrometers*.

17.2. WATER AND THE HYDROSPHERE

The *hydrosphere* (the Greek prefix *hydro* means water) is the great mass of water that surrounds the crust of the earth. Water is one of a few substances that, at the temperatures normal on the surface of the earth (which range between about −50 and 50°C), exists in three different states: liquid, gas, and solid. Liquid water makes up the oceans, seas, and lakes, flows in rivers, and underground streams. Solid water (ice) occurs in the polar masses, in glaciers, and at high altitudes, and gaseous water (moisture) is part of the atmosphere (O'Toole 1995). Liquid and solid water cover over 70% of the surface of the earth.

Water is the most abundant substance on the surface of the earth and the major constituent of all living organisms. Life on earth would not be possible and the biosphere would not exist without water. Water plays a key role in the sustenance of living organisms, since it is essential for all living processes, transporting nutrients to wherever they are required in their bodies, and removing from them their waste products. Thus, since it plays a fundamental role in all biological processes, water is also the most important substance for living organisms on the surface of the earth. All this is because water is an outstanding substance that has exceptional properties. It is the only substance that occurs on the surface of the earth not only in three different states – as a gas, a liquid, and a solid – but also as ordinary and heavy water. It is also unique in its molecular architecture, which accounts for many of its properties, as detailed in Textbox 72.

TEXTBOX 72

WATER

Pure water is colorless, odorless, and tasteless. The earth is pretty much a closed system, neither gaining nor losing much water, with very little of the earth's water escaping into outer space; thus, the same water that existed on the planet millions of years ago is still here. Water is, however, continually changing its form between water vapor, liquid water and ice, and moving around through, below, and above the surface of the earth (see Fig. 86).

Water is also so abundant and so familiar to everybody that it is difficult to appreciate that it has many outstanding properties; some of its chemical and physical properties are remarkably different from those of most other substances with which it would be expected to share certain properties. For example:

- Water is an excellent solvent that dissolves many materials that are insoluble in most other liquids having a composition similar to that of water.
- Water melts at 0°C and boils at 100°C; both these temperatures are much higher than the melting and boiling temperatures of other substances that have a compositions similar to that of water.
- When most liquids cool down, their volume gradually contracts. Water also contracts when it cools down, but only until the temperature reaches 4°C, when it is at it lowest volume, expanding as the temperature decreases or increases.
- Most substances are more dense in the solid than in the liquid state; solid water (ice) is lighter than liquid water, on which it floats.

These exceptional properties, exclusive to water, result from the structure of its molecule and the nature of the bonds (links) between its constituent atoms. The molecule of water is made up of two atoms of hydrogen bonded to one of oxygen, so that its formula is H_2O. The composition gives no clue as to its unusual properties, however. It is the special architecture of the molecule, the special way in which the hydrogen and oxygen atoms are bonded to each other, as illustrated in Figure 85, which accounts for the unusual behavior and properties of water.

The atom of oxygen in the molecule of water is bonded to the two of hydrogen in such way that the angle between the two oxygen-hydrogen bonds is under 180°. This means that the two atoms of hydrogen are located just at one side of the oxygen, as illustrated in Figure 85. Although the molecule has, overall, no net electric charge, the two hydrogen ends of the molecule have slight positive electric charges, while the oxygen has a feeble but definite negative charge. Molecules having separate positive and negative electric charges are known as *polar molecules*, and it is the *polarity* of the water molecules that explains, for example, why water is a good solvent of many substances. The opposite ends of the polarized molecules of water also attract each other and form *intermolecular bonds*, known as *hydrogen bonds*. The hydrogen bonds link between the molecules and account, for example, for the relatively high boiling point of water and for several other of its unusual properties.

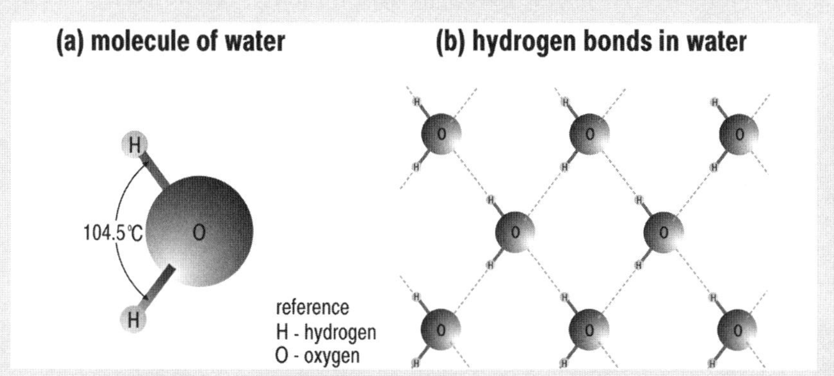

FIGURE 85 Water. A molecule of water is made up of two atoms of hydrogen bonded to one atom of oxygen **(a)**. The three atoms in the molecule are not aligned; because there are electrons on the oxygen atom (which are not involved in the bonding between oxygen and hydrogen), the hydrogen atoms form an angle, as shown in the figure. Since the electrons have negative electric charges, they repel each other; moreover, their repulsive force coerces the hydrogen atoms to come close together and form an angle of about 105° between them. Molecules of liquid water are linked to each other by "hydrogen bonds" **(b)**; these are weak attractive forces between the positively charged nuclei of the hydrogen atoms and the negatively charged electrons on the oxygen atoms. The distance between the hydrogen and oxygen atoms within each molecule are much shorter and the bonding force between them much stronger than between hydrogen-bonded molecules.

Isotopes in Water and Water Fractionation

The greatest number of water molecules (99.7%) are composed of the most abundant isotopes of hydrogen (hydrogen-1) and oxygen (oxygen-16). A small proportion, however, include in their composition other, less common isotopes of these elements: either hydrogen-2 (also known as *deuterium*), hydrogen-3 (also known as *tritium*), and oxygen-17 or oxygen-18. Thus, although most of the oxygen (99.76%) in the water of seas and rivers is oxygen-16, about 0.20% is oxygen-18 and about 0.04% is oxygen-17. The molecules in which there are heavier isotopes, which are therefore known as *heavy-water molecules*, have chemical properties identical to those of ordinary water; since they are heavier, however, the physical properties heavy-water molecules are slightly (but critically) different from those of ordinary water: they boil at a higher temperature and have a greater density than ordinary water molecules. The difference in physical properties between the heavier an lighter molecules results, for example, in the natural separation between ordinary and heavy water molecules. Molecules of heavy water are slower to evaporate than those of ordinary water: they evaporate at a slightly higher temperature than do ordinary water molecules. As a consequence of this difference, the water of the seas and oceans, from which for millions of years ordinary water has been evaporating at a slightly faster rate than has heavy water, is rich in the latter. Sea and ocean waters are, therefore, rich in the heavy isotopes of hydrogen and oxygen. The natural isotopic separation between ordinary and heavy-water molecules provides a tool, often used in archaeological studies, for establishing the provenance of materials and determining ancient temperatures (see Textbox 47).

Although water is a definite chemical compound and its chemical composition therefore does not vary in any way, people talk about various forms of water, such as drinking water, *rainwater*, *spring*, *lake*, and *river water*, *seawater*, and even *wastewater*. This is due to the exceptionally good solvent properties of water, which make it practically impossible to find pure water in nature. *Rainwater*, for example, one of the purest forms of natural water, contains gases dissolved from the atmosphere; *surface* and *groundwater*, from springs, streams, and lakes, usually contains solids dissolved from the surrounding soil and rocks. Over millions of years the concentration of the solids in ground and surface water has built up; since all flowing water ultimately runs into the seas and oceans, the water of seas and oceans contains, on the average, about 3.5% (by weight) of dissolved solids. The difference between the various types of water rests, therefore, not in the water itself, but on the nature and amount of gases and/or solids dissolved or suspended in the water. *Spring water*, for example, contains few solids dissolved from the surrounding ground and is usually suitable for human and animal consumption. *Groundwater* (water in the pores and crevices in the ground, in either soil or rock) also contains solids dissolved from the surrounding ground. *Hard water* contains relatively large amounts of dissolved lime. *Seawater* contains a wide range of dissolved solids such as common salt (sodium chloride), salts of potassium and magnesium, and many others. *Lake* and *river waters* have much less dissolved salts than seawater, but usually contain suspended small particles of matter of vegetable and mineral origin. Table 100 lists some of the substances generally dissolved, or suspended, in natural waters.

TABLE 100 Substances Often Dissolved or Suspended in Natural Waters

Water	Dissolved or suspended substances	Source
Rainwater	Carbon dioxide, nitrogen, oxygen, dust	Air pollutants
Groundwater, lakes, rivers, seas, and oceans	Sand (silica) and soil particles; chlorides, bicarbonates, and sulfates, mainly of calcium, sodium, magnesium, and iron ions; organic	Rocks and soil, microorganisms, plant and animal

TEXTBOX 73

HARD WATER: SCALE AND INCRUSTATIONS

Rain and groundwater naturally dissolve atmospheric carbon dioxide and, while doing so, they turn into aqueous solutions of carbonic acid, a weak acid:

$$\underset{\text{carbon dioxide}}{CO_2} + \underset{\text{water}}{H_2O} = \underset{\text{carbonic acid}}{H_2CO_3}$$

In the ground and when in contact with rocks, the acid solution reacts with limestone, and the result is a solution of calcium bicarbonate:

$$H_2CO_3 + \underset{\substack{\text{limestone, insoluble in}\\ \text{water (calcium carbonate)}}}{CaCO_3} = \underset{\substack{\text{calcium bicarbonate}\\ \text{(the } hardness \text{ in } hard \text{ water)}}}{Ca(HCO_3)_2}$$

If the water then evaporates, the dry calcium bicarbonate decomposes, recreating calcium carbonate, which precipitates and forms hard deposits and incrustations while carbon dioxide and water are released into the atmosphere:

$Ca(HCO_3)_2$	=	$CaCO_3$	+ H_2O +	CO_2
dissolved calcium bicarbonate		reprecipitated solid calcium carbonate		carbon dioxide (gas)

Hard Water. *Hard water*, as groundwater from areas rich in lime is known, contains relatively large amounts of dissolved salts, such as calcium bicarbonate and magnesium bicarbonate (see Textbox 73). The higher the amount of the dissolved solids the water contains, the higher its hardness. The evaporation of water from hard-water solutions in caves results in the formation of insoluble deposits of calcium and/or magnesium carbonate, and sometimes of other metal carbonates and sulfates, that precipitate in the form of *speleothems*. These are usually columnar deposits that either hang from the ceiling of caves (generally known as *stalactites*) or project upward from the floor (known as *stalagmites*). Consolidated precipitates of a composition rather similar to that of the solid that makes up speleothems are also formed, generally as unsightly and usually damaging concretions, on archaeological remains buried in places where hard groundwater flows.

Efflorescence. The solvent properties of water also causes *efflorescence*, a phenomenon whereby soluble or slightly soluble substances migrate from the interior of porous solids to the surface, where they precipitate. Efflorescence is an important factor in the decay and disintegration of many rocks, and of human-made porous materials such as ceramics, and even of some types of glass. On archaeological objects, efflorescence generally occurs mostly as a white, powdery, but sometimes consolidated accretion on the surface of the objects. Calcite, a form of calcium carbonate, is one of the most common substances to effloresce on archaeological ceramics.

Efflorescence usually takes place when groundwater penetrates within porous solids, where it *leaches* (dissolves) soluble salts from the solids. When the water with the leached solids eventually evaporates, the solution migrates toward the surface; if the water continues to evaporate, the dissolved salts are redeposited, forming small crystals just below and on the surface of the objects. The forces generated by the crystallization of the efflorescent salts below the surface (in the bulk of the solid), as well as differences between the thermal expansion of the solid and the now-crystallized salts, may bring about the breakdown of the solids and eventually, their total disintegration into small pieces. Another source of efflorescence that often damages porous archaeological objects are salts dissolved in hard groundwater that flows into a burial site and into the objects there. These salts usually begin to effloresce soon after removal of the objects from their burial place to drier surroundings; unless such objects are thoroughly washed and the salts removed immediately after excavation, they may cause, on drying, disintegration of the objects.

Freezing Water. Water expands when it cools down below 4°C and freezes into *ice*. When water trapped in the crevices and pores of rocks of solids such as pottery, brick, and wood freezes, therefore, the expansion of the water generates pressure on the surrounding matter. The pressure generated is extremely high, many times higher than the atmospheric pressure and sufficiently high to disrupt the crevices and the pores within those materials. In other words, water freezing within voids and crevices breaks up solid materials, a process generally known as *freeze wedging* (see Fig. 45). Moreover, when freezing occurs, moisture in the air is strongly attracted to the freezing ice and, as it cools down, it first liquefies and then freezes. Thus the total volume of ice within the crevices and open voids in solids increases while freezing; this contributes to intensifying the stress caused by the freeze wedging process on the surrounding solid, which may split up and break objects into fragments.

The Water Cycle. The evaporation of water from land and water surfaces, the transpiration from plants, and the condensation and subsequent precipitation of rain cause a cycle of transportation and redistribution of water, a continuous circulation process known as the *hydrologic cycle* or *water cycle* (see Fig. 86). The sun evaporates fresh water from the seas and oceans, leaving impurities and dissolved solids behind; when the water vapor cools down, it condenses to form clouds of small droplets that are carried across the surface of the earth; as the clouds are moved inland by the wind and are further cooled, larger droplets are formed, and eventually the droplets fall as rain or snow. Some of the rainwater runs into natural underground water reservoirs, but most flows, in streams and rivers, back to the seas and oceans, evaporating as it travels.

The purity of the water changes constantly during the water cycle. As rain falls through the air, for example, the water dissolves some atmospheric gases such as oxygen, carbon dioxide, and in industrial regions also such air pollutants as sulfur dioxide and oxides of nitrogen. Still more carbon dioxide

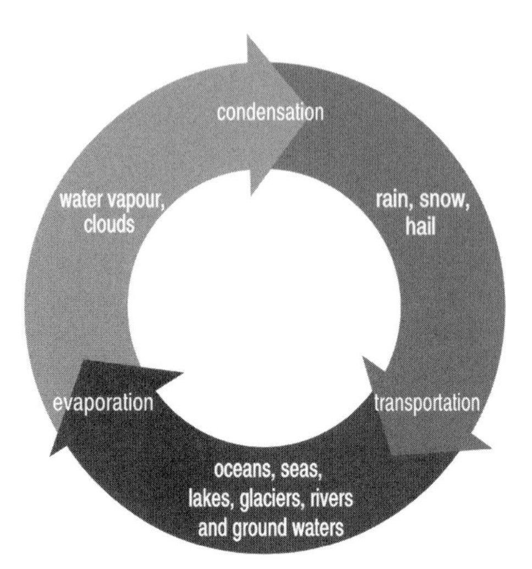

FIGURE 86 The water cycle. The water on the earth is constantly being cycled between the oceans, the atmosphere, and land. This cycling is a very important process that helps sustain life in the planet. Most of the water is in the oceans, seas, lakes, rivers, streams and wetlands. The heat of the sun warms the water and turns it into vapor, which goes into the air, where it becomes clouds. Under warm conditions the clouds become oversaturated with water, which, as the temperature decreases, precipitates as either rain, snow, or hail (ice) onto the surface of the earth. Some of the precipitated water evaporates again, but some soaks into the ground and stays there until plants absorb it, or penetrates deep into the soil or rock and becomes groundwater; most of the precipitated water, however, runs again into streams, rivers and lakes, and finally into the seas and oceans.

is dissolved when water flows on the surface of the earth or soaks through the soil and carbonaceous rocks. Carbon dioxide dissolved in water reacts with the water to form carbonic acid; thus, "pristine" rainwater is usually acidic, having a pH value below 7, generally between 5 and 6, which makes it a solvent of minerals in rocks and soil (see Textbox 48). When this occurs in lime areas, hard water is formed. Eventually most water flows back to seas and oceans, carrying with it all the impurities previously dissolved. The sea is, therefore, a vast store of water-dissolved solids.

17.3. POLLUTION

The term *pollution* refers to materials or energy in the atmosphere, water, and/or ground, that are, or may be, detrimental to inanimate matter and objects and/or to living organisms on the surface of the planet. Polluted air, it follows, is air that includes gases, small droplets of liquids, and/or particulate matter that may be harmful to materials as well as to living organisms. Some air pollutants are produced by natural processes such as volcanic eruptions, but many have entered the atmosphere since primeval times, and they still do so, as a result of human activities, as, for example, when organic matter is burned, or wastes are disposed of. Since prehistoric times humans have created air, water, and soil pollutants when lighting fires to clear land, dumping wastes on the ground, tossing them into waterways, or burning them. Fires pollute the air with smoke, and the cleared and exposed land becomes vulnerable to dust storms, which further pollute the atmosphere with particles of sand and clay. Dumping wastes on the ground or in water pollutes the soil as well as surface and underground water. Prehistoric and even later populations were relatively small, however, and widely spread out over large areas. The land clearing activity of ancient peoples was of limited extension, and their wastes were mostly food scraps and other substances that were easily broken down by natural decay processes. Ancient human-caused pollution was therefore limited, and widely dispersed, and caused little, if any, damage to the environment. The growth of pollution therefore, started when large numbers of people began living together in cities. As the cities grew, so did the extent of pollution and, with it, the amount of pollutants released from wastes and refuse. During the first century C.E., for example, Roman writers were already referring to the "stink" and "heavy air" of Rome.

At still later times, beginning at the end of the eighteenth century, large urban populations began to burn fossil fuel such as coal (which contains sulfur) for their home and working activities. This resulted in the release, into the atmosphere, of relatively large amounts of polluting sulfur dioxide, smoke, and other dangerous discharges. Environmental problems caused by pollution became even more serious and widespread in the early nineteenth century, during the period known as the *Industrial Revolution*, which was characterized by the development of large factories, the widespread use of coal as an energy source, and the overcrowding of cities with factory workers who used coal for their domestic activities. In modern society, which has since then been using increasing amounts of *fossil fuels*, including not only coal but also oil and natural gas, the high levels of air pollution have not only become detrimental to the quality of life of living things, they also affect relatively stable metals and many types of stone, such as sandstone, limestone, and marble, damaging and in many cases even destroying antiquities and archaeological structures. Ancient statues, for example, and other archaeological structures made of marble are slowly but extensively being disfigured, and even obliterated, under the aggressive and continuous attack of airborne pollutants and acid rain. Moreover, many archaeological objects are complex assemblages of organic and/or inorganic materials, each of which reacts to the environment according to its chemical and physical properties; if there are incompatibilities between the materials and their environment, pollutants usually worsen the existing incompatibilities (Camuffo et al. 2000; Hicks et al. 1991).

Air Pollutants

Among the most abundant air-polluting gases are ozone, hydrogen sulfide, the oxides of some nonmetallic elements such as sulfur and nitrogen, and several hydrocarbons.

Ozone. *Ozone*, an *allotrope* of oxygen (see Textbox 19), is a powerful air pollutant, an oxidizing and bleaching agent that reacts easily with many organic and biological substances and negatively affects living processes. Ozone is abundant mainly in the higher layers of the atmosphere, where most of it is created; however, small amounts of the gas reach the lower layers, near the surface of the earth, and are extremely active in the decay and breakdown of archaeological materials such as animal skins, textiles, and dyes (Sherwood 1995; Shaver et al. 1983).

Sulfur Dioxide. *Sulfur dioxide* and *nitrogen oxides*, which are produced in modern society when coal, gas, and oil are burned in cars, power plants, and factories, react with water vapor in the air to form acids that negatively affect organic materials and even metals and stone; when dissolved in airborne rainwater, the oxides of sulfur and nitrogen are the main cause of the formation of acid rain (see below).

Particulate Pollutants. *Smoke*, *ash*, *viruses*, *pollen*, *sand*, and in contemporary industrial society also *coal* and *cement dust*, are generally known as *particulate pollutants* (they occur as extremely small solid particles suspended in the atmosphere). The combination of air, pollutant gases, small liquid droplets, and particulate matter constitutes what is known as *smog*, which, since the second half of the eighteenth century, has beset antiquities, damaging and disintegrating even those made of stone and metals.

Water Pollutants

Acid Rain. *Acid rain* is a broad term used to refer to acid fallout of the atmosphere in the form of rain, snow, and ice (see Textbox 74) (Bubernick and Record 1984). As acid rainwater flows on the surface of the earth, it affects rocks, the environmental condition of plants and animals on the ground, and practically all material structures built by humans; acid rain constitutes, therefore, a severe pollutant that affects large parts of the world. Carbon dioxide and other air pollutants that enhance the acidity of rainwater have long been released into the atmosphere by such natural processes as the activity of volcanoes, animals, and soil bacteria. But all these occurs in relatively low amounts and therefore rarely have adverse effects on either living organisms or inanimate materials. On the contrary, the natural acidity of rainwater caused by the dissolution of carbon dioxide (see text above), is beneficial to plants and animals. It has been only since the Industrial Revolution, and in particular during the twentieth century, when fossil fuels became used in ever-increasing quantities, that acid rain has affected the environment in general, and ancient materials in particular. Fossil fuels contain a relatively high proportion of compounds of sulfur and nitrogen that are released into the atmosphere as the fuel burns; when dissolved in rainwater these compounds form strong acids that are extremely harmful to the environment.

TEXTBOX 74

ACID RAIN

Fossil fuels, such as coke and petrol, are natural fuels created during past geologic periods from the remains of living organisms. They consist mainly of a mixture of combustible *hydrocarbons*, organic substances composed of hydrogen and carbon derived from the remains of the ancient organisms. Since the nineteenth century fossil fuels constitute the main energy resource of modern society. Although their main components are hydrocarbons, a small proportion of most fossil fuels, generally between 6% and 9% of their total weight, are compounds of sulfur and nitrogen. When fossil fuel burns, the compounds of sulfur and nitrogen in the fuel are oxidized by oxygen in the air and are converted into oxides, mainly the gases sulfur dioxide and nitrous oxide, which are released into the atmosphere. There they constitute the main air pollutants of modern times. Moreover, in the atmosphere, both sulfur dioxide and nitrous oxide are further oxidized and then dissolve in water droplets in the air, forming the very strong sulfuric and nitric acids. Sulfur dioxide, for example, is oxidized to sulfur trioxide, which, when dissolved in water droplets in the atmosphere and in rainwater gives rise to sulfuric acid:

S sulfur (in fuel)	+	O_2 oxygen (in the air)	=	SO_2 sulfur dioxide (a gas)
$2\ SO_2$ (in the air)	+	O_2	=	$2SO_3$ sulfur trioxide (a gas)
SO_3 (in atmospheric water)	+	H_2O	=	H_2SO_4 sulfur acid

Under similar conditions, nitrogen compounds in the fuel react as follows:

$2N$ nitrogen (in fuel)	+	$2O_2$ in the air	=	N_2O nitrous oxide (a gas)
$2N_2O$	+	O_2	=	$2NO$

Ultimately, these gases are converted into nitric acid, HNO_3.

Sulfuric and nitric acids are extremely strong. They lower the pH of rainwater much below that resulting from the dissolution of atmospheric carbon dioxide, rendering it strongly acid (the pH of rainwater occasionally reaches pH values as low as 2.1). The solution of sulfuric and nitric acids in rainwater is generally known as *acid rain*. Acid rain is corrodes most metals and alloys, dissolves many minerals and rocks, and decomposes a wide range of materials. At present times, sulfuric and nitric acids together account for 95% of the acids in acid rain. Although it is customary to speak primarily of acid rain, all forms of water precipitation, including fog, snow, sleet, and hail, have been shown, during the late twentieth century, to contain increasing concentrations of acids. A more appropriate collective term referring to all of these forms of acid-polluted downfall is therefore *acid precipitation* (Schmitz 1996).

Acid rain harms the environment in a number of ways: it dissolves many rocks and metals, alters the composition of soils, groundwaters, and lakes, and alters the environmental conditions of living organisms. Acid rain is also particularly harmful to ancient objects and structures, as it plays an important role in their deterioration and sometimes total destruction. Unprotected limestone, marble, and sandstone, all of them widely used in ancient times for building and making statuary, are disintegrated by acid rain, which slowly disfigures and eventually destroys ancient structures and statues made from any of these types of stones (Bubernick and Record 1984).

17.4. INTERACTION OF MATERIALS WITH THE ENVIRONMENT

Most materials are affected by physical, chemical, and/or biological factors in their environment that may alter their composition and cause either decay or conversion to other materials. It is not always possible, however, to make a clear-cut separation between the different forms of interaction of materials with the environment; some chemical agents may affect materials physically, physical agents may cause their chemical alteration, and, under certain circumstances, the same agent may induce physical or chemical changes and even both simultaneously. Water, for example, may erode and wash away particles from solid matter, thus affecting it physically without chemically altering its composition; but it may also dissolve the washed-away particles and later even cause their chemical alteration. Two or more physical and/or chemical agents may act simultaneously, often in ways that are interrelated, and the effect of one may even exacerbate that of the other, thus causing an overall intensification of the decay process. Thus it is customary nowadays not to classify the agents of decay into separate (physical or chemical) categories, but to consider an all-embracing category of *physicochemical agents of decay*. Some naturally occurring substances such as ozone and water vapor, in addition to human-generated atmospheric and water pollutants, are among the most active physicochemical agents affecting antiquities.

Temperature Effects

Changes in temperature alter most of the physical properties of materials and affect the rates of chemical reactions. Thus heat and cold may advance or inhibit the deterioration of antiquities (see Textbox 75).

TEXTBOX 75

THE THERMAL PROPERTIES OF SOLIDS

When a solid is heated, the heat is transferred throughout its mass by *conduction*: the heat energy passes from one part of the solid to another without the solid undergoing any mass changes. *Heat conduction* is the flow of energy from a spot of higher temperature to one of lower temperature by the interaction of the atoms in the space between them. The rate at which heat is transferred is determined by the *thermal conductivity* of the solid, an intrinsic property of every material that describes the rate at which heat flows within a body of the material at a given temperature difference. Most metals and alloys have *high thermal conductivity* and are said, therefore, to be *good conductors* of heat. The nonmetals, whether inorganic or organic, on the other hand, generally have *low thermal conductivity* and are *bad conductors* (but *good insulators*) of heat. Stone, brick, pottery, glass, and all forms of living matter, such as wood and textile fibers, for example, are good insulators of heat, shielding from either heat or cold. It is because of their low heat-conducting properties that these materials are used for in building constructions and making clothes. Table 101a lists the thermal conductivities of solids frequently found in archaeological sites.

Temperature changes also cause dimensional changes in materials. When a material is heated or cooled, its length changes by an amount pro-

TABLE 101a Thermal Properties of Some Archaeological Materials

Thermal Conductivity (at 25°C)

Material	Thermal conductivity[a]
Metals and Alloys	
Silver	4.3
Copper	4.0
Gold	3.2
Brass	0.8
Iron	3.2
Steel	0.5
Lead	0.4
Nonmetals	
Obsidian (at 0°C)	1.35
Pottery	1.2
Sodalime glass	0.9
Mortar cement	0.5
Wood (at 20°C)	0.03–0.3

[a] Expressed as the amount of heat (measured in watts) that passes, per second, through one cubic meter of the material when the temperature at opposite sides of the cube is kept at a difference of one degree centigrade, at the temperature stated.

portional to the original length and the change in temperature. The *coefficient of thermal expansion* of a material is the fractional change in length (or in volume) that a body of a material undergoes for a unit change in temperature. Most materials expand when heated and contract when cooled, and Table 101b lists the values of the coefficients of linear expansion of some materials. Some materials, however, have abnormal coefficients of thermal expansion at particular temperature ranges, contacting when heated and expanding when cooled at particular temperature ranges. Water, for example, expands when heated and contracts when cooled. When it cools down to about 4°C, it begins to expand, however, this expansion peaks (reaches a maximum) as the temperature approaches the freezing point (see Textbox 72). Although the expansion is relatively small, it is sufficient to exert pressure on surrounding the matter; freezing water trapped in the pores or crevices of stone or pottery, for example, often causes the breakup and disintegration of these materials into small pieces.

TABLE 101b Thermal Properties of Some Archaeological Materials

Coefficient of Linear Expansion (at 25°C)

Material	Expansion coefficient[a] $N \times 10^{-6}$
Metals and Alloys	
Lead	29
Brass	20
Silver	19
Copper	17
Gold	14
Iron	12
Steel	10–17
Nonmetals	
Marble	4
Brick	5– 7
Glass	5–11
Mortar cement	7–14
Wood	6–11

[a]Expressed as the change in length per degree centigrade.

Thermal Conductivity. Materials that have low thermal conductivity, such as wood, stone, ceramic, and glass, are generally referred to as *bad thermal conductors* [see Table 101a]. When heated or cooled, such materials heat up or cool down not uniformly but differentially by way of temperature gradients throughout their volume. Consequently, when an object made from a bad thermal conductor is heated or cooled, different portions of it will expand, or contract, at different times and at different rates; the temperature gradients throughout the volume create, internal mechanical stresses that the object may or may not endure. Moreover, as a result of the mechanical stresses set up by repeated heating and cooling cycles, the shape of such objects becomes distorted and, if the thermal cycling is repeated over long periods of time, the objects may crack, break down, and shatter into small fragments.

Thermal Expansion. When heated, most materials change size, expanding when heated and shrinking when cooled, the extent of the change is expressed by their *coefficient of thermal expansion* (the fractional change in length or in volume of a material per degree of temperature, which is characteristic for every material) [see Table 101b]. Because of the differences in the coefficients of thermal expansion, when objects or structures assembled from two or more materials undergo temperature changes, some of the components are compressed by the changes, whereas others are forced apart; the stresses generated by such changes, repeated time and again, often result in the physical breakdown of such structures.

Chemical Effects of Temperature. Changes in temperature also affect the chemical properties of materials. The rate at which most chemical reactions take place, for example, is roughly doubled when the temperature of the reactants increases by 10°C. Consequently, any increase in temperature intensifies the rate at which most materials react with substances in the environment such as oxygen, water, and atmospheric and soil pollutants, and hastens their chemical degradation.

Sunlight

Sunlight is often the single natural cause for the degradation of archaeological materials of organic origin. Although only a very narrow portion of the solar spectrum, that of ultraviolet radiation, causes most deterioration, its effects on organic materials are generally deleterious and often devastating. Solar ultraviolet radiation accelerates many chemical reactions that, in the absence of radiation, occur at much slower rates. The effectivity of *photochemical degradation reactions*, as such reactions are known, is determined mainly by the nature of the exposed material, the ambient environmental conditions, and the intensity of the solar radiation. The more intense the ultraviolet radiation, the faster the rate of degradation. Common textile yarns such as cotton, wool, and silk, when exposed to solar radiation, undergo very fast photochemical degradation, disintegrating in relatively short periods of time; most dyestuffs are readily bleached by the photochemical activity of sunlight.

Oxygen and Ozone

Oxygen is the most universal chemical agent of decay. Less plentiful under ordinary conditions but far more destructive is the unstable and highly reactive allotrope of oxygen, *ozone* (see Textbox 19). The degradation of materials brought about by oxygen and ozone are quite similar, although ozone is a much more aggressive oxidizing agent. Under ordinary environmental conditions, for example, few materials are affected by oxygen in the absence of other agents. In general it can be said that for materials to be oxidized by atmospheric oxygen requires the presence of at least one other "activating agent". Iron, for example, is not oxidized (does not rust) when exposed to dry oxygen; moisture is an essential concomitant in the conversion of iron to rust. Ozone, however, attacks many materials directly. Iron is oxidized by ozone, and so is silver, which is also unaffected by oxygen alone; when exposed to ozone, however, silver is quickly covered with a coating of silver oxide. Most textile yarns, dyes, and paint layers are rapidly oxidized by ozone, particularly when illuminated by sunlight.

Water

Water is a most important agent of deterioration. Its action is sometimes chemical, sometimes physical, but mostly physicochemical; it interacts with materials simultaneously, affecting them mechanically as well as chemically. In its various forms water affects many and most diverse materials unfavorably. Corrosion cannot take place without water (see Textbox 42). Textiles subjected to periodical wetting and drying are weakened; parchment exposed to a humid environment for long periods of time breaks down completely, decaying into a viscous gelatine commonly known as *parchment size*. Also the adhesion of paints to solid surfaces is damaged by the action of excessive moisture; initially the paint layer forms blisters that eventually break and scale off.

Variations in the relative amounts of atmospheric moisture affect wood, subjecting it to dimensional changes. When moisture is taken up or given off by wood, a *moisture gradient* is established across the wood (since not all the fibers of the wood swell or shrink at the same rate); the moisture gradient creates internal stresses, and these often result in warping of the wood. Usually, the more moisture in the environment, the more serious its detrimental effects on wooden objects, and the more rapidly they decay. The strength and rigidity of wood is also greatly weakened by continuous soaking in water, which results in the formation of very soft and flabby waterlogged wood.

Large quantities of flowing water, in the form of rain, snow, hail, or river or flood streams, exert strong erosive action on practically all materials. Water also brings about the physical breakdown of materials through the *frost wedging* effect (see Fig. 45), which becomes active when water in the internal voids and crevices of materials cools down; since below 4°C water expands before freezing, the dimensional expansion of water during frost wedging has a sort of localized explosive action that can reduce porous stone, ceramics, and wood to dust (see Textbox 45).

Water can contribute to the decay and breakdown of some materials not only by its presence but also by its absence; most organic materials require some optimum moisture content to preserve their mechanical properties; papyrus, skin, parchment, and leather that are too dry, for example, are extremely stiff, brittle, and friable.

Air Pollutants

The atmosphere is a reservoir of aggressive air pollutants that may cause serious deterioration to archaeological objects; some of the most active pollutants are *sulfur dioxide, hydrogen sulfide,* and *ozone.*

Sulfur Dioxide. *Sulfur dioxide* is a dangerous air pollutant because of its corrosive properties, which are causal factors in the decomposition of most organic and some inorganic materials. Moreover, sulfuric acid, formed when sulfur dioxide is oxidized and reacts with moisture, is one of the harshest solvents of a wide range of materials: it dissolves for example all of the metals known in antiquity, with the exception of gold, converting them into metallic salts. Sulfuric acid also dissolves many common stones, such as limestone and marble, which are widely used for building and statuary (see Textbox 74). Sulfur dioxide can, however, also act as a reducing agent, and as such, it causes textiles, particularly cotton, to be rapidly broken down, and leather to change its texture, become brittle, and eventually break down completely.

Hydrogen Sulfide. *Hydrogen sulfide* is a foul-smelling gas that is released into the atmosphere from volcanoes as well as in the course of decay of animal tissues. As an air pollutant, it reacts with almost all metals, with the exception of gold, forming a dark-colored corrosive layer of metal sulfide, commonly known as *tarnish*, which discolors the exposed surface of most metals.

Soil Pollutants

Contaminants in the soil, mainly acids and salts, are powerful agents of decay. All soils are either acid or alkaline and rarely, if ever, neutral (see Textbox 48). Acid, as well as alkaline groundwaters, readily attack most buried materials of either inorganic or organic origin, altering their properties and, over long periods of time, causing their disintegration. Moreover, in the presence of water, salts in the soil undergo *hydrolysis*, the process by which water and the salts interact to form acid or alkaline solutions. Many of the salts generally abundant in soils are hydrolyzed by water and form electrically charged *ions* (cations or anions) that enhance the corrosion of buried metals and alloys. The ions formed when the salts are hydrolized either react with the metals directly or facilitate their corrosion (see Textbox 42). Quite often, excavated archaeological objects are covered by a whitish, sometimes relatively hard efflorescent incrustation on their surface; such a coating is generally formed when the objects have undergone successive cycles of soaking with hard groundwater followed by drying (see above, *efflorescence*). As the soil and objects dry, the salts in the groundwater precipitate and crystallize on the surface, forming defacing layers, and often causing total breakdown of the surface.

When salts in groundwater precipitate and crystallize within the cavities of buried materials such as pottery, cement, and wood, they may generate internal pressures sufficient to disrupt these materials and turn them into gravel. Salts are also active in blistering and scaling painted surfaces on a variety of materials.

17.5. DETERIORATION OF SOME ARCHAEOLOGICAL MATERIALS

Pottery

Well-fired pottery, fired at temperatures above 850°C, is a very stable material; in fact, it is practically inert and indestructible. If the firing temperature is low, however, say below 600°C, an inferior material is obtained that rain and groundwater may soften and even cause to disintegrate. The penetration of salts into well-fired pottery can, however, cause serious deterioration. For example, if groundwater containing dissolved salts infiltrates into the pores of pottery, when the water evaporates, the salts crystallize at and just under the surface of the pottery (where the rate of evaporation and of salt crystallization is fastest), setting up internal stresses within the bulk. Recurrence of the infiltration and evaporation cycle over long periods of time results in the crystallization of more salts, their efflorescence, and the consequent generation of strong stresses in the surface layer of the pottery, which may then crumble. In extreme cases, when the crystallization of the salts occurs deep in the bulk, the process may generate internal pressures sufficient to reduce the entire pottery to dust.

Glass

Contrary to appearance and to apparent common experience, many types of *glass* (particularly ancient glass) are, to a certain extent, soluble in water, and therefore susceptible to deterioration by water. In ancient soda, potash, and early sodalime glass in particular, the deterioration is due to the *leaching out* (preferential dissolution) of the glass modifiers by moisture, rain, and groundwater. The most common glass modifier used in antiquity, soda, was often used in a much higher proportion than that required for making the glass; this made the glass partially soluble in water and therefore very susceptible to deterioration. Also potash, another glass modifier frequently used in the past, renders the glass vulnerable to water. Acids in groundwaters and particularly in acid rain may accelerate the leaching of the modifiers and thus accelerate the decay of the glass. When any of these types of glass becomes wet, the water selectively dissolves the excess, or part of the modifiers, from the surface layer; after drying, all that remains on the surface of the glass is a relatively thin layer of silica, depleted of modifier in varying degrees. Seasonal cycles of soaking, leaching, and drying of the glass, repeated over long periods of time, result in the formation of successive layers of silica, which build up in onion-skin fashion on the inner core of the remaining glass. The superpositioned layers of silica cause the iridescent color effects often seen on the surface of ancient glass objects.

If a very large excess of modifier was introduced into the glass at the time of manufacture, part of it may remain uncombined within the body of the glass and lead to another form of glass deterioration known as *sweating* or *weeping glass*. Glass modifiers, especially potassium oxide, are strongly alkaline and very *hygroscopic*, interacting with water from their surroundings. Under highly humid conditions the *hydrolysis* (interaction with water) of the glass modifiers results in the formation, within the glass, of a solution of potassium hydroxide, a very strong alkali that migrates to the surface, where it forms beads. There the beads absorb carbon dioxide from the atmosphere and form potassium carbonate, which further attacks the silica skeleton of the glass and further hastens its deterioration. Moreover, the potassium carbonate also absorbs moisture, promoting the movement of even more alkali to the surface. This form of deterioration continues as long as the glass is in a humid environment, and unless it is transferred to drier surroundings, it eventually disintegrates completely. In alkaline surroundings, such as those that exist when glass is buried in calcareous soil, the deterioration of glass is even faster, since the alkaline surrounding reacts and dissolves the silica that makes up the main structural framework of glass. Thus prolonged burial of glass in calcareous soils weakens glass to such an extent that any slight pressure may be sufficient to cause it to crumble into small fragments.

Metals

See Section 5.9, in Chapter 5.

Wood

Wood is an *anisotropic* material that undergoes uneven dimensional changes and, under extreme variations of environmental conditions, becomes distorted and warped (see Chapter 10). Exposed to the atmosphere, wood is also susceptible to the mechanical forces of wind and rain, and the effects of solar radiation; the latter, in particular, causes discoloration initially, and then photochemical degradation, which often results in the wood's total decomposition. Wood is also prone to consumption by bacteria, fungi, insects, and rodent animals (Unger et al. 2001).

Waterlogged Wood. *Wood* submerged in water or embedded in mud or peat for long periods of time becomes saturated with water; *waterlogged wood*, as it is generally called, holds the maximum amount of water that, at ambient temperature, can be held in its cells and in the intercellular spaces of the wood. Although the lignin structure of wood generally does not decay and preserves its original shape, the decay of at least part of the cellulose enlarges the voids within the wood, which becomes very frail. As long as waterlogged wooden objects are wet, they preserve their shape; if dried, however, the lignin structure of the waterlogged wood collapses, and so does the shape of wood-made objects, usually beyond repair. Appropriate conservation treatment may prevent this form of disintegration (Mühlethaler 1973).

Skin, Hide, and Leather

Skin is unstable to varying environmental conditions and deteriorates readily under humid conditions or through biological activity, or both. Basically, the decay of much ancient skin and hide results from *hydrolysis*, that is, the reaction of the protein fibers in the skin with water; in extreme cases, the hydrolysis of skin and hide may cause their total dissolution, and quite often, under humid and hot environmental conditions, nothing remains to indicate that skin or hide was once there.

Leather, a human-made, more complex material than skin (see Chapter 11), is generally stable to decay processes; differently tanned leathers vary, however, in their resistance to breakdown. Heat and exposure to oxygen and ozone in the air may, over long periods, cause leather to crumble to a powdery residue. Sunlight has relatively little direct effect on leather, although it accelerates the effects of high humidity and water, as well as the oxidation of the leather by oxygen, ozone, and sulfur dioxide in the air. Sulfur dioxide is easily absorbed by leather and in time causes the leather to disintegrate by a process known as *red rot*: after it is absorbed, the sulfur dioxide is further oxidized and, under humid conditions, converted to sulfuric acid, which dissolves and eventually disintegrates the leather, leaving only dusty red remains.

Fibrous Matter

The decay of yarn, textiles, and cordage is caused mainly by physicochemical processes, such as the activity of sunlight, environmental oxygen and moisture, and pollutants (ozone and sulfur dioxide for example). Often their decay is accelerated by biological processes such as the activity of bacteria, fungi, insects, and rodents. The most evident effects of physicochemical deterioration are loss of strength and change in the appearance of the fibers. Sunlight, mainly ultraviolet radiation, weakens them and initiates the chemical breakdown of the fibers, mainly through oxidation processes. Excessive moisture causes swelling of the fibers, which is always accompanied by softening; desiccation, on the other hand, leads to their hardening and becoming brittle and frail. Wool seems to be the natural fiber most resistant to decay; cotton and linen come second, whereas silk is most easily degraded.

18 AUTHENTICATION OF ANTIQUITIES

Ancient tools and artifacts have been appreciated and sought after by collectors since, at least, as far back as the prime times of ancient Egypt. The demand usually exceeds supplies, and forgers set to work and dispense the necessities; the intense yearning of collectors for works of earlier periods has since promoted a lucrative trade in careful restorations, meticulous reproductions, and fakes (Cohon 1996; Howard 1992). An early manuscript known as *The Stockholm Papyrus*, probably the "laboratory notebook" of an ancient Egyptian "chemist," for example, deals with a variety of ways for making base metals resemble precious metals and for faking gemstones (Caley 1927). Later during the early centuries C.E., Roman sculptors made replicas of esteemed Greek works from earlier times to satisfy the cravings of demanding collectors. An early reference to the forgery of antiquities was found in the verses of the Roman fabulist Phaedrus, who lived during the first century B.C.E., when he wrote:

> Ut quidam artifices nostro faciunt saeculo, qui pretium operibus maius inveniunt, novo si marmori adscripserunt Praxitelem suo, trito Myronem argento.
>
> (Some artists of our times obtain a higher price for their works by inscribing on their new marble the name of Praxiteles or the name of Myron on worn-out silver.)

The active trade in false ancient tools and artifacts, deceptive reproductions of sculptures, and fictitious precious stones, continued during the Middle Ages; stylistic criteria were as yet unknown, and *forgery*, in the modern sense of the word, concerned nobody. The Renaissance brought a renewed appreciation of antiquity, ancient craftmanship, and the works then created, that has endured ever since; continuing generations of plagiarists and counterfeiters labor incessantly to supply the world with fictitious antiquities (Rieth

Archaeological Chemistry, Second Edition By Zvi Goffer

1970; Muscarella 2000). The tradition of simulating and forging antiquities has became continuous; only the type of objects that attract the attention of collectors, and therefore also of fakers and frauds, varies from time to time and from place to place.

During Renaissance times, works of art from classical times were highly appraised by collectors, and Michelangelo, for example, early in his life, made a marble statue that he buried in the ground for a time to give it an ancient appearance and then sold it as a classical sculpture (Condivi 1928). During the seventeenth century, when the ancient Egyptian civilization was rediscovered, it stimulated a great deal of fraudulent activity, which extended into the eighteenth century, to make imaginative copies and variations of Egyptian artifacts, sculptures, and jewelry. Later on, during the eighteenth century, the unearthing of Roman mosaics that had been buried for many centuries prompted numerous crafters to compose new, unfeigned "Roman mosaics" (Herbig 1933). During the eighteenth century Roman catacombs began to be excavated on a large scale and antique gilded glass objects, which had been in favor with early Christians, became such an antiquities market favorite that it prompted one of the founders of modern archaeology to write: "This technique (of gilded glass) has been reinvented in Rome a few years ago. I had the opportunity of examining some specimens which are very well done; one (workman) has utilized this device to fool foreigners" (Caylus 1759).

During the nineteenth and twentieth centuries, public interest in antiquities continued to grow, and created an increasing demand for artifacts of the ancient world that ingenious forgers and deceivers were only too ready to satisfy, supplying the market with "antiquities" of their own manufacture. Some, like clever businesspeople, proceded to create a demand for previously unheard-of objects that deceived even some of the greatest archaeologists and museum curators in their moments of unfortunate credulity. In the second half of the nineteenth century, for example, a cunning businessman from Jerusalem invented and sold to the Berlin Museum the now infamous "Moabite pottery," which, although still kept in the basement pottery collection at the museum, was never displayed to the public (Yahuda 1944; Clermont-Ganneau 1885). Early in the twentieth century, Sir Arthur Evans, the excavator of Knossos, acquired the "Ring of Nestor," which he believed to be a genuine Minoan antiquity but on later chemical analysis proved to be of modern origin. Later, many terracotta sculptures claimed in the antiques market to be West African antiquities were revealed by a wide-ranging investigation to be fakes, traded and sold at exorbitant prices, many of which had been acquired by some of the world's most prestigious museums. Even at the beginning of the twenty-first century, a burial box with the inscription

"James, son of Joseph, brother of Jesus," claimed to have contained his bones, was shown to be a recent forgery (Mayell 2003).

A common type of forgery consists not of newly made objects, but of the remains of old ones, which serve as the foundation for additional new work. The advantages of this method are obvious: since the basic materials of this type of forgeries have all the marks of age and authenticity, the objects seem to have been restored, and the borderline between restoration and outright forgery is difficult to draw. Occasionally, enthusiastic restorers create new styles by working derivatively from small fragments; many "ancient" bronzes and ceramics are faked in this way, sometimes supplemented by additional parts (Woolley 1962; Schmitt 1959).

When ancient objects are removed from their original archaeological location, they lose much of their initial value to the scholar, since site, position, and surroundings convey a great deal of information about an object and also enable its authentication. Authenticating isolated objects relies on the fact that antiquities carry within themselves evidence of the time and place of their manufacture (Jones 1992; Jaffé and Van der Tweel 1979).

In archaeology and art the term *authenticity* refers to the positive identification of objects, their origin and attributes, and describes the relative integrity of the object in relation to its original creation. *Authentication* is, therefore, the process of determining whether objects are what they are asserted to be. Nowadays, archaeological, artistic, and scientific guidelines generally complement one another in determining the authenticity of objects, and the authentication of antiquities is generally based on collaborative archaeological, artistic, and scientific studies. The scientific examination of antiquities has therefore become not merely desirable but absolutely essential for the positive identification of antiquities. The information derived from archaeological, stylistic, and aesthetic considerations is generally reinforced by scientific evidence (Bibliothèque Nationale 1988). During the second half of the twentieth century the results of many physical and chemical investigations on antiquities revealed that such studies are of great assistance for establishing the authenticity or otherwise of antiquities.

18.1. TECHNICAL AND SCIENTIFIC METHODS OF AUTHENTICATION

Many material properties lend themselves to the authentication of objects; the composition and the physical and chemical properties of materials and their decay products provide objective criteria for establishing the authenticity of archaeological objects. Contradictions between the claimed origin or age of an object and the composition and properties of the material from which it is made imply that the object could not have been made either at the place or during the period assigned to it. Finding aluminum to be a component of a supposedly ancient metallic object, for example, would clearly brand the object as a modern forgery, since aluminum metal was first made only at the end of the nineteenth century. Similarly, the detection of titanium white in a layer of paint would clearly label the painting as recent, or recently restored, since titanium white (titanium dioxide) is a pigment that was first used only during the early twentieth century (Fleming 1975). In the following pages are discussed some technical and scientific tools by which material evidence as to the authenticity, or otherwise, of ancient objects can be established.

Material Evidence

Surface Erosion Marks. Surface smoothing and finishing operations always leave distinctive erosion marks on the surface of objects that can sometimes be seen by the naked eye and are clearly discerned under amplification. It is particularly difficult to remove the marks left by carving and abrasion or by burnishing tools used for finishing surfaces in such a manner that they cannot be detected even after a long period of use. Studying the nature of surface erosion marks under a low-magnification microscope, for example, generally allows one to differentiate between ancient and modern worked surfaces: random or subparallel marks usually indicate that the surface was produced with the help of pointed tools or by other crude means, while parallel abrasion marks mostly indicate that a surface was finished using modern tools such as rasps, files, or cutting and polishing wheels.

Cracks. Cracking is a frequent form of deterioration of ancient paint layers, and cracks on painted surfaces are widely regarded as definite evidence of age and authenticity. The cracks are usually the result of chemical and physical changes in the paint layers, which gradually lose elasticity and become increasingly stiff and unable to adapt to the expansion and contraction of the support backing the layer of paint. Cracks in painted surfaces are often caused by cycles, repeated over long periods of time, of relatively high and low temperatures. Forgers often induce the stiffening of paint layers by submitting freshly painted surfaces to short, repeated cycles of high and low temperatures; this treatment hardens and stiffens the layer (or layers) of paint that cannot follow the dimensional changes of the substrate and therefore initially crack and eventually break up. To enhance the ancient appearance of such paint layers, the cracks are usually also filled with dark dust. Differentiating between natural cracks and modern, forged ones is not easy, especially since forgers usually know the general outline of cracking patterns, which are different and characteristic for every substrate. Sometimes sharp or pointed instruments are used to make sham crack patterns, although such cracks are not sufficiently uniform to escape detection, particularly when observed with magnifying instruments.

Patina. The surface of many materials, such as stone, metals, and alloys that remain exposed or buried for long periods of time often undergoes chemical alteration, which results in the formation of a layer of decay products generally known as *patina*. Some of the characteristics of the patina, such as its composition, are related to the place of exposure or burial, while others, such as, its thickness, is generally determined by the length of time that the surface covered by the patina has been exposed or buried. Studying the nature and structure of layers of patina contributes, therefore, to ascertaining the genuineness of the layers as well as of the object it covers, whether ancient or otherwise. In fact, examination of most ancient metals and alloys, minerals, glass, ivory, and some ceramics generally begins with a study of their patina. Nevertheless, a false patina should not be considered as prima facie evidence of forgery, since a layer of false patina is often applied by conservators to genuine antiquities from which the original patina may have been removed during cleaning. A false patina should, however, be removed before an object is either authenticated or denounced as a fake (Craddock 1992; Lewin and Alexander 1967, 1968).

Metal Casting Techniques. Many ancient cast metal objects were made by the *cire perdue* (lost wax) casting process, which involves pouring molten metal into a one-piece mold and letting it solidify; modern fakes are usually cast in two halves that are then joined. A casting fin, or a fine line of filed solder on a cast object, usually reveals that the casting is modern.

Scientific Examination

Ultraviolet Examination. When the surface of many materials is illuminated by *ultraviolet light*, it emits *fluorescence* (see Textbox 22). Since no two materials exhibit the same form of fluorescence, the inspection of surfaces under ultraviolet light is often appropriate for determining the nature of the surface; differences between an original surface, repaired areas, patches, and repaints are normally clearly shown by differences in fluorescence; the technique is not suitable, however, for detecting entire fakes. Objects made of marble that are from different sources or that have undergone different weathering processes, for example, even though appearing similar in daylight, emit different fluorescence. Thus, areas of contrasting fluorescence observed on the surface of marble sculptures illuminated by ultraviolet light may reveal that parts of the marble were either replaced during restoration or faked. Also, differences between the fluorescence of genuine and manipulated patina on bronze objects, and of different pigments of the same color on painted surfaces, provide clues for differentiating between them and identifying their nature.

Radiography. *Penetrating radiation* is absorbed to varying degrees by different materials, and this property makes them useful for examining (*radiographing*) solid objects (see Textbox 11). The shape of wrapped or encased objects, as well as the internal structure and condition of solid objects, can usually be revealed using conventional X-rays radiography. Extremely large objects, or objects to which X-rays equipment is not accessible, can be radiographed using the *gamma-rays radiography* technique, while *neutron radiography* is useful for revealing the structure of objects made of organic materials. The information provided by radiographs could otherwise be obtained only by damaging the specimens examined. Consequently, radiographs are of much help for establishing the internal structure of ancient objects, authenticating questionable ones, and detecting forgeries. Thus have been radiographed a wide variety of archaeological objects and structures, including metal casts and mummies. The X-rays radiography of a small mummy found in the tomb of an Egyptian high priestess from the twenty-first dynasty, for example – which until radiographed was believed to be that of her daughter – revealed that it was actually of a baboon.

Occasionally, because of the large dimensions of some objects and/or the high density of the material from which they are made, X-rays may not be of sufficient energy to penetrate them, and their radiographs may provide no useful information. In such cases the gamma rays of highly energetic radioactive isotopes provide penetrating radiation of sufficient energy to reveal hidden details and/or unexpected material components.

Chemical Analysis. The chemical composition of ancient objects is important for their authentication. The nature as well as the relative amounts of major, minor, and trace elements in any object are of use for determining the authenticity or otherwise of ceramics, glass, or alloys. A wide range of analytical techniques, depending on the nature of the material studied, have been used for this purpose, including X-rays fluorescence analysis, mass spectrometry, atomic absorption spectroscopy, and neutron activation analysis, to mention only a few of the most widely used methods. Of special importance for examining small or highly valued objects that because of their rarity or cultural value should not be damaged, are the nondestructive techniques, which enable one to examine and characterize objects without destroying any visible surface blemishes, material or living (see Textbox 11).

Neutron activation analysis, for example, a nondestructive analytical technique (see Textbox 10), was used to characterize each one of a collection of gold coins, including some suspected as fakes; analyzing the elemental composition of only an extremely small area of the surface of each coin and then calculating the relative weight ratios of silver and copper to that of gold in the coin showed that most coins fell into one of two well-defined groups, each group having definite and characteristic weigh ratios; those suspected as fakes had however very different weight ratios of these metals, which clearly distinguished them from the genuine coins (Keisch 1972).

Isotopic Evidence. The study of the *isotopic ratios* of the stable isotopes of carbon and oxygen in the patina of bronze objects has been used to characterize natural patinas and recognize artificially made ones. This technique involves comparing the weight relationship between the amounts of the isotopes oxygen-18 and oxygen-16 or those of carbon-13 and carbon-12 in the patina. Determining these weight ratios in malachite, in genuine patina of ancient bronze, and then comparing the results with those obtained in suspected patina from a similar origin provided sufficient evidence to establish the authenticity or otherwise of suspected patina: in artificial patina the isotopic ratios of both these elements was well outside the range of values found in ancient layers (Smith 1978).

Dating. Radiocarbon dating generally provides reliable dates of organic materials. As for the age of objects: ages determined with the radiocarbon dating method are not a convincing proof of antiquity, since the method yields information only about the age of the materials and not the time when the materials were worked. Modern forgers have therefore learned to use ancient wood for their work, just as they have learned to use ancient materials for most of their work. It is not likely, however, that forgers before the 1950s, when the method of radiocarbon dating was developed, would have taken the precaution of using appropriate ancient materials.

Thermoluminescence measurements on a few milligrams of material scraped from fired-clay objects, such as pottery and figurines, can readily establish whether objects made of these materials are genuine antiquities, recent copies, or fakes. Chemical tests on what appeared to be a broken and restored terracotta statuette, purporting to be of Etruscan origin (from approximately the fifth century B.C.E.), for example, revealed that it could have been made in Italy but gave no hint as to its age. Although thermoluminescence measurements indicated that all the pieces that made up the statue were of the same age (ca. 2400 years old), each piece had formerly been part of different statues, and the "reassembled, pastiche statue" was a modern forgery.

18.2. SOME AUTHENTICATION STUDIES

The Piltdown Man

An illuminating example of a longstanding authentication controversy settled by the use of scientific methods is the by now historical case of the Piltdown Man (Spencer 1990; Weiner 1981). Fragments of a human braincase and a jawbone, together with teeth of various mammals and a tool carved from an elephant tusk, were reportedly found at Piltdown in Sussex, England, in 1912; a thorough anthropological examination, it was claimed, revealed that the human bones belonged to a 500,000-year old, hitherto unknown fossil ancestor of human beings named *Evanthropus dawsoni* or Piltdown Man.

Some anthropologists at the time could not accept, however, that both the braincase and the jawbone belonged to a single individual, since the jawbone appeared to belong to an ape while the braincase was more human. The controversy that hence arose occupied the scientific world for over 45 years.

Fluorine analysis of the Piltdown bones made in 1949, showed that neither the braincase bones nor the jawbone contained more than minute traces of fluorine, whereas other fossil bones from the same gravel bed contained a great deal of it. This reduced the possible age of the skull to below 50,000 years. A more refined analytical technique revealed, however, that the amount of fluorine in the braincase was sufficient to account for its being ancient, but that the jawbone and teeth contained no more fluorine than did modern bones (Oakley and Hoskins 1950). Other chemical analyses confirmed that the jawbone and teeth contained the same amount of nitrogen as did modern species. The braincase, on the other hand, contained much less. Used as a crosscheck against the fluorine tests, this provided quite conclusive evidence that the jawbone was modern; additional tests that followed, confirmed that the jawbone really was modern. Electron microscopy revealed preserved fibers of organic tissues in the jawbone, in contrast to the braincase, where there were no traces of such fibers. So the jawbone was exposed as a bogus fossil of undoubted modern origin (probably from a partly grown female orangutan, a conclusion supported by independent anatomical investigations) deceitfully buried at the Piltdown site.

X-rays analysis of the mineral structure of the fragments of the braincase revealed that the phosphate mineral component of the bones had partly converted to gypsum, a mineral found in neither human nor other mammal bones, nor in the Piltdown gravel. It had previously been shown, however, that if a partially fossilized bone were stained with an acid solution of an iron salt (which tints it brown), part of the bone would convert to gypsum. Thus the Piltdown skull seemed to have been stained in this way, probably to match the gravel in which it was planted, hence providing an explanation of the otherwise inexplicable presence of gypsum. It was therefore concluded that the dark brown fragments of the braincase were also fraudulently introduced into the Piltdown gravel.

The supposed human remains were accompanied by flint implements and animal teeth, all of which apparently supported the early date attributed to the find. The flint implements were of a reddish color, matching that of local flints, but spectrographic analysis revealed that they had been artificially stained with chromium and iron salts; below the layer of stain there was a white crust, whereas local flints were brown throughout. The animal teeth also seemed to have been "planted" at the burial site so as to suggest an early date; the red-brown color of the teeth had also been artificially stained: chromium and iron had been found when they were analyzed. A hippopotamus tooth, previously supposed to be contemporary with the tool carved from the elephant's tusk, was also stained with chromium and iron salts and contained little fluorine.

When the Piltdown Man finds were finally dated by the radiocarbon dating method, the braincase was proved to be probably less than 700 years old and the jawbone even younger, about 500 years old. Thus a large body of scientific evidence established, beyond any doubt, that the Piltdown Man was a bogus human ancestor, and that the whole group of teeth and implement were of different chronologic and geographic origin; they had all been fraudulently introduced into the Piltdown gravel.

In conclusion, thorough investigation revealed that the Piltdown find contained the following:

a human skull dated as medieval, about 620 years old
an orangutan jawbone, around 500 years old, probably from Sarawak
an hippopotamus tooth, genuine fossil, probably from Malta or Sicily
an elephant tooth, genuine fossil, probably from Tunisia
a canine tooth

An interesting postscript to this extraordinary story was the disclosure that the Piltdown Man seems to have been a trap set by one eminent scientist to another.

Francis Drake's Brass Plate

An inscribed thick plate of brass attributed to the landing, in 1579, of Francis Drake on the coast of California, is retained in safekeeping at the University of California, Berkeley. Since its discovery, in the San Francisco Bay area in 1936, however, there have been doubts about the authenticity of the plate, although an early chemical study had apparently confirmed its authenticity. Regardless of this initial study, doubts about the origin of the plate persisted, and a new study, based on the composition of the brass as determined by neutron activation, X-rays fluorescence, and atomic absorption analysis was initiated to reevaluate the earlier authentication of the plate. The results of this study were then compared with the composition typical of brass from Drake's time as well as from modern brass, and it was then concluded that the plate was probably made during the latter part of the nineteenth century or the early years of the twentieth century (Hedges 1979).

The Greek Bronze Horse from the Metropolitan Museum

A bronze statuette of a horse just over 40 cm tall, owned by the Metropolitan Museum of New York, was considered for many years to be 2400 years old, and one of the finest examples of early Greek art. A magnetic study indicated the presence of iron inside the figure of the horse, prompting radiographing of the statuette. The X-ray radiograph indeed disclosed inside an iron armature and a casting fin around the sculpture. The presence of the casting fin and the use of iron wire inside, intended to provide hollow space within the object, were accepted as combined evidence that the statuette had been cast by a technique apparently not practiced until much later than the period to which the statuette had been assigned. As a consequence of this study, the horse was exposed as a modern forgery.

The authenticity of the horse was still supported by some, however, on the grounds that examples of genuine Greek bronzes containing iron armatures are known (Blumel 1969). Also, the composition and structure of the bronze used to make the casting and the nature and extent of the corrosion layer were said to be consistent with the statuette being authentic. The statuette had a solid core made up of sand, clay, calcite, and mineral inclusions that would have been heated during casting operations. Such a solid is suitable for thermoluminescence dating and could provide evidence that might answer the question regarding the statuette's real age; particularly since it was also determined that at the time the statuette was cast the core would had been heated sufficiently to remove all stored geologic thermoluminescence in the temperature range of interest (Zimmerman et al. 1974).

Previous radiographic work on the horse would, however, have given the core an artificial radiation dose that would have interfered with a reliable thermoluminescent date. As luck would have it, a sizable sample of the core material had been removed before most of the radiographic work on the statuette was carried out, and the thermoluminescence dating results proved that the horse was, in fact, made in antiquity. Its thermoluminescence age was determined as being somewhere within the 2000–2500 years B.P. range; this is sufficiently accurate to prove that the horse is not a modern forgery, but extends over too wide a period of time to prove with certainty whether the statuette is a classical Greek sculpture of the fifth century B.C.E. or a more recent creation, from the Hellenistic or Roman period (Lefferts et al. 1981).

The Shroud of Turin

The Shroud of Turin is an ancient piece of sepia-yellow linen cloth, in a remarkably good state of preservation, over 1 m wide and about 4 m long, which bears a faint frontal image and a dorsal image of a human being. Stains of a dark red or rust color, with the appearance of blood, are seen in areas consistent with the New Testament's account of the scourging and crucifixion of Christ. Many millions of people around the world believe that the image on the cloth is that of Jesus of Nazareth, and that the shroud was used to wrap his body after he was crucified and died. For this reason the shroud has become probably the most intensively studied object in the history of humankind; a large number of scientists have jointly dedicated thousands of years to its examination. The shroud was first exhibited in France in 1357; it then traveled through many places until 1578, when it was taken to Turin, Italy, where it is presently kept in the Chapel of the Shroud. Over the years the shroud has become the object of mass veneration by believers but of scorn and contempt on the part of skeptics.

The scientific evidence from the many studies done since 1975 seem to indicate that the shroud is a real burial cloth with the image and bloodstains of a crucified human; no stains seem to have been made by an artist or faker (Iannone 1998; Heller 1983). Radiocarbon dating of the linen cloth at three different laboratories yielded, however, dates that are in exceptionally good agreement with each other, varying between 646 ± 31 and 750 ± 30 years before present. The linen of the shroud is, therefore, medieval, originating sometime between 1260 and 1390 C.E. and not at the beginning of the Christian era (Gove 1996; Damon et al. 1989). Some scientists, however, dispute these findings, claiming that a fire, during the 16th century, may have introduced extra carbon into the cloth and thus have made the date appear more recent than it really is.

Many more studies have been initiated in attempts to explain the discrepancy between the belief of millions of people and the evidence provided by the radiocarbon dating technique on the age of the shrouds's cloth. Such studies, however, have resulted only in still further studies and a persistent controversy on the real nature of the relic.

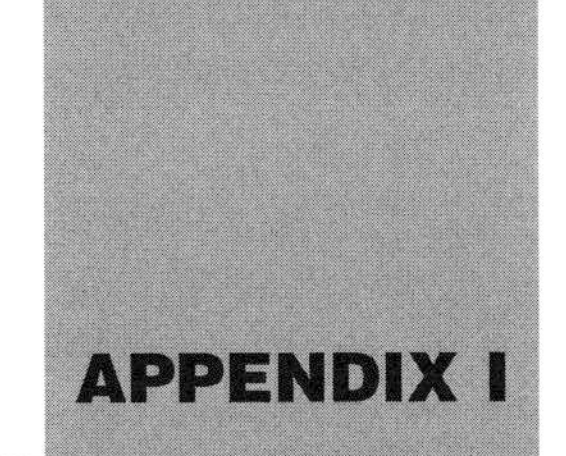

THE CHEMICAL ELEMENTS

APPENDIX I

The chemical *elements* are substances that cannot be decomposed into simpler substances by chemical means, but combine with other elements to form new substances. All *substances* and *materials* other than the elements are made up from the combination of two or more elements. Thus, either singly or in combination with others, the elements constitute all matter. Well over 100 elements occur in nature, but only 92, listed below, occur in the planet earth.

The smallest unit having the chemical properties of the element are the *atoms*. All atoms are made up from a number of *elementary particles* known as the *protons, neutrons,* and *electrons*. The *protons* and *neutrons* make up an *atomic nucleus* at the center of the atom, while the *electrons*, distributed in *electron shells*, surround the atomic nucleus. The atoms of each element are identical to each other but differ from those of other elements in *atomic number* (the number of protons in the atomic nucleus) and *atomic weight* (their weighted average mass) as listed in the table below.

Atomic number	Symbol	Name	Weight	Atomic number	Symbol	Name	Weight
1	H	Hydrogen	1.0	12	Mg	Magnesium	24.3
2	He	Helium	4.0	13	Al	Aluminum	26.9
3	Li	Lithium	6.9	14	Si	Silicon	28.1
4	Be	Beryllium	9.0	15	P	Phosphorus	30.9
5	B	Boron	10.8	16	S	Sulfur	32.1
6	C	Carbon	12.0	17	Cl	Chlorine	35.5
7	N	Nitrogen	14.0	18	Ar	Argon	39.9
8	O	Oxygen	16.0	19	K	Potassium	39.1
9	F	Fluorine	19.0	20	Ca	Calcium	40.1
10	Ne	Neon	20.2	21	Sc	Scandium	44.9
11	Na	Sodium	23.0	22	Ti	Titanium	47.9

Archaeological Chemistry, Second Edition By Zvi Goffer

Atomic number	Symbol	Name	Weight	Atomic number	Symbol	Name	Weight
23	V	Vanadium	50.9	58	Ce	Cerium	140.1
24	Cr	Chromium	52.0	59	Pr	Praseodymium	140.9
25	Mn	Manganese	54.9	60	Nd	Neodymium	144.2
26	Fe	Iron	55.8	61	Pm	Promethium	[145]
27	Co	Cobalt	58.9	62	Sm	Samarium	150.4
28	Ni	Nickel	58.7	63	Eu	Europium	151.9
29	Cu	Copper	63.5	64	Gd	Gadolinium	157.2
30	Zn	Zinc	65.4	65	Tb	Terbium	158.9
31	Ga	Gallium	69.7	66	Dy	Dysprosium	162.5
32	Ge	Germanium	72.6	67	Ho	Holmium	164.9
33	As	Arsenic	74.9	68	Er	Erbium	167.3
34	Se	Selenium	78.9	69	Tm	Thulium	168.9
35	Br	Bromine	79.9	70	Yb	Ytterbium	173.0
36	Kr	Krypton	83.8	71	Lu	Lutetium	174.9
37	Rb	Rubidium	85.5	72	Hf	Hafnium	178.5
38	Sr	Strontium	87.6	73	Ta	Tantalum	180.9
39	Y	Yttrium	88.9	74	W	Tungsten	183.9
40	Zr	Zirconium	91.2	75	Re	Rhenium	186.2
41	Nb	Niobium	92.9	76	Os	Osmium	190.2
42	Mo	Molybdenum	95.9	77	Ir	Iridium	192.2
43	Tc	Technetium	97.9	78	Pt	Platinum	195.1
44	Ru	Ruthenium	101.1	79	Au	Gold	196.9
45	Rh	Rhodium	102.9	80	Hg	Mercury	200.6
46	Pd	Palladium	106.4	81	Tl	Thallium	204.3
47	Ag	Silver	107.9	82	Pb	Lead	207.2
48	Cd	Cadmium	112.4	83	Bi	Bismuth	208.9
49	In	Indium	114.8	84	Po	Polonium	[209]
50	Sn	Tin	118.7	85	At	Astatine	[210]
51	Sb	Antimony	121.8	86	Rn	Radon	[222]
52	Te	Tellurium	127.6	87	Fr	Francium	[223]
53	I	Iodine	126.9	88	Ra	Radium	[226]
54	Xe	Xenon	131.3	89	Ac	Actinium	[227]
55	Cs	Cesium	132.9	90	Th	Thorium	232.0
56	Ba	Barium	137.3	91	Pa	Protactinium	[231]
57	La	Lanthanum	138.9	92	U	Uranium	238.0

The atomic weight of atoms of the same element may differ, and atoms of the same element, having an equal atomic number but different atomic weights are said to be *isotopes*. Some elements occur in only one isotopic form, whereas others have many isotopes. Some isotopes, known as *radioactive isotopes*, are unstable: they undergo *radioactive decay* and transmute into other elements.

A convenient way of listing all the elements is the *periodic table*, in which they are grouped according to their chemical properties. As early as in 1869, when only 63 elements were known, a Russian chemist, Dimitri Mendeleev, devised a periodic table based on the periodic pattern of their properties. He noticed that there were yet unknown elements with portended properties for which he left empty spaces in the table. The elements for which Mendeleev predicted their properties were discovered within a few years of the publication of the table, fitting exactly into the empty spaces. The modern periodic table (see Fig. 87) includes 112 elements, only 92 of which are naturally found in the earth. As seen in the figure, the modern table has seven rows, called *periods*, and eight columns, called *groups*. In the horizontal periods, the elements are listed in order of increasing atomic number. Elements listed in the vertical groups have similar chemical properties. The position of an element on the periodie table thus provides useful information about its structure and chemical properties. Four families of elements (identified in the table by shade) that share many properties can be recognized: *metals*, *metalloids*, *nonmetals* and *noble gases*.

The *metals* are typically shiny and good conductors of heat and electricity; most of them are *ductile* and *malleable*, so that their shape can be easily changed and formed into thin wires or sheets. When exposed to the environment, most metals, except the noble metals, *corrode*, gradually wasting

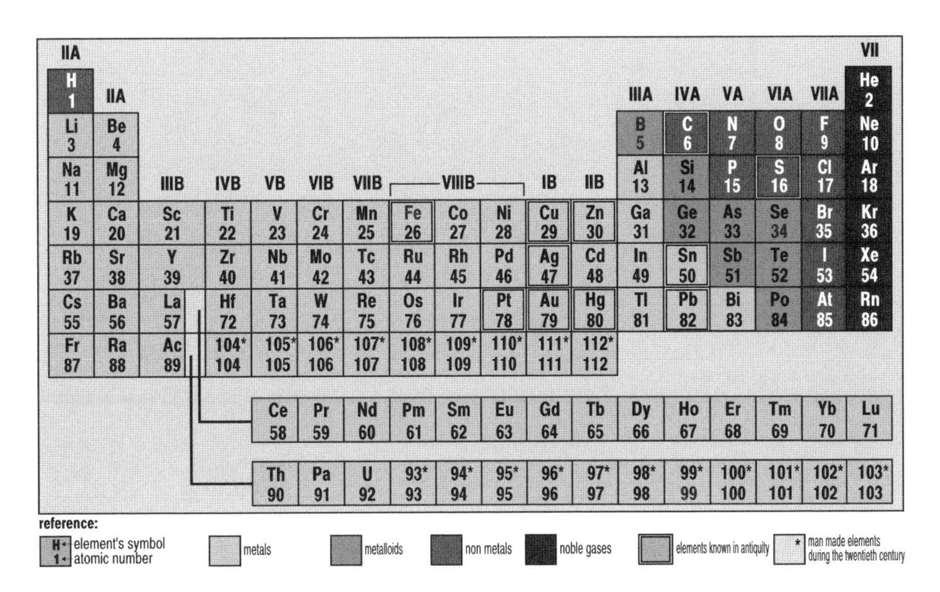

FIGURE 87 The periodic table of the elements.

away. The atoms of the metals tend to lose one or more negatively charged electrons and form positively charged particles (ions), known as *cations*.

The *metalloids*, also known as *semimetals*, have the characteristics of both the metals and the nonmetals: they conduct electricity and heat better than do the nonmetals, but not as well as the metals do, and they can be shiny or dull; their properties are therefore between those of the metals and the nonmetals.

The *nonmetals* are poor conductors of electricity and heat, and their surfaces generally are dull. Their atoms tend to acquire one or more electrons and become negatively charged ions, known as *anions*. Cations and anions combine with each other, forming the myriad of compounds that make up most of the universe.

The *noble gases* have a tendency for neither acquiring or losing electrons and do not react chemically.

Sixty-seven elements, almost three-quarters of the 92 that occur naturally in the earth, are metals; 7 are metalloids, 12 are nonmetals, and 6 are noble gases. Only 8 metals and 2 or 3 nonmetals were recognized in antiquity.

CHRONOMETRIC DATING METHODS
SELECTION CRITERIA

Dating methods are used for placing ancient materials, objects or events, within a timescale. Two basically different types of dating methods are recognized: *relative* and *absolute dating*. *Relative dating* methods reveal the temporal order of a sequence of materials, objects or events, disclosing whether these occurred before, contemporarily or after other materials, objects or events. *Absolute*, or *chronometric dating* methods reveal the age, measured in calendar years, of materials, objects or events. Chronometric dating methods (see Fig. 88) make use of a variety of physical or chemical measurements to ascertain the time when events occurred or when materials and objects were made, used, or altered. Chronometric dates are not exact dates, but numerical age estimations and are generally expressed as a range of dates qualified by an indication of the uncertainty of the dates.

When selecting a method of dating, the nature of of the material to be dated and the estimated age to be determined must be considered. Only wood, for example, can be dated by dendrochronology; all organic materials, can be dated with the radiocarbon and many with the amino acid racemization methods. Coral, bones, teeth, and shell can also be dated using the electron spin resonance method. Pottery and other burned-clay materials that contain crystalline solids are usually dated using the thermoluminescence technique. Glassy minerals, such as mica and obsidian, can be dated by employing the fission track method. The potassium–argon method is used to date volcanic rock and ash and the uranium series method, for dating such materials as travertine, marl, coral, mollusc shells, bone, and teeth, which are formed through precipitation processes (Schwarcz 2002; Smart and Frances 1992).

As for the estimated age: two chronometric methods, radiocarbon and dendrochronological dating, enable the accurate dating of materials or/and events in the timespan between the present and 10,000 years ago. The

(a): Feasibility of dating methods

dating method	bone	burnt flint	glass	igneous rock	obsidian	pottery	shell	tooth	wood
amino acid racemization	+					+	−		
dendrochronology									+
electron spin resonance		−		−			−	−	
fission track		+	+	+					
obsidian hydration					+				
potassium argon				+					
radiocarbon						−	+		+
thermoluminescence	+								
uranium series	−		+					−	

note: (+) feasible (−) occasional inconsistent results

(b): Time range of dating methods

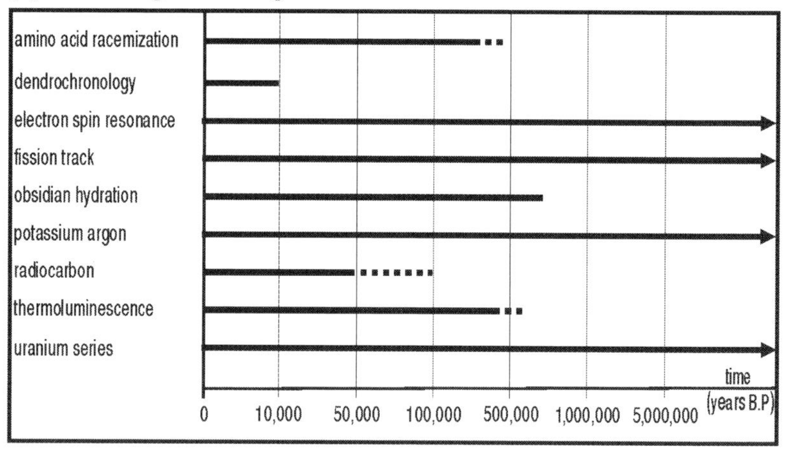

FIGURE 88 Dating methods. Shortly after the discovery of radioactivity, at the beginning of the twentieth century, it was found that the decay of radioactive elements could be used to keep track of time. Many of the dating techniques developed since then are, therefore, based on radioactive decay phenomena, but others, such as the hydration of obsidian, amino acid racemization, and dendrochronology, are based on other physical, chemical, or biological phenomena.

radiocarbon method is also applicable for dating materials older than 10,000 and as much as about 50,000 years. The amino acid racemization, electron spin resonance, fission track, and thermoluminescence methods are effective for dating materials within the time range 10,000–100,000 years B.P. For materials older than 100,000 years, potassium–argon, uranium series, and fission track dating methods are the most widely used; refinements in the amino acid racemization, thermoluminescence, and electron spin resonance methods, however, have also made them sometimes applicable for this early time range.

The evolution of humans and society can be broadly divided into three periods: (1) before one million years before present, when early hominids developed into anatomically modern humans; (2) from one million to about 10,000 years before present, when modern humans developed and dispersed through the world; and (3) between 10,000 years before present and the present time, the period when humans settled, social communities emerged, and organized societies developed. Following this classification, the relevance of the dating methods for the study of these periods is listed below.

Dating method	Period
Electron spin resonance	1,2,3
Fission tracks	1,2,3
Obsidian hydration	1,2,3
Potassium argon	1,2,3
Uranium series	1,2,3
Amino acid racemization	2,3
Radiocarbon	2,3
Thermoluminescence	2,3
Dendrochronology	3

SYMBOLS, CONSTANTS, UNITS, AND EQUIVALENCIES

Symbols

Property, unit, or term	Symbol and/ or value	Property, unit, or term
Magnitude		Length
Mass	m	Meter (m)
Time	t	Kilometer (km) = 1000 m
Velocity	v	Centimeter (cm) = 0.01 m
Force	F	Millimeter (mm) = 0.001 m
Wavelength	λ	Micrometer (μm) = 10^{-6} m
Frequency	ν	Nanometer (nm) = 10^{-9} m
Others		Ångstrom unit (Å) = 10^{-10} m
Equal	=	Mass and weight
Greater than	>	Gram (g)
Smaller than	<	Kilogram (kg) = 1000 g
Plus or minus	±	Milligram (mg) = 0.001 g
Constants		Microgram (μg) = 10^{-6} g
Speed of light	3×10^{8} m/sec	Pressure
Speed of sound (in air)	343 m/sec	Atmosphere = 760 mm mercury
Mass of electron	9.1×10^{-31} kg	Frequency
Mass of proton	1.7×10^{-27} kg	Hertz (Hz) = 1 cycle/second
Mass of neutron	1.7×10^{-27} kg	

Archaeological Chemistry, Second Edition By Zvi Goffer

Large and Small Numbers: Expression, Units, and Prefixes

Expression		
Exponential	Decimal	Prefix (and symbol)
10^{12}	1,000,000,000,000	tera (T)
10^{9}	1,000,000,000	giga (M)
10^{6}	1,000,000	mega (M)
10^{3}	1,000	kilo (k)
10^{2}	100	hecto (h)
10^{1}	10	deca (da)
10^{-1}	0.1	deci (d)
10^{-2}	0.01	centi (c)
10^{-3}	0.001	milli (m)
10^{-5}	0.000001	micro (μ)
10^{-9}	0.000000001	nano (n)
10^{-12}	0.000000000001	pico (p)

GLOSSARY

A Abbreviation for *adenine.*

Å Abbreviation for *Ångstrom unit.*

AAR dating Abbreviation for *amino acid racemization dating.*

AAS Acronym for *atomic absorption spectroscopy.*

abaca A strong *bast fiber* derived from Manila hemp (*Musa textilis*) plants.

abiotic Nonliving, not of a biological nature.

abrasion The mechanical process of wearing down a solid surface.

abrasive A more or less finely comminuted hard and brittle *solid* that, by scraping or rubbing, wears away the surface of other solids.

absolute age See *age, absolute.*

absolute date See *date, absolute.*

absolute dating See *dating, absolute.*

absorption The process of absorbing (taking up) something, such as some form of *energy* by *matter* or of gaseous or liquid matter by a solid.

absorption spectrum See *spectrum, absorption.*

accelerator mass spectrometer (AMS) See *spectrometer, accelerator mass.*

accelerator mass spectrometry radiocarbon dating See *dating, accelerator mass spectrometry, radiocarbon.*

accretion The increase in the volume of a solid body by the addition of new matter during *crystallization* and/or *precipitation* processes.

accuracy The extent to which a measurement is free from *errors.*

achromatin Together with *chromatin,* one of the two portions of the *nuclei* of living *cells.*

acid (1) One of a large class of substances that, when dissolved in water, have some common properties, such as a sour, bitter taste, and react with *metals* and *bases* to form *salts*; (2) a *solution* whose *pH* is less than 7; the stronger the acid the lower its pH. Acid solutions whose pH values varies between 2.5 and 0 are said to be

Archaeological Chemistry, Second Edition By Zvi Goffer

strong acids, while those with *pH* values that vary between 6 and 4 are said to be *weak acids*. The most widely used acids in antiquity were the juices of some fruits, sour milk, and *vinegar*.

acid rain Rainwater that has a *pH* value below 5.0.

activation analysis See *neutron activation analysis*.

additive See *temper*.

additive color See *color, additive*.

adenine (A) A *nitrogenous base*, one of the constituents of *DNA* and *RNA*.

adenosine triphosphate (ATP) A relatively stable *nucleotide* whose high-energy *molecule* is used by living organisms to fuel *chemical reactions* within their *cells*.

adhesive A *substance* or preparation used to bond objects together or to *size* surfaces; two basic types of adhesives were used since prehistoric times and until the beginning of the twentieth century: *inorganic* adhesives known as *stony cements* and *organic* adhesives, which include *glues* and *gums*.

adjective dye See *dye, mordant*.

adobe The Spanish word for a brick or cement made of a mixture of *clay*, straw, and water.

adsorption The take-up of *molecules* of a *gas* or *liquid* on the surface of a *solid*.

aerobic An environment or a process associated with free oxygen; see *anaerobic*.

aerosol A *suspension* of solid particles or liquid droplets in a *gas*.

affresco See *fresco*.

agate A semiprecious *gemstone* that consists of banded colored layers of the mineral *chalcedony*.

agate glass See *glass, agate*.

age, absolute The age of an object, feature or event, determined by scientific evidence; it is usually expressed in years *before present* (B.P.).

age, relative The age of an object, feature, or event, expressed in a timescale related to that of other objects, features, or events; see *absolute date*.

aggregate A mass of *mineral* grains, such as of *gravel* or *sand*, consolidated into solid *rock* by a *cementing* material.

aging The change of the *physical* and/or *chemical properties* of *materials* with time; the aging process is accelerated by increasing temperatures.

air-dry wood See *seasoned wood*.

alabaster A relatively compact, soft, and *translucent* form of *gypsum*, widely used for carving decorative objects; see *oriental alabaster*.

alabaster, oriental A soft and translucent form of *calcite* or *aragonite*, also known as *onyx marble*, used (in ancient Egypt) for carving decorative objects; see *alabaster*.

alcohol One of a large group of *organic substances*, which include in their *molecules* a *hydroxyl* (OH^-) group of *atoms*; the term is commonly used to refer to *ethyl alcohol*.

alcohol, ethyl See *ethyl alcohol.*

alpha recoil-track dating See *dating, alpha recoil track.*

alga A class of one-celled, vegetable organisms.

alkali (1) One of a large class of substances also known as *bases* that, when dissolved in water, form *solutions* having some common properties such as a bitter taste, being *caustic,* and reacting with *acids* to form *salts;* (2) an aqueous solution whose *pH* value is more than 7; the stronger the alkali, the higher its pH. Alkalies whose solutions have high *pH* values, between 10 and 14, are said to be *strong alkalies,* while those whose solutions have low *pH* values, between just over 7 an up to 10 are said to be *weak alkalies.*

alkaline A *solution* having *pH* greater than 7.

alkaline glaze See *glaze, alkaline.*

allotrope One of the forms of a *chemical element* with characteristic *physical properties,* different from those of the other allotropic forms of the same element.

alloy A *mixture* of two or more *elements* that has *metallic properties* and one of the elements, at least, is a *metal.*

alloy, cast An *alloy* that is suitable for *casting.*

alloy, ferrous An *alloy* composed mainly of iron and smaller amounts of carbon; *steel* and *wrought iron,* for example, are ferrous alloys.

alloy, nonferrous See *nonferrous metals and alloys.*

alloying The process of making *alloys* by mixing other *metals* or *nonmetals* with a molten metal.

alluvial A *sediment* deposited by a stream or river.

alpaca (1) The thin, long, and strong hair *fibers* of alpaca (*Lama pacos*) mammals; (2) an *alloy* (also spelled *alpacca*) composed mostly of copper (about 60%); the remainder is nickel (less than 20%), zinc (less than 20%), and tin (about 5%); it has been used as a substitute for silver.

alpacca See *alpaca.*

alpha decay A process of *radioactive decay* during which *alpha particles* are emitted from the *nuclei* of unstable *isotopes.*

alpha particle A subatomic *particle* emitted from the nuclei of some *radioactive isotopes;* they consists of two *protons* and two *neutrons* bearing a positive electric charge.

alpha recoil track dating See *dating, alpha recoil track.*

alpha tin (α tin) A brittle *allotrope* of *tin,* stable below –18°C.

alum The common name of a group of *minerals* composed of mixed *sulfates* of aluminum and another metal (such as potassium); they have been used for *tanning* animal *skins* and *hides* and as *mordants;* for *dying yarns* and textile fabrics.

aluminate A complex *anion* composed of an *atom* of aluminum and two of oxygen, to which the formula AlO_2^- is assigned; it is a constituent of many *minerals* and *sedimentary rocks.*

amalgam An *alloy* of mercury and one or more other *metals;* thus silver amalgam or gold amalgam.

amalgam gilding See *gilding, amalgam.*

amalgamation Making an *amalgam.*

amber A *translucent, fossilized resin* that varies in color from yellow to brown, derived from ancient trees, appreciated as a *gemstone.*

ameba A one-celled, generally microscopic living organism.

amiantus See *asbestos.*

amino acid One of a group of *organic substances*, the building blocks of the *proteins*, whose *molecules* include both *amino* and *carboxyl* groups of *atoms.*

amino acid, essential An *amino acid* required for the formation of *proteins* in vertebrate animals, but not produced by their bodies and therefore acquired in their diet.

amino acid, nonessential An *amino acid* produced by the body of vertebrate animals and therefore not required in their diet.

amino acid racemization dating (AAR dating)

amino group An *anion* found in *amino acids;* it consists of a nitrogen *atom* bonded to two hydrogen *atoms* ($-NH_2$).

amorphous A *solid* that has no *crystal structure.*

amplification The production of multiple copies of a segment of *DNA.*

amplitude The maximum displacement of a *wave* from its mean value.

AMS Acronym for *accelerator mass spectrometer.*

AMS ^{14}C dating Abbreviation for *accelerator mass spectrometry radiocarbon dating.*

AMU Acronym for *atomic mass unit.*

anaerobic An environment or a process in which there is no free oxygen; see *aerobic.*

analog A device or system that presents information in the form of continuously varying quantities, as opposed to a *digital* device or system.

analysis, chemical The identification of *substances* and their distinct compositions.

analysis, elemental The identification of the chemical *elements* and determination of their relative amounts in a *substance.*

analysis, gravimetric The generic name for *classical methods* of *quantitative analysis* based on the measurement of weight; see *volumetric analysis.*

analysis, qualitative The identification of a solid, liquid, or gaseous *substance* or a mixture of substances.

analysis, quantitative The determination of the precise relative amounts of the components of a sample, based on its *chemical* and/or *physical properties.*

analysis, volumetric The generic name for *classical methods* of *quantitative analysis* based on the measurement of the volume of *solutions*; see *gravimetric analysis.*

angora The fine, long, and straight *hair fiber*, also known as *mohair*, derived from the angora goat.

Ångstrom unit (Å) A unit of length, one ten-millionth of a millimeter or one ten-thousandth of a *micrometer* long.

anhydrous Without water.

animal fibers *See fibers, animal.*

anion An *ion* having a negative electric charge.

anisotropic *Matter* that exhibits different property values when tested along different directions; see *isotropic.*

annealing A *thermal process* for the removal of internal stresses and/or the hardness from objects made of *glass, metals,* or *alloys;* it entails heating the object, holding it for some time at a relatively high temperature, and finally cooling to ambient temperature at a relatively slow rate.

annual growth ring A growth layer of *wood* produced by a tree in a single year, as seen on a transverse section of a trunk or branch; see *dendrochronology.*

anode The positive terminal on an electric source or circuit; see *cathode.*

anthropogenic Produced by or related to human activity.

anthroposoil See *archaeological soil.*

antibody A *protein* produced by the body of animals and circulated in the bloodstream in response to foreign substances to help the body fight disease.

antimoniate A complex *anion* composed of *atoms* of antimony and oxygen and assigned the *formula* SbO_3^- or SbO_4^{3+}.

antimony oxide A *glass opacifier.*

antimony red A *mineral* composed of antimony pentasulfide, used as a red *pigment* and *stain.*

antiparallel Two objects that lie along a common line but in opposite directions to each other.

aplastic material See *temper.*

aqueous Watery.

aragonite A *mineral* composed of calcium carbonate, of which it is a *polymorph;* see *calcite.*

Ar–Ar dating Abbreviation for *argon–argon dating.*

archaeology The study of the human past based on its material remains.

archaeomagnetic dating See *dating, archaeomagnetic.*

archaeometallurgy Study of the *metals* and *alloys* used in antiquity and the ancient *metallurgical* processes.

argentarium An *alloy* of lead and tin containing 50% tin; it was used by the Romans for *tinning.*

argon-40 An *isotope* of the *inert,* gaseous *element* argon produced by the *radioactive decay* of the isotope *potassium-40,* used for *dating minerals* and *rocks.*

argon–argon dating (Ar–Ar dating) See *dating, argon–argon (Ar–Ar dating).*

arsenical bronze See *arsenical copper.*

arsenical copper An ancient *alloy* of copper and arsenic also known as *arsenical bronze.*

artifact An object used, made, or modified by human beings, such as *lithic, ceramic, metallic,* and *organic* artifacts.

asbestos A fire-resistant *mineral fiber,* also known by its Latin name, *amianthus.*

ash The *inorganic, noncombustible* residue left after *organic matter* has been burned.

ash, bone See *bone ash.*

ash glaze An ancient *glaze* in which *soda* or *potash* are used as *flux.*

asphalt A natural mixture of a smooth, brown to black *organic thermoplastic* solid and comminuted *minerals;* the organic solid is composed of a mixture of *hydrocarbons* that soften and become liquid at about below 60°C.

asphalt rock A *rock,* generally *sandstone* or *limestone,* naturally impregnated with *asphalt.*

assay A procedure for determining the amount of *precious metals* in *alloys;* three assaying techniques were known several centuries B.C.E.: *fire assay, specific gravity assay,* and *touching.*

asterism A starlike luminous effect that reflects light in some *gemstones,* like star *sapphires* and star *garnets.*

astringent A *substance* that causes living tissues to contract.

asymmetric carbon atom An *atom* of *carbon bonded* to four different *atoms* or *atomic groups* that cannot, therefore, be superimposed on its mirror image.

atmosphere (1) The gaseous envelope that surrounds the earth; (2) a unit of pressure equal to the force exerted by the weight of a vertical column of air in the atmosphere on an area of the surface of the earth at sea level; (3) the term is also used to refer to the gases in a heating device, such as a *kiln* or *furnace,* during *firing* or *smelting* operations; see *oxidizing atmosphere, reducing atmosphere.*

atmosphere, oxidizing See *oxidizing atmosphere.*

atmosphere, reducing See *reducing atmosphere.*

atmospheric pressure The force exerted by the *atmosphere* on a surface; at the sea level of the earth the average atmospheric pressure is one atmosphere (see *atmosphere, barometer*).

atom The smallest amount of an *element* that retains the properties of the element.

atom, radioactive See *radioactive atom.*

atomic absorption spectroscopy (AAS) See *spectroscopy, atomic absorption.*

atomic group A group of *atoms* held together by chemical *bonds.*

atomic mass The *mass* of an *atom* expressed in *atomic mass units.*

atomic mass unit (AMU) An internationally accepted unit of *mass* (weight) used to express the mass of *particles* and *atoms;* it is based on arbitrarily assigning a mass of 12 to the isotope carbon-12.

atomic number The number of *protons* in an atomic *nucleus;* each *element* has a characteristic atomic number.

atomic weight The relative weight of an *atom*, measured in *atomic mass units*; it is determined by adding the number of *protons* and *neutrons* in the atomic *nucleus*.

atomic weight unit The mass of one *proton* or *neutron*.

ATP Acronym for *adenosine triphosphate*.

austenite A solid *solution* of carbon in iron that is formed at high temperatures (above 750°C) in *cast iron* and in some *steels*.

authigenesis A *diagenetic* process characterized by the formation of new *minerals* within the voids or between the grains and particles of *sediments* or buried *material* remains.

autolysis The process whereby *cells* and *tissues* of dead organisms are decomposed by *autolytic enzymes* contained in them.

autolytic enzyme An *enzyme* in the *cell's* wall that causes the breakdown of the cell after death.

autoradiography The *imaging* of an object or part of an object using *energetic radiation*, such as *gamma rays*, emitted by *radioactive elements* within the object.

background radiation See *radiation, background*.

ball clay A *plastic secondary clay*, usually rich in *organic matter*.

barilla *Alkaline ash* derived from *calcined* Mediterranean marsh plants, either *Salsola soda* or *S. kali*; *S. soda* yields *soda*, *S. kali* yields *potash*.

bark The outer, protective layer of dead *cells* that surrounds the *wood* of the trunk and branches of trees.

bark-tanned leather *Leather tanned* with *tannins* derived from the *bark* of plants.

barometer An instrument for measuring *atmospheric pressure*.

barwood A *dye*; see *camwood*.

basalt A dark *igneous rock* composed mainly of a mixture of *silicate minerals* rich in magnesium and iron.

base (1) See *alkali*; (2) one of the main constituents of *DNA* and *RNA* known as *nitrogenous bases*, namely *adenine*, *cytosine*, *guanine*, *thymine*, and (only in RNA) *uracil*.

base, nitrogenous One of the nitrogen-containing *bases* – *adenine*, *cytosine*, *guanine thymine*, and *uracil* – that make up *DNA* and *RNA*.

base metal See *metal, base*.

basepair (bp) Two chemically bonded *nitrogenous bases*.

base sequence The order of *nitrogenous bases* in a *molecule* of *DNA*.

bast The flexible *phloem* tissue that makes up the stem of some plants; see *bast fibers*.

bast fiber See *fiber, bast*.

batch A mixture of a number of *materials* or objects that are handled together during a process.

batch processing The fabrication of *materials* or objects in *batches*.

batik A method of *dyeing cotton* textiles, original to Southeast Asia, in which *wax* is used to prevent the *dye* from penetrating *fibers* in wax-coated areas.

bating In *leather* technology, the removal of residual *dehairing* chemicals and non-leathermaking substances; an ancient bating process that entails immersing the skins in an *infusion* of *fermenting* chicken manure results in soft and supple skins; see *scudding* and *puering*.

beam A unidirectional flow of *particles*, such as *electrons* or *neutrons*, or *radiation* as, for example, *visible light* or *gamma rays*.

B.C.E. Before the common era, that is, before Christ.

bedrock The consolidated *rock* underlying loose matter, such as *sediments* or *soil* on the earth's surface.

before present (B.P.) A *date* measured in calendar years before the present time. By convention, when dating with *radiocarbon*, the year 1950 of the Gregorian calendar is assigned as the present.

benefication See *ore dressing*.

beta decay A process of *radioactive decay* during which *beta particles* are emitted from the *nuclei* of unstable *isotopes*.

beta particle A (negatively charged) *electron*, or (positively charged) *positron* emitted from an atomic *nucleus* undergoing *radioactive decay*.

beta particle radiography See *radiography*.

beta tin (β tin) A gray, powdery *allotrope* of tin that exists at temperatures lower than 18°C; also known as *tin pest* or *tin disease*.

billet A block of *forged metal*.

billon An *alloy* of copper with silver or gold, in which the proportion of copper is above 50% (by weight).

binder A *substance* or mixture of substances, also known as *binding medium* or *film former*, that acts cohesively, attaching and consolidating grains or particles of other materials into solid units; binders are basic components of *cements*, *adhesives*, *paints*, and *inks*.

binding medium See *binder*.

binocular microscope See *microscope, binocular*.

biochemical substance See *substance, biochemical*.

biochemistry Study of the composition and changes of *living matter* created by, or derived from living organisms.

biodegradation The *decay* of *organic matter* by living organisms such as fungi and bacteria.

biogenic Created by living organisms.

biological Related to living organisms.

biology Study of living organisms.

biomarker A *molecule* or *substance* whose composition or basic structure suggests an unambiguous, direct link to a biological precursor substance, or to changes in the precursor as a result of injury, disease, or interaction with the environment.

biomass The total *mass* of all living organisms in a particular *habitat*.

biomineral A *mineral* created by living processes.

biomineralization The formation of *minerals* by living organisms.

biomolecule The *molecule* of a *substance* that is a constituent of living organisms.

biooxidation An *oxidation* process brought about by living processes.

biopolymer A *polymer* created by living processes.

biosphere The zone of the earth, including land, water, and air, that can support life.

biostratinomy The physical, chemical, and biological changes that affect the remains of living organisms between the time of death and their burial in the ground; see *diagenesis.*

biotic Pertinent to the living components of the *environment,* such as animals, plants, fungi, and bacteria.

biscuit See *bisquit.*

bisque See *bisquit.*

bisquit Unglazed, once-fired *porcelain,* also known as *bisque* or *biscuit.*

black core (pottery) A dark-to-black coloration at the center of *fired pottery,* due to elementary *carbon* derived from only partly burned *organic matter* in the *clay mixture.*

black and red glaze A type of ancient Greek *pottery* decoration with red figure motifs on a black background.

blast furnace See *furnace, blast.*

bleaching Removal or modification of the intrinsic color of *dyes* in dyed *materials,* such as *yarns* and *textiles;* see *crofting.*

bleeding The loss and flow of *dyes* from dyed *yarns* and *textiles,* when in contact with water.

blend yarn See *yarn, blend.*

blister copper See *copper, blister.*

bloom A porous, spongy mass of iron mixed with *slag;* see *bloomery iron.*

bloomery An early type of *furnace* for *reducing* iron *ore* without *melting,* at temperatures below the *melting point* of iron.

bloomery iron See *iron, bloomery.*

blowpipe A metal tube, generally made of iron, used for blowing *glass.*

blue vitriol The archaic name for copper sulfate; see *vitriol.*

body See *ceramic body.*

bog A spongy, poorly drained type of *soil* that consists mostly of *peat,* a high percentage of *organic matter,* and *acid* water.

boiling point The temperature at which a *liquid* turns into *gas.*

bond The (physical) attractive force that joins *atoms* together into *molecules;* four main types of bonds are recognized: *covalent* and *ionic* bonds, a combination of these two, and *van der Waals* forces.

bond, covalent The chemical *bond* (link) between two *atoms* that share *electrons.*

bond, hydrogen A chemical *bond* (link) resulting from the attraction between positively charged *hydrogen ions* and negatively charged *electrons* from other atoms, such as those of oxygen.

bond, ionic The *bond* (link) between two opposite charged *ions* resulting from the attractive force between their electric charges.

bone The rigid *composite material* that makes up the skeleton of vertebrate animals; it consists of a mixture of *organic* matter, mainly *collagen,* impregnated with a *mineral* component, calcium carbonate phosphate.

bone ash *Calcinated bone.*

bone china A type of *porcelain* made in Europe similar to Chinese porcelain, using *bone ash* as a *flux.*

borax A *mineral* composed of *hydrated* sodium borate, also known as *tinkal;* it has been used as a *flux* in *brazing* operations.

bp Acronym for *basepair;* often used to express the length of *DNA* segments.

B.P. Acronym for *before present; radiocarbon* dates are, by convention, related to the year 1950, which has been specified as "present."

brass An *alloy* consisting of copper (50–90%) and zinc (50–10%), in which small amounts of other *elements,* such as lead, may also be present.

brazing A *metallurgical* technique, also known as *hard soldering,* for permanently joining together *metallic* parts using a *brazing alloy.*

brazing alloy Generic name for *alloys,* also known as *hard or hot solders,* which are used for joining *metals and alloys* at temperatures above 400°C and below the *melting point* of the *base metal* in the alloy.

brick Small, generally rectangular block used for building *masonry;* in antiquity it was made of a mixture of mud and straw, twigs, and/or dung that was either dried (see *mud brick*) or fired (see *fired brick*).

brick, fired *Brick* fired to high temperature, generally above 600°C.

brick, mud *Brick* dried by exposure to the atmosphere and the heat of the sun.

brin The common name used to refer to the two adjacent *filaments* of *silk* produced by the silkworm.

brittleness The property of *hard* and *rigid* solids to break without undergoing deformation.

brocade A firm, textured *textile* fabric with a complex design.

bronze An *alloy* of copper (80–95%) and tin (20–5%) to which minor amounts of other elements, such as zinc and lead, may have been added.

bronze disease A particular form of the *corrosion* of *bronze,* very damaging to excavated archaeological bronze objects, resulting from the interaction of chloride *ions* in bronze *patina* with moisture and oxygen in the air.

brown rot See *rot, brown.*

buffing The roughing, smoothing, or polishing of a surface with an *abrasive* or a sharp tool.

bull's-eye glass An ancient type of flat *glass*, also known as *crown glass*; it was made by blowing a bubble of hot glass, cutting the bubble open and then rotating it in order to spread it, by centrifugal force, into a flat disk with a "bull's-eye" at the center, where the *blowpipe* was attached.

bullion (1) *Crude* lead or copper containing appreciable amounts of one or more *precious metals,* silver and/or gold; (2) refined silver or gold.

burl A natural distortion in the *grain* of *wood.*

burlap A coarse, heavy, plain-weave *fabric* made from *jute.*

burned brick See *fired brick.*

burning (1) The *oxidation* process resulting from the reaction between any oxidizable *material* as, for example, *organic* substances and oxygen in the air; (2) in *ceramic* technology, *firing pottery* at temperatures above 850°C.

burnishing Smoothing or polishing the surface of *pottery* or a *metal*, by rubbing it with a hard tool, such as a smooth stone.

butt The best part of an animal *skin* or *hide*, from around the neck, back, and flanks.

C Abbreviation for *cytosine.*

C-14 Abbreviation for *carbon-14.*

C-14 dating Abbreviation for *radiocarbon dating.*

C3 plants Plants as, for example, cotton and wheat, in which, during the *photosynthesis* process, a stable compound consisting of three bonded *carbon atoms* is first formed during the fixation of carbon derived from carbon dioxide in the plants; over 95% of the plant species on the earth are C3 plants; see *C4* and *CAM* plants.

C4 plants Plants as, for example, maize, in which during the *photosynthesis* process, a stable compound consisting of four bonded *carbon atoms* is first formed during the fixation of carbon derived from carbon dioxide, in the plant; less than 1% of the earth's plant species can be characterized as C4 plants; see *C3* and *CAM plants.*

cable *Cordage* over 2.5 cm in diameter.

calcareous A *material* composed of, or containing mainly calcium carbonate.

calcination The process of heating *mineral ores* or *metal concentrates* to high temperature, generally about 600°C, to drive off volatile components such as chemically combined water, carbon dioxide, and/or sulfur dioxide; see *roasting.*

calcite A *mineral*, composed of calcium carbonate, the most common *polymorph* of calcium carbonate on the surface of the earth; it is the main component of *limestone*, most *marble*, and the *shell* of many sea animals; see *aragonite.*

calico A plain *weave*, *textile* cloth.

CAM plants Abbreviation for *crassulacean acid metabolism plants.*

cambium The layer of *cells* between the *phloem* and the *wood* of a tree.

cameo A small relief carving on a different colored background, made from either *onyx* or two-color layered *glass.*

camwood Generic name for a group of natural *dyes*, also known as *redwood*, that include *barwood* and *sanders wood*, derived from the *wood* of *pterocarpus* and *baphia* trees.

carat A unit to measure the of weight of *gemstones*, equivalent to 0.2 g; see *karat*.

carbohydrate A substance belonging to the large class of naturally occurring *substances* of *biological* origin that are composed of only carbon, hydrogen, and oxygen, with the latter two elements occurring in the proportion 2:1, as in water; see *sugar*, *starch*, *cellulose*.

carbohydrate, complex A *carbohydrate* whose *molecules* consist of more than two *simple carbohydrate* molecules joined together.

carbohydrate, simple A *carbohydrate* whose *molecules* consist of either just one or two joined *carbohydrate molecules*.

carbon A very common *element* on the surface of the earth and the main constituent of *organic matter*.

carbon, asymmetric See *asymmetric carbon atom*.

carbon-14 (C-14) A naturally occurring *radioactive isotope* of *carbon* used for *dating* carbon-containing archaeological *materials*; see *radiocarbon dating*.

carbon-14 dating See *dating, radiocarbon*.

carbon cycle The circulation of *atoms* of *carbon* through the *atmosphere*, *hydrosphere*, *biosphere*, and *lithosphere* as a result of chemical changes.

carbon steel *Steel*, also known as "ordinary" steel, containing up to 2% carbon.

carbonate A complex *anion* composed of *atoms* of carbon and oxygen and assigned the *formula* CO_3^{2-}; it is a constituent of many *sedimentary minerals* and *rocks*.

carbonization The conversion of *organic matter* into carbon that takes place when it is heated to high temperature (above 450°C) in the absence of oxygen, that is, under *reducing* conditions.

carboxyl group The group of COOH *atoms* that occurs in *organic acids*.

carburizing A *metallurgical* technique for introducing carbon into the surface of a low-carbon *steel*; see *case hardening*.

case hardening A *thermal process* for producing a hard surface in *steel*; it entails *carburizing* and *quenching* operations.

casein A *protein* in the milk of mammals; it has been used, since early antiquity, as a *binder*, *size*, and *adhesive*.

cashmere The soft *hair fibers* of the *cashmere* (*Capra hircus*) goat.

cast alloy See *alloy, cast*.

cast iron See *iron, cast*.

casting The process of shaping objects from *materials* that can be brought into the liquid state, such as *metals*, *alloys*, *glass*, and *plaster of Paris*; it entails pouring the liquid into a *mold* where it is allowed to solidify; see *open-mold casting*, *hollow sand casting*, *false core casting*, *lost wax casting*, and *die casting*.

casting, die See *die casting*.

casting, false core See *false core casting.*

casting, hollow See *hollow casting.*

casting, hollow sand See *hollow sand casting.*

casting, lost wax See *lost wax casting.*

casting, open-mold See *open-mold casting.*

catalysis The acceleration, or repression, of a *chemical reaction* by a substance, known as a catalyst, that is chemically unaltered by the reaction.

catechu A natural *dye* derived from several trees, the main variety is *Acacua catechu,* natural to India.

catenation The *bonding* of *atoms* of the same *element* into chains or rings.

cathode The negative terminal of an electric source or circuit; see *anode.*

cathodic protection The protection of a *metal* or *alloy* from *corrosion* due to its contact with a *baser* metal or alloy.

cation An *ion* with a positive electric charge.

caustic A *basic substance* that irritates and disintegrates living tissues.

cell (1) In biology, the basic structural unit of all living organisms; (2) in chemistry, a receptacle in which an electric current in a circuit flows from one metal terminal known as the *cathode,* through a chemical *solution* to a second terminal known as the *anode;* (3) a small, limited space.

cellulose The main constituent of *wood* and the most abundant biological *material* on the surface of the earth; it is a *biopolymeric carbohydrate* that consists of linearly joined *glucose molecules.*

cellulose fibers Vegetable *fibers,* such as *cotton, linen,* and *jute,* which are composed mainly of *cellulose.*

Celsius (°C) A scale for measuring temperatures, also known as the *centigrade scale,* where the *freezing point* of water is 0°C and the *boiling point* 100°C.

cement A natural or artificial fluid or semifluid *substance,* or mixture of substances, that hardens to act as an *adhesive* for binding solid surfaces together.

cement, hydraulic A type of waterproof *stony cement* that sets even under water; see *pozzolana.*

cement, stony A soft and *plastic adhesive* mixture of an *inorganic substance* such as *plaster of Paris,* or a mixture of inorganic materials, as, for example, *clay* and *lime cement* with water, that on drying *sets* into a hard and stable *cementing* solid; see *adobe.*

cementation (1) The deposition of *mineral matter* between *rock* fragments; (2) a *diagenetic* process that involves the *precipitation* of solids in the empty space between the particles of *sediments,* binding them into *rock.*

cementite The common name for a very hard and brittle *compound* of iron and carbon, namely, iron carbide.

ceramic An *inorganic, nonmetallic* solid made by processing earthy *materials* such as *clays* at high temperature; see *burned brick, pottery, porcelain,* and *glass.*

ceramic body The mixture of *clay* and *nonclay materials*, also called a *clay paste*, that has properties suitable for making *pottery*.

ceramic materials Inorganic *materials* such as *pottery*, *burned brick*, and *porcelain* that have been made, since antiquity, mainly by *firing clay*.

chain reaction (1) In chemistry, a series of *reactions* in which one product of a *reaction* is a *reactant* in the following one; (2) in physics, a multistage *nuclear reaction* in which *neutrons* expelled from a *radioactive*, decaying *atom* cause the *decay* of others; the decay chain is interrupted when a product of the decay is a *stable isotope*.

chalcedony A semitranslucent form of the *mineral quartz*; see *agate*.

chalkos Ancient Greek name for the *alloys* of copper (*brass* and *bronze*).

chamois leather A soft *leather* that was made since very early antiquity by *oil-tanning* the skin of the chamois goat antelope (*Rupricarpa rupicapra*) and at present, the skin of sheep.

chamoising The ancient name for *oil tanning*.

chamotte See *grog*.

charcoal An impure, very porous form of *carbon*; it is an excellent *fuel* that has been made since early antiquity by burning *organic matter*, mainly *wood*, in a *reducing atmosphere*.

chemical analysis See *analysis, chemical*.

chemical compound See *compound*.

chemical element See *element*.

chemical equation A symbolic representation of the relationship between the *reactants* and the *products* of a *chemical reaction*.

chemical formula See *formula*.

chemical group Two or more *atoms bonded* together as a single unit and forming part of a *molecule*.

chemical properties Those properties of *substances* and *materials* related to their interaction with other substances.

chemical reaction A chemical change that takes place when the *atoms* or *molecules* of two or more *substances* interact with each other and become rearranged, forming new substances whose properties differ from those of the original ones.

chemical symbols A worldwide conventionally accepted form of chemical notation that makes use of symbols, made up of one or two letters, to represent the chemical *elements*.

chemical weathering See *weathering, chemical*.

chemistry The *science* that deals with the properties of *substances* and how substances react with one another.

China clay See *kaolin*.

China grass See *ramie*.

chiral An object that cannot be superimposed on its mirror image.

chiral carbon atom See *asymmetric carbon*.

chiral molecule An *asymmetric molecule* that cannot be superimposed on its mirror image.

chirality The property of an object that cannot be superimposed on its mirror image; see *isomerism, optical.*

chloride A *chemical compound* that contains negatively charged chloride (Cl^-) *ions.*

chromatin One of the two portions of the *nuclei* of living *cells* (the other is *achromatin,* in which chromatin is embedded) that make up the chromosomes; it consists of a mixture of threadlike *proteins* and *DNA.*

chromatography Generic name for chemical techniques for separating and characterizing the components of a *mixture.*

chromatography, gas A *chromatographic* technique applicable to gases or liquids that can be brought into the gaseous state.

chromatography, gas–mass spectrometry A technique of *chemical analysis* that combines *gas chromatography* and *mass spectrometry.*

chromophore From the Greek word for "color bringer;" a *molecule* or group of *atoms* that absorbs part of *visible white light* and, therefore, imparts *color* to the *substance* of which it is a component or to which it is attached.

chromosome A threadlike structure made up of a linear, end-to-end sequence of *genes* in the nucleus of the *cell,* which transmits the *genetic code* of living organisms; see *chromatin.*

chronometric date See *absolute date.*

chronometric dating See *dating absolute.*

cinnabar A mineral composed of mercury sulfide, used as a red *pigment* and main *ore* of mercury.

cinder A mixture of *slag* and *charcoal* formed during *smelting* operations.

cire perdue casting The French name for the *lost wax casting* method.

classical methods of chemical analysis See *analysis, gravimetric* and *analysis, volumetric.*

clastic A *rock* or *sediment* made up mainly of broken fragments of *minerals* or older rocks that have been transported some distance from their places of origin.

clay (1) *Mineral* particles smaller than 0.004 mm in diameter; (2) a *sediment* composed of *clay minerals.*

clay, primary *Clay* formed in situ by the natural *weathering* of *igneous rocks;* also called *residual clay;* see *secondary clay.*

clay, secondary A natural *clay* that was moved by such natural agencies as water streams or wind from its place of origin, before it became deposited at a site; also called *sedimentary clay.*

clay body A mixture of *clay* and *minerals* used for making *ceramic* objects.

clay minerals Fine-particled *sediments,* such as *kaolinite, montmorillonite,* and *halloisite,* composed of aluminum silicate *minerals* derived from the *chemical weathering* of *felspathic rocks.*

clay mixture A mixture of *clay, temper,* and water of a suitable consistency to be shaped into objects.

clay paste See *ceramic body.*

cleavage The property of *crystals* to easily split at particular planes of weakness within the *crystal lattice.*

closed system A *material* system that can exchange *energy* but not *matter* across its boundaries.

cloth A two-dimensional, planar structured *textile material,* made up of interwoven *yarn.*

cloth, grass See *grass cloth.*

coal A *carbonaceous sedimentary rock* formed by the decomposition and compaction of vegetable matter.

cochineal A natural red *dye* derived from the dried dead bodies of *Coccus cacti* insects.

code, genetic See *genetic code.*

coefficient of thermal expansion See *thermal expansion coefficient.*

coinage metals See *metal, coinage.*

coir The *hard fibers* derived from the outer husk of the coconut used for making *cordage.*

cold working The structural deformation of a piece of *metal* or *alloy,* such as bending, piercing, and forging, carried out at a temperature below that at which the metal or alloy *crystallizes;* cold working brings about the hardening and strengthening of metals, properties that can be eliminated by subsequent *annealing.*

collagen A linear *protein,* a main component of *bone,* cartilage, *skin,* and ligaments in the body of vertebrate animals.

colloid A mixture of finely divided solid particles dispersed in a liquid, usually water.

color (1) The visual sensation by which the eye distinguishes between differences in the *spectral* distribution of *light;* (2) one of the constituents into which *visible light* can be separated; (3) a *substance,* such as a *pigment* or a *dye,* used for coloring others.

color, additive A *color* created by the mixture of two or more colored beams of *light,* that is, light of different *wavelengths.*

color, complementary See *complementary colors.*

colorant A substance, either a *dye* or a *pigment,* that endows *color* to other *materials;* also known as *coloring matter.*

colorimeter An instrument for measuring *color* in a numerical scale.

colorimetry The science that deals with the definition and measurement of undefined colors in terms of standard colors by either visual or instrumental methods, such as *photoelectric* or *spectrophotometric* techniques.

coloring matter See *colorant*.

combustible A substance that burns with a flame.

combustion The process of *burning*.

common salt Sodium chloride.

compaction A *diagenetic* process that involves the reduction in volume of the pores and voids in buried *materials* a result of the rearrangement of their constituent particles.

complementary colors Two *colors* that, when combined with each other, produce gray.

complex carbohydrate See *carbohydrate, complex*.

composite material A solid *material* that is made up of a mixture of two or more different chemical substances that do nor react with but reinforce each other.

compound A *substance* composed of two or more *elements* combined in a particular weight ratio and *bonded* to each other in a specific, definite manner.

compound, chemical See *compound*.

concentrate, metal See *metal concentrate*.

concentration A measure of the ratio between the amount of *solute* and the amount of *solvent* in a *solution*. Also, see *ore dressing*.

conchiolin A complex hard *protein*, the *organic* component of the shells of molluscs; it is closely related to *keratin*.

conchoidal A characteristic form of the smooth, curved, shell-shaped fracture of solids such as *flint* and *obsidian*.

concrete A fluid, plastic mixture of *cement*, water, and particles (larger than 5 cm in diameter) of an inert *material* such as *sand* or *gravel*, that *sets* and hardens; see *mortar*.

concretion A compact mass of *mineral* matter deposited (by the *precipitation* of solids from aqueous *solution*) on exposed or buried shell, bone, fossils, or objects; concretions are usually of different composition from the rock or object on which they are formed.

conductivity (thermal and/or electrical) The property of a material to conduct heat, electricity, or both; thus *thermal conductivity* and *electrical conductivity*.

conductor A *material* through which heat (*thermal conductor*) or electricity (*electrical conductor*) flows.

conglomerate A *sedimentary rock* composed of gravel, pebbles, cobbles, or boulders *consolidated* into a solid mass.

conifer A usually evergreen type of tree belonging to the order *Coniferae* that produces *wood* known as *softwood*.

consolidated Forming a solid mass as, for example, *sediment* particles held together into *rock* by a *mineral cement*.

consolidation The compaction of *sediments* or *soils* into *rock*.

copper, blister An impure form of (ancient) copper; the name is derived from blisters formed when gases released as bubbles from the molten metal during *smelting* were trapped and formed blisters in the solidifying, upper surface.

copper vitriol Blue copper sulfate pentahydrate; see *vitriol*.

copperas See *vitriol*.

coprolites *Fossilized* feces.

cordage Stringlike flexible *materials*, such as *cable, rope*, and *string*, usually made up of *cordage fibers*.

cordage fibers The hard and coarse *fibers*, also known as *hard fibers*, derived mainly from the leaves of such plants as *sisal* and *coir*.

core (1) A relatively large, cylindrical sample of material(s) obtained by drilling soil, rock, ice or wood; (2) a massive stone from which flakes can be removed to make lithic tools; (3) in geology, the innermost and most dense layer of the earth.

core, deep-sea See *deep-sea core*.

core forming An ancient technique for making *glass* vessels by trailing hot, viscous, molten glass so as to form a layer around a removable core; the latter is usually made of *sand* or animal dung mixed with a binder such as *clay*.

corium See *dermis*.

corrosion The generic name for natural chemical processes that result in the chemical deterioration and degradation of *metals* and *alloys*.

corrosion, localized A *corrosion* process that is confined to small areas on the surface of metals bodies but may penetrate deep into the bulk, causing the formation of cracks and holes.

corrosion, uniform A *corrosion* process that expands over large areas forming a layer of corrosion products that often covers the entire surface of metallic objects.

corundum A transparent or semitransparent, often multicolored hard *mineral* (only *diamond* is harder) composed of aluminum oxide; see *ruby, sapphire*.

cosmic radiation See *cosmic rays*.

cosmic rays Beams of *penetrating radiation* or electrically charged *subatomic particles* from outer space that bombard the outer layers of the earth.

cotton The natural *cellulose fiber* obtained from a variety of plants belonging to the botanical genus *Gossypium*, Malvacea family.

covalent bond See *bond, covalent*.

crackle See *crazing*.

craquelle See *crazing*.

craquelure See *crazing*.

crassulacean acid metabolism (CAM) plants Plants such as the cacti and other succulents; see C3 and *C4 plants*.

crazing The hairlike surface cracks that develop in layers of *glaze, enamel, varnish*, and *paint* as a result of tensile stresses between the layer and the substrate; it is also known as *crackle, craquelure*, or *craquelle*.

crisseling See *crizzling.*

crizzling The diminished transparency of *glazes* after exposure to the environment that results in the formation of a very fine form of surface *crazing:*

crofting An ancient method of *bleaching cellulose fibers,* such as *cotton* and *linen,* by successive stages of rinsing and exposure to sunlight.

crosscut A cut made across the *grain* of wood.

crown glass See *bull's-eye glass.*

crucible A hollow vessel, made mainly of a *refractory material,* used for melting *metals, alloys,* or *glass.*

crust In geology, the rocky, relatively low-density, outermost layer of the earth.

cryptocrystalline A *crystalline substance* with a sub*microscopic crystal structure,* too fine to be resolved under an *optical microscope;* see *microcrystalline.*

crystal A homogeneous solid made up of an ordered, uniform, three-dimensional, repetitive arrangement of *ions, atoms,* or *molecules.*

crystal lattice A three-dimensional, regularly spaced array of ions, *atoms,* or *molecules* that make up *crystalline solids.*

crystal structure See *crystal lattice.*

crystalline solid *Solid matter* that has a *crystal structure.*

crystallization The formation of *crystals* from a *melt, solution,* or *gas.*

cullet Broken *glass,* often from former glass production wastes, recycled in a glass melting *furnace.*

cupel A shallow cup or *crucible* made of *bone ash;* see *cupellation.*

cupellation A *metallurgical* process for separating remains of lead from silver or gold by the oxidation, at high temperature, of the lead to its oxide, in a *cupel.*

Curie temperature The temperature above which a magnetic *material* loses its magnetism.

curing (1) The temporary preservation of animal *skins* and *hides* by either drying, freezing, or treating them with salt–see *salting;* (2) the hardening, or *setting,* of *paints, cements,* and *adhesives.*

currying The application of *oil* or *fat* to *leather* to make it pliable and impervious to water.

cuticle See *epidermis.*

cutis See *dermis.*

cyclic compound Any one of a myriad of *organic compounds* whose *molecules* include at least one closed ring of *atoms.*

cyclotron A device for accelerating electrically charged *particles* to high energies in a circular path; it is used to produce *isotopes.*

cytoplasm The content of a living *cell,* excluding the *nucleus.*

cytosine (C) A *nitrogenous base,* one of the constituents of *DNA* and *RNA.*

DNA Acronym for *deoxyribonucleic acid.*

DNA, mitochondrial (mt-DNA) A double-stranded molecule of *DNA* that controls the development and functioning of the *mitochondrion* containing it; mt-DNA is passed along female lines.

DNA double helix Two interlocking helixes joined by *hydrogen bonds* between pairs of *organic bases* that make up a *molecule* of *DNA.*

DNA replication The process that takes place in the *nuclei* of living *cells,* by which *DNA molecules* create exact copies of themselves.

DNA sequence The linear assembly of *nucleotides* along a *DNA* strand.

dahllite A *mineral* composed of carbonated hydroxyapatite, the main constituent of *bone.*

damascening A method for decorating *steel* sword blades with delicate surface patterns by *etching* the surface so as to emphasize local differences in the composition of the steel.

damask A firm, textured *textile fabric* similar to, but lighter than *brocade.*

Damascus steel A *hard* and *resilient* decorated *steel* used for sword blades; see *damascening.*

date, absolute Also known as *numerical date* or *chronometric date*; the time of occurrence of past events, or of the formation or alteration of ancient objects, measured in a definite timescale as, for example, the number of calendar years *before present*; see *relative date.*

date, chronometric See *absolute date.*

date, numeric See *absolute date.*

date, relative A *date* that expresses the age of ancient materials and objects or the time of occurrence of past events in a timescale related to that of other events or objects; it specifies only whether the material, object, or event is earlier, contemporary, or later than another; see *absolute date.*

dating The placement of ancient objects or past events in a timescale to which the human calendar seldom can be applied.

dating, absolute The placement of past events or the formation or alteration of ancient *materials* and objects in a definite timescale, based on scientific evidence as, for example, the rate of *decay* of *radioactive isotopes.*

dating, accelerator mass spectrometry radiocarbon A method of *radiocarbon dating* based on counting, in an *accelerator mass spectrometer*, the number of *radiocarbon atoms* in a sample.

dating, alpha recoil track A *radiometric dating* method based on the fact that if the *concentration* of *alpha emitting radioisotopes* and the volume *density* of *alpha recoil tracks* in a suitable *mineral* are known, it is possible to *date* the mineral.

dating, amino acid racemization (AAR dating) A method of *dating* animal remains, based on the fact that, following the death of an organism, the *amino acids* in the dead body undergo *racemization.*

dating, archaeomagnetic A method of *dating* based on the study of the movement of the magnetic north pole of the earth around the rotational north pole during the past.

dating, chronometric See *absolute dating.*

dating, cross A *dendrochronological* procedure for matching variations in the width of *growth rings* and other *growth rings* characteristics among trees that have grown in the same region.

dating, electron spin resonance (esr dating) A *radiometric* method of *dating*, based on the fact that *energetic radiation* causes *electrons* to separate from their *atoms* and then become trapped in *defects* in the *crystal lattice* of *minerals.*

dating, fission track A *radiometric dating* method based on the fact that when the *isotope uranium-238 fissions* there is a release of *alpha particles* that create damage trails known as *fission tracks* through glassy materials such as *obsidian* and other *vitreous* materials; the number of tracks is directly proportional to the time since the material cooled from a molten state.

dating, numeric See *dating, absolute.*

dating, obsidian hydration A method of *dating obsidian*; based on the fact that newly formed surfaces of obsidian absorb water from the environment and progressively develop a chemically altered (*hydrated*) outer layer whose thickness is time-dependent.

dating, oxidizable carbon ratio A new and as yet unexplored technique for dating organic carbon in soil.

dating, potassium argon An *absolute dating* technique based on determining the proportions between the amounts of the *radioactive isotope* potassium-40 and its *decay* product, argon-40, in rocks; it is particularly important for dating rocks associated with human remains.

dating, radiocarbon A *radiometric* method of *dating* based on the fact that the amount of *carbon-14* in the dead remains of organisms steadily decreases with time.

dating, radiometric The common name for scientific techniques of *dating* based on the known *half-lives* of particular *radioactive isotopes*, or on the rate of occurrence of other cumulative changes in *matter* resulting from *radioactive decay* processes.

dating, relative The placement of ancient *materials*, objects, or past events on a timescale related to that of other materials, objects, or events.

daughter isotope See *isotope, daughter.*

debitage Waste flakes or chunks resulting from the manufacture of *lithic artifacts.*

decay (1) The deterioration of *material*s because of their exposure to the *environment* and particularly to *oxidizing* agents, erosion, and/or variations in temperature and humidity—see *diagenesis, weathering*; (2) see *radioactive decay.*

decay, alpha See *alpha decay.*

decay, beta See *beta decay.*

decay series A sequence of *nuclides* that are formed and subsequently transformed by *radioactive decay* until a *stable isotope* that ends the series is formed.

decolorizer, glass See *glassmaker's soap.*

decomposition The breakdown and disintegration of materials; see *decay.*

deep-sea core A *core* drilled from the seabed that provides a coherent record of *paleotemperatures.*

defect An imperfection in the *crystal lattice* of a *solid.*

deflocculation See *peptization.*

dehairing An ancient chemical or mechanical process for the removal of *hair* from animal *skin*; see *liming, scudding.*

dehydration The removal of water.

denaturation A term used particularly in reference to *proteins* and *nucleic acids* to refer to the alteration, caused by heat or *acid* conditions, of their structure, without breaking major internal *bonds.*

dendrochronology A method of *dating wood* by comparing sequences of the *wood*'s *annual growth rings* with those of *wood* of known *ages.*

density The *mass* (or weight) per unit volume of a substance; see *specific gravity.*

dentine The yellow-white, hard material, also known as *ivory*, which makes up the bulk of the tooth; its composition is similar to that of bone but is harder and denser than bone.

deoxyribonucleic acid (DNA) A *nucleic acid* that carries the *genetic* information of living organisms; its *molecule* consists of two spiraling strands made up of long chains of *deoxyribonucleotides bonded* to each other.

deoxyribonucleotide The building block of the long *molecules* of *DNA*; it consists of bonded *deoxyribose*, a *nitrogenous base* and a *phosphate group*; see *ribonucleotide.*

deoxyribose A *simple carbohydrate*, one of the basic components of *DNA*, having one less *hydroxyle* group than *ribose.*

dermis The middle layer of animal *skin*, also known as *corium* or *cutis*; see *epidermis, hypodermis.*

detector A *material* or device that is sensitive to *radiation* and produces a response signal or *image* suitable for viewing, measurement, or analysis.

detergent A substance that enhances the cleaning properties of *water.*

devitrification Loss of the properties characteristic of *glass* by its conversion into a *crystalline material.*

dezincification A form of *corrosion* of *brass* whereby zinc is gradually lost from the *alloy*, ultimately resulting in the formation of spongy copper.

diagenesis The physical, chemical, and biological changes that affect *sediments* and organic remains after they become buried in the ground; see *biostratinomy.*

diamond A *gemstone* that consists of pure, crystallized carbon; it is the hardest naturally *material* occurring in the crust of the earth.

diatom A microscopic, unicellular organism that has *siliceous* cell walls, found in the seas and oceans and in *fossil* deposits.

diatomaceous earth See *diatomite.*

diatomite A *sedimentary rock* also known as *diatomaceous earth*, formed mainly from the accumulations of *diatoms*.

dichromatic glass *Glass* that shows different *colors* depending on whether it is viewed under *transmitted* or *reflected light*.

die A device for shaping *materials* by *casting*, *extrusion*, or *stamping*.

die casting Shaping a *metal* or *alloy* in a *die* made of a harder material.

diffraction The breakup of a *beam* of *particles* or *radiation* that encounters an obstacle or edge, into a series of alternatively low and high *intensities* or into a *spectrum* of *wavelengths*.

digital Term used for describing a device or system that presents information in the form of discrete digits, as opposed to a continuous, *analog* device or system.

diluent A substance that, when added to a solution, increases its volume or dilutes (decreases the concentration of) the *solute*.

dipping A technique for *plating* objects made of *base metals* or *alloys*, by dipping them in a bath of a molten *precious metal*.

direct dye A *dye* that colors *fibers* without a *mordant*.

dispersion A *heterogeneous mixture* consisting of finely divided *particles* of a *material* within a bulk of solid, liquid, or gaseous matter, such as *fog* or *paint*.

display A (usually electronic) device that provides an output visible to the human eye.

dissociation The splitting of *molecule*s into smaller molecules, *atoms*, or *ions*.

dolomite A *mineral* composed of calcium magnesium carbonate.

Doppler imaging See *imaging, Doppler*.

dosimeter A portable instrument for measuring *ionizing radiation*.

double helix The common name for the *structure* of *DNA molecule*, which is composed of two interlocking helices in an *antiparallel* arrangement, joined together by paired *bases*.

drenching An ancient process for the removal of residual *dehairing* chemicals from treated animal *skins* by immersing the skins in a bath of barley, bran, or the husks of cereals in water; see *bating*, *puering*.

dressing See *ore dressing*.

dross See *slag*.

dry rot A type of *brown rot* caused by the dry rot fungus (*Serpula destruent* or *Merulius lacrymans*).

dry wood *Seasoned wood* or *timber*.

drying oil See *oil, drying*.

ductility The property of *metals* and *alloys* to be drawn into thin sheets or wires without breaking.

dye A *soluble organic substance* used to impart color to others, mainly *fibrous, materials*; see *pigment.*

dye bath A solution in water of *dyes, mordant* (if required), and any other *dyeing* assistants required for *dyeing.*

dye, fast A *dye* that provides a *fast color.*

dye, fugitive A *dye* that changes color or fades.

dye, mordant A *dye* that becomes attached to fibrous materials, through *mordants.*

dye, polygenetic A dye that, depending on the method of *dyeing* and the conditions during the dyeing process, yields different hues and shades.

dye, vat A *dye insoluble* in water that can be used for *dyeing* only after preliminary chemical processing to make it *soluble.*

dyeing The process of imparting *color* to *fibrous materials* such as *yarn, textiles,* skin, or *leather,* by immersion in a *dye bath*; see *staining.*

dyewood Generic name for natural *dyes* derived from the colored *wood* of trees.

early wood See *wood, early.*

earthenware A type of colored *pottery,* usually not *glazed,* that is *fired* at relatively low temperature (below 900°C).

ecology Study of the interaction of living organisms with one another and with their physical surroundings.

efflorescence A fine powdery deposit on the surface of some *rocks, burned brick,* and *pottery,* that results from the *leaching* of one or more of the components from the bulk of the *material.*

Egyptian faience A *non-ceramic material* made from a mixture of *quartz, lime, soda, alumina,* and *feldspar,* which is fired and covered with a layer of usually blue, *alkaline glaze.*

elasticity The property of some materials to be deformed and resume their original shape.

electric conductivity See *conductivity.*

electric conductor A *material* through which an electric current flows; most *metals* and *alloys* are good electrical conductors; see *insulator.*

electric insulator See *insulator.*

electrochemical cell See *electrolytic cell.*

electrochemical circuit An electric circuit that includes an *electrolytic cell.*

electrochemical potential A numerical value that expresses the relative tendency of *metals* and *alloys* to *corrode*; the *noble metals* have low electrochemical potentials. The more easily corroded is a metal, the higher the value of its electrochemical potential.

electrode An *electric conductor* in and *electrochemical cell*; there are two different electrodes: a positive *anode* and a negative *cathode.*

electrolysis The production of chemical changes in an *electrolyte* by the passage of an electric current through an *electrochemical cell.*

electrolyte A solution in water of any *substance* that undergoes *ionization*, dissociating into positively and negatively charged *ions*.

electrolytic cell The assembly of two *electrodes* – a positive *anode* and a negative *cathode* – between which there is an *electrolyte* in which *electrolysis* may take place.

electromagnetic radiation (EMR) A form of *energy* such as *visible light, infrared, ultraviolet*, *X*, and *gamma rays*, that moves at the speed of light; all forms of *EMR* are originated by the interaction of electrical and magnetic energy; see *periodic process*.

electron A *subatomic particle* that bears the smallest electric charge known.

electron capture A *radioactive decay* process in which an *electron* is captured by, and merges with an atomic *nucleus*; the electron combines with a *proton* to form a *neutron* that stays within the nucleus. The mass number of the resulting isotope is unchanged, but its *atomic number* is decreased by one.

electron microprobe See *microprobe, electron*.

electron microscope See *microscope, electron*.

electron spin resonance (ESR) dating See *dating, electron spin resonance*.

electrum (1) A naturally occurring *alloy* of gold and silver usually containing over 20% silver; (2) synthetic (human-made) silver-gold alloys varying in composition from almost pure gold to almost pure silver.

element A *substance* that consists of only one type of *atom* and cannot, therefore, be separated into other *substances*.

element, major An essential component of a *material*, also referred to as a *main* or *matrix element*, that determines the nature and properties of the *material*; see *minor element, trace element*.

element, minor A nonessential element of a component of a *material* that occurs within its basic *matrix* structure in a concentration below 1%; see *major element, trace element*.

element, primordial An *element* created at the time of the creation of the earth.

element, trace A nonessential component of a *material* that occurs in very low concentration, usually expressed in *ppm* or *ppb*.

elemental analysis See *analysis, elemental*.

elementary particle See *particle, elementary*.

elutriation A technique for separating smaller and larger particles of finely divided *materials* such as clay, by subjecting them to an upward flow of a fluid, such as air or water; the flowing fluid causes different sized particles to settle at different rates and therefore to separate.

emerald A green, *precious gemstone* that consists mainly of *metal silicates*.

emery A dark brown variety of the *mineral corundum*.

emission spectrum See *spectrum, emission*.

empirical formula See *formula, empirical*.

EMR Acronym for *electromagnetic radiation*.

emulsifier A substance that promotes the formation of a stable mixture (*emulsion*) of two liquids that are regularly immiscible with each other.

emulsion A mixture of two normally unmixable liquids in which very small droplets of one are dispersed in the other.

enamel (1) *Glaze* used for coating *metals*; (2) the outer layer of the teeth.

enameling A technique for coating *metallic* surfaces with a thin layer of *enamel*.

enantiomer An *optical isomer*; either one of two *molecules* that have the same *composition* but different *structures* and cannot be superimposed on each other; see *chirality*.

encaustic A *painting* technique in which the *binder* is molten beeswax.

end grain The section of *wood* exposed when *wood* is *crosscut*.

energetic radiation energetic *Electromagnetic radiation* as, for example, *X-rays* and *gamma rays*.

energy The potential to do work by the utilization of physical or chemical resources. There are various forms of energy, such as potential energy, kinetic energy, heat energy, chemical energy, electrical energy, and atomic energy; some of these can be interchanged into others.

energy, kinetic The *energy* of motion, associated with moving objects.

engobe A *slip* used for coating the surface of *pottery* before *firing* so as to give it a particular texture and color.

environment The combined physical, chemical, and biological factors affecting the existence of objects or the life of organisms at a particular place in the earth; it includes the composition of the *atmosphere*, *hydrosphere*, and *soil*, measures of temperature and *radiation*, and the activity of living organisms.

enzyme A *protein* produced by living *cells* that *catalyzes* the buildup as well as the *decay* of many biological *substances*.

epidermis The outer layer of animal *skin*; see *dermis*, *hypodermis*.

epimerization The process in which the configuration of a *molecule* containing two or more *asymmetric carbon atoms* is converted into its *optical isomer*; see *racemization*.

equation, chemical See *chemical equation*.

erosion The wearing away, dissolution, breakdown, and/or transportation of *solid matter*; natural erosion is mostly caused by flowing gas (wind) or water (rain and water steams).

error The deviation of an experimental measurement from its correct or true value; see *accuracy*.

esparto A coarse, long *fiber* derived from esparto grass (*Stipa tenacissima*).

ESR dating Abbreviation for *electron spin resonance dating*.

essential amino acid See *amino acid, essential*.

etching The controlled dissolution of the surface of solids (such as of *metals*, *alloys*, or *minerals*) with a suitable *reagent* that selectively dissolves part of the surface *atoms*, revealing the *structure* of the solid.

ethanol See *ethyl alcohol.*

ethyl alcohol The *alcohol* in alcoholic beverages.

eutectic A mixture of two or more *substances*, such as *metals*, that melts at the lowest possible temperature of all mixtures of the same substances.

excited atom An *atom* that has more *energy* than do regular, ordinary atoms.

extrusion An ancient process for shaping *ductile* and *malleable metals* by forcing them to flow through a shaped opening (or *die*).

fabric (1) *Cloth* made of *knitted, woven*, or *felted fibers*; (2) the physical *structure*, that is, the distribution and arrangement, of the constituents of *solid* matter.

fabric, knit *Fabric* made by interlocking loops of one or more *yarns.*

fabric, nonwoven A web structure, such as that of *felt*, made by entangling animal *fibers* without *weaving* or *knitting.*

fabric, woven *Fabric* made by interlocking loops of one or more *yarns.*

fading The reduction in the intensity of the *color* of *materials* as a result of exposure to *light* and/or to *oxydizing* agents.

faience (1) A type of *glazed earthenware* made originally in Faenza, Italy, and later also elsewhere in Europe; (2) the term is also loosely, although erroneously, used to refer to quartz-cored objects made, since antiquity, mostly in Egypt; see *Egyptian faience.*

faience, Egyptian See *Egyptian faience.*

fast color A *color* resistant to *fading*, not impaired by exposure to air and/or *light.*

fast dye See *dye, fast.*

fat A *lipid* solid at ambient temperatures.

fatty acid An *organic acid* generally having rather long carbon chains, which living organisms combine with *glycerol* to produce *lipids* (*triglycerides*), namely, *oils* and *fats.*

feldspar The most common and widespread group of *minerals* on the surface of the earth; it is composed mainly of silicates of the *elements* calcium, sodium, potassium, barium, and even others; the silicon of the silicates, is often replaced by aluminum or boron, thus giving rise to aluminates or borates.

feldspathic rock A type of *igneous rock* that contains a high proportion of *feldspar.*

felt See *fabric, nonwoven.*

fermentation The conversion, usually caused by *yeast* in the absence of oxygen, of *sugar* into *ethyl alcohol* and *carbon dioxide.*

ferrite A soft and *ductile ferrous alloy* that contains minimal amounts (below 0.01%) of carbon, although it may also contain other *elements.*

ferrous alloy See *alloy, ferrous.*

fertilizer A substance that adds nutrients to *soil* for the purpose of increasing the growth of crops or other form of vegetation.

fiber An elongated form of solid *material* whose length is significantly greater than its width or thickness.

fiber, animal A *fiber* derived from animals, mainly *hair*, *wool*, and *silk*; the main components of animal fibers are *proteins* such as *fibroin* and *keratine*.

fiber, bast *Cellulose fiber* derived from the stems of plants.

fiber, cellulose See *cellulose fibers*.

fiber, hair *Fiber* derived from animals such as *alpaca*, *angora*, *cashmere*, *guanaco*, *llama*, *mohair*, and *vicuña*.

fiber, hard See *cordage fibers*.

fiber, inorganic *Fiber* derived from *minerals* or *rocks*, such as *asbestos*.

fiber, protein A *fiber* derived from animals that consists mainly of one of the fibrous *proteins*, either *keratin* or *fibroin*.

fiber, soft The generic name for either vegetable or animal *fibers* that, because of their soft texture, are used for making *yarns* and *textiles*.

fiber, staple A *fiber*, usually of short length, in its natural, unprocessed state.

fiber, vegetable *Fiber* derived from plant tissues, such as *cotton*, *jute*, or *linen*.

fiber, wood See *wood fiber*.

fibril A small *fiber* or a subdivision of a fiber.

fibroin A fibrous *protein*, the main component of *silk*.

fibrous material Matter, mostly of *organic* origin, in the form of *fibers*.

filament A slender, very long threadlike material.

filigree A technique for decorating flat *metal* surfaces, usually of gold or silver, with thin wire of the same metal; see *granulation*.

filler A comminuted solid added to a mixture or preparation such as a *cement*, *paint*, *adhesive*, or potter's *clay*, to improve some particular property of the mixture.

film former See *binder*.

fine silver See *silver, fine*.

fineness, fiber See *fiber fineness*.

fire assay An ancient technique of *assaying* that entails heating a sample of a *precious metal alloy* to a high temperature in a *cupel*.

fire clay A *clay* that does not fuse when heated to high temperatures and that is, therefore, used for making *crucibles* and *molds*.

fire polish A brilliant *glass* surface obtained by heating a glass object, after *glass blowing*, at the mouth of the glass *furnace*.

fired brick See *brick, fired*.

fired clay A mixture of *clay* and *temper* that has been *fired* into a *ceramic material*.

firing The process of heating to high temperatures so as to melt the raw *materials* of *glass*, or to convert *clay* into *ceramic materials*.

fission A *nuclear reaction* in which a *radioactive isotope* splits into two lighter isotopes while *energy* is released.

fission, nuclear See *fission*.

fission, spontaneous A form of *radioactive decay* in which an unstable *nucleus* (usually of a heavy *element*) *fissions* without external stimulation.

fission track A *submicroscopic* trail of damage in *crystalline* and/or *vitreous* materials, created by *nuclear particles* ejected from *radioactive isotopes* of *uranium impurities* in the material.

fission track dating See *dating, fission track.*

flashing A technique for coating the outer surface of *glass* objects with a layer of *glass* of another color by dipping a partly blown object into molten glass.

flax The *bast fiber* obtained from the *flax* plant (*Linum usitatissimum*); it is used to make *linen yarn* and *fabrics.*

fleshing The removal of flesh and/or fatty tissue from the *hypodermis* of animal *skin.*

flint A *mineral* composed of almost pure *silica*, which may also include some (3–5%) trapped water.

flint glass See *lead glass.*

flocculation The agglomerating of small *colloidal* particles, as those of *clay* suspended in water, into *aggregates* that are too heavy to remain suspended and therefore *precipitate.*

fluid A *gas* or *liquid* that has the property of flow.

fluidity The property of a liquid, or liquid *solution*, to flow freely without the hindrance of friction.

fluorescence A form of *luminescence* emitted by some *substance*s when they are irradiated with *energetic radiation*; see *phosphorescence.*

flux A solid *material* that, when added to another solid, promotes the *fusion* of other solids by lowering the *melting point* of the mixture.

fog A *dispersion* of extremely small droplets of water or particles of ice in the air, near the surface of the earth.

food chain The movement of *energy* through the nutrition levels of living organisms; it generally begins with plants and ends with carnivore and detritivore animals.

foraminifera Microscopic organisms (protozoa) that live mainly in marine environments and produce *shells* rich in *calcite*; the *sedimentation* and *lithification* of their shells makes up much of the *sedimentary rocks* of the earth.

forge welding See *welding, forge*

forging *Shaping* a *metal* or *alloy*, generally at relatively high temperatures but below its *melting point*, by beating or hammering.

former (1) One of the components of *paints* – see *film former*; (2) the constituent of *glass* whose *atoms* make up the basic structural network of the glass; the former in all ancient glass is *silica.*

formula See *formula, chemical.*

formula, chemical A conventionally agreed-on combination of *chemical symbols* and numbers that express the composition of a *compound.*

formula, empirical The simplest chemical *formula* for a *compound* that expresses the relative ratios of different *atoms* rather than their total number in a *molecule* of the compound.

formula, molecular A *chemical formula* that expresses the type and number of different *atoms* in the *molecule* of a *compound.*

formula, structural The graphical representation of a *molecule* showing the arrangement of the *atoms* and the *bonds* between them.

fossil The body remains or the impression, outline, or track of an ancient animal or plant after its *organic matter* is *mineralized* or removed.

fossil fuel See *fuel, fossil.*

foundry A workshop where *metal castings* are produced.

Fourier transform infrared (FTIR) A *spectroscopic* technique of *analysis* based on the use *of infrared radiation.*

fractionation The separation of the components of a *mixture;* see *isotopic fractionation.*

fractionation, isotopic See *isotopic fractionation.*

freeze drying A method of preserving fragile *waterlogged* objects by freezing them to very low temperatures (usually about –25°C) and then removing from them water by the process of *sublimation.*

freezing (1) The change in state of *matter* from liquid to solid that occurs with cooling; (2) a process for *curing* animal *skins* practiced mostly in northern cold countries.

freezing point The temperature at which a liquid turns into a solid; it is identical in value to the *solidifying* or *melting point.*

frequency The number of repetitions, or cycles per unit time, of a *wave motion.*

fresco A technique of *painting* walls also known as *affresco,* in which the wall's plaster serves as the *binder* as well as the *support* of the *paint* layer.

frit *Glass* crushed to small particles, resembling coarse sand.

frost wedging A process of physical *weathering* resulting when water *freezing* within cracks or voids in a *solid* exerts sufficient force to rupture the solid.

FTIR Acronym for *Fourier transform infrared.*

fuel An energy rich *material* used for heating. When fuel is heated in air, it ignites and continues to burn with the evolution of heat until exhausted. The fuels most widely used in antiquity include *wood, charcoal,* dung, *shale, lignite,* and *peat.*

fuel, fossil An energy-rich *organic substance* such as *coal,* natural oil, and natural gas in the earth's crust, formed by the fossilization of ancient organic materials, mostly of vegetable origin; now used as fuel.

fugitive dye See *dye, fugitive.*

fuller's earth See *fullers clay.*

fuller's clay A *clay,* generally *montmorillonite,* used for *fulling* cloth.

fulling An ancient finishing process for *woven* or *knitted cloth* that entails washing the cloth and beating it in the presence of *fuller's clay* and finally rinsing and drying; fulled cloth becomes compact and firm and its *yarn* and weave are obscured, giving the appearance of *felt*.

fundamental particle See *elementary particle*.

fur The *skin* with *hair* or *wool*, of some animals; often also referred to as *pelt*.

furnace An enclosed structure in which *materials* such as *mineral ores, metals*, or *glass* can be heated to high temperature so as to alter their physical properties, chemical composition, or both; see *kiln* and *oven*.

furnace, blast A *furnace* used for *smelting* metal *ores*.

fusion See *melting*.

fusion joining A method for joining two pieces of *metal* or *alloy* by heating to, or just below, the *melting point*.

fustic A natural yellow *dye* derived from the *wood* of *Morus tinctoria* trees.

G Abbreviation for *guanine*.

gall An excrescent growth, also known as *gall nut* or *nut gall*, on certain oak trees from Europe or sumac from China and Japan; they provide a source of black *dye*.

gall nut See *gall*.

galvanic series See *electrochemical potential*.

gamma decay A process of *radioactive decay* during which *gamma rays* are emitted from the *nuclei* of unstable *isotopes*.

gamma radiation See *gamma rays*.

gamma rays Highly penetrating *electromagnetic radiation* with *wavelengths* shorter than 10^{-10} (0.0000000001) m.

gamma-ray radiography See *radiography*.

gangue The valueless components of *metalliferous ores*.

garnet Common name for a group of hard, complex *silicate minerals*, some varieties of which are appreciated as a semiprecious *gemstones*; powdered garnet is an excellent *abrasive* material.

gas One of the three states of *matter, solid, liquid*, and *gas*, that, at constant temperature has neither a stable volume nor a definite shape.

gas chromatography (GC) See *chromatography, gas*.

gas chromatography–mass spectrometry (GC-MS) See *chromatography, gas–mass spectrometry*.

gather In *glass* technology, a blob of hot, soft *glass* on the end of a *blowpipe*, preparatory for shaping glass objects.

GC Acronym for *gas chromatography*.

GC-MS Acronym for *gas chromatography–mass spectroscopy*.

Geiger counter A *detector* of *radiation* named after his developer, Hans Geiger.

gelatine A purified form of *glue*.

gelatinous A smooth, jellylike *substance* or *mixture* of substances with a relatively high *viscosity*.

gemstone A *mineral* or other natural *material*, such as a natural *glass* or an *organic substance*, that is beautiful, rare, and durable enough to be used for personal adornment or for the embellishment of decorative objects, such as a *diamond*, *sapphire*, or *amber*.

gemstone, semiprecious A *gemstone* that is used for personal adornment or for the embellishment of decorative objects, such as turquoise and jade, but that is not extremely valuable; the term is ambiguous and rather confusing, since there is no strict boundary between *precious* and semiprecious gemstones.

gene The fundamental unit of heredity; it consists of section of *DNA* located in a particular position in a *chromosome*.

genetic code The heredity instructions, embodied in the sequence of *nucleotides* of *DNA* (and *RNA* in some viruses), that determine the nature of the *proteins* synthesized by living organisms.

genetics The study of *heredity*, how particular traits or qualities of living organisms are transmitted from parents to offsprings.

genetics, molecular See *molecular genetics*.

genome All the *genetic material* in the *chromosomes* of a particular organism.

geochronology The study of the (*absolute* and/or *relative*) *dating* of objects, formations, or events through the use of scientific measurements; see *chronometric dating*.

geomorphology The study of origin and development of present-day landforms, including their classification, description, nature, and relationships to other structures.

gilding The process of coating a *material*, such as a *base metal*, *wood*, *glass*, or *paper*, with a very thin layer of gold.

gilding, amalgam A process for the *gilding* copper and copper *alloys* by applying a thin layer of gold *amalgam* to a surface of copper or copper alloy and then heating the coated surface to drive off the mercury.

glass A uniform *solid* that, after *melting*, cooled in such a way as to remain in an *amorphous* state, lacking long-range internal order.

glass, agate Decorative *opaque glass* made to imitate the *gemstone agate* by joining several different-colored types of *glass* before shaping.

glass, flint See *lead glass*.

glass, lead A type of glass also known as *crystal glass* or *flint glass*, which contains a large proportion of lead oxide.

glass, luster A type of *glass* decorated with a thin surface layer of a *metal oxides*.

glass, opal A *translucent* and sometimes *iridescent* type of *glass*, made to imitate *opal*.

glass, soda *Glass* made from only two components, *silica* and *soda*.

glass, sodalime *Glass* made from three main components, *silica*, *soda*, and *lime*.

glass blowing The process of shaping *glass* objects by inflating a *gob* of hot, soft glass gathered at the end of a *blowpipe*.

glass decolorizer See *glassmaker's soap*.

glassmaker's soap Popular name for the mineral *pyrolusite*, composed of manganese oxide; it is used as a *glass* decolorizer, to counteract the dark green color imparted mainly by iron *impurities* in the silica used for making glass.

glaze A *vitreous* surface coating on *pottery* and other *ceramic materials.*

glaze, alkaline A *glaze* in which the *flux* is an *alkaline material.*

glaze, salt An *alkaline glaze* formed, for example, when common *salt* is heated in a *kiln.*

glazing (1) A technique for coating *ceramic* surfaces with a thin layer of *glaze;* (2) a thin layer of *glue* or *fat* on *fur* or *leather.*

glow temperature See *incandescence.*

glucose A simple *carbohydrate,* commonly known as grape sugar.

glue An *organic, adhesive* substance derived from animal tissue; see *gum.*

glycerol An *organic compound,* a basic component of the *lipids;* it is an *alcohol* whose *molecule* contains three atoms of carbon.

glycine A non*essential amino acid,* the main component of *collagen.*

glycoside One of a group of natural compounds derived from vegetable tissues, in which *carbohydrates,* particularly *glucose,* are major components.

gob A lump of molten *glass* of appropriate *viscosity* for *glass blowing.*

gold leaf An extremely thin (less than 0.0001 mm thick) leaf of gold, used for *gilding;* it has been made, since early antiquity, by hammering a piece of the metal.

grain (1) A small, hard piece of *solid matter;* (2) the natural pattern of pores and wrinkles that creates the texture on the outer surface of a *skin* or *hide;* (3) the top, outer side of animal *skin* or *hide* that has been split into two or more layers; (4) the appearance, direction, arrangement, and size of *wood fibers.*

granite An *igneous rock* that is made up of a mixture of mainly *feldspar* and *quartz,* with minor quantities of other *minerals* such as *biotite* and *mica.*

granulation A technique for decorating flat *metal* surfaces, mostly of silver or gold, by attaching to it minute spheres of the same metal; see *filigree.*

grass cloth A loose, *open-weave fabric* made of reads or coarse *vegetable fibers.*

gravel *Stone* grains varying between 6 and 76 mm in diameter.

gravimetric analysis See *analysis, gravimetric.*

gravimetry See *analysis, gravimetric.*

green copperas An archaic name for ferrous sulfate, one of the best known iron *mordants;* see *green vitriol, vitriol.*

green vitriol An archaic name for ferrous sulfate; see *green copperas, vitriol.*

greenware In *ceramic* technology, objects made of *clay* that have been shaped and dry-hardened but have not yet been *fired.*

grog Crushed or ground *pottery,* also known as *chamotte,* used as a non*clay additive* when making pottery; see *temper.*

ground A usually white fluid coating applied to solid surfaces before *painting.*

groundwater Water beneath the surface of the earth in the pores of *soil, sand,* or *gravel* and in fractures within *rock* formations.

group, chemical See *chemical group.*

growth ring See *annual growth ring.*

guanine (G) A *nitrogenous base*, one of the constituents of *DNA* and *RNA.*

gull nuts See *galls.*

gum An *organic, adhesive substance* derived from vegetable matter; see *glue.*

gypsum A *monomineral sedimentary rock*, composed of *hydrated* calcium sulfate.

habitat The *environment* where a living organism naturally lives.

habutai A soft, lightweight type of raw *silk fabric* from Japan and China.

hackling See *scutching.*

haematite An iron *ore* composed of ferric oxide; the term is often also used to refer to *ochre.*

hair The slender, threadlike structures that grow out of the *skin* of mammals; see *wool.*

hair fiber See *fiber, hair.*

half-life The time required in a process to reduce some quantity, such as the number of *atoms* or *molecules*, to one half its original value.

halloysite A *clay mineral.*

hammer welding See *forge welding.*

hammering Beating a sheet of *metal* into a desired shape.

hard fiber See *cordage fibers.*

hard solder See *solder, hard.*

hard water Water with a high concentration of calcium and/or magnesium *ions*; see *water, hard.*

hardening In *metallurgy*, enhancing the *hardness* of *metals* and *alloys*, usually by a *thermal process.*

hardness The property of *solids* to be resistant to scratching, abrasion, and permanent deformation; see *Mohs scale.*

hardness, water See *water hardness.*

hardwood The *wood* of trees of the class *Angiospermae*, which have broad leaves; the name has no relation to the actual hardness of the wood; see *softwood.*

hearth (1) A site in which humans made fire; (2) the lowest part of a metallurgical *furnace.*

heartwood The inner core of a tree trunk or branch, composed of dead *cells* and differentiated from the enveloping *sapwood* by its darker color.

heat A form of *energy.*

heat capacity The amount of *heat* required to change the temperature of a *material* by one degree.

hemicellulose The common name for a group of *carbohydrates* that, associated with *cellulose* and *lignine* make up *wood.*

hemp A *coarse bast fiber* obtained from the herbaceous hemp (*Cannabis sativa*) plant.

henequen A strong *hard fiber* derived from agave plants (of the family *Agavaceae*).

heredity (1) The *genetic* constitution of a single living organism; (2) the transmission of the genetic constitution of living organisms from one generation to another.

Hertz A unit of measurement of *frequency*, equivalent to one *wave cycle* per second.

heterogeneous matter See *matter, heterogeneous.*

hide The *skin* of large animals such as cows and horses.

hollow sand casting See *casting, hollow sand.*

homogeneous matter See *matter, homogeneous.*

honan A Chinese *cloth* made from *wild silk.*

hormone A *protein* produced in the body and circulated in the bloodstream that controls the activity of certain cells and of many life processes.

hot forming See *hot working.*

hot solder See *solder, hard.*

hot soldering See *brazing.*

hot working The structural deformation of a piece of metal, *alloy*, or *glass*, such as bending, piercing, and *forging*, at high temperature, but below the *melting point* of the *material.*

humic acid Generic name for *organic* substances in the *soil* that are soluble in *alkaline* solutions but insoluble in *acid* solutions; see *humus.*

humic substance *Organic* matter in the *soil* whose exact nature is unknown, made up of the *decay* products of biological matter.

humidity The amount of water *vapor* in the *atmosphere.*

humus Dark, partially soluble *matter* resulting from the decomposition of *organic substances* in the *soil*; see *humic substance.*

hydrated Containing or *bonded* to water.

hydration The incorporation of one or more *molecules* of water into a *compound*, which is converted into a *hydrated* compound.

hydration dating See *dating, obsidian hydration.*

hydraulic cement See *element, hydraulic.*

hydrocarbon An *organic substance* composed only of *atoms* of the *elements hydrogen* and *carbon* arranged in chains or rings.

hydrogen bond See *bond, hydrogen.*

hydrogen ion An *atom* of hydrogen devoid of its only *electron* and thus bearing a positive electric charge.

hydrogen ion concentration See *pH.*

hydrolysis The decomposition of *substances* by their interaction with water.

hydrophilic Literally, having affinity for water; substances or materials that are wetted by, absorb, mix with, or easily dissolve in water; see *hydrophobic.*

hydrophobic Literally, averse to water, or water-repelling; substances or materials that are not wetted by, do not mix, and do not dissolve in water; see *hydrophilic.*

hydroxide A *compound* that, when dissolved in water, becomes *ionized* and forms *hydroxyl anions.*

hydroxyapatite A constituent of *bone*; see *dahllite.*

hydroxyl ion An *anion* composed of bonded hydrogen and oxygen atoms (OH^-).

hygrometer An instrument for measuring *atmospheric humidity.*

hygroscopic A *substance* that easily absorbs water.

hypodermis The innermost layer of animal skin; see *dermis, epidermis.*

igneous rock See *rock, igneous.*

imaging The creation of visual representations of measurable properties of objects, living organisms, or physical phenomena, by either drawing, painting, or by the use of technics such as *photography, radiography,* and *ultrasound.*

imaging, Doppler An *imaging* technique that takes advantage of the change in *wavelength* of *electromagnetic radiation* emitted or reflected from moving objects to create an *image* of the objects.

imaging, magnetic resonance An *imaging* technique based on the use of *nuclear magnetic resonance.*

impurity A foreign *element* or *substance,* incorporated in relatively low concentration in a host *material*; some impurities have little effect on the *physical properties* of the host material while others may significantly alter them.

incandescence The emission of *visibly light* by a material due to its high temperature.

inclusion See *temper.*

incrustation An irregular crust on the surface of a *solid.*

index of refraction See *refractive index.*

indigo A blue *vat dye* derived from plants of the botanical genus *Indigofera.*

inert An *element* or *compound* that does not react chemically.

inert gas See *noble gas.*

infrared *Electromagnetic radiation* with *wavelengths* ranging between 7.5×10^{-7} and 7.5×10^{-3} m.

ink A liquid preparation used for writing, consisting of a mixture of a *pigment* or another *colorant*, a *binder*, and a *vehicle* (usually water).

inorganic *Matter* that does not contain the *element* carbon, with the exception of the oxides of carbon (carbon monoxide and carbon dioxide) and compounds containing *carbonate ions.*

inorganic substance See *substance, inorganic.*

insoluble Not *soluble*; a *substance* that does not dissolve in another.

insulation The property of a *material* to reduce or prevent the flow of heat or electricity.

insulator A *material* that reduces or precludes the flow of heat (*thermal insulator*) or electricity (*electrical insulator*).

intensity See *light intensity, radiation intensity.*

interference (of waves) The constructive or destructive interaction between *waves* traveling through a medium.

ion An *atom* or group of atoms that, because of the gain, or loss, of *electrons,* has a negative (see *anion*) or positive (see *cation*) electric charge.

ion, hydroxyl See *hydroxyl ion.*

ionic bond See *bond, ionic.*

ionization A process in which *ions* are created from *atoms* or *molecules* by the addition (when *anions* are formed) or subtraction (when *cations* are formed) of *electrons.*

ionizing radiation See *radiation, electromagnetic.*

iridescence The property of very thin transparent colorless films, to *scatter visible light* and exhibit a variety of rainbowlike changing colors, as displayed by *mother of pearl* and *weathered glass.*

iron, bloomery An early form of iron also known as *bloom,* produced mostly in a solid state as iron sponge, without melting either the *ore* or the *metal.*

iron, cast An *alloy* of iron and carbon containing more than 0.9% (generally between 2–4%) carbon, which is very brittle and not *malleable,* either when hot or when cold; see *casting.*

iron, pig The brittle product of *smelting iron,* suitable for *casting.*

iron, wrought A *ferrous alloy* suitable for *forging;* it is composed of *iron,* carbon (less than 3%) and small amounts of *slag.*

iron sponge See *bloomery iron.*

iron vitriol See *green vitriol.*

irradiation The exposure of something, *matter,* for example, to *radiation.*

isomers *Molecules* that have the same chemical *formula* but different structure, because the *atoms* in their molecules are arranged differently.

isomer, optical One of two substances whose *molecules* have the same chemical formula but different structures, each molecule containing one or more *asymmetric (chiral) carbon atoms* and, therefore, not superimposable on each other; also known as *stereoisomer;* see *enantiomer.*

isotope One of the different forms of *atoms* of the same *element,* which have the same number of *protons* and therefore share the same *atomic number,* but differ in the number of *neutrons* in their *nuclei.*

isotope, daughter The generic name for *isotopes* formed as a result of the *radioactive decay* of other isotopes known as the *parent isotopes.*

isotope, parent *A radioactive isotope* that *decays* into a different isotope known by the generic name *daughter isotope.*

isotope, primordial An *isotope* created at the time of the creation of the earth.

isotope, radioactive An *isotope* that undergoes *radioactive decay.*

isotope, radiogenic See *radiogenic* isotope.

isotope, stable An *isotope* that does not undergo *radioactive decay.*

isotopic fractionation The change in *isotopic ratio* that sometimes occurs when a *substance* undergoes a chemical or certain type of physical process.

isotopic ratio The relative amount of the different *isotopes* of an *element* in a *substance.*

isotropic *Matter* having identical properties in different directions; see *anisotropic.*

ivory The yellow-white, hard material, also known as *dentine,* that makes up the bulk of the tooth; the word is traditionally applied to the material that constitutes the tusks of the elephant, mammoth, and walrus, which is used for making decorative and ornamental objects.

jade The common name for either one of two white to green *minerals;* appreciated as gemstones: nephrite and the rarer jadeite.

jasper The colloquial name for some fine-grained varieties of banded *quartz;* it is usually semi*translucent,* dark red or dull green and is appreciated as a *gemstone.*

joining, fusion See *fusion joining.*

jute A *bast fiber* derived from various annual plants of the botanical species *Corchorus capsularis.*

kaolin A *clay mineral* composed mainly of *kaolinite;* also known as *china clay,* widely used for making *ceramic materials.*

kaolinite A *clay mineral* composed of aluminum silicate.

kapok A lightweight type of *fiber,* also known as *silk–cotton,* derived from the seeds of *Ceiba pentandra* trees.

karat A form of expressing the purity of *gold alloys,* as parts by weight of gold in 24 parts of *alloy;* fine gold is 24 karat; see *carat.*

kelp The ash of burned seaweed, especially wracks; in antiquity it was a source of *soda,* used for making *glass.*

kenaf A *bast fiber* derived from the stems of the kenaf (*Hibiscus cannabinus*) plant.

keratin A complex, fibrous *protein,* the main component of *hair* and *wool,* the horns and nails of mammals, and the beaks of birds.

kiln A *furnace* used for drying, *firing,* or *burning materials.*

kiln, periodic A *kiln* used in repeated cycles of operations such as loading, heating, holding at peak temperature, cooling, and unloading.

kinetic energy See *energy, kinetic.*

knit fabric See *fabric, knit.*

knitting Interlocking stands of *string* or *yarn* into garments.

K–Ar dating Abbreviation for *potassium–argon dating.*

lacquer An *organic* coating preparation composed of a film forming material dissolved in a suitable *solvent,* which dries by solvent evaporation.

lake An artificial type of *pigment,* made since antiquity, by causing a *dye* to *precipitate* on a porous, usually white, powdered *mineral.*

lanolin A *fat* that naturally covers *wool fibers*.

laser A devices that creates highly concentrated and directional beams of *light*; the word *laser* is an acronym for *light amplification by stimulated emission of radiation*.

laser-induced breakdown spectroscopy (LIBS) See *spectroscopy, laser-induced breakdown*.

late wood See *wood, late*.

lattice, crystal See *crystal structure*.

lava Molten *magma* released from a volcano.

leaching The removal of *soluble* components from *solid matter* by percolating water.

lead glass See *glass, lead*.

leaf fibers See *hard fibers*.

leaf, metal See *metal leaf*.

leather The generic name for *tanned* animal *hide* or *skin*.

leather hard The condition of a *ceramic body* from which most of the *water of plasticity* has evaporated.

leuko compound A colorless (and *reduced*) transient form of a *vat dye* from which the original *color* of the dye can be regenerated by *oxidation*.

levigation A method for separating fine from coarse *particles*, such as those of *clay*, by mixing a powdered *solid* with a *liquid*, usually water, and then allowing the particles to settle; the coarser particles settle first.

LIBS Acronym for *laser-induced breakdown spectroscopy*.

light A form of *electromagnetic radiation* that makes things visible; see *visible light*.

light intensity The amount of *light* per unit area falling on a surface.

light interference The interaction between *light waves* traveling in the same medium.

light reflection The turning back of a beam of *light* when striking the surface of a *medium* that it does not penetrate.

light scattering The physical process in which light is deflected and haphazardly dispersed as a result of collisions with particles.

light transmission The passing of *light* through a *medium*.

light, transmitted Light that passes through a transparent or *translucent* medium.

light, visible *Electromagnetic radiation* with *wavelengths* ranging between 4.0×10^{-7} and 7.5×10^{-7} m, that is, extending from red, through orange, yellow, green, blue, indigo, to violet.

lignin A *biopolymer* that, together with *cellulose*, is a major constituent of *wood* and straw.

lignite A brownish black sort of *coal* formed from the decay of ancient *wood* in which the texture of the original *wood* is often recognizable; it is used as *fuel*.

lime Calcium oxide.

lime cement A *stony cement* that, when freshly prepared from a mixture of *slaked lime, gravel* or *sand*, and water, forms a *plastic* paste that *sets* into a hard and stable solid; it is used mainly for building; see *plaster of Paris*.

lime, quick See *quicklime*.

limestone A *sedimentary, monomineral rock* composed of calcium carbonate.

liming A process for the removal of the *hair* from animal *skin* using *lime;* see *scudding, dehairing*.

linear expansion See *thermal expansion*.

linen The generic name for *yarn* and *fabric* made from *flax fibers*.

lipid The chemical name for *oils* and *fats*; see *triglycerides*.

liquid One of the three states of matter, *gas*, liquid, and *solid*; at constant temperature liquids have a definite volume but no fixed shape and flow.

liquid, supercooled A *liquid* that has been cooled below its *freezing point* and becomes *solid*, without *crystallizing*.

liquor A liquid *solution*.

listing See *selvedge*.

lithification A *diagentic* process whereby, over long periods of time, buried materials are converted into *rock*.

lithosphere The rigid outermost layer of the earth, composed of *rocks* similar to those exposed at the surface.

localized corrosion See *corrosion, localized*.

loess A very fine-grained, loose-textured mixture of *quartz, clay*, and other *minerals* that is blown by the wind and becomes a constituent of many *soils*.

log A cut section of the trunk of a tree.

logwood A natural red *dye* derived from the wood of *Haematoxylon campechianum* trees.

longitudinal wave See *wave, longitudinal*.

loom An implement for *weaving yarn* or *thread* into *cloth*.

lost wax casting An ancient technique of *metal casting*, also known as by its French name, *cire perdue casting*, that entails the following steps: (1) making a wax pattern in the shape of the desired object; (2) surrounding the pattern with a mold of a refractory *material*; (3) melting and discarding the wax; (4) casting the metal into the empty mold.

lumber Cut *wood*, also known as *timber*.

luminescence The emission of light by a *material* as a result of its irradiation with *energetic radiation*, such as *ultraviolet light, X-rays*, or *gamma radiation*; see *fluorescence* and *phosphorescence*.

luminescence, optically stimulated *Thermoluminescence* stimulated by a beam of *laser* light.

luster glass See *glass, luster*.

lye A strong *alkaline* solution.

lysosome A saclike *organelle* inside a *cell* in which *enzymes* break down *substances* that the *cell* does not need.

maceration The softening of *tissues* steeped or soaked in a liquid, usually water.

macromolecule A very large *molecule*, generally formed from many *monomers* bonded together.

macroscopic Something that can be seen by the naked eye.

madder A natural *dye* derived from the roots of the madder (*Rubia tinctorum*) plant.

magma Molten *rock* generated beneath or within the surface of the earth.

magmatic rock See *igneous rock.*

main element See *major element.*

magnetic resonance imaging See *imaging, magnetic resonance.*

major element See *element, major.*

malleability The mechanical property of *solids*, only *metals* and *alloys* in antiquity, to undergo *plastic* deformation when hammered, pounded, or pressed into different shapes, without rupture.

mantle The layer of the earth below the *crust* and above the *core*, which consist mainly of dense iron- and magnesium-rich *minerals*.

marble A *metamorphic* type of *rock* composed of *calcium carbonate* that is widely used for building and making statuary.

marble, onyx See *oriental alabaster.*

marl A natural mixture of grains of *secondary clay* and of *calcareous rocks,* such as *calcite* or *dolomite.*

martensite An extremely hard and brittle microconstituent of *steel*, formed when steel is *quench*-cooled above a certain temperature.

marver A slab of *stone* on which hot, molten *glass* on a *blowpipe* is rolled for shaping.

mass The total amount of *matter* in an object; it is usually expressed, but not measured, in units of weight.

mass number The total number of *protons* and *neutrons* in an atomic *nucleus.*

mass spectrometer See *spectrometer, mass.*

mass spectrometry See *spectrometry, mass.*

masonry A structure made from *stones* or *bricks* bonded with *mortar.*

material A specific type of *matter*; materials may be either *homogeneous substances* or *mixtures,* or *heterogeneous mixtures.*

matrix (1) The major components of a *substance*—see *major element*; (2) a physical environment in which *particles* are embedded.

matter Anything that has *mass* and occupies space.

matter, heterogeneous *Matter* made up of two or more visibly different components.

matter, homogeneous *Matter* of uniform *composition* and *structure.*

matter, organic See *organic matter.*

maturity (of clay) The appropriate conditions created by the complex of temperature and time at which a *clay mixture* acquires specific properties such as *plasticity, hardness*, and *color*.

medium A physical environment as, for example, a *solid* or *liquid*.

melting The physical process of a *solid* becoming a *liquid*; see *melting point*.

melting point The temperature at which a *solid* liquefies; see *freezing point*.

mercaptan Generic name for foul-smelling *organic substances* consisting of, among other components, a sulfur-containing group of *atoms*.

messenger RNA See *RNA, messenger*.

metabolism The whole range of physical and chemical processes by which living organisms convert food into *energy*.

metal One of a large group of *elements* that are mostly solid at ambient temperatures, have a shiny surface, and are good conductors of electricity and heat; only eight metals were known in antiquity.

metal, base (1) The major component of an alloy; (2) the opposite of *noble metal*, that is, a *metal* such as iron, which is easily *corroded*.

metal, coinage Generic name for such *metals* as copper, silver, and gold, and *alloys* of these metals, as for example *bronze* and *electrum*, whose chemical, physical, and economic qualities make them suitable for coinmakings.

metal concentrate A metallic *ore* that has been *dressed*.

metal, native A *metal* that occurs naturally in irregularly shaped masses, uncombined with other *elements* and can be worked by hammering, cutting, and/or *annealing*; see *nugget*.

metal, noble Generic name for the *metals* that are not *corroded* under any environmental conditions as, for example, gold and platinum; see *base metal*.

metal, precious See *noble metal*.

metal, virgin See *native metal*.

metal leaf An extremely thin plate or leaf of a *malleable, precious metal* such as gold.

metallic materials Generic name for *metals* and *alloys*.

metalliferous Bearing or producing a *metal*.

metallography The study of the composition, structure, and properties of *metals* and *alloys*, as revealed by the *microscopic* examination of their polished and *etched* surfaces.

metalloid From the Greek words *metallon* (metal) and *eidos* (sort), an *element* whose properties are intermediate between those of the *metals* and the *nonmetals*; antimony, arsenic, boron, germanium, polonium, silicon, and tellurium are metalloids.

metallurgy The science and technology of *metals* and *alloys*, including their extraction from *mineral ores*, working, and use.

metamorphic rock See *rock, metamorphic*.

metastable An apparently stable condition that may be perturbed to become unstable.

microcrystalline A *crystalline substance* with a very fine *crystal structure* visible only under an *optical microscope*; see *cryptocrystalline*.

microlith A very small and usually long, narrow, and sharp lithic tool (often less than 2 cm long); it was used as a tip of arrows or, placed edge-to-edge along with others in a piece of wood, bone, or antler, as knife or saw.

micrometer (1) A unit of length, formerly known as a *micron*, equivalent to one-millionth (10^{-6} or 0.000001) of a meter; (2) an instrument for measuring very small linear distances.

micromorphology The *microscopic* study of *sediments* and *soils*, as well as of synthetic (human-made) *materials* such as *brick*, *mortar*, and *ceramics*.

micron See *micrometer*.

microorganism Extremely small organism that can be seen only under a *microscope*.

microprobe An instrument in which a stable and well-focused beam of electrically charged *particles*, such as *electrons* or *ions*, is used to analyze the composition of *solids* at the *microscopic* level.

microprobe, electron An instrument for *imaging* and *analyzing* solids at the *microscopic* level.

microscope An instrument that provides enlarged images of objects that are too small to be seen by the naked eye; the mayor types of microscope are based on the use *beams* of either *visible light* (*optical microscopes*), *electrons* (*electron microscopes*), or *X-rays* (*X-ray microscopes*) that interact with an object and create an enlarged image of the object.

microscope, binocular A *microscope* with two eyepieces for viewing a magnified image of an object with two eyes.

microscope, electron A type of *microscope* in which an enlarged image of an object is generated by bombarding the object with a beam *of electrons*; there are two main types of electron microscopes: the *transmission electron microscope* (TEM) and the *scanning electron microscope* (SEM).

microscope, transmission electron (TEM) An *electron microscope* that provides extremely large magnifications, making it possible to see magnified images of objects as small as a few *angstrom* units in size and to get much better resolution than with an *optical microscope*.

microscope, optical A *microscope* in which an enlarged image of an object is produced by visible *light*.

microscope, scanning electron (SEM) An *electron microscope* in which an *electron beam* is scanned (moved systematically) in successive parallel lines across the surface of a *sample* and provides a lifelike, three-dimensional image of the surface.

microscope, X-ray A *microscope* in which *X-rays* create enlarged images of objects, much clearer than images obtained with an *optical microscope*.

microscopic Related to objects too small to be seen by the naked eye.

microscopy The science of the uses and applications of *microscopes*.

microstructure The microscopic *morphology* of a *substance* or *material.*

microwear Microscopic *use wear.*

microwaves See *waves, micro.*

migration In *dye* technology, the seepage of a *dye* from one part of a colored *material* to another.

mineral A natural *inorganic solid* that has a definite *chemical composition* and a characteristic *crystal structure.*

mineral, primary A *mineral* that has not been chemically altered since it *crystallized* from molten lava; see *secondary mineral.*

mineral, secondary A *mineral* formed as a result of the decomposition or *weathering* of *primary minerals,* or of the *reprecipitation* of the products of decomposition of primary minerals.

mineral dressing See *ore dressing.*

mineral fiber See *fiber, inorganic.*

mineral ore See *ore, mineral.*

mineral rock See *monomineral.*

mineral tanning See *tanning, mineral.*

mineralization A process whereby *organic matter* in the remains of dead organisms is partly or wholly replaced by *inorganic* matter and therefore converted into *mineral* matter.

mineraloid A natural *material* as, for example, *obsidian* or *opal* that, although of definite composition, lacks the typical *crystal structure* characteristic of the *minerals.*

mining The extraction and exploitation of *mineral ores.*

minor element See *element, minor.*

mitochondria *Organelles* in the *cytoplasm* of living cells that contain *enzymes* and DNA.

mitochondrial DNA (mt-DNA) See *DNA, mitochondrial.*

mixture A blend of two or more *substances* that retain their characteristic properties and can be separated from each other by physical means.

modifier In *glassmaking,* an *alkaline* metal oxide, such as *soda* and *potash* that, when added in small quantities to *silica,* acts as a *flux.*

Mohs scale of hardness An empirical scale that grades the *hardness* of *minerals* from 1 (softest, *talc*) to 10 (hardest, *diamond*).

mohair See *angora.*

moisture Water, in the form of *gas* or *vapor,* dispersed throughout a gas or *solid;* thus also water in the *atmosphere.*

mokume A technique of *metal* and *alloy* surface decoration intended to imitate wood grain patterns, practiced since antiquity in Japan.

mold A hollow recipient of a required shape into which a liquid, such as molten *metal, alloy,* or *glass,* or fresh, runny *cement,* is poured to acquire its shape when the liquid solidifies.

molding To form a material into a particular shape in a *mold.*

molecular biology Study of the composition, structure, and function of the substances in, and those associated with, living organisms.

molecular formula See *formula, molecular.*

molecular genetics The branch of *genetics* involving the study of how *genetic* information is encoded within *DNA* and how *biochemical* processes in the *cell* translate the genetic information into the characteristics of an organism.

molecule The smallest unit into which a *compound* can be divided and still retain it *chemical properties;* it consists of *bonded atoms* of one or more *elements.*

monochromatic *Light* of a single color or *radiation* of single *wavelength.*

monomer A substance made up of relatively small identical *molecules* that can bond to each other to form very large molecules known as *polymers.*

monomineral A *rock* made up of just one *mineral.*

montmorillonite A *clay mineral;* the main component of *fuller's clay.*

mordant A *substance* used in *dyeing,* which binds *dyes* to *textile fibers* and other *fibrous materials.*

mordant dye See *dye, mordant.*

morphology The physical shape, form, and structure of objects, *organisms,* and *materials.*

mortar A *stony cement* used in *masonry* to bind blocks of *stone* or *bricks* together and to fill gaps between them; it consists of a mixture of a *cement,* small grains of stone (such as *lime* or *sand*), and water, that sets on drying into a strong and stable solid; see *concrete.*

Mössbauer spectroscopy See *spectroscopy, Mössbauer.*

mother of pearl A smooth *iridescent substance* in the inner layer of the shell of some molluscs.

MRI Acronym for *magnetic resonance imaging.*

mt-DNA Acronym for *mitochondrial DNA.*

mud brick See *brick, mud.*

mud cement See *cement, mud.*

muffle A wall inside a *kiln* where *pottery* is fired, that prevents the flames of the fire from reaching the *pottery.*

mullite A silicate of aluminum formed when *clay* is heated above 1000°C.

mummification The natural or artificial preservation of the body of dead animals.

mummy The naturally or artificially preserved body of a dead animal.

muslin The generic name for a large group of plain-*weave* cotton *fabrics.*

NAA Acronym for *neutron activation analysis.*

nanometer A unit of length equivalent to one-thousandth millionth (10^{-9} or 0.000000001) of a meter.

nap A soft, fluffy effect achieved in *leather* and/or *textile fabrics* by *brushing* or *buffing.*

napa The surface or top *grain* of a *hide*.

native metal See *metal, native*.

natron A *mineral* composed mainly of hydrated sodium carbonate.

natural glass See *obsidian*.

NDT Acronym for *nondestructive testing*.

necrolysis The decomposition and dissolution of the remains of an organism after death.

neutral solution An *aqueous* solution whose pH value is 7.

neutralization The chemical process for rendering a solution *neutral*, neither *acid* nor *alkaline*.

neutron An electrically uncharged *subatomic particle* in the *nucleus* of an *atom*.

neutron activation analysis (NAA) An analytical technique for identifying and evaluating the relative amounts of *elements* in a sample; it is based on bombarding the sample with *neutrons* so as to create *radioactive isotopes* and then measuring the characteristic *radiation* emitted by each isotope.

neutron radiography See *radiography*.

nitrate A complex *anion* composed of *atoms* of nitrogen and oxygen and assigned the *formula* NO_3^-.

nitrogenous base See *base, nitrogenous*.

NMR Acronym for *nuclear magnetic resonance*.

noble gas A gaseous *element* also known as *inert gas* as, for example, helium and argon, which is chemically inert and does not combine with other elements.

noble metal See *metal, noble*.

nonclay additive See *temper*.

nondestructive testing (NDT) One of a range of a *physical* and *chemical* techniques for determining the properties of *materials* and/or the structural integrity of material assemblies without permanently altering them or their properties.

nondrying oil See *oil, nondrying*.

nonessential amino acid See *amino acid, nonessential*.

nonferrous alloys See *nonferrous metals and/or alloys*.

nonferrous metals and/or alloys Generic name for all *metals* and *alloys* with the exception of iron and the *ferrous alloys*.

nonmetal A *nonmetallic element* such as *sulfur* or *carbon*.

nonplastic material See *temper*.

nonwoven fabric See *fabric, nonwoven*.

nuclear magnetic resonance (NMR) A *physical technique* used to study *physical, chemical*, and *biological* properties of matter by determining the detailed, three-dimensional structure of *molecules*; it requires the use of a strong magnet that interacts with the natural magnetic properties of *atomic nuclei*.

nuclear particle See *particle, nuclear*.

nuclear reaction A *reaction* that involves changes in the *nuclei* of *atoms*, such as *decay* and *fission*.

nuclear reactor A device in which *nuclear reactions* can be initiated, maintained, and controlled.

nuclear transformation A *radioactive decay* process whereby the *isotope* of an *element* is transformed into an isotope of another element.

nuclear transmutation See *nuclear transformation*.

nucleic acid A class of complex *biogenic* substances comprising two basic types: *ribonucleic acid* (RNA) and *deoxyribonucleic acid* (DNA), both consisting of very large *molecules*, among the largest known.

nucleon A *nuclear particle*, either a *proton* or a *neutron*.

nucleoside In *nucleic acids*, a *base* bound to a *sugar*; see *nucleotide*.

nucleotide The basic unit from which *nucleic acids* are built up; it consists of a *nitrogenous base*, a *phosphate group*, and a *sugar molecule* (*deoxyribose* in *DNA* and *ribose* in *RNA*).

nucleus (1) The central part of an *atom* that consists mainly of *neutrons* and *protons* and makes up nearly all of the atom's mass; (2) the site of a living *cell* that contains *chromatin*, where RNA is synthesized and *DNA* is *replicated*.

nuclide An *isotope*; an *atom* with a particular combination of *protons* and *neutrons* in its *nucleus*.

nugget A lump of *native metal*, such as gold, silver, or copper, uncombined with other *elements*.

numerical date See *absolute date*.

nut gall See *gall*.

nutrient A chemical *element*, *compound*, or food required by a living organism to live, grow, or reproduce.

obsidian A black or dark volcanic *glass* that is formed when *lava* cools so rapidly that *crystals* do not have time to grow.

obsidian hydration dating See *dating, obsidian* hydration.

ocher A yellow, red, or brown earthy *mineral* composed of hydrated ferric oxide.

OCR dating Abbreviation for *oxidizable carbon ratio dating*.

OES Acronym for *optical emission spectrometry*.

oil A liquid *lipid*.

oil, drying An *oil*, such as *linseed oil*, which, when exposed to air, is *oxidized* and hardens ("dries"), forming a tough solid.

oil, nondrying An *oil* that neither is *oxidized* nor *hardens* when exposed to air.

oil tanning See *tanning, oil*.

onyx A banded type of *agate* that incorporates straight, parallel layers of different colors.

onyx marble See *oriental alabaster*.

opacifier A *material* that, when incorporated into *glass, glaze,* and *enamel,* decreases their *transparency;* tin and antimony oxide *minerals* were used in antiquity as opacifiers.

opal A usually colorless or white *mineraloid* that consists of *hydrated silica;* it is appreciated as a *semiprecious gemstone,* sometimes showing *iridescent* changing colors.

opal glass See *glass, opal.*

opaque A *material* that is not *transparent* such as opaque *glass, glaze* or *enamel;* these were produced in the past by adding tin or antimony oxide *minerals* to molten glass.

open firing A method of *firing pottery* without using a *kiln.*

open-flame kiln See *kiln, open flame.*

open system A *material system* in which *energy* as well as *matter* flow across its boundaries.

opening material See *temper.*

optical emission spectrometry (OES) See *spectrometry, optical emission.*

optical isomer See *isomer, optical.*

optical microscope See *microscope, optical.*

optically stimulated luminescence (OSL) See *luminescence, optically stimulated.*

ore A natural mixture of *minerals* from which *metals* or a *nonmetallic elements* can be profitably extracted.

ore, mineral The *mineral* in an *ore* that contains a useful *metal* or other *element.*

ore benefication See *ore dressing.*

ore concentrate That part of an *ore* in which a required *metal* or other *element* is highly concentrated.

ore concentration See *ore dressing.*

ore dressing The process, also known as *ore benefication, ore concentration,* and *ore preparation,* for separating *metal* containing *minerals* from the *gangue,* after *mining,* but before *smelting.*

ore preparation See *ore dressing.*

organelle A distinctive organized structure in the *cytoplasm* of a living *cell.*

organic See *organic matter.*

organic acid An *organic substance* whose *molecule* includes a *carboxyl* group of *atoms.*

organic matter *Matter* in which *carbon atoms* are linked together by carbon–carbon *bonds.*

organic substance See *substance, organic.*

orichalcum An *alloy* of copper and zinc used by the Romans for the production of coins.

oriental alabaster See *alabaster, oriental.*

orpiment A yellow *mineral* composed of arsenious trisulfide, used as a yellow *pigment.*

orthoclase A *mineral* of the *feldspar* group.

OSL Acronym for *optically stimulated luminescence.*

oven A *furnace* used at relatively low temperatures.

overglaze *Pigment* applied to the surface of *pottery* after *glazing.*

oxidation A type of *chemical reactions* in which an *atom* or *atomic group* combines with oxygen or one or more *electrons* are removed from it.

oxide A substance composed by the chemical combination of *oxygen* with another *element.*

oxidizable carbon ratio (OCR) dating See *dating, oxidizable carbon ratio.*

oxidizing atmosphere A smoke-free *atmosphere* in which there is abundant oxygen.

oxygen isotope ratio The weight ratio of the *stable isotopes* of oxygen in such *materials* as seawater, glacier ice, and *fossil* shells; this measurement provides a reliable indicator of past temperatures and climatic conditions.

ozone An *allotrope* of *oxygen,* which occurs as *molecules* composed of three atoms of oxygen; it is an air *pollutant* and a powerful *oxidizing* agent.

paint A colored preparation, mainly a *dispersion* of particles of a *pigment* in a liquid *vehicle* and a *binder,* which is applied to the surface of materials for decorative or protective purposes.

paktong A silver-colored *alloy* of copper, nickel, and zinc that was used in China and India (where it was known as *tutenag*) many centuries before it was known to Western civilizations.

paleoclimate The climate prevailing in a region in archaeological or geologic times.

paleomagnetism The magnetism of *igneous rocks* that is used for their *dating.*

paleontology The branch of science concerned with the study of *fossils,* ancient living organisms and their evolution.

paleosol A *soil* formed or altered by human inhabitation or human activity.

paleotemperature The temperature prevailing in a region during a certain archaeological or geologic era.

palynology The study of *pollen.*

parchment *Cured* and dried animal *skin* used mainly for writing; see *rawhide, vellum.*

parent isotope See *isotope, parent.*

parent, radioactive See *radioactive parent.*

particle An extremely small, minute portion of matter; the term is usually reserved for portions of matter measuring about 50 μm.

particle, alpha See *alpha particle.*

particle, beta See *beta particle.*

particle, elementary A *particle* that cannot be subdivided into smaller units and has negligible dimensions but a definite weight.

particle, fundamental See *elementary particle.*

particle, nuclear A *particle* in the *nucleus* of an *atom,* such as the *proton* and the *neutron;* see *nucleon.*

particle, subatomic An *elementary particle* that may be a component of an *atom.*

particle size distribution Classification of the *particles* of *soil* samples; it is often expressed as a mass percentage.

parts per billion (ppb) A unit used for expressing the concentration of *trace elements;* it is equivalent to one gram per one thousand tons.

parts per million (ppm) A unit used for expressing the concentration of *trace elements;* it is equivalent to one gram per one ton.

paste See *ceramic body.*

patina A usually green layer or *incrustation* of *corrosion* products that appears on exposed copper or *bronze* surfaces.

PCR Acronym for *polymerase chain reaction.*

pearlite A mixture of *ferrite* and *cementite* that occurs in such *ferrous alloys* as *cast iron* and *steel.*

peat A naturally dark deposit of partially decomposed vegetable matter accumulated in very wet *anaerobic* environments that prevent further decay.

pelt See *fur.*

penetrating radiation *Energetic radiation* including X- and *gamma rays,* which penetrates solid materials such as *brick,* rock and *metals.*

peptide A *compound* made up of two or more *amino acids bonded* in a chain.

peptide bond The chemical *bond* linking between *amino acids* in *peptides* and *polypeptides.*

peptization The process of preparing *colloids,* also known as *deflocculation.*

percolation The movement of *water* through porous *solids.*

periodic kiln See *kiln, periodic.*

periodic process A process that changes regularly: as time passes, the process advances from one stage to another.

periodic table The classified arrangement of all the chemical *elements* in order of their increasing *atomic numbers.*

pewter An *alloy* of tin and lead known in antiquity, which contains 75–95% tin and 25–5% lead.

pH A measure of *hydrogen ion concentration* (i.e., of the acidity or alkalinity) of water solutions. A pH number of 7 indicates that a solution is neutral; lower than 7, that it is *acid;* and higher than 7, that it is *alkaline.*

pH meter An electrochemical instrument for the measurement of the *pH* of *aqueous solutions.*

phase Any physically separate part of a material; a *homogeneous* component of a material system that has a characteristic chemical composition. A pure *metal,* for example, is composed of only one phase; most rocks however, usually include two or more phases.

phloem The food-conducting tissue of plants; see *bast*.

phosphate An *anion* consisting of an atom of phosphorus bonded to four oxygen atoms, assigned the *formula* PO_4^{3-}.

phosphate group A *phosphate anion* often attached to biological *molecules*, such as *proteins*, *carbohydrates*, and *lipids*.

phosphorescence The *luminescence* emitted by a material after it is irradiated with *energetic radiation* such as *ultraviolet light*, *X-rays*, or *gamma radiation*; see *fluorescence*.

photometry Measurement of the characteristics of *visible light*, such as its *spectral distribution*, *wavelength*, or *intensity*.

photosynthesis The natural process by which plants, algae, and some bacteria use the sun's light energy to create *organic molecules* from atmospheric carbon dioxide and environmental water.

phylogenetics The study of the evolutionary relationships among organisms.

phylogeny See *phylogenetics*.

physical weathering See *weathering, physical*.

phytoliths Literally, "plant stones"; microscopic particles of *inorganic matter*, mostly *silica*, created by plants in vegetable tissues.

pig iron See *iron, pig*.

pigment An intensely colored and finely powdered *solid*, insoluble in most liquids, that is used to impart color to other materials; see *dye*.

placer A deposit of *gravel* or *sand* in the bed of a stream, containing valuable *metal* or *mineral* grains.

plagioclase Generic name for *feldspar minerals* composed of *silicates* containing considerable amounts of sodium and calcium.

plain weave See *weave*.

plaster A more or less uniform coating of *cement* on walls, ceilings, and floors.

plaster of Paris A *stony cement* that, when freshly prepared from a mixture of calcium sulfate and water, forms a paste that *sets* into a hard and stable solid; it is used mainly for building and making *molds* for *casting* other *materials*.

plastic A supple and pliable *material*, such as wet *clay*, that can be easily shaped by the application of external *stresses* and that retains the new shape when the stresses are removed.

plastic deformation The deformation of a *solid* by an applied external *stress* beyond its elastic limit, which creates irreversible effects.

plastic strain A deformation on a *solid* body that disappears when an external stress is removed.

plastic properties See *plasticity*.

plasticity The property of some *semisolid* materials, such as wet *clay* and *molten glass*, to retain deformations caused by applied external *strains* after these are removed.

plasticity, water of See *water of plasticity.*

plate (lithospheric) A slab of rigid, moving *crust* in the uppermost *mantle* of the earth.

plating Formation of an adherent layer of (a usually *precious) metal* on an object made from another, usually *base,* metal; see *dipping, metal leaf, amalgamation, gilding.*

pollen The male sex spores of flowering plants.

pollution The contamination of the *environment, atmosphere,* water, and/or *soil,* with undesirable substances and/or factors that are detrimental to living organisms or to *inorganic mater.*

polygenetic dye See *dye, polygenetic.*

polymer A *substance* whose *molecules* are very large and are formed by the bonding together of many (more than five) *monomers.*

polymerase Any *enzyme* that *catalyzes* the *polymerization* of small, *monomeric molecules,* into large *polymers.*

polymerase chain reaction (PCR) A technique employed to rapidly *amplify* segments of *DNA.*

polymerization A type of chemical reaction in which *monomers* combine to form *polymers.*

polymorph Any one of various *crystalline* forms of the same *substance.*

polypeptide A *biopolymer* formed by *amino acid monomers;* polypeptide *molecules* are larger than those of the *peptides,* but smaller than the molecules of *proteins.*

porcelain A highly *vitrified* and *translucent ceramic* material made when *kaolin clay* that is *fired* at very high temperatures (above 1350°C).

porosity The volume of the voids within a porous solid; it is usually expressed as a percent of the total volume of the solid.

positron A *subatomic particle* having the same mass as an *electron* but a positive electric charge.

positron decay A form of *radioactive decay* whereby a *proton* emits a *positron* and is converted to a *neutron.*

positron emission A form of *radioactive decay* involving the emission of a *positron* from an unstable *atomic nucleus.*

potash The common name for the *basic* substance potassium carbonate; it is used in *glassmaking* as a *modifier,* an alternative to *soda.*

potassium–argon dating (K–Ar dating) See *dating, potassium-argon* (*K–Ar dating*).

potential, electrochemical See *electrochemical potential.*

pottery A *ceramic* material made of *fired clay;* see *terracotta* or *earthenware;* the term is often also used in a loose sense to refer to most types of ceramic objects made of *clay.*

pozzolana A *siliceous* and *aluminous* material that combines with *lime* to form *hydraulic cement* that *sets* even under water; natural pozzolanas, derived primarily from volcanogenic materials, were used in antiquity, particularly by the Romans, to prepare *hydraulic cement.*

ppb Acronym for *parts per billion.*

ppm Acronym for *parts per million.*

precious metal See *metal, precious.*

precious stone See *gemstone.*

precipitate A *solid* that separates, as extremely small *particles*, out of a *liquid solution.*

precipitation Formation of a *precipitate* as a result of either a chemical *reaction* or the evaporation of water.

primary clay See *clay, primary.*

primary mineral See *mineral, primary.*

primordial element See *element, primordial.*

primordial isotope See *isotope, primordial.*

properties, chemical See *chemical properties.*

properties, physical See *physical properties.*

protein One of a large group of complex *organic compounds* that consist of many *amino acids* joined together by *peptide bonds*; they are essential to all living organisms, making up the main component of the *cells protoplasm.*

protein fiber See *fiber, protein.*

proton A *subatomic particle* in the *nuclei* of *atoms* that has the same mass as the *electron* but with a positive electric charge.

protoplasm The entire contents of a *cell*, including the *nucleus*; see *cytoplasm.*

provenance The geologic or geographic source of a material or object.

provenience See *provenance.*

PSD Acronym for *particle size distribution.*

puering The removal of residual *dehairing* chemicals from treated animal skin; an ancient puering process entails immersing the skins in an *infusion* of *fermenting* dog manure and results in supple, stretchy, and soft skins, softer than those obtained by *bating*; see *drenching.*

pulled wool See *wool, pulled.*

pumice A light-colored, frothy volcanic *rock*, formed by the expansion of gas in solidifying molten *lava.*

putrefaction The decomposition, also known as *rot*, of *organic matter*, particularly *proteins*, which is accompanied by the evolution of foul-smelling compounds, such as *mercaptanes* and ammonia.

pyrolusite A *mineral* composed of manganese oxide, used as a black *pigment* and as *glassmaker's soap.*

pyrolysis A process of chemical decomposition brought about by heat.

pyrometallurgy The use of fire and heat to extract and recover *metals* from *ores.*

qualitative analysis See *analysis, qualitative.*

quantitative analysis See *analysis, quantitative.*

quantize To produce an output in discrete steps; relating to a *quantum.*

quantum The smallest discrete amount of *energy.*

quartz A transparent or *translucent mineral* composed of silicon dioxide; see *silica.*

quenching The *metallurgical* process of hardening *metals* and/or *alloys* by first heating them to high temperature and then rapidly cooling them, usually by immersion in a cold liquid; see *thermal processing* and *case hardening.*

quercitron A natural (*polygenetic*) *dye* derived from the inner bark of the oak tree (*Quercus tinctoria*).

quicklime *Calcinated limestone*; it is composed of calcium oxide.

RNA Acronym for *ribonucleic acid.*

RNA, messenger *RNA* that serves as a template for the synthesis of a *protein* or of *DNA.*

RNA, ribosomal *RNA* that encodes information necessary to *synthesize proteins.*

RNA, transfer One of a class of *RNA molecules* that transfer to the *ribosomes amino acids* required for the *synthesis* of *polypeptides.*

racemic mixture A mixture of equal amounts of the *optical isomers* of a substance.

racemization The process of conversion of an *optical isomer* to its *enantiomer* and formation of a *racemic mixture*; see *epimerization.*

radar Acronym for *ra*dio *d*irection and *r*anging, the common name for a number of systems used for the location and/or *imaging* of either buried or distant objects; it is based on the use of *electromagnetic radiation* with *wavelengths* ranging between 10 and 10^2 cm, that is, high-frequency *radio waves.*

radiation A form of *energy*, such as *ultraviolet radiation, gamma rays*, or *radio waves.*

radiation, background *Ionizing radiation* in the *environment*, such as *cosmic radiation* and radiation derived from *radioactive isotopes* in the *soil* and *atmosphere.*

radiation, electromagnetic *Electromagnetic radiation* as, for example, *gamma radiation* and *X-rays* that, when incident on matter, can cause the *ionization* of electrically neutral *atoms.*

radiation, ionizing See *ionizing radiation.*

radiation, penetrating See *penetrating radiation.*

radiation, solar *Electromagnetic radiation* that originates in the sun; most solar radiation that reaches the earth ranges in *wavelength* between the near *infrared* and *ultraviolet*, including *visible light.*

radiation intensity The amount of *radiation* per unit area falling upon a surface.

radio waves See *waves, radio.*

radioactive Undergoing *radioactive decay*; an unstable *nuclide* that *decays* to a different nuclide and emits *radioactivity radiation.*

radioactive atom See *radioactive isotope.*

radioactive daughter See *daughter isotope.*

radioactive decay See *radioactivity.*

radioactive element See *radioactive isotope.*

radioactive isotope See *isotope, radioactive.*

radioactive nuclide See *radionuclide.*

radioactive parent A *radioactive nuclide* that *decays* into another *nuclide* that may be either *stable* or *radioactive.*

radioactivity The process of disintegration of the *nuclei* of unstable, *radioactive isotopes*; it is always accompanied by the emission of *radiation* and/or *subatomic particles.*

radiocarbon See *carbon-14.*

radiocarbon calibration The process of converting *radiocarbon* dates to calendar dates.

radiocarbon date A *date* determined by measuring the relative amount of *radiocarbon* in a *material*; see *radiocarbon dating.*

radiocarbon dating See *dating, radiocarbon.*

radiocarbon dating, accelerator mass spectrometry See *dating, accelerator mass spectrometry radiocarbon.*

radiogenic Created or generated by *radioactive decay.*

radiography The generic name for *nondestructive* techniques for *imaging* the internal structure and morphology of objects based on the use of *penetrating radiation*; thus *beta particle, gamma,* and *X-rays* and *neutron* radiography.

radioisotope A *radioactive isotope.*

radiometric A physical process or technique based on measuring the *rate of disintegration* of *radioactive matter.*

radiometric dating See *dating, radiometric.*

radionuclide A *nuclide* that undergoes *radioactive decay.*

radio waves See *waves, radio.*

rain, acid See *acid rain.*

ramie A *bast fiber* derived from the ramie plant, also known as *China grass*, of the nettle (Urticaceae) family.

rancidity The spoiled, foul smell and taste of *lipids* as a result of *oxidation* and/or *hydrolysis* processes.

rate The quantitative measurement of a magnitude or frequency in relation to a unit of time, such as kilometers per hour, particles per second, or *atoms* per minute.

rattan A *hard fiber* derived from climbing palm plants of the genera *Calamus* and *Daemonothops.*

rawhide *Hide* of great strength used for making such objects as saddles and lashing; it is prepared by stretching and dry-*curing* the hide of large animals, such as cows and buffalo.

raw silk See *silk, raw.*

reactant One of the initial *substances* in a *reaction.*

reaction, chemical See *chemical reaction.*

reaction, nuclear See *nuclear reaction.*

recovery The process of extracting valuable materials from *rocks, minerals* and *ores.*

recrystallization (1) Repeated *crystallization;* (2) a *diagenetic* process that involves the recrystallization of *minerals* after the dissolution by *groundwater* of *primary* or *secondary minerals* and the subsequent *precipitation* of new minerals.

reducing atmosphere A smoke-laden *atmosphere* in which there is practically no *oxygen.*

reduction A *chemical reaction* in which a *substance* loses *oxygen* and/or gains *hydrogen* or *electrons.*

redwood See *camwood.*

refining The removal of *impurities* from a *substance.*

reflected light See *light reflection.*

refraction The change in direction of a beam of *light* as it passes from one medium (e.g., air or water) into another, such as *glass.*

refractive index The numerical value of the extent to which a beam of *light* is deflected from its original path when passing from a *vacuum* into a given *substance.*

refractory material A solid *material* (i.e., *pottery* or *brick*) that withstands high temperatures without changes in shape or in chemical composition.

relative age See *age, relative.*

relative date See *date, relative.*

relative dating See *dating, relative.*

remote sensing A technology for acquiring information about distant objects or phenomena, without physical contact with them; it is based on the use of *electromagnetic radiation* either reflected from or radiated by objects of interest.

replication See *DNA replication.*

residual clay See *primary clay.*

resin A *solid* or semisolid *organic substance,* generally of high *molecular weight,* with no definite *melting point;* until the twentieth century all resins were of natural origin, most being exudations from plants.

respiration The process by which living organisms use oxygen for their life processes so as to make *substances* required for maintenance and growth, and give off carbon dioxide.

retting A bacterial process for softening and causing the *decay* of woody tissue associated with vegetable *fibers,* such as *flax* and *hemp.*

rhyolite A low-*density* volcanic *rock,* also known as *lava,* that contains a high proportion of *silica.*

ribonucleic acid (RNA) A *nucleic acid* in the *cell's cytoplasm* that transfers information from *DNA* to the *proteinmaking* sites of the *cells;* its *molecule* is very similar to that of DNA, although it is single-stranded and consists of a chain of *ribonucleotides* instead of DNA's deoxyribonucleotides.

ribonucleotide The building block of the long *molecules* of *RNA*; it consists of bonded *ribose*, a *nitrogenous base*, and a *phosphate group*; see *deoxyribonucleotide*.

ribose A *simple carbohydrate*, a basic component of *RNA*; see *deoxyribose*.

ribosomal RNA (rRNA) See *RNA, ribosomal*.

ribosome An *organelle* within a *cell*, in which *amino acids* combine into *polypeptides*.

rigidity The property of solids to resist deformation by external forces.

ring, annual growth See *annual growth ring*.

roasting The process of *oxidizing* mineral *ores* by heating them in an abundant supply of air; see *calcination*.

rock A naturally formed, firm, and generally coherent mass or aggregate of one or more *minerals* or *mineraloids* that makes up a large part of the crust of the earth; see *igneous*, *sedimentary*, and *metamorphic* rock.

rock, igneous *Rock* formed when *magma* or *lava* cool down and solidify.

rock, metamorphic *Rock* whose physical properties have been altered by such geologic processes as high pressure and/or heating to high temperature but below their *melting point*.

rock, secondary See *sedimentary rock*.

rock, sedimentary A type of *rock*, also known as *secondary rock* such as *limestone* and shale, that is formed by the *lithification*, mostly by pressure, of *sediments*.

rock varnish A natural *accretion* on *rocks* that consists mainly of magnesium and iron oxides mixed with *clay minerals* and *organic matter*.

rope *Cordage* 0.8–2.5 cm thick.

rosin A *wax* derived from coniferous trees that has been used as a *flux* for *soldering* operations as well as a *varnish*.

rot (1) The *decay* of *organic matter* (see *putrefaction*); (2) the softening, change in texture and color, and loss of strength and mass of *wood* as a result of fungal activity.

rot, brown A type of *wood*-destroying fungus that decomposes *cellulose* but leaves the *lignin* in the wood almost unaltered and results in a powdery mass of decayed wood of varying shades of brown; see *white rot*.

rot, soft A slimy or mushy form of *wood decay* caused by fungi or bateria.

rot, white A type of *wood*-destroying fungus that attacks both *cellulose* and *lignin*, leaving a usually white, but also light brown or yellow, spongy mass of *wood decay* products; see *brown rot*.

row edge See *selvedge*.

ruby A precious red *gemstone*, a variety of *corundum*.

rust The usually reddish brown, brittle, and porous product of the oxidation of iron and *ferrous alloys*; it is composed mainly of hydrated ferric oxide.

saccharide *Sugar*, from the Latin word *saccharum* (sugar), a *carbohydrate*.

safflower A natural yellow *coloring matter* derived from the flowers of safflower (*Carthamus tinctoria*) plants.

salt (1) A *compound* formed when *hydrogen ions* in an *acid* are replaced by *metal* ions; (2) the common name for naturally occurring sodium chloride (table salt) that has been used, since very ancient times, as a seasoning and food preservative.

salt glaze See *glaze, salt.*

salting A process for *curing* natural animal *skins* and *hides*, by strewing them with dry common *salt* on the flesh side and then air-drying them.

sample A small part of something that is representative of the whole.

sampling The selection of a *sample.*

sand Small grains of *rock*, composed mostly of *silica* but also of *limestone* and other rocks, between 0.06 and 2 mm in diameter; *silica* sand is the most common form of silica on the crust of the earth and the almost universal *former* of all ancient *glass.*

sanders wood See *camwood.*

sandstone A *sedimentary rock* made up of *sand* cemented together and consolidated into a *solid mass.*

sap The fluid water *solution* in fresh *wood.*

sapphire A precious *gemstone*, a variety of *corundum*; sapphires occur in a range of colors, but the most common color is blue.

sapwood The *wood* made up mainly of living *cells* that forms the layer beneath the *inner bark* of the trunk and branches of trees.

saturation The point at which a *solution* has reached the limit of concentration for dissolving a *solute.*

scanning electron microscope (SEM) See *microscope, scanning electron.*

scattering The haphazard deviation of a *wave* or *particle* that encounters an obstacle on its path.

science The systematic study and organized knowledge of the material world.

scouring A *textile* cleaning process for freeing *wool* from dirt, grease, and sweat, usually before *dyeing.*

scudding The mechanical removal of *hair* from animal *skins* with a blunt tool, in order to leave their surface as free as possible of hair; also known as *dehairing*; see *liming.*

scutching The separation of *bast fibers*, such as *flax* and *hemp*, from decayed woody tissue produced by the *retting* process.

seasoned wood See *wood, seasoned.*

seasoning The process of drying and exposing to the *environment*, for a long time, fresh, wet, recently cut *wood* so as to bring it into equilibrium with its surroundings.

secondary clay See *clay, secondary.*

secondary mineral See *mineral, secondary.*

secondary rock See *sedimentary rock.*

sediment Particles (of *inorganic* or/and *organic matter*) deposited by some process.

sediment, clastic See *clastic sediment*.

sedimentary clay See *secondary clay*.

sedimentary rock See *rock, sedimentary*.

selfedge See *selvedge*.

selvedge Each side of a *woven fabric* and an actual part of the *warp* of the *fabric*; also called *listing*, *rowedge*, *selfedge*, and *selvedge*.

SEM Acronym for *scanning electron microscope*.

semimetal See *metalloid*.

semiprecious gemstone See *gemstone, semiprecious*.

sensor A device that detects a signal (e.g., *electromagnetic radiation*) and converts it into a another signal (numerical data or an image) that can be displayed and recorded.

sequencing An analytical procedure for determining of the arrangement of *nucleotides* in *nucleic acids*, or of *amino acids* in *proteins*.

seriation A technique for the *relative dating* of artifacts based on establishing their chronological order by their stylistic similarities.

sericin The *waxy protein* secreted by the *silkworm* to bind together the *brins* of *silk fibers*.

sericite A white, fine grained *mineral* composed of silicates of various metals that occurs as minute flakes.

setting The process of solidification of a such materials as *cements* and *adhesives*; see *curing*.

shale A *sedimentary rock* formed from a mixture of *clay* and very small grains of *quartz* pressed into thin sheets.

shammy See *chamois leather*.

shantung A *plain-weave fabric* made of *silk* in Shantung, China.

signature, isotopic See *isotopic signature*.

silica Silicon dioxide.

silicate A negative *ion* that includes *silica* and *oxygen* as its two components, usually assigned the *formula* SiO_3^{2-}, a constituent of many *igneous minerals* and *rocks*.

siliceous Term used to refer *rocks* and *minerals* that contain a high proportion of *silica*.

silk The *filament* made by the silkworm from *fibroin*, a *protein*, for the construction of its cocoon; it is the only natural *fiber* that occurs in the form of extremely long *filaments*.

silk, raw *Silk* from which a natural coating of *silk wax* has not been removed.

silk, wild See *tusah*.

silk cotton See *kapok*.

silt Extremely small *mineral* grains measuring between 0.004 and 0.06 mm in diameter, coarser than those of *clay* but finer than *sand*.

silver, fine An *alloy* of silver and another *metal*, usually copper, that contains at least 99.9% silver.

simple carbohydrate See *carbohydrate, simple.*

sine (sin) In *trigonometry*, the sine of an *acute* angle in a right-angled triangle is the *ratio* of the length of the side opposite the given angle to the length of the *hypotenuse.*

sintering A process in which finely divided particles of *solid* matter such as *ceramics, metals*, or *alloys* agglomerate, when heated, into larger particles or into a solid mass.

sisal The *hard fiber*s derived from the agave plant (*Agave sisalana*), used mainly for making *cordage.*

size A liquid preparation, usually a *solution* or *suspension* of an *adhesive* in water, that is applied as a thin coating to porous surfaces to seal the pores and/or to protect the surface.

skin (1) The outer covering of the body of animals; (2) the skin of young or small animals; see *hide.*

slag The waste product, also known as *dross*, of *metal smelting* operations, which consists of the valueless components of *metalliferous ores* left after removal of the metal.

slake To disintegrate a *substance* (e.g., *quicklime*) by the addition of water.

slaked lime Calcium hydroxide formed by the *slaking* action of water on *quicklime;* it is one of the components of *lime cement.*

slip A *fluid, homogeneous suspension* of *clay* particles in water; slips are used to cover the surface of *pottery* so as to endow *pottery* objects with a smooth texture and, if *pigments* are added to the slip, also with color.

slurry A fluid, *heterogeneous suspension* of solid particles in water.

smelting The basic process by which *metals* are produced from metalliferous *ores;* it is a *pyrometallurgical* process that involves the chemical *reduction* and *recovery* of the metals contained in the ores.

SMOW Acronym for *Standard Mean Ocean Water*, used as a chemical and isotopic reference standard, for example, when studying *oxygen isotope ratios.*

soap A cleaning agent made by the reaction of an *oil* or *fat* with an *alkali* such as *soda.*

soda The common name used to refer to a number of *alkaline substances* that include sodium in their composition, but mainly to sodium carbonate.

soda glass See *glass, soda.*

sodalime glass See *glass, sodalime.*

soft fiber See *fiber, soft.*

soft rot See *rot, soft.*

soft solder See *solder, soft.*

softening point (of glass) The temperature at which heated *glass* becomes sufficiently soft and fluid for working and shaping; see *working point* and *working range*.

softwood The *wood* of *coniferous* trees; the term has no relation to the actual *hardness* of the wood; see *hardwood*.

soil The loose mixture of small particles of *clay minerals, sand,* and *decayed organic matter* on the surface of the earth, above *bedrock*.

soil horizon A layer of *soil* that differs from other layers above or below it because of its texture, composition, and/or color.

soil profile A vertical section through the *horizons* of a *soil*.

solar radiation See *radiation, solar*.

solder The generic name for *alloys* composed mainly of lead (50–70%) and tin (50–30%), which *melt* at relatively low temperatures and are used for joining *metal* parts.

solder, hard An *alloy* that softens at a relatively high temperature (above 400°C), used for *hot soldering* or *brazing*.

solder, hot See *solder, hard*.

solder, soft The generic name for *alloys* composed mainly of lead (50–70%) and tin (50–30%), which *melt* at relatively low temperatures (below 400°C) and are used for joining metal parts; see *soldering, soft*.

soldering A *metallurgical* operation for joining *metal* parts using *solder; see solder, hot* and *solder, soft*.

soldering, hot See *brazing*.

soldering, soft A *metallurgical* operation for joining *metal* parts using a *soft solder*.

solid One of the three states of *matter, gas, liquid* and solid; at constant temperature solids have a definite and stable shape and volume.

solute A gaseous, liquid, or solid *substance* dissolved in a *solvent* and forming a *solution*.

solution (1) A *homogeneous mixture* of one or more *solutes* in a *solvent*; (2) a *diagenetic* process that involves the dissolution of matter in sediments or in the remains of living organisms, by *groundwater*.

solvent A *substance* that dissolves others.

sorbite A fine *dispersion* of *cementite* in *ferrite*, formed when *steel* is *tempered*.

sound An audible *longitudinal wave* propagated through a medium that displaces *molecules* from their equilibrium state.

sour (1) *Acid*; (2) having an acid taste.

souring The primary method of *neutralizing cloth*, by removing *alkalis* that remain after *bleaching*; it entails dipping the cloth in a solution of a mild *acid*, such as *sour* milk in water.

spall Splinter or chip; the term is sometimes also used to refer to relatively large chunks, flakes, or flake fragments of *obsidian* that were traded between the natural resources of obsidian and the workshops where the obsidian was shaped into tools.

specific gravity A comparison, at a specified temperature, of the weight of a volume of matter with that of an equal volume of water; it is a measure the *density* of matter.

specific gravity assay An ancient method of *assaying*, based on the fact that the *specific gravity* of *alloys* is progressively reduced when *precious metals* are *alloyed* with increasingly larger relative amounts of lighter *metals*.

spectrometer An instrument for measuring the *intensity* of *electromagnetic radiation* as a function of *frequency* (or *wavelength*).

spectrometer, accelerator mass An instrument for accelerating beams of *isotopes* to high velocities in order to sort them in terms of their different masses and measure their relative abundance; it is used to measure the number of *radiocarbon atoms* in a sample dated by the *accelerator mass spectrometry radiocarbon dating* method.

spectrometer, mass An instrument that distinguishes between *isotopes* or *ions* in terms of their different masses.

spectrometry Measurement of the characteristics – *wavelength*, spectral distribution, and *intensity*—of all forms of *electromagnetic radiation*; see *spectroscopy*.

spectrometry, mass An *analytical* technique for identifying the chemical constitution of substances by separating gaseous *ions* according to their different mass and electric charge; see *mass spectrometer*.

spectrometry, optical emission A *spectroscopic* method of *analysis* for identifying the composition of a sample, based on the fact that when *atoms* are heated to high temperature, they emit *electrons* and release characteristic *light* of specific *wavelength* that can be measured.

spectrophotometry Measurement of the characteristics (*wavelength*, spectral distribution, and *intensity*) of *visible* light.

spectroscopy Study of the characteristics (*wavelength*, *intensity*, and spectral distribution) of all forms of *electromagnetic radiation* (see *spectrometry*) and measurement of these properties.

spectroscopy, atomic absorption A *spectroscopic* technique of *analysis* based on measurement of the *wavelength* and *intensity* of *visible light* that is absorbed by a *substance* being analyzed.

spectroscopy, laser-induced breakdown An emerging *spectroscopic* technique for determining the *elemental* composition of a *sample* regardless of whether it is a solid, liquid, or gas, which provides a portable, quick, and easily adaptable facility that can be used in the field.

spectroscopy, Mössbauer A *spectroscopic technique* of *analysis* based on the use of *gamma radiation*; in archaeological studies it yields information on the *firing* conditions used in making *pottery*.

spectroscopy, X-rays fluorescence A *spectroscopic* technique of *analysis* based on irradiating a sample with a beam of *X-rays*; the X-rays excite *electrons* of characteristic *wavelengths*, associated with *atoms* on the surface of the sample; the nature of the *elements* can be calculated from the wavelength of the radiation emitted and the relative amounts of these elements in the surface from the *intensity* of the *radiation*.

spectrum (1) The range of *wavelengths* or *frequencies* of any form of *electromagnetic radiation*; (2) the range of *colors* of *visible light*, which is perceived as a sequence of red, orange, yellow, green, blue, indigo, and violet.

spectrum, absorption The *spectrum* of *electromagnetic radiation* absorbed by specific *atoms* or *molecules*.

spectrum, emission The *spectrum* of *electromagnetic radiation* emitted by *elements* or *molecules* excited by an external source of *energy* such as *heat* or/and *electric discharge*.

speculum The Latin name for an *alloy* of copper and tin containing about 40% of the latter *metal* that, when polished, has very high optical *reflectivity*; it was used by the Romans for making mirrors.

speleothem A *mineral* deposit formed in caves when calcium carbonate, or some other *mineral*, precipitates from drops of water; see *stalactite, stalagmite*.

spelter An alternative name for the *metal zinc*.

spinning The process of twisting *fibers* together to make *yarn*.

splitting Reduction of the thickness of a *skin* or *hide* by shaving a layer with a sharp instrument.

spontaneous fission See *fission, spontaneous*.

spring wood See *early wood*.

stabilizer A *substance* that, when added to others, increases their stability and makes then less liable to *decay*.

stable isotope See *isotope, stable*.

staining The process of imparting *color* to *textile fibers* with *pigments*; see *dyeing*.

stalactite A *speleothem* that hangs downward from the roof or wall of a cave; see *stalagmite*.

stalagmite A *speleothem* that grows upward from the floor or wall of a cave; see *stalactite*.

stamping A technique for transferring to a sheet of a *ductile metal* an impression from a *die* by the application of pressure.

standard deviation In statistics, a measure of the dispersion of a group of measurements around the average or mean value.

stannic oxide See *tin oxide*.

staple fiber See *fiber, staple*.

starch A *biopolymer* made up by most plants to store energy; it consists of linearly joined *glucose monomers*; see *cellulose*.

state (of matter) One of the three conditions in which *matter* exists: *solid, liquid,* and *gaseous*.

steatite A gray to green soft *mineral,* two varieties of which are soapstone and talc.

steel A *ferrous alloy* that contains small amounts of *carbon* and often other additional *elements.*

steel, Damascus See *Damascus steel.*

stereoisomers See *optical isomers.*

stochastic (A process) determined by random variables.

stone Broken-up or cut *rock.*

stone, precious See *gemstone.*

stone, semiprecious See *gemstone, semiprecious.*

stoneware A type of *ceramic* material made of *clay* heated above 1200°C that is therefore highly *vitrified,* hard, and strong.

stony cement See *cement, stony.*

strain A dimensional change in a *solid* body caused by the application of an external *stress.*

strain, plastic See *plastic strain.*

strata Layers of *rocks, sediments,* soils, or archaeological debris.

stratigraphy The study of *strata* (layers) of *rock, soil,* or archaeological debris including their origin, characteristics, and correlation with one another.

stratosphere The outermost layer of the earth's *atmosphere,* above the *troposphere.*

stress A pressure or force exerted on a *solid* body.

string *Cordage* about 1 mm thick.

strong acid See *acid.*

structural formula See *formula, structural.*

structure See *crystal structure.*

stucco Very fine *plaster* used to smooth, decorate, or ornament the surface of the interior walls of building constructions.

stuffing (leather) A process for imparting special properties to leather by working into it *wax* or *fat.*

subatomic particle See *particle, subatomic.*

subduction A geologic process during which a *plate* in the earth's *lithosphere* slides down below another plate.

sublimation The direct conversion, by heat, of a *solid substance* into a *gas* without formation of a *liquid.*

subsoil The layer of *soil* that lies below the *topsoil.*

substance *Homogeneous matter* made up of one or more *elements,* which has a specific, well-defined composition; see *compound.*

substance, biochemical A *substance* created by living organisms.

substance, biogenic A *substance* created by living organisms.

substance, inorganic *Homogeneous matter* that does not contain the *element* carbon, with the exception of the oxides of carbon (carbon monoxide and carbon dioxide) and compounds containing *carbonate ions.*

substance, organic *Homogeneous organic matter* that has a specific composition.

suede *Oil-tanned leather* finished with a velvetlike *nap* on the flesh side.

sueding The process of raising *fibers* on the surface of *oil-tanned leather* to give a velvet nap effect.

suet Hard animal *fat* surrounding the kidneys.

sugar See *carbohydrate.*

sulfate A complex *anion* composed of *atoms* of sulfur and oxygen and assigned the *formula* SO_4^{2-}; it is a component of many *minerals* and *sedimentary rocks.*

sulfide A *chemical compound* that contains negatively charged sulfide (S^{2-}) *ions.*

sulfur A *nonmetallic element.*

sumac The *tannin*-rich extract from small trees or shrubs of the genus *Rhus,* such as *Rhus coriaria* of southern Europe, which is used for *tanning* animal skins and as a *mordant* for *dyeing.*

summer wood See *wood, late.*

supercooled liquid See *liquid, supercooled.*

surface soil See *topsoil.*

surface treatment Alteration of the surface of *pottery* by *burnishing* or by coating it with a of *slip* or *glaze.*

surfactant A *substance* that, when added to a liquid, reduces the surface tension of the liquid.

suspension A mixture of very fine, nonsettling particles of a *solid* within a *liquid* or *gas.*

sweating An ancient process for the removal of *hair* or *wool* from animal *skins* by a process of controlled *putrefaction.*

symbol, chemical See *chemical symbol.*

synthesis The combination of constituent chemical units into a single *compound.*

system A group of interrelated and interacting *matter* and phenomena.

system, closed See *closed system.*

system, open See *open system.*

T Abbreviation for *thymine.*

tailings See *gangue.*

tallow Hard *fat* derived from animal carcasses, widely used in the past for making *soap* and candle wax.

tannin Generic name for bitter-tasting, astringent *organic substances* of vegetable origin that have been used since early antiquity for *tanning* animal skins.

tanning The production of *leather* from animal *skins;* see *mineral tanning, oil tanning,* and *vegetable tanning.*

tanning, mineral A process for making *leather* by treating animal *skins* with *mineral salts* such as *alum*.

tanning, oil A process for making *leather*, by treating animal *skins* with animal *oil* or oily materials derived mainly from fish or marine animals.

tanning, vegetable The process for making *leather* by treating animal skins with *tannins* derived from plants.

tanning agent See *tannin*.

tapa A *nonwoven fabric* used for writing in the South Pacific Ocean islands; it is made from the bark of mulberry trees, of the botanical genus *Morus*.

taphonomy Study of the processes that affect the remains of living organisms from the moment of death until they become *fossilized*; see *biostratinomy* and *diagenesis*.

tarnish A form of *decay* of the exposed surface of some *metals*, such as copper and silver, when they interact with *sulfur* compounds (usually *sulfides*) in the environment that is accompanied by changes in the *color* of the surface.

tattoo An implantation under the skin of a *pigmented* indelible design, usually of a decorative or symbolic nature.

tawing A very ancient *mineral tanning* process in which use was made of *alum* and *common salt*.

teasel The prickly flowerhead of plants of the family *Dipsacaceae* or a wire device, used to raise a *nap* on *textile fabrics*.

teaseling A process for making *textile fabrics* soft and rise a *nap* on their surface with a *teasel*.

TEM Acronym for *transmission electron microscope*.

temper An inert *material* added to *ceramic bodies* to alter their *plasticity* and *firing* properties; also known as *additive*, *inclusion*, *filler*, and *grog*, as well as *aplastic*, *nonplastic*, *nonclay*, or *opening material*.

temperature A measure of the average energy of *atoms* and *molecules*; the higher the temperature, the more energy the particles have.

tempering (1) A *thermal process* for softening hard and brittle *steel* by heating it to a given temperature, usually within the range 150–650°C, and then cooling at an appropriate rate; (2) in *ceramic* technology, the addition of *temper* to *clay* when preparing a *ceramic body*.

template (1) A pattern or gauge; (2) in *molecular biology*, a *molecule* (e.g., of *DNA*) that serves as a pattern for the generation of another molecule (of *RNA* in the example above).

tephra General term for *mineral* grains of all sizes ejected into the atmosphere during volcanic eruptions and deposited on the surface of the earth.

tephrochronology Study of *tephra* deposits to determine their *age* and age relationships.

terracotta A type of *pottery*, generally *unglazed*, made from *clay* and coarse *tempers* and generally *fired* at temperatures well below 900°C.

tertiarium A lead–tin *alloy* containing 33% *tin*, used by the Romans mainly for *tinning*.

tesserae Small, usually roughly square pieces of *stone* or *glass*, used to make decorative mosaics.

textile *Cloth woven* from *yarns* and used for making clothing and other coverings.

texture The distinctive appearance or feel of a solid resulting from the size and shape of the grains or particles of which it is composed.

thermal capacity See *heat capacity*.

thermal conductivity See *conductivity*.

thermal conductor See *conductor*.

thermal expansion The expansion in size undergone by most materials when heated.

thermal expansion coefficient The fractional change in length (linear expansion) or in volume (volume expansion) of a material when heated, per degree of temperature.

thermal insulator See *insulator*.

thermal shock A mechanical *stress* created within a *solid* body by a fast change in *temperature*.

thermal processing Cyclic thermal operations, such as heating and cooling, to modify the structure and *physical properties*, although not the *chemical composition*, of objects made from *metals*, *alloys*, or *glass*; see *case hardening*, *quenching*, *tempering*.

thermoluminescence The emission of *light* by *solids* heated to rather high temperatures, albeit below those at which they become *incandescent*.

thermoluminescence dating See *dating, thermoluminescence*.

thermoplastic A *material* that can be repeatedly softened when heated and hardened when cooled without undergoing chemical changes.

thermoremanent magnetism Magnetism acquired by *igneous rocks* and burned *clay* as they cool below their *Curie temperature*.

thermotropy The change of *color* due to changes in *temperature*.

thickener A *substance* added to a liquid to increase its *viscosity*.

thixotropy The property of some *viscous solutions* or *suspensions* of solids in liquids, such as *paints* and *ceramic slips*, to become more *viscous*, that is, less *fluid*, when left undisturbed, and less viscous and more fluid when shaken or agitated.

thread A fine string of made from two or more *yarns* twisted together.

threading A process for making *metal filaments* by slipping a hot metal wire through a small hole in a hot bar.

thymine (T) A *nitrogenous base*, one of the constituents of *DNA*.

timber See *lumber*.

tin disease See *beta tin*.

tin oxide Also known as *stannic oxide* used as a glass *opacifier*.

tin pest See *beta tin*.

tinkal See *borax*.

tinning A process for coating objects made of *metals* or *alloys*, generally copper, *brass*, or *bronze*, with a thin layer of tin or a lead–tin alloy; the ancient Romans used two main types of tinning alloys: a better grade known as *argentarium* and an inferior grade, *tertiarium*.

tissue *Cells* in living organisms that are organized into a structure with a specific function.

TL Acronym for *thermoluminescence*.

TL dating Abbreviation for *thermoluminescence dating*.

touching A method of *assaying*; comparing the color of a streak made by an *alloy* of unknown composition on a *touchstone*, with that left by a *precious alloy* of standard composition reveals the composition of the *assayed* alloy.

touchstone A small tablet of a dark *stone*, such as black jasper or flinty slate, which has a finely grained surface used for *assaying*; see *touching*.

toughness The property of a *material* to resist sudden stress, bending, or pulling without fracturing.

trace A very small amount; see *trace element*.

trace element See *element, trace*.

transcription The *biochemical* process by which *RNA* is *synthesized* using a *DNA template*, in the *nuclei* of *cells*.

transfer RNA See *RNA, transfer*.

transformation, nuclear See *nuclear transformation*.

translucence The property of some materials to transmit *visible light* so that objects viewed through them cannot be seen distinctly; see *transparency*.

transmission electron microscope (TEM) See *microscope, electron transmission*.

transmitted light See *light, transmission*.

transmutation, nuclear See *nuclear transformation*.

transparency The property of some materials to transmit *visible light* so that objects viewed through it can be seen distinctly; see *translucence*.

transversal wave See *wave, transversal*.

travertine A deposit of *limestone* formed in caves and around hotsprings, where *groundwater* saturated with carbonate salts cools down on exposure to air.

tree-ring dating See *dendrochronology*.

triglycerides Generic name for the *lipids*, which expresses their composition, since their *molecules* consist of three molecules of *fatty acids* bonded to one of a *glycerol*.

troposphere The layer of the *atmosphere* closest to the surface of the earth.

tuff A compacted or consolidated deposit of volcanic dust and ash.

tumbago An *alloy* of copper, silver, and gold used by pre-Columbian Indians; it usually contains about 50% copper, 33% gold, and 12% silver.

turquoise A porous blue-green to gray *mineral* composed of a hydrous phosphate of aluminum and copper; the blue variety is appreciated as a *semiprecious gemstone.*

tussah *Silk* derived from moths other than *Bombyx mori,* sometimes called *nonmulberry silk moths,* which are not reared in captivity.

tutenag See *paktong.*

tuyere A duct or pipe in the wall of a *blast furnace,* through which air may be forced into the furnace to produce a high-temperature, *oxidizing* fire.

U Abbreviation for *uracil.*

ultrasound waves See *waves, ultrasound.*

ultraviolet *Electromagnetic radiation* with *wavelengths* ranging between 10^{-8} and 400×10^{-7} m.

unconsolidated Loose, unattached, as, for example, loose dust, gravel, or sand, which has no matrix or *mineral cement* binding the grains.

underfired Term used in *ceramic* technology to refer to *pottery fired* below a required temperature and therefore faulty and flawed.

underglaze A *pigment* applied to the surface of *pottery,* before *glazing* with a transparent *glaze.*

uniform corrosion See *corrosion, uniform.*

uracil (U) A *nitrogenous base,* one of the constituents of *RNA.*

uranium series dating See *dating, uranium series.*

use wear Alteration to the surface and bulk of artifacts caused by use.

use wear analysis A technique for determining the function and/or use of ancient artifacts by analyzing alterations to their surface and bulk caused by use.

vacuum A region in space devoid of *matter.*

valence The number of *bonds* that an *atom* forms with other atoms when in a *compound.*

van der Waals forces Weak attractive *forces* bonding (linking) between *atoms* and/or *molecules* due to the opposite electric charges that they bear.

vapor The *gaseous* form of *substances* that are normally *liquids* or *solids.*

varve Thin layer, or layers, of *sediments* formed annually on the beds of lakes; varves are of use for reconstructing past environmental conditions and for *dating.*

vat A vessel used for the *dyeing* of *textiles.*

vat dye See *dye, vat.*

vegetable tanning See *tanning, vegetable.*

vehicle The fluid component (mostly water) of *ink, paint,* or *glaze,* that assists in applying the *pigments* or glaze to a surface.

vellum Animal *skin* used mainly for writing; it is made mostly from the dried and *cured* but *untanned* skins of calves; see *parchment.*

veneer A very thin, decorative sheet of a fine *wood* that is often attached to the surface of objects made from ordinary wood.

verdigris An artificial green *pigment* prepared in antiquity by exposing sheets of copper to *vinegar vapor*.

vicuña The fine and very soft *hair fiber* derived from the vicuña (*Lama vicugna*), a mammal native from the Andes mountains of South America.

vinegar A naturally occurring *acid* formed when fruitjuices exposed to the atmosphere are oxidized, and consists of a solution of acetic acid in water; it was one of the few acids known in antiquity.

virgin metal See *native metal*.

viscosity The resistance of a liquid *solution* or *suspension* to flow, due to the high friction between its *molecules*.

visible light See *light, visible*.

vitreous A *glasslike, amorphous material*.

vitrification The formation of *glass*.

vitriol (1) The archaic and confusing name for the *compounds* of three different *metals*, namely, the sulfates of copper (*blue vitriol*), iron (*green vitriol*), and zinc (*white vitriol*); all these occur naturally as *minerals* but have also been artificially made since antiquity; the outdated word *copperas* is also used to refer to these compounds and particularly to iron sulfate (green copperas); (2) the ancient name for sulfuric acid.

volumetric analysis See *analysis, volumetric*.

volumetry See *analysis, volumetric*.

ware Term widely used to refer to *pottery* objects of any kind and in any manufacturing stage.

warp The *yarn* that runs along the length of *woven fabrics*.

warping The distortion of *timber*, caused by differential swelling or shrinkage.

water, hard *Groundwater* that contains relatively large amounts of dissolved *lime*.

water hardness The concentration of calcium (Ca^{2+}) and/or magnesium (Mg^{2+}) ions in *groundwater*.

waterlogged *Saturated* with water.

water of plasticity Water that, when mixed with *clay*, endows the mixture with *plastic properties*.

wave A periodic disturbance that propagates through space, characterized by its *wavelength*, *frequency*, and *amplitude*.

wave cycle See *wavelength*.

wave motion See *wave*.

wave, longitudinal A *wave*, also known as a compression or pressure wave, such as those of *sound*, in which the periodic disturbance is along the direction traveled by the wave.

wave, transversal A *wave*, such as those of *electromagnetic radiation*, in which the periodic disturbance is across the direction traveled by the wave.

wavelength The distance between successive wave crests or wave valleys in all forms of *waves*, whether *electromagnetic* or *sound waves*; see *frequency*.

waves, electromagnetic A form of *energy* that propagates through space or *matter* in the form of *transversal waves* resulting from the oscillation of an electric field and a magnetic field at right angles to one another.

waves, micro *Electromagnetic radiation* with *wavelengths* ranging between 1×10^{-3} and 0.1 m.

waves, radio *Electromagnetic radiation* having *wavelengths* longer than 0.1 m.

waves, ultrasound *Sound waves* of very high *frequency* that are often used for *imaging* and *nondestructive resting*.

wax One of a number of natural materials of either *mineral*, vegetable, or animal origin, such as *bitumen*, *rosin*, and *beeswax*, which are brittle solids at ambient temperatures but soften, become pliable, and can be easily molded when warm.

wax printing See *batik*, *encaustic*.

weak acid See *acid*.

weathering The *decay* of materials caused by their exposure to the *environment*; weathering may be caused by either *physical* or *chemical* processes.

weathering, chemical The *weathering* of *materials* into others as a result of *chemical reactions* occurring on exposure to the *environment*; see *decay*, *diagenesis*.

weathering, physical The *weathering* of solid matter into smaller pieces caused by eroding fluids, such as air and water and the disrupting effects of climatic changes such as freezing.

weave The interlacing of *warp* and *weft* that forms *textile fabrics*; in a *plain weave*, for example, each *weft yarn* passes successively over and under single *warp* yarns in alternating rows; see *weaving*.

weaving The technology of making *cloth* by interlacing, in a *loom*, usually two (but sometimes more) sets of *yarn*, one known as the *weft* and the other set known as the *warp*, in a direction perpendicular to the weft; see *weave*.

weft The *yarn* in woven *textiles*, also known as *woof*, that run across the length of *woven fabrics*; see *weaving*.

welding The process of joining two or more *metal* parts by applying heat, pressure, or both, with or without a *filler metal*; the parts are thus joined by either *fusing* or *forging*.

welding, forge The process for joining two or more *metal* parts by striking (hammering) or pressing; when carried out by hammering, the process is also known as *hammer welding*.

welding, hammer See *welding, forge*.

wetland The land of marshes, ponds, wet meadows and *bogs*, that is inundated by water for varying periods of time during each year and supports vegetation which is adapted to saturated soil conditions.

white rot See *rot, white.*

white vitriol The archaic name for zinc sulfate; see *vitriol.*

whiting A white *pigment* that consists of calcium carbonate; it was usually obtained in antiquity by crushing *limestone* or seashells.

wicking *Cordage,* usually thin, used as candlewicks.

wild silk See *tussah.*

wood The hard, inner part of the trunks, branches, and stems of trees, inside the *bark*; it is a *composite material* made up of *cellulose fibers, hemicellulose,* and *lignin.*

wood, early The relatively light layer of annual rings *wood* formed at the beginning of the growing season; also known as *spring wood;* see *late wood.*

wood, late The relatively dense layer of *annual rings* in *wood,* also known as *summer wood,* formed during the latter part of the growing season (the summer); see *early wood.*

wood, seasoned Wood, also known as *air-dry wood,* in which the *moisture* is in equilibrium with that of the surrounding air.

wood fibers The common name for the narrow and relatively long *cellulose fibers* in *wood.*

wood rays Strands of tissue that transport *sap* in a radial direction across the *grain* of *wood.*

woof See *weft.*

wool A term generally used to refer to the soft *hair fibers* derived from the fleece of sheep or lambs.

wool, pulled *Wool* pulled from the *skin* of dead sheep.

wool, sheared Wool sheared from live sheep.

work hardening Modification of the structural properties of a *metal* or *alloy,* by mechanical operations such as hammering.

working point In *glass* technology, the temperature at which cooled molten glass becomes *viscous* enough to be worked and shaped; see *softening point.*

working range (1) In *glass* technology, the range of temperatures between the *working point* and the *softening point,* at which glass can be shaped by blowing; the working range for most types of ancient glass ranges between 600 and 900°C; (2) in *ceramic* technology the term is used to refer to the variable amount of water, dependent on the nature of the *clay* used, that endows a mixture of *clay* and water with *plastic* properties.

woven fabric *Cloth* made by interlacing long *threads* in two directions.

wrought A *metallic* object or part of an object shaped by beating or hammering.

wrought iron See *iron, wrought.*

X-rays *Electromagnetic radiation* with *wavelengths* ranging between 10^{-10} and 10^{-8} cm.

X-rays diffraction A *physical method* for determining the *structure* of *crystalline* solids by exposing the solids to *X-rays* and then studying the varying intensity of the *difracted* rays due to *interference* effects.

X-rays fluorescence spectroscopy (XRF) See *spectroscopy, X-ray fluorescence.*

X-rays microscope See *microscope, X-ray.*

X-rays radiography See *radiography.*

XRF Acronym for *X-rays fluorescence spectroscopy.*

yarn A continuous strand made up of either *fibers* or of long *filaments* bound together, suitable for *weaving, knitting,* and making *cordage.*

yarn, blend *Yarn* made by *spinning* together two or more different types of *fibers.*

yeast A type of fungus that causes the *fermentation* of *carbohydrates.*

zinc vitriol White zinc sulfate heptahydrate; see *vitriol.*

zoogenic Related to or produced by the activity of animals.

BIBLIOGRAPHY

Abraham, M. (2004), Ion beam analysis in art and archaeology, *Nuclear Instrum. Meth. Phys. Research B* **219–220**, 1–6.

Abrahams, D. H. and A. M. Edelstein (1967), A new method for the analysis of ancient dyed textiles, in Levey, M. (ed.), *Archaeological Chemistry – a Symposium*, Univ. Pennsylvania, Philadelphia, pp. 15–27.

Aceto, M. and E. Marengo (2004), The application of ICP-MS and multivariate analysis to provenance studies of pottery samples, *Sci. Tech. Cultural Heritage* **13**, 63–70.

Adam, J. P. (1994), *Roman Building: Materials and Techniques*, Batsford, London.

Adan-Bayewitz, D., F. Asaro, and R. D. Giauque (1999), Determining pottery provenance: Application of a new high-precision X-ray fluorescence method and comparison with instrumental neutron activation analysis, *Archaeometry* **41**, 1–24.

Adovasio, J. (1970), The origin, development and distribution of Western archaic textiles, *Tebiwa* **13**(2), 1–40.

Ainswoth Mitchel, C. and T. C. Hepworth (1924), *Inks, Their Composition and Manufacture*, Griffin, London.

Aitken, M. J. (1999), Archaeological dating using physical phenomena, *Reports Progress Phys.* **62**, 1333–1376.

Aitken, M. J. (1998), *An Introduction to Optical Dating: The Dating of Quaternary Sediments by the Use of Photon-Stimulated Luminescence*, Oxford Univ. Press, Oxford, UK.

Aitken, M. J. (1997), Luminescence dating, in Taylor, R. E. and M. J. Aitken (eds.), *Chronometric Dating in Archaeology*, Advances in Archaeological and Museum Science Series, Vol. 2, Plenum, New York.

Aitken, M. J. (1985), *Thermoluminescence Dating*, Academic Press, London.

Aitken, M. J. (1974), *Physics and Archaeology*, Oxford Univ. Press, Oxford, UK.

Aitken, M. J. and S. Stokes (1997), Climatostratigraphy, in Taylor, R. E. and M. J. Aitken (eds.), *Chronometric Dating in Archaeology*, Advances in Archaeological and Museum Science Series, Vol. 2, Plenum, New York.

Archaeological Chemistry, Second Edition By Zvi Goffer

Aitken, M. J., C. B. Stringer, and P. A. Mellars (eds.) (1992), *The Origin of Modern Humans and the Importance of Chronometric Dating*, Princeton Univ. Press, Princeton, NJ.

Akroyd, T. N. W. (1962), *Concrete: Properties and Manufacture*, Pergamon, Oxford.

Alberts, B., A. Johnson, J. Lewis, M. Raff, K. Roberts, and P. Walter (1998), *Essential Cell Biology: An Introduction to the Molecular Biology of the Cell*, Garland, Oxford, UK.

Alexander, J. (1923), *Glue and Gelatine*, Chemical Catalog Company, New York.

Alfassi Z. (ed.) (1990), *Activation Analysis*, Chemical Rubber Company, Boca Raton, FL.

Allan, R. O. (ed.) (1989), *Archaeological Chemistry*, Vol. 4, Advances in Chemistry Series, American Chemical Society (ACS), Washington, DC.

Allsopp, D. and K. J. Seal (1986), *Introduction to Biodeterioration*, Arnold, London.

Almond, M., S. Eversfield, and S. Atkinson (2005), A window on the past: Scientific study and chemical analysis in archaeological research, *Chemistry & Industry* **14**, 14–17.

Ambrose, S. H. (2006), A tool for all seasons – Laser ablation carbon isotope analysis and ancient diets, *Science* **314**, 930–931.

Ambrose, S. H. (1993), Isotopic analysis of paleodiets: Methodological and interpretive considerations, in Sanford, M. K. (ed.), *Investigations of Ancient Human Tissue: Chemical Analyses in Anthropology*, Gordon & Breach, Langhorne, pp. 59–130.

Ambrose, S. H. (1987), Chemical and isotopic techniques of diet reconstruction in eastern North America, in Keegan, W. F. (ed.), *Emergent Horticultural Economies of the Eastern Woodlands*, Southern Illinois Univ., Carbondale, Center for Archaeological Investigations, Occasional Papers, Vol. 7, pp. 78–107.

Ambrose, S. H. and M. A. Katzenberg (eds.) (2001), *Biochemical Approaches to Palaeodietary Analysis*, Kluwer, New York.

Ambrose S. H. and J. Krigbaum (2003), Bone chemistry and bioarchaeology, *J. Anthropol. Archaeol.* **23**, 193–199.

Ambrose, W. R. (2001), Obsidian hydration dating, in Brothwell, D. R. and A. M. Pollard (eds.), *Handbook of Archaeological Sciences*, Wiley, New York, pp. 82–92.

Ammen, C. W. (1979), *Lost Wax Investment Casting*, McGraw-Hill, New York.

Anderson, B. W. (1974), *Gem Testing*, Newness-Butterworth, London.

Andreani, C., V. C. Nunziante, G. Cinque, G. Gorini, A. Granelli, and M. Martini (eds.) (2006), Atomic and nuclear techniques for the diagnostics and the preservation of archaeological artifacts, *J. Neutron Res.* **14**(1), Special Issue.

Andrefsky, W. (2005), Lithic studies, in Maschner, H. D. G. and C. Chippindale (eds.), *Handbook of Archaeological Methods*, Altamira Press, Walnut Creek, CA, Chapter 19.

Andrefsky, W. (1998), *Lithics: Macroscopic Approaches to Analysis*, Cambridge Univ. Press, Cambridge, UK.

Andrews, C. (1998), *Egyptian Mummies*, British Museum, London.

Anglos, D. (2001), Laser-induced breakdown spectroscopy (LIBS) in art and archaeology, *Appl. Spectrosc.* **55**(6), 186A–205A.

Angus, T. (1976), *Cast Iron: Physical and Engineering Properties*, Butterworth, London.

Anheuser K. (2000), Amalgam tinning of Chinese bronze antiquities, *Archaeometry* **42**, 189–200.

Anheuser, K. (1997), The practice and characterization of historic fire gilding techniques, *J. Mining, Metals Mater. Soc.* (Nov.), 58–62.

Anheuser, K. and J. P. Northover (1994), Silver plating on Roman and Celtic coins from Britain – a technical study, *B. Numismatic J.* **64**, 22–32.

Anovitz, L. M., J. Michael Elam, L. R. Riciputi, and D. R. Cole (1999), The failure of obsidian hydration dating: Sources, implications and new directions, *J. Archaeol. Sci.* **26**, 735–752.

Arem, J. E. (1987), *Color Encyclopedia of Gemstones*, Van Nostrand-Reinhold, New York.

Armiento, G., D. Attanassio, and R. Platania (1997), Electron spin resonance study of white marbles, *Archaeometry* **39**, 309–319.

Aronson, J. L., T. J. Schitt, R. C. Walter, M. Taieb, J. J. Tiercelin, D. C. Johnson, C. W. Naeser, and A. E. M. Nairn (1976), New geochronologic and palaeomagnetic data from the hominid bearing hadar formation of Ethiopia, *Nature* **267**, 323–327.

Arpino, P., J. P. Moreau, C. Oruezabal, and E. Flieder (1977), Gas chromatographic-mass spectrometric analysis of tannin hydrolysates from the ink of ancient manuscripts (XI to XVI centuries), *J. Chromatogr.* **134**, 433–439.

Arrhenius, O. (1935), *Markundersökning och Arkeologi*, Fornvännen, Stockholm, pp. 65–76.

Asahina, T., F. Yamasaki, and K. Yamasaki (1958), in Exterman, R. C. (ed.), *Int. Conf. Isotopes in Scientific Research*, Vol. 2, Pergamon, London, pp. 528–532.

Asaro, F., F. H. Stross, and R. L. Burger (2002), *Breakthrough in Precision (0.3 percent) of Neutron Activation Analyses Applied to Provenience Studies of Obsidian*, paper LBNL-51330, Lawrence Berkeley National Laboratory, Berkeley, CA.

Ashok, R. (1993), *Artists Pigments*, National Gallery of Art and Oxford Univ. Press, Oxford.

Ashurst, J. (1991), Mortars and stone buildings, in Ashurst, J. and F. G. Dimes (eds.), *Conserv. Build. Decor. Stone* **2**, 90–106.

Ashurst, J. (1983), *Mortars, Plasters and Renders in Conservation, a Basic Guide*, I.C.R.O.M., Rome.

Ashurst, J. and F. G. Dimes (eds.) (1991), *Conservation of Building and Decorative Stone*, Vol. 2, Butterworth-Heinemann, London.

Asimov, I. (2002), *Atoms: Journey Across the Subatomic Cosmos*, Penguin, New York.

Asimov, I. (1974), *Building Blocks of the Universe*, Abelard-Schuman, New York.

Asimov, I. (1962), *The Search for the Elements*, Basic Books, New York.

Asquith, L. (1977), *Chemistry of Natural Protein Fibers*, Plenum, New York.

Astruc, L., R. Vargiolu, and H. Zahouani (2003), Wear assessments of prehistoric instruments, *Wear* **255**(1), 341–347.

Attas, M. (1986), *A Review of Neutron Activation Analysis*, Reactor Technology Notes, no. 5.

Aufderheide, A. C. (2003), *The Scientific Study of Mummies*, Cambridge Univ. Press, Cambridge, UK.

Avery, D. (1982), The iron bloomery, in Wertime, T. A. and S. F. Wertime (eds.), *Early Pyrotechnology*, Smithsonian Institution, Washington, DC.

Ayala, F. J. (1995), The myth of Eve: Molecular biology and human origins, *Science* **270**, 1930–1936.

Azaroff, L. V. (1968), *Elements of X-Ray Crystallography*, McGraw-Hill, New York.

Bacci, M. (2000), UV, VIS, NIR, FT-IR and FORS Spectroscopy, in Ciliberto, E. and G. Spoto (eds.), *Modern Analytical Methods in Art and Archaeology*, Chemical Analysis Series, Vol. 155, Wiley, New York, pp. 321–362.

Bachmann, H. G. (1982), *The Identification of Slags from Archaeological Sites*, Historical Metallurgical Publications, Institute of Archaeology, London Univ., London.

Bada, J. L. (1990), Racemization dating, *Science* **248**, 539–540.

Bada, J. L. (1985), Amino acid racemization dating of fossil bones, *Ann. Rev. Earth Planet. Sci.* **13**, 241–268.

Bada, J. L. (1972), Dating of bones using the racemization of isoleucine, *Earth Planet. Sci. Lett.* (July).

Baer, N. S. and N. Indictor (1974), Chemical investigations of ancient Near Eastern archaeology ivory artifacts, in Beck, C. W. (ed.), *Archaeological Chemistry*, Advances in Chemistry Series, Vol. 1, ACS, Washington, DC, pp. 241–251.

Bahn, P. G. (1992), The making of a Mummy, *Nature* **356**, 109.

Bahn, P. G. and K. Everett (1993), Iceman in the cold light of day, *Nature* **362**, 11–12.

Baillie, M. G. L. (1995), *Tree rings in archaeology*, Univ. Chicago Press, Chicago.

Bakas, Th. H., N. H. Gangas, I. Sigalas, and M. J. Aitken (1980), Mössbauer study of Glozel tablet, *Archaeometry* **22**, 69–80.

Baker, J. T. (1974), Tyrian purple: An ancient dye, a modern problem, *Endeavour* **33**(118), 11–17.

Baldwin, H. and L. B. Marman Jr. (1981), *What Is Water?*, U.S. Department of Interior, Geological Survey, Washington, DC.

Balfour-Paul, J. (ed.) (2000), *Indigo*, Fitzroy Dearborn, London.

Bandyopadhyay, D. (2006), Study of materials using Mossbauer spectroscopy, *Int. Materials Rev.* **51**, 171–208.

Bar Yosef, O. (2000), The impact of radiocarbon dating on Old World archeology: Past achievements and future expectations, *Radiocarbon* **42**(1), 23–39.

Barba, L. A. and A. Ortiz (1992), Análisis químico de pisos de ocupación: Un caso etnográfico de Tlaxcala, *Latin Am. Antiq.* **3**, 63–82.

Barber, D. J. and I. C. Freestone (1990), An investigation of the origin of the colour of the Lycurgus Cup by analytical transmission electron microscopy, *Archaeometry* **32**, 33–45.

Barbetti, M. F. (1976), *J. Archaeol. Sci.* **3**, 137–151.

Bareham, T. (1994), Bronze casting experiments, *J. Hist. Metal. Soc.* **28**(2), 112–116.

Barfield, L. H. (1994), The Iceman reviewed, *Antiquity* **68**, 10–26.

Barilaro, D., G. Barone, V. Crupi, and D. Majolino (2005), Characterization of archaeological findings by FTIR spectroscopy, *Spectroscopy* **20**, 18–22.

Barker, H. (1973), Scientific Criteria in the Authentication of Antiquities, *Application of Science in Examination of Works of Art: Proc. Seminar, June 15–19, 1970, Museum of Fine Arts, Boston*, Museum of Fine Arts, Boston.

Barnes, J. W. (1988), *Ores and Minerals: Introducing Economic Geology*, Open Univ., Milton Keynes.

Barnett, J. R., S. Miller, and E. Pearce (2006), Colour and art – A brief history of pigments, *Optics Laser Tech.* **38**, 445–453.

Barrandon, J. N., J. P. Callu, and C. Brenot (1977), The analysis of Constantinian coins by non-destructive californium analysis, *Archaeometry* **19**, 173–186.

Barrett, T. (1988), *Japanese Papermaking*, John Weatherill, New York.

Barrio, M. J. (2003), Evaluacion critica de los principios en arqueometria, conservacion y restauracion de los vidrios arqueologicos, *Patina* (Epoca 2) **12**, 53–64.

Baruchel J., J. Y. Buffiere, E. Maire, P. Merle, and G. Peix (eds.) (2000), *X-Ray Tomography in Material Science*, Hermes, Kogan Page, London.

Batt, C. (2005), *Analytical Chemistry in Archaeology*, Cambridge Univ. Press, Cambridge, UK.

Baxter, M. J. and C. E. Buck (2000), Data analysis and statistical analysis, in Ciliberto, E. and G. Spoto (eds.), *Modern Analytical Methods in Art and Archaeology*, Chemical Analysis Series, Vol. 155, Wiley, New York, pp. 681–746.

Baxter, M. S. and A. Walton (1970), *Nature* **225**, 937–939.

Bayer, G. and H. G. Wiedermann (1983), Papyrus: The paper of ancient Egypt, *Anal. Chem.* **55**, 1220A–1230A.

Bayley, J. (1998), The production of brass in antiquity with particular reference to Roman England, in Craddock, P. T. (ed.), *Two Thousand Years of Zinc and Brass*, London.

Bealer, A. W. (1969), *The Art of Blacksmithing*, Funk & Wagnalls, New York.

Béarat, H., M. Fuchs, M. Maggetti, and D. Paunier (eds.) (1997), *Roman Wall Paintings: Materials, Techniques, Analysis and Conservation*, Institute of Mineralogy and Petrography, Fribourg Univ., Fribourg, Switzerland.

Béarat, H. and T. Pradell (1997), Contribution of Mössbauer spectroscopy to the study of ancient pigment and paintings, in Béarat, H., M. Fuchs, M. Maggetti, and D. Paunier (eds.), *Roman Wall Paintings: Materials, Techniques, Analysis and Conservation*, Institute of Mineralogy and Petrography, Fribourg Univ., Fribourg, Switzerland, pp. 239–256.

Beck, C. W. (1986), Spectroscopic investigations of amber, *Appl. Spectrosc. Rev.* **22**, 57–110.

Beck, C. W. (ed.) (1974), *Archaeological Chemistry*, Advances in Chemistry Series, Vol. 1, ACS, Washington, DC.

Beck, C. W. (1970), Amber in archaeology, *Archaeology* **23**, 7–11.

Beck, C. W. and S. Shennan (1991), *Amber in Prehistoric Britain*, Oxbow, Oxford, UK.

Beck, L., S. Bosonnet, S. Reveillon, D. Eliot, and F. Pilon (2004), Silver surface enrichment of silver-copper alloys: A limitation for the analysis of ancient silver coins by surface techniques, *Nuclear Instrum. Meth.* (*B*) **226**(1–2), 153–162.

Bekefi, G. and A. H. Barrett (1987), *Electromagnetic Vibrations, Waves and Radiation*, MIT Press, Cambridge, MA.

Bell, L. A. (1983), *Papyrus, Tapa, Amate and Rice Paper*, Liliacea, McMinnville.

Bellhouse, D. R. (1980), Sampling studies in archaeology, *Archaeometry* **22**, 123–132.

Bellot-Gurlet, L., S. Pages-Camagna, and C. Coupry (eds.) (2006), Raman spectroscopy in art and archaeology II, *J. Raman Spectr.* **37**.

Bergsoe, P. (1937), *The Metallurgy and Technology of Gold and Platinum among pre-Columbian Indians* (transl. F. C. Reynolds), Copenhagen.

Berke, H. (2007), The invention of blue and purple pigments in ancient times, *Chem. Soc. Rev.* **36**, 15–30.

Berke, H. (2002), Chemistry in ancient times: The development of blue and purple pigments, *Angewandte Chemie. Int. Ed.* **41**(14), 2483–2487.

Berthelot, M. (1901), *Comptes Rendus* **132**, 9.

Bertrand, L., J. Doucet, P. Dumas, A. Simionovici, A, G. Tsoucaris, and P. Walter (2003), Microbeam synchrotron imaging of hairs from ancient Egyptian mummies, *J. Synchrotron Rad.* **10**(5), 387–392.

Berzero, A., V. Carmella-Crespi, and P. Cavagna (1997), Direct gamma-ray spectrometric dating of fossil bones: Preliminary results, *Archaeometry* **39**, 189–203.

Bethell, P. H., R. P. Evershed, and L. J. Goad (1993), The investigation of lipids, in Lambert, J. B. and G. Grufoe (eds.), *Organic Residues by Gas Chromatography / Mass Spectrometry; Applications to Palaeodietary Studies*, Archaeology at the Molecular Level, Springer, Berlin, pp. 227–255.

Bethell, P. H. and I. Máté (1989), The use of phosphate analysis in archaeology, in Henderson, J. (ed.), *Scientific Analysis in Archaeology*, Oxford Univ. Committee Monograph, Vol. 19, pp. 1–29.

Bibliothèque Nationale (1988), *Vrai or Faux?*, Paris.

Bicchieri, M., A. Sodo, G. Piantanida, and C. Coluzza (2006), Analysis of degraded papers by non-destructive spectroscopic techniques, *J. Raman Spect.* **37**, 1186–1192.

Bicchieri, M., M. Nardone, A. Sodo, M Corsi, G. Cristoforetti, V. Palleschi, A. Salvetti, and E. Tognoni (2000), The characterization of historical pigments: A crucial problem in the connotation of ancient manuscripts, *Proc. 23rd Int. Conf. Lasers*, pp. 803–806.

Bieber, A. M., D. W. Brooks, G. Harbottle, and E. V. Sayre (1976), Application of multivariate techniques to analytical data on Aegean ceramics, *Archaeometry* **18**, 59–74.

Bienkiewicz, K. (1983), *Physical Chemistry of Leather Making*, Krieger, Malabar.

Biers, W. R., K. O. Gerhardt, and R. A. Braniff (1994), *Lost Scents: Investigation of Corinthian Plastic Vases by Gas Chromatography – Mass Spectrometry*, Univ. Pennsylvania Museum, Philadelphia.

Bigazzi, G., Z. Yegingil, T. Ercan, M. Oddone, and M. Ozdogan (1994), Provenance studies of prehistoric artifacts in eastern Anatolia: Interdisciplinary research results, *Mineralogica et Petrographica Acta* **37**, 17–36.

Billmeyer, F. W. and M. Saltzman (1981), *Principles of Color Technology*, Wiley, New York.

Bimson, M. and I. C. Freestone (eds.) (1992), *Early Vitreous Materials*, British Museum, London.

Bird, J. N. and M. S. Weathers (1977), Native iron occurrence in Dikko Island, Greenland, *J. Geol.* **85**, 359–371.

Biró, K. T. (2005), Non-destructive research in archaeology, *J. Radioanal. Nuclear Chem.* **265**, 235–240.

Biró, K. T., I. Pozsgai, and A. Vlader (1986), Electron beam micro-analysis of obsidian samples from geological and archaeological sites, *Acta Archaeologica Academia Scientiarium Hungaricae* **38**, 257–278.

Bishay, A. (ed.) (1974), *Recent Advances in Science and Technology of Materials*, Vol. 3, Plenum, New York,

Bishop, R. L. and M. J. Blackman (2002), Instrumental neutron activation analysis of archaeological ceramics: Scale and interpretation, *Acc. Chem. Research* **35**(8), 603–610.

Bitossi, G., R. Giorgi, M. Mauro, B. Salvadori, and L. Dei (2005), Spectroscopic techniques in cultural heritage conservation: A survey, *Appl. Spectrosc. Rev.*, **40**(3), 187–228.

Blackwell B. A. and H. P. Schwarcz (1995), The uranium series disequilibrium dating methods, in Rutter, N. W. and N. R. Catto (eds.), *Dating Methods for Quaternary Deposits*, Geotext 2, Geological Association of Canada.

Bläuer-Böhm, C. and E. Jägers (1997), Analysis and recognition of Dolomitic lime mortars, in Béarat, H., M. Fuchs, M. Maggetti, and D. Paunier (eds.), *Roman Wall Paintings: Materials, Techniques, Analysis and Conservation*, Institute of Mineralogy and Petrography, Fribourg University, Fribourg, Switzerland, pp. 223–235.

Block, B. P., W. H. Powell, and W. C. Fernelius (1990), *Inorganic Chemical Nomenclature, Principles and Practice*, Oxford Univ. Press, Oxford, UK.

Bloom, A. L. (1969), *The Surface of the Earth*, Prentice-Hall, Englewood-Cliffs, NJ.

Bloomfield, V. A., D. M. Crothers, I. Tinoco, and J. Hearst (2000), *Nucleic Acids: Structures, Properties and Functions*, University Science Books, Sausalito, CA.

Blumel, C. (1969), Zur Echtheitsfrage des Antiken Bronzepferdes im Metropolitan Museum in New York, *Archaologischer Anzeiger* 208–216.

Bohn, H. (1979), *Soil Chemistry*, Wiley, New York.

Bonnin, A. (1924), *Tutenag and Paktong*, Oxford Univ. Press, London.

Born, H. (1985), *Archäeologische Bronzen: Antike Kunst, Moderne Technik*, Reiner, Berlin.

Born, W. (1936), Der Scharlach, *Ciba Rundschau* 218–228.

Bose, P. R., Y. Sankarnarayanan, and S. C. Sengupta (1963), *The Chemistry of Lac*, Indian Lac Research Inst., Ranchi, India.

Bossom, M. (1992), *Encaustic Art*, Arts Encaustic International, London.

Bourriau, J. D., P. T. Nicholson, and P. J. Rose (2000), Pottery, in Nicholson, P. T. and I. Shaw, *Ancient Egyptian Materials and Technology*, Cambridge Univ. Press, Cambridge, UK, pp. 121–147.

Bowen, E. J. (1946), *The Chemical Aspects of Light*, Oxford Univ. Press, Oxford, UK.

Bowen, R. (1966), *Paleotemperature Analysis*, Elsevier, Amsterdam.

Bowie, S. H. V. and C. F. Davidson (1955), *Bull. Br. Museum (Natural History and Geology)* **2**, 276.

Bowman, H. R., F. Asaro, and I. Perlman (1973), Composition variations in obsidian sources and the archaeological implications, *Archaeometry* **15**, 123–132.

Bowman, H. R., F. Asaro, and I. Perlman (1972), *Lawrence Berkeley Laboratory Report L.B.L. 661*, Berkeley, CA.

Bowman, H. R., F. H. Stross, F. Asaro, R. L. Hay, R. F. Heizer, and H. V. Michel (1984), The Northern Colossus of Memnon: New slants, *Archaeometry* **26**, 218–229.

Bowman, S. G. E. (ed.) (1991), *Science and the Past*, British Museum, London.

Bowman, S. G. E. (1990), *Radiocarbon Dating*, British Museum, London.

Bowman, S. G. E. (1982), TL Studies of burnt clays, *PACT* **6**, 351–361.

Bowman, S. G. E. and G. de G. Sieveking (1983), Thermoluminescence dating of burnt flint from Combe Grenal, *PACT* **9**, 253–268.

Boynton, R. S. (1980), *Chemistry and Technology of Lime and Limestone*, Wiley, New York.

Brachert, T. (1985), *Patina*, Calway, Munich.

Bradley, R. (1999), *Paleoclimatology*, Academic Press, New York.

Bradley, S. (ed.) (1997), *The Interface Between Science and Conservation*, British Museum, London.

Branch, N., M. Canti, P. Clark, C. Turney, and A. Hodder (2005), *Environmental Archaeology: Theoretical and Practical Approaches*, Arnold, London.

Bray, C. (2000), *Ceramics and Glass: A Basic Technology*, Society of Glass Technology, London.

Brill, R. H. (1979), A small glass factory in Afghanistan, *Glass Art Soc. J.* 26–27.

Brill, R. H. (ed.) (1971), *Science and Archaeology*, MIT, Cambridge, MA.

Brill, R. H. et al. (1990), *Scientific Investigations of Ancient Glasses and Lead-Isotope Studies*, Collected Papers, Lexis Nexis, New York.

Brill, R. H., I. L. Barnes, and E. C. Joel (1984), Lead isotope studies of early Chinese glasses, in Brill, R. H. and J. H. Martin (eds.), Scientific research in early Chinese glass, *Proc. Inte. Symp. Glass at Beijing*, Corning Museum of Glass, Corning, NY.

Brill, R. H. and H. P. Hood (1961), A new method for dating ancient glass, *Nature* **189**(4738), 12–14.

Brill, R. H. and J. H. Martin (eds.) (1984), Scientific research in early Chinese glass, *Proc. Int. Symp. Glass at Beijing*, Corning Museum of Glass, Corning, NY.

Brill, T. B. (1980), *Light – Its Interaction with Art and Antiquities*, Plenum, New York.

Brimhall, G. (1990), The genesis of ores, *Sci. Anal.* (May), 48–55.

Brocardo, G. (1982), *Mineral Gemstones, an Identification Guide*, Hyppocrene, New York.

Broecker, W. S., J. L. Kulp, and C. S. Tucek (1956), *Science* **124**, 154.

Bromelle, N. S. (1984), *Adhesives and Consolidants*, IIC, London.

Bronitsky, G. (ed.) (1989), *Pottery Technology*, Westview, San Francisco.

Bronitsky, G. and R. Hamer (1986), Experiments in ceramic technology: The effects of various tempering materials on impact and thermal-shock resistance, *Am. Antiquity* **51**, 89–101.

Brothwell, D. (1987), *The Bog People and the Archaeology of People*, Harvard Univ., Cambridge, MA.

Brothwell, D. (1981), *Digging Up Bones*, Oxford Univ. Press, London.

Brothwell, D. and E. Higgs (1963), *Science in Archaeology*, Thames & Hudson, London.

Brothwell, D. R. and A. M. Pollard (eds.) (2001), *Handbook of Archaeological Sciences*, Wiley, New York, pp. 449–460.

Brown, H. B. and J. O. Ware (1958), *Cotton*, McGraw-Hill, New York.

Brown, P. W. and J. R. Clifton (1979, 1978), Adobe, *Stud. Conserv.* **24**, 23–39; **23**, 139–146.

Brown, T. and K. Brown (1992), Ancient DNA and the archaeologist, *Antiquity* **66**, 10–23.

Browning, B. L. (1970), The nature of paper, in Winger, H. W. and R. D. Smith (eds.), *Deterioration and Preservation of Library Materials*, Chicago Univ. Press, Chicago, pp. 18–38.

Brunello, F. (1973), *The Art of Dyeing in the History of Mankind*, Neri Pozza, Vicenza.

Bryant, V. M., Jr. (1986), Prehistoric diet: A case for coprolite analysis, in Shafer, H. J. (ed.), *Ancient Texans: Rock Art and Lifeways along the Lower Pecos*, San Antonio Museum Association, Texas Monthly Press, Austin, pp. 132–135.

Bryant, V. M., Jr. (1974), The role of coprolites in archaeology, *Bull. Texas Archaeol. Soc.* **45**, 1–28.

Bryant, V. M., Jr. and G. W. Dean (1975), The coprolites of man, *Sci. Am.* **232**, 100–109.

Bubernick, D. and F. Record (1984), *Acid Rain Information Book*, Noyes, Ridge Park.

Buchwald, V. F. and H. Wivel (1998), Slag analysis as a method for the characterization and provenancing of ancient iron objects, *Mater. Charact.* **40**, 73–96.

Buckley, S. (2003), Resurrecting the past – The hidden power of chemistry, *Chem. Rev.* **13**(2), 24–28.

Buckley, S. A. and R. P. Evershed (2001), Organic chemistry of embalming agents in Pharaonic and Graeco-Roman mummies, *Nature* **413**, 837–841.

Budd, P., R. Haggerty, A. M. Pollard, B. Scaife, and R. G. Thomas (1996), Rethinking the quest for provenance, *Antiquity* **70**, 168–174.

Budd, P. and B. Ottaway (1991), *The Properties of Arsenical Copper Alloys: Implications for the Development of a Neolithic Metallurgy*, Oxbow Monograph 9, Oxford.

Budd, S. M. (1975), ESCA examination of tin oxide coatings on glass surfaces, *J. Non-Cryst. Solids* **19**, 55–64.

Budden, S. (ed.) (1991), *Gilding and Surface Decoration*, UK Institute of Conservation, London.

Burgess, D. (1990), *Chemical Science and Conservation*, Macmillan, London.

Burgio, L., R. J. H. Clark, T. Stratoudaki, M. Doulgeridis, and D. Anglos (2000), Pigment identification in painted artworks: A dual analytical approach employing LIBS (laser-induced breakdown spectroscopy) and Raman microscopy, *Appl. Spectrosc.* **54**(4), 463–469.

Burkett, M. E. (1979), *The Art of the Felt Maker*, Abbot Art Gallery, Kendal.

Burmester, A. (1983), Far Eastern lacquers, *Archaeometry* **25**, 73–81.

Burnham, A. J. and R. I. Macphail (1995), *Archaeological Sediments and Soils: Analysis, Interpretation and Management*, Institute of Archaeology, London Univ., London.

Burns, G. (1992), Eco-archaeometry: Interdisciplinary applications in Egypt, *Interdisc. Sci. Rev.* **17**, 81–90.

Burroughs, W. J. (2005), *Climate Change in Prehistory: The End of the Reign of Chaos*, Cambridge Universtiy, Cambridge, UK.

Burton, J. J. (1996), Trace elements in bone as palaeodietary indicators, in Orna, M. V. (ed.), *Archaeological Chemistry – Organic, Inorganic and Biochemical Analysis*, ACS, Washington, DC, pp. 327–333.

Bushong, S. C. (2000), *Computerized Tomography*, McGraw-Hill, New, York.

Cahill, J., K. Reinhard, D. Tarler, and P. Warnock (1991), Scientists examine remains of ancient bathroom, *Bibl. Archaeol. Rev.* **27**, 64–69.

Caley, E. R. (1967), The origin and manufacture of orichalcum, in Levey, M. (ed.), *Archaeological Chemistry, a Symposium*, Univ. Pennsylvania, Philadelphia, pp. 59–73.

Caley, E. R. (1964), *Analysis of Ancient Metals*, Macmillan, New York.

Caley, E. R. (1962), *Analysis of Ancient Glasses*, Corning Museum of Glass, Corning.

Caley, E. R. (1946), Ancient Greek Pigments, *J. Chem. Educ.* **23**, 314–315.

Caley, E. R. (1927), The Stockholm Papyrus, *J. Chem. Educ.* **4**, 979–1002.

Caley, E. R. and D. T. Easby (1964), New evidence of tin smelting and the use of metallic tin in pre-Conquest Mexico, *Actas del XXXV Congreso Internacional de Americanistas*, Mexico, pp. 507–517.

Callen, E. O. (1969), Diet as revealed in Coprolites, in Brothwell, D. R. and E. S. Higgs (eds.), *Science in Archaeology*, Thames & Hudson, London.

Callen, E. O. (1967), Analysis of Tehuacan Coprolites, in Byers, D. S. (ed.), *The Prehistory of the Tehuacan Valley*, London.

Calnan, C. and B. Haines (eds.) (1991), *Leather: Its Composition and Changes with Time, Proc. 1996 Conf.* UK Leather Conservation Centre, Northampton.

Camufo, D., C. Daffara, and M. Sghedoni (2000), Archaeometry of air pollution: Urban emission in Italy during the 17th century, *J. Archaeol. Sci.* **27**, 685–690.

Cannon, J. and M. Cannon (1994), *Dye Plants and Dyeing*, Royal Botanic Gardens, London.

Cano, R. J. (2000), Biomolecular methods, in Ciliberto, E. and G. Spoto (eds.), *Modern Analytical Methods in Art and Archaeology*, Chemical Analysis Series, Vol. 155, Wiley, New York, pp. 241–254.

Cano, R. J. (1996), Analyzing ancient DNA, *Endeavour* **20**(4), 162–167.

Cariati, F. and S. Bruni (2000), Raman spectroscopy, in Ciliberto, E. and G. Spoto (eds.), *Modern Analytical Methods in Art and Archaeology*, Chemical Analysis Series, Vol. 155, Wiley, New York, pp. 255–275.

Carnot, A. (1893), Sur une application de l'analyse chimique pour fixer l'age d'ossements humaines préhistoriques, *Comptes Rendus de l'Academie des Sciences* **115**, 337–339, 243–246; **114**, 1189–1192.

Caroll, D. (1970), *Rock Weathering*, Plenum, New York.

Carrillo y Gariel, A. (1946), *Técnica de la Pintura de Nueva España*, Imprenta Universitaria, Mexico, pp. 42–48.

Carter, G. F. (1993), Chemical and discriminant analysis of Augustan Asses, *J. Archaeol. Sci.* **20**, 101–115.

Carter, G. F. (ed.) (1978), *Archaeological Chemistry* (Vol. 2), Advances in Chemistry Series, ACS, New York.

Carvalho, D. N. (1971), *Forty Centuries of Ink*, Franklin, New York.

Cassar, M., G. V. Robins, R. A. Fletton, and A. Alstin (1983), Organic components in historical non-metallic seals investigated using C-13-NMR spectroscopy, *Nature* **303**, 238–239.

Cassey, J. and R. Reece (1988), *Coins and the Archaeologist*, Seaby, London.

Catling, H. W., A. E. Blyn Stoyle, and E. E. Richards (1961), *Archaeometry* **6**, 1.

Catling, D. M. and J. Grayson (1982), *Identification of Vegetable Fibers*, Chapman & Hall, London.

Cattaneo, C., K. Gelsthorpe, P. Phillips, and R. J. Sokol (1993), Blood residues on stone tools, *World Archaeol.* **87**, 365–372.

Cauet, B. (1999), *L'Ordans L'Antiquite, de la Mine a l'Object* (Gold in Antiquity, from the Mine to the Object), Federation Aquitania, Bordeaux.

Cauvin, M. C. (ed.) (1998), *L'Obsidienne au Proche et Moyen Orient: Du Volcan à l'Outil*, Archaeopress, Oxford, UK.

Cavalli-Sforza, L. L., P. Menozzi, and A. Piazza (1994), *The History and Geography of Human Genes*, Princeton Univ. Press, Princeton, NJ.

Caylus, C. de (1759), *Recueil d'Antiquites* **3**, 195–199.

Cechak, T., J. Gerndt, C. M. Kubelik, L. Musilek, and M. Pavlik (2000), Radiation methods in research of ancient monuments, *Appl. Rad. Isotopes* **53**(4–5), 565–570.

Celoria, F. (1971), Archaeology and dyestuffs, *Sci. Archaeol.* **6**, 15–39.

Chabas, A. and D. Jeannette (2001), Weathering of marbles and granites in marine environment: Petrophysical properties and special role of atmospheric salts, *Environ. Geol.* **40**(3), 359–368.

Chaplin, R. E. (1971), *The Study of Animal Bones from Archaeological Sites*, Seminar, London.

Charters, S., R. P. Evershed, P. W. Blinkhorn, and D. Denham (1995), Evidence for the mixing of fats and waxes in archaeological ceramics, *Archaeometry* **37**, 113–127.

Chawla, K. K. (1998a), *Fibrous Materials*, Cambridge Univ. Press, Cambridge, UK.

Chawla, K. K. (1998b), *Composite Materials, Science and Engineering*, Springer, New York.

Chenciner, R. (2001), *Madder Red: A History of Luxury and Trade*, Curzon, Richmond.

Cherry, J. F. and A. B. Knapp (1991), Quantitative provenance studies and Bronze Age trade in the Mediterranean: Some preliminary reflections, in Gale, N. H. (ed.), *Studies in Mediterranean Archaeology*, Ästroms, Jonsered, Vol. 40, pp. 92–111.

Chirnside, R. C. and P. M. C. Proffit (1963), *J. Glass Stud.* **5**, 18.

Chowdhury, S. N. (1981), *Muga Silk*, Directorate of Sericulture and Weaving, Cauhati, Assam.

Christian, G. D. (1994), *Analytical Chemistry*, Wiley, New York.

Christie, W. W. (1989), *Gas Chromatography and Lipids: A Practical Guide*, The Oily Press, Ayr.

Ciliberto, E. (2000), Analytical methods in art and archaeology, in Ciliberto, E. and G. Spoto (eds.), *Modern Analytical Methods in Art and Archaeology*, Chemical Analysis Series, Vol. 155, Wiley, New York, pp. 1–10.

Ciliberto, E. and G. Spoto (eds.) (2000), *Modern Analytical Methods in Art and Archaeology*, Chemical Analysis Series, Vol. 155, Wiley, New York.

Cipollaro, M., U. Galderisi, and G. Di Bernardo (2005), Ancient DNA as a multidisciplinary experience, *J. Cell. Physiol.* **202**(2), 315–322.

Clark, A. J. (1996), *Seeing Beneath the Soil: Prospecting Methods in Archaeology*, Batsford, London.

Clark, G. R. (1974), Growth lines in some invertebrate skeletons, *Annu. Rev. Earth Planetary Sci.* **2**, 77–79.

Clark, J. E. (1981), Multifaceted approach to the study of Meso-American obsidian trade, *Proc. 47th Mtg. Society for Analytical Archaeology*, San Diego.

Clark, R. J. H. (2007), Raman microscopy as a structural and analytical tool in the fields of art and archaeology, *J. Molec. Struct.* 834–836, 74–80.

Clark, R. J. H. (2002), Pigment identification by spectroscopic means: An arts – science interface, *Comptes Rendus Chimie* **5**(1), 7–20.

Clark, R. J. H. (1995), Raman microscopy: Applications to the identification of pigments on medieval manuscripts, *Chem. Soc. Rev.* **24**, 187–194.

Clark, R. J. H., C. J. Cooksey, M. A. M. Daniels, and R. Whithnall (1993), Indigo, woad, and Tyrian Purple: Important vat dyes from antiquity to the present, *Endeavour* (New Series), **17**(4), 191.

Clark, R. J. H. and P. J. Gibbs (1997), Non-destructive *in situ* study of ancient Egyptian Faience by Raman microscopy, *J. Raman Spectrosc.* **28**, 99–103.

Clark, R. J. H. and K. Huxley, K. (1996), Raman spectroscopic study of the pigments on a large illuminated Qur'an, circa thirteenth century, *Sci. Technol. Cult. Heritage* **5**(2), 95–101.

Clark, S. P. (1968), *Structure of the Earth*, Prentice-Hall, Englewood Cliffs, NJ.

Clayton, P. (1998), *Desert Explorer*, Zerzura, Cornwall.

Cleg, W. (1998), *Crystal Structure Determination*, Oxford Univ. Press, Oxford, UK.

Clermont-Ganneau, C. (1885), *Les Grandes Fraudes Archéologiques à Palestine*, Paris.

Clifton, J. R. (1977), Technology and conservation of art, *Architect. Antiquities* **77**, 30–37.

Clow, P. (1982), *The Prevention of Bronze Disease in Archaeological Specimens*, thesis, Portsmouth Polytechnic, Portsmouth.

Cobean, R. H., E. A. Perry, K. K. Turekian, and D. P. Kharkar (1971), *Science* **174**, 666.

Coburn, A., E. Dudley, and R. Spence (1990), *Gypsum Plaster: Its Manufacture and Use*, Intermediate Technology Publications, London.

Cochran, W. G. (1977), *Sampling Techniques*, Wiley, New York.

Cockburn, T. A., E. Cockburn, and T. A. Reyman (1998), *Mummies, Disease and Ancient Cultures*, Cambridge Univ. Press, Cambridge, MA.

Coghlan, H. H. (1951), *Notes on the Prehistoric Metallurgy of Copper and Bronze in the Old World*, Pitt Rivers Museum, Univ. Oxford, Oxford, UK.

Cohen-Ofri, I., L. Weiner, E. Boaretto, G. Mintz, and S. Weiner (2006), Modern and fossil charcoal: Aspects of structure and diagenesis, *J. Archaeol. Sci.* **33**, 428–439.

Cohey, J. M. D., R. Bouchez, and N. V. Dang (1979), Ancient techniques, *J. Appl. Phys.* **50**, 7772–7777.

Cohon, R. (1996), *Discovery and Deceipt, Archaeology and the Forgery Craft*, Nelson-Atkins Museum, Kansas City.

Collins, J. (1978), *New York Times*, Nov. 7.

Collins, M. J., C. M. Nielsen–Marsh, J. Hiller, C. I. Smith, J. P. Roberts, R. V. Prigodich, T. J. Wess, J. Csapo, A. R. Millard, and G. Turner–Walker (2002), The survival of organic matter in bone – a review, *Archaeometry* **44**(3), 383–394.

Collinson, J. D. and D. B. Thompson (1989), *Sedimentary Structures*, Unwin-Hyman, London.

Colman S. and K. L. Pierce (2002), Classifications of quaternary geochronologic methods, in Noller, J. S., J. M. Sowers, and W. R. Lettis (eds.), *Quaternary Geochronology Methods and Applications*, American Geophysical Union, Washington, DC, pp. 2–5.

Colombini, M. P., M. Orlandi, F. Modugno, E. L. Tolppa, M. Sardelli, L. Zoia, and C. Crestini (2007), Archaeological wood characterization by PY/GC/MS, GC/MS, NMR, and GPC techniques, *Microchemical J.* **85**, 164–173.

Colombo, L. (1995), *I Colori Degli Antichi*, Nardini, Firenze.

Colson, I. B., J. F. Bailey, M. Vercauteren, and B. C. Sykes (1997), The preservation of ancient DNA and bone diagenesis, *Ancient Biomol.* **1**(2), 109–117.

Comyn, K. (1996), *Adhesion Science*, Royal Society of Chemistry, Cambridge, UK.

Condamin, J., F. Fromenti, M. O. Metais, M. Michel, and P. Blond (1976), The application of chromatography to the tracing of oil in ancient Amphorae, *Archaeometry* **18**, 195–201.

Condivi, A. (1928), *Michelangelo*, d'Ancona, Roma, p. 59.

Connan, J. (1996), Chemical sleuths on ancient trails – molecular archeology reveals secrets of past civilizations, *Sci. Spectra* **5**, 30–35.

Conyers, L. B. (2004), *Ground-Penetrating Radar for Archaeology – Geophysical Methods for Archaeology*, AltaMira Press, Walnut Creek, CA.

Cook, S. F. and R. F. Heizer (1965), *Studies on the Chemical Analysis of Archaeological Sites*, Publications in Archaeology, Vol. 2, Univ. California, Berkeley.

Cooksey, C. J. (2001), Tyrian Purple: 6,6′-Dibromoindigo and related compounds, a review, *Molecules* **6**, 736–739.

Coopper, D. (1988), *The Art and Craft of Coin Making*, Cooper, Epsom.

Copley, M. S., H. A. Bland, P. Rose, M. Horton, and R. P. Evershed (2005), Gas chromatographic, mass spectrometric and stable carbon isotopic investigations of organic residues of plant oils and animal fats employed as illuminants in archeological lamps from Egypt, *Analyst* **130**, 860–871.

Cornwall, I. W. (1958), *Soils for the Archaeologist*, Phoenix House, London.

Corset, J., P. Dhamelincourt, and J. Barbillat (1989), Raman microscopy, *Chem. Britain* (June), 612–616.

Cotte, M., P. Dumas, G. Richard, R. Breniaux, and Ph. Walter (2005), New insight on ancient cosmetic preparation by synchrotron-based infrared microscopy, *Anal. Chim. Acta* **553**, 105–110.

Coupry, C. (1987), *Proc. 8th Int. Council of Museums Commission for Conservation*, Sydney.

Courty, M. A., P. Goldberg, and R. MacPhail (1990), *Soils and Micromorphology in Archaeology*, Cambridge Univ. Press, Cambridge, UK.

Coutts, P. J. F. (1970), Bivalve growth patterning as a method of seasonal dating in archaeology, *Nature* **226**, 874.

Cowell, N. R. and S. G. E. Bowman (1983), Provenancing and dating of flint, in Phillips P. (ed.), *The Archaeologist and the Laboratory*, Council of British Archaeological Research, Vol. 58, pp. 36–40.

Cox, B. (1938), Henna, *The Analyst* **63**, 397.

Craddock, P. T. (1998), *Two Thousand Years of Zinc and Brass*, British Museum, London.

Craddock, P. T. (1997), The detection of fake and forged antiquities, *Chem. Indust.* **13**, 514–519.

Craddock, P. T. (1992), A short history of the patination of bronze, in Jones, M. (ed.), *Why Fakes Matter – Essays on Problems of Authenticity*, British Museum, London, pp. 63–70.

Craddock, P. T. (1990), *Early Metal Mining and Production*, Edinburgh Univ., Edinburgh.

Craddock, P. T. (1987), The early history of zinc, *Endeavour* **11**, 183–191.

Craddock, P. T. (ed.) (1980), *Scientific Studies in Early Mining and Extractive Metallurgy*, British Museum, London.

Craddock, P. T. and M. J. Hughes (eds.) (1985), *Furnaces and Smelting Techniques in Antiquity*, British Museum, London.

Craddock, P. T. and N. D. Meeks (1987), Iron in ancient copper, *Archaeometry* **29**, 187–204.

Craig, H. and V. Craig (1972), Greek marbles: Determination of provenance by isotope analysis, *Science* **176**, 401–403.

Crane, E. (1983), *The Archaeology of Beekeeping*, Duckworth, London.

Creighton, T. E. (1983), *Proteins: Structures and Molecular Properties*, Freeman, San Francisco.

Creaser, C. S. and M. C. Davies (1988), *Analytical Applications of Spectroscopy*, Royal Society of Chemistry, London.

Crestini, C., N. M. N. El Hadidi, and G. Paleschi (2000), Towards the understanding of oxidative degradation processes in archaeological wood, in *Book of Abstracts, 219th American Chemical Society National Mtg.*, San Francisco, March 26–30.

Crew, P. (2001), Iron in archaeology: The European Bloomery smelters by Radomir Pleiner (review article), *J. Hist. Metallurgy Soc.* **35**(2), 99–102.

Cronyn, J. M. (2001), The deterioration of organic materials, in Brothwell, D. R. and A. M. Pollard (eds.), *Handbook of Archaeological Sciences*, Wiley, New York, pp. 627–647.

Cronyn, J. M. (1990), *The Elements of Archaeological Conservation*, Routledege, London.

Cullity, B. D. (2001), *Elements of X-Ray Diffraction*, Prentice-Hall, New York.

Currie, C. K. (1995), Altered soils: a need for a radical revision in policy, in Beavis, J. and K. Barker (eds.), *Science and Site: Evaluation and Conservation, Proc. Conf.*, Bornemouth Univ., Bornemouth, pp. 90–106.

Dahlgren, B. (1990), *La Grana Cochinilla*, UNAM, Mexico.

Damon, P. E., D. J. Donahue, B. H. Gore, A. L. Hatheway, A. J. T. Jull, T. W. Linick, P. J. Secrel, L. J. Toolin, C. R. Bronk, E. T. Hall, R. E. M. Hedges, R. Housley, I. A. Law, C. Perry, G. Bonani, S. Trumbore, W. Woelfli, J. C. Ambers, S. G. E. Bowman, M. N. Leese, and M. S. Tite (1989), Radiocarbon dating of the Shroud of Turin, *Nature* **337**, 611–615.

Dana, E. D. (1949), *Minerals and How to Study Them*, Wiley, New York.

Daniels, V. (ed.) (1988), *Early Advances in Conservation*, British Museum, London.

Danti, C., M. Matteini, and A. Moles (eds.) (1990), *Le Pitture Murali: Tecniche, Problemi, Conservazione*, Centro DI, Firenze.

Davey, N. (1961), *A History of Building Materials*, Phoenix House, London.

David, A. R. (2000), Mummification, in Nicholson, P. T. and I. Shaw (eds.), *Ancient Egyptian Materials and Technology*, Cambridge Univ. Press, Cambridge, UK.

Davidson, C. F. and S. H. U. Bowie (1955), *Bull. Br. Museum Natural His. Geol.* **2**, 27.

Davidson, D. A. and M. L. Shackley (eds.) (1976), *Geoarchaeology, Earth Sciences and the Past*, Westview, Boulder, CO.

Davies, A. M. C. and C. S. Creaser (1991), *Analytical Applications of Spectroscopy* II, Royal Society of Chemistry, London.

Davison, C. C., R. D. Giauque, and D. Clark (1971), *Man* **6**(4), 645.

Dawkins, J. M. (1956), Zinc and saltpeter, *Notes on the Early History of Zinc*, Zinc Development Association, New York.

Dayton, J. A. and A. Dayton (1986), Uses and limitations of lead isotopes in archaeology, *Proc. 24th Int. Archaeometry Symp.*, pp. 13–41.

de Hoffmann, E. and V. Stroobant (2001), *Mass Spectrometry: Principles and Applications*, Wiley, New York.

de Korosy, F. (1975), *Naturwissenschaften* **62**, 484–491.

de Maria y Campos, T. and T. Castelló Yturbide (1989), *Historia y Arte de la Seda en Mexico*, Fomento Cultural, Mexico.

de Michele, V. (ed.) (1996), *Silica '96, Proc. Mtg. on Libyan Desert Glass and Related Desert Events*, July 18, Bologna Univ., Bologna.

de Michele, V. (1998), The Lybian desert glass scarab in Tutankhamen's pectoral, *Sahara* **10**, 107.

de Niro, M. J. (1987), Stable isotopes in archaeology, *Am. Sci.* **75**, 190.

de Niro, M. J. and S. Epstein (1978), Influence of diet on the distribution of carbon isotopes in animals, *Geochimica and Cosmochimica Acta* **42**, 341–351; 495–506.

de Niro, M. J. and S. Weiner (1988), Chemical, enzymatic and spectroscopic characterization of collagen and other organic fractions from prehistoric bones, *Geochimica and Cosmochimica Acta* **52**, 2197–2206.

Deal, M. (1990), Exploratory analysis of food residues from prehistoric pottery and other artifacts from eastern Canada, *SAS Bull.* **13**(1), 6–12.

Dean, A. M. (1998), The molecular anatomy of an ancient adaptive event, *Am. Sci.* **86**(1), 5–17.

Dean, J. S. (1997), Dendrochronology, in Taylor, R. E. and M. J. Aitken (eds.), *Chronometric Dating in Archaeology*, Advances in Archaeological and Museum Science Series, Vol. 2, Plenum, New York.

Deer, L., B. L. Howie, and A. Zussman (1992), *An Introduction to Rock Forming Minerals*, Longmans, London.

Del Pozzo, G. and J. Guardiola (1989), Mummy DNA fragment identified, *Nature*, **339**(6224), 431–432.

Delhaye, M., B. Guineau, and J. Vezin (1984), *Le Courrier du C.N.R.S.*, Centre National de la Recherche Scientifique, Paris.

Delibrias, G. and J. Labeyrie (1965), in Charters, R. M. (ed.), *Proc. 6th Int. Conf. Radiocarbon Dating*, CONF 650652, p. 634.

Delly, J. (1988), *Photography through the Microscope*, Eastman Kodak, Rochester, New York.

Delmonte, J. (1980), *Origin of Materials and Processes*, Technomic, Lancaster, PA.

Desautels, P. E. (1971), *The Gem Kingdom*, Random House, Toronto.

Desch, H. E. and J. M. Dinwoodie (1996), *Timber: Structure, Properties, Conservation and Use*, Macmillan, London.

Devlin, T. M. (2002), *Textbook of Biochemistry*, Wiley-Liss, New York.

de Vries, H. and K. P. Oakley (1959), Radiocarbon dating of the Piltdown skull and jaw, *Nature* **184**, 224–225.

de Wilde, W. P. and W. R. Blain (1990), *Composite Materials*, Springer.

Dimroth, O. and W. Scheurer (1913), Kermes, *Analitische Chemie* **399**, 43–61.

Dirksen, D. and G. von Bally (eds.) (1997), *Optical Technologies in the Humanities*, Springer, Berlin.

Dixon, J. E., J. R. Cann, and C. Renfrew (1968), Obsidian and the origins of trade, *Sci. Am.* **218**(3), 38–46.

Doerner, M. (1984), *The Materials of the Artist*, Harcourt Brace, New York.

Domanski, M., J. A. Webb, and J. Boland (1994), Mechanical properties of stone artifact materials and the effect on heat treatment, *Archaeometry* **36**, 177–208.

Donaghue, D. N. M. (2001), Remote sensing, in Brothwell, D. R. and A. M. Pollard (eds.), *Handbook of Archaeological Sciences*, Wiley, New York, pp. 555–564.

Donkin, R. A. (1977), Spanish Red: An etnographical study of cochineal and the Opuntia Cactus, *Trans. Am. Phil. Soc. (NS)*, **67**(5), 5–84.

Donovan, S. K. (ed.) (1991), *The Processes of Fossilization*, Belhaven, London.

Doremus, R. H. (1994), *Glass Science*, Wiley, New York.

Doucet, J. (2003), New synchrotron radiation-based imaging techniques and archaeology, *Phys. Chem.* **117** (Molecular and Structural Archaeology: Cosmetic and Therapeutic Chemicals) 179–192.

Doumet, J. (1980), *A Study on the Ancient Purple Color*, Imprimerie Catholique, Beirut.

Dowsett, M. and A. Adriaens (2004), The role of SIMS in cultural heritage studies, *Nuclear Instrum. Meth.* (*B*) **226**(1–2), 38–52.

Doxiadis, E. (1995), *The Mysterious Fayum Portraits; Faces from Ancient Egypt*, Abrams, London.

Dram, J. C. T., T. Calligaro, and J. Salomon (2000), Particle induced X-ray emission, in Ciliberto, E. and G. Spoto (eds.), *Modern Analytical Methods in Art and Archaeology*, Chemical Analysis Series, Vol. 155, Wiley, New York, pp. 135–166.

Drayman-Weisser, T. (2000), *Gilded Metals: History, Technology and Conservation*, Archetype, London.

Drever, J. I. (1985), *Chemistry of Weathering*, Kluwer, New York.

Driessen, L. (1944), Über Eine Charakteristische Reaktion des Antiken Purpurs auf der Faser, *Melliand Textilberichte* **25**, 66–69.

Drooker, P. and L. D. Webster (2000), *Beyond Cloth and Cordage – Archaeological Textiles Research in the Americas*, Univ. Utah Press, Salt Lake City.

Dunnell, R. (1996), Evolutionary theory and archaeology, in O'Brien, M. (ed.), *Evolutionary Archaeology*, Univ. Utah, Salt Lake, pp. 30–67.

Dunnell, R. (1993), Why archaeologists don't care about archaeometry, *Archaeomaterials* **7**, 161–165.

Dunnell, R. (1971), *Systematics in Prehistory*, The Free Press, New York.

Duplessy, J. C., L. D. Labeyrie, C. Lalou, and H. Nguyen (1970), Continental climatic variations between 130,000 and 90,000 years BP, *Nature* **226**, 631–632.

Duval, A. R., Ch. Eluere, and L. P. Hurtel (1989), The use of the scanning electron microscope in the study of gold granulation, *Proc. Archaeometry Symp.* (in Athens), Amsterdam.

Eagland, D. (1988), Adhesives and adhesive performance: The silent background, *Endeavour* **12**, 183–191.

Earl, B. and A. Adriens (2000), Initial experiments on arsenical bronze production, *J. Mining Metals Mater. Soc.* (March), 14.

Eastaugh, N., V. Walsh, T. Chaplin, and R. Siddall (2005), *Pigment Compendium: Optical Microscopy of Historical Pigments*, Butterworth-Heinemann, London.

Eastlake, C. L. (1960), *Materials for the History of Oil Painting*, Dover, New York.

Easton, B. K. (1971), Textile bleaching, *Ciba Geigy Rev.* 3–97.

Edmonds, M. (2001), Lithic exploitation and use, in Brothwell, D. R. and A. M. Pollard (eds.), *Handbook of Archaeological Sciences*, Wiley, New York, pp. 461–470.

Edwards, H. G. M. and K. J. M. Chalmers (eds.) (2005), *Raman Spectroscopy in Archaeology and Art History*, Royal Society of Chemistry, London.

Eglinton, G. and G. B. Curry (eds.) (1991), *Molecules through Time: Fossil Molecules and Biochemical Systematics*, The Royal Society, London.

Ehlers, E. G. and H. Blatt (1982), *Petrology: Igneous, Sedimentary and Metamorphic*, Freeman, San Francisco.

Ehrenreich, R. M. (1995), Archaeometry into archaeology, *J. Archaeol. Meth. Theory* **2**, 1–6.

El Ashry, H., Y. El Kilany, A. El Engebawy, and M. A. El Tarabouls (2003), Some chemical data on *Cyperus papyrus L.* the ancient papermaking material, *Egyptian J. Chem.* **46**(1), 181–185.

El-Kammar, A., R. G. V. Hancock, and R. O. Allen (1989), Human bones as archaeological samples: Changes due to contamination and diagenesis, in Allan, R. O. (ed.), *Archaeological Chemistry*, Advances in Chemistry Series, Vol. 4, American Chemical Society, Washington, DC.

Elster, H., E. Gil-Av, and S. Weiner (1991), Amino acid racemization of fossil bone, *J. Archaeol. Sci.* **18**, 605–617.

Emery, K. F., L. E. Wright, and H. Schwarcz (2000), Isotopic analysis of ancient deer bone: Biotic stability in collapse period maya land use, *J. Archaeol. Sci.* **27**(6), 537–550.

Emiliani, C. (1958), Ancient temperatures, *Sci. Am.* 54–63.

Emiliani, C. and N. J. Shackleton (1974), The Brunhes epoch: Isotopic paleotemperatures and geochronology, *Science* **183**, 511–514.

Emsley, J. (2001), *Nature's Building Blocks: An A-Z Guide to the Elements*, Oxford Univ. Press, Oxford, UK.

Encaust (1984), Pictura Resevata: Über Punisches Wachs, *Maltechnik* **90**(3), 44–53.

Engel, M. H. and P. E. Hare (1985), Gas liquid chromatography of amino acids and their derivatives, in Barrett, G. C. (ed.), *Chemistry and Biochemistry of the Amino Acids*, Chapman & Hall, London, pp. 462–499.

Engh, T. A. (1992), *Principles of Metal Refining*, Oxford Univ. Press, Oxford, UK.

Engle, A. (1991), *From Myth to Reality – Readings in Glass History*, Phoenix, Jerusalem.

Entwistle, J. A. and P. W. Abrahams (1997), Multi-element analysis of soils and sediments from Scottish historical sites. The potential of inductively coupled plasma-mass spectrometry for rapid site investigation, *J. Archaeol. Sci.* **24**, 407–416.

Epstein, S., R. Buchsbaum, H. A. Lowenstam, and H. C. Urey (1953), Revised carbonate water isotopic temperature scale, *Bull. Geol. Soc. Am.* **64**, 1315–1322.

Erhardt, D. (2005), Analysis of organic archaeological samples, *Abstracts of Papers, 230th ACS National Meeting, Washington, DC*, August 28–September 1, 2005.

Ericson, J. E., J. D. McKenzie, and R. Berger (1976), Physics and chemistry of the hydration process in obsidian, in Taylor, R. E. (ed.), *Advances in the Study of Obsidian Glass Studies*, Noyes, Park Ridge, pp. 46–62.

Ericson, J. E., M. West, C. H. Sullivan, and H. W. Krueger (1989), The development of maize agriculture in the Viru Valley, Peru, in Price, T. D. (ed.), *The Chemistry of Prehistoric Human Bone*, C.U.P, Cambridge, pp. 68–104.

Eriksen, E. and S. Thegel (1966), *Conservation of Iron Recovered from the Sea*, Tojhusmuseets Skrifter & Copenhagen.

Ernst, W. G. (1969), *Earth Materials*, Prentice-Hall, Englewood Cliffs, NJ.

Evans, A. V. (1987), *An Introduction to Ore Geology*, Blackwell, Oxford, UK.

Evernden, J. F. and G. H. Curtis (1965), The potassium-argon dating of late Cenozoic rocks in East Africa and Italy, *Current Anthropol.* **6**, 343–385.

Evershed, R. P. (2000), Biomolecular analysis by organic mass spectroscopy, in Ciliberto, E. and G. Spoto (eds.), *Modern Analytical Methods in Art and Archaeology*, Chemical Analysis Series, Vol. 155, Wiley, New York, pp. 177–240.

Evershed, R. P. (1993), Biomolecular archaeology and lipids, *World Archaeol.* **25**(1), 74–93.

Evershed, R. P. and P. H. Bethell (1996), *Application of Multi-molecular Biomarker Techniques to the Identification of Faecal Material in Archaeological Soils and Sediments*, ACS Symposium Series, Vol. 625, pp. 157–172.

Evershed, R. P., R. Berstan, F. Grew, M. S. Copley, A. J. H. Charmant, E. Barham, H. R. Mottram, and G. Brown (2004), Archeology: Formulation of a Roman cosmetic, *Nature* **432**(7013), 35–36.

Evershed, R. P., S. N. Dudd, M. S. Copley, R. Berstan, A. Stott, W. Andrew, H. Mottram, S. A. Buckley, and Z. Crossman (2002), Chemistry of archaeological animal fats, *Acc. Chem. Research* **35**(8), 660–668.

Evershed, R. P., S. N. Dudd, M. J. Lochheart, and S. Jim (2001), Lipids in archaeology, in Brothwell, D. R. and A. M. Pollard (eds.), *Handbook of Archaeological Sciences*, Wiley, New York, pp. 351–358.

Evershed, R. P., C. Heron, and L. J. Goad (1990), Analysis of organic residues of archaeological interest by high temperature gas chromatography and high temperature gas chromatography-mass spectrometry, *Analyst* **115**, 1339–1342.

Evershed, R. P. and N. Tuross (1996), Proteinaceous material from potsherd and associated soils, *J. Archaeol. Sci.* **23**, 429–436.

Evershed, R. P., S. J. Vaughan, S. N. Dudd, and J. S. Soles (1997), Fuel for thought? Beeswax in lamps and conical cups from Late Minoan Crete, *Antiquity* **71**, 979–985.

Falcone, R., A. Renier, and M. Verita, M. (2002), Wavelength-dispersive X-ray fluorescence analysis of ancient glasses, *Archaeometry* **44**(4), 531–542.

Fano, U. and L. Fano (1973), *Physics of Atoms and Molecules*, University of Chicago Press.

Faraday Society (1953), *Nature and Structure of Collagen*, Butterworth, London.

Farnsworth, M. (1951), Red madder dye, *J. Chem. Educ.* **28**, 72–78.

Farnsworth, M., C. S. Smith, and J. L. Rodda (1949), Metallographic examination of a sample of metallic zinc from ancient Athens, *Hesperia* (Suppl. 8), 126–129.

Farnsworth, M. and A. Wisely (1958), *Am. J. Archaeol.* **62**, 165–169.

Farquar, R. M. and V. Vitali (1989), Lead isotope measurements and their application to Roman artifacts from Carthage, *Museum Applied Science Center for Archaeology Papers in Science and Archaeology*, Vol. 6, pp. 39–45.

Faure, G. (1986), *Principles of Isotopic Geology*, Wiley, New York.

Feathers, J. K. (2003), Use of luminescence dating in archaeology, *Meas. Sci. Technol.* **14**, 1493–1509.

Feathers, J. K., M. Berhane, and L. May (1998), Firing analysis of South-Eastern Missouri Indian pottery using iron Mössbauer spectroscopy, *Archaeometry* **40**, 59–70.

Fedorovich, E. F. (1965), *Sovetskaia Arkheologyia* **4**, 124–127.

Feinberg, W. (1983), *Lost Wax Casting*, Interim Technology, London.

Feist, W. C. and D. N. S. Hon (1984), Chemistry of weathering and protection, in Rowell, R. (ed.), *The Chemistry of Solid Wood*, Advances in Chemistry Series, Vol. 207, ACS, Washington, DC, pp. 401–451.

Feller, R. L. (ed.) (1986), *Artists Pigments: A Handbook of their History and Characteristics*, Vol. 1, Cambridge Univ. Press, Cambridge, UK.

Feltz, A. (1993), *Amorphous Inorganic Materials and Glasses*, VCH. Verlag GmbH, New York.

Fernandes, P. (2006), Applied microbiology and biotechnology in the conservation of stone cultural heritage materials, *App. Microbio. Biotech.* **73**, 291–296.

Ferraro, J. R. and K. Nakamoto (1994), *Introduction to Raman Spectroscopy*, Academic Press, New York.

Fester, G. A. (1962), *Chymia* **8**, 21–25.

Fester, G. A. (1953), Madder, *Isis* **44**, 13–17.

Fiedel, S. (1996), Blood from stones? Some methodological and interpretative problems in blood residue analysis, *Archaeol. Sci.* **23**(1), 139–147.

Fieser, L. (1930), Alizarin, *J. Chem. Educ.* **7**, 2609–2611.

Figiel, L. (1991), *On Damascus Steel*, Atlantis, New York.

Fine, L. W. (1980), *Chemistry Decoded*, Oxford Univ. Press, New York.

Fink, C. G. and C. G. Eldridge (1925), *The Restoration of Ancient Bronzes and Other Alloys*, Metropolitan Museum of Art, New York.

Fink, C. G. and E. P. Polushkin (1938), *Plate of Brass: Evidence of the Visit of Francis Drake to California in the Year 1579*, California Historical Society Special Publication, Vol. 14, San Francisco.

Finlayson-Pitts, B. G. and J. N. Pitts (1986), *Atmospheric Chemistry*, Wiley, New York.

Fiori, F., G. Giunta, A. Hilger, N. Kardjilov, and F. Rustichelli (2006), Non-destructive characterization of archaeological glasses by neutron tomography, *Physica B, Condensed Matter*, Part 1, 385–386; Part 2, 1206–1208.

Fitzhugh, E. W. (1986), Lead read and minium, in Feller, R. L. (ed.), *Artists Pigments: A Handbook of Their History and Characteristics*, Vol. 1, National Gallery of Art, Washington, DC.

Fleisher, R. L. (1998), *Tracks to Innovation, Nuclear Tracks in Science and Technology*, Springer, Berlin.

Fleischer, R. L., P. B. Price, and R. M. Walker (1975), *Nuclear Tracks in Solids: Principles and Applications*, Univ. California Press, Berkeley.

Fleischer, R. L., P. B. Price, and R. M. Walker (1965a), *Application of Fission Tracks and Fission Track Dating to Anthropology*, General Electric Report 65-RI-3876M, pp. 1–12.

Fleischer, R. L., P. B. Price, and R. M. Walker (1965b), Fission track dating of a mesolithic knife, *Nature* **205**, 1138–1140.

Fleischer, R. L., P. B. Price, and R. M. Walker (1965c), Fission track dating of Bed I, Olduvay Gorge, *Science* **148**, 72–74.

Fleming, S. J. (ed.) (1980), *The Egyptian Mummy: Secrets and Science*, University Museum Handbook Series, Vol. 1, Univ. Pennsylvania, Philadelphia.

Fleming, S. J. (1975), *Authenticity in Art: The Scientific Detection of Forgery*, Institute of Physics, London.

Flieder, E., R. Barroso, and C. Oruezabal (1975), *Analyse des Tannins Hydrolisables Susceptibles d'Entrer Dans la Composition des Encres Ferro-Galliques*, International Council of Museums Committee for Conservation Report 75-5-12, Venice.

Flieder, F., E. Delange, A. Duval, and M. Leroy (2001), Papyrus: The need for analysis, *Restaurator* **22**(2), 84–106.

Florian, M. L. E. (ed.) (1990), *Archaeological Wood*, Advances in Chemistry Series, Vol. 225, ACS, Washington, DC.

Forbes, R. J. (1997a), *Metallurgy in Antiquity*, Brill, Leiden.

Forbes, R. J. (1997b), *Studies in Ancient Technology: The Fibres and Fabrics of Antiquity – Washing, Bleaching, Fulling and Felting – Dyes and Dyeing – Spinning – Sewing, Basketry and Weaving*, Brill, Leiden.

Fortey, R. (1991), *Fossils, the Key to the Past*, Harvard Univ., Cambridge, MA.

Fouqué, N. (1869), *Revue des Deux Mondes* **83**, 923–931.

Fox, R. (1948), *Vat Dyestuffs and Vat Dyeing*, Chapman & Hall, London.

Fox, R. and W. H. Powell (eds.) (2001), *Nomenclature of Organic Compounds, Principles and Practice*, Oxford Univ. Press, Oxford, UK.

Frankel, R., S. Avitsur, and E. Ayalon (1994), History and technology of olive oil, in E. Ayalon (ed.), *The Holy Land*, Olearius, Arlington and Tel Aviv.

Franklin, A. D., J. S. Olin, and T. A. Wertime (eds.) (1978), *The Search for Ancient Tin*, Smithsonian Institution, Washington, DC.

Franks, F. (1983), *Water*, Royal Society of Chemistry, London.

Fraquet, H. (1987), *Amber*, Butterworth, London.

Freemantle, M. (1996), Historic stone monuments pose challenge to conservation scientists, *Chem. Eng. News* (April 15), 20–23.

Freestone, I. C. (2005), The provenance of ancient glass through compositional analysis, in Vandiver, P. B., J. L. Mass, and A. Murray (eds.), *Materials Issues in Art and Archaeology VII* (Symposium, November 30–December 3, 2004, Boston, Massachusetts), *Materials Research Society Symposium Proceedings*, Vol. 852, Materials Research Society, Warrendale, Pennsylvania.

Freestone, I. C. (2001), Post depositional changes in archaeological ceramics and glasses, in Brothwell, D. R. and A. M. Pollard (eds.), *Handbook of Archaeological Sciences*, Wiley, New York, pp. 615–626.

Freestone, I. C. (1991), Looking into glass, in Bowman, S. (ed.), *Science in the Past*, British Museum, London.

French, C. A. I. and Ch. French (2002), *Geoarchaeology in Action: Studies in Soil Micromorphology and Landscape Evolution*, Routledge, London.

Friedlander, P. (1909), Tyrian Purple, *Berichte der Deutscher Chemischer Geseltshaft* **42**, 765–770.

Friedman, F. D., G. Borromeo, and M. Leveque (eds.) (1998), *Gifts of the Nile: Ancient Egyptian Faience*, Thames & Hudson, London.

Friedman, I. and W. D. Long (1976), Hydration rate of obsidian, *Science* **191**, 347–352.

Friedman, I. and J. Obradovich (1981), Obsidian hydration dating of volcanic events, *Quatern. Research* **16**, 37–47.

Friedman, I. and R. L. Smith (1960), A new dating method using obsidian, *Am. Antiquity* **25**, 476–493.

Friedman, I. and F. W. Trembour (1978), Obsidian: The dating stone, *Am. Sci.* **66**, 44–54.

Friedman, I., F. W. Trembour, and R. E. Hughes (1997), Obsidian hydration dating, in Taylor, R. E. and M. J. Aitken (eds.), *Chronometric Dating in Archaeology*, Advances in Archaeological and Museum Science Series, Vol. 2, Plenum, New York.

Friend, J. N. A. (1926), *Iron in Antiquity*, Charles Griffin & Co., London.

Friesen, P. L. (1995), *Natural Fibers*, Arbidar, New York.

Frink, D. S. (1995), Application of the oxidizable carbon ratio (OCR) dating procedure and its implications for pedogenic research, in *Pedological Perspectives in Archaeological Research*, Soil Science Society Special Publication, Vol. 44, Soil Science Society of America, Madison, WI.

Fritts, H. C. (1976), *Tree Rings and Climate*, Academic Press, New York.

Fry, G. A. (1985), Analysis of fecal matter, in Gilbert, R. I. and J. H. Mielke (eds.), *The Analysis of Prehistoric Diets*, Academic Press, Orlando, FL, pp. 127–154.

Gaines, A. M. and V. L. Handy (1975), Mineralogical alteration of Chinese tomb jades, *Nature* **253**, 433–434.

Gale, N. (1981), Mediterranean obsidian source characterization by strontium isotope analysis, *Archaeometry* **23**, 41–51.

Gale, N. and Z. Stos-Gale (2000), Lead isotope analysis applied to provenance studies, in Ciliberto, E. and G. Spoto (eds.), *Modern Analytical Methods in Art and Archaeology*, Chemical Analysis Series, Vol. 155, Wiley, New York, pp. 503–584.

Gale, N. and Z. Stos-Gale (1996), Lead isotope methodology: The possible fractionation of lead isotope compositions during metallurgical processes, in S. Demirci, A. M. Özer, and G. D. Summers (eds.), *Proc. Archaeometry Symp.* Ankara, 1994, pp. 287–299.

Gan, F., H. Cheng, and Q. Li (2006), Origin of Chinese ancient glasses – Study on the earliest Chinese ancient glasses, *Science in China, Ser. E: Tech. Sci.* **49**, 701–713.

Gani, M. S. J. (1997), *Cement and Concrete*, Chapman & Hall, London.

Garcia Gimenez, R., R. Vigil de la Villa, M. D. Petit Dominguez, and M. I. Rucandio (2006), Application of chemical, physical and chemometric analytical techniques to the study of ancient ceramic oil lamps, *Talanta* **68**(4), 1236–1246.

Garland, G. D. (1979), *Introduction to Geophysics: Mantle, Core and Crust*, Holt, Reinhart, Winston, Toronto.

Garland, M. (2002), *The Alluvial Sapphire Deposits of Western Montana*, Ph.D. thesis, Univ. Toronto.

Garrison, E. G. (2001), Physics and archaeology, *Phys. Today* **54**(10), 32–36.

Garza-Valdes, L. A. and B. Stross (1992), Rock varnish analysis, in Vandiver, P. J., D. Druzik, G. S. Wheeler, and I. C. Freestone (eds.), *Materials Issues in Art and Archaeology III, a Symposium in San Francisco, California, April 27–May 1*, Materials Research Society Symposium Proceedings, Pittsburgh, PA, Vol. 267, pp. 891–900.

Gaxiola, G. M. and J. E. Clark (coordinators) (1989), *La Obsidiana en Mesoamerica*, a symposium report, INAH, Mexico.

Gebhard, R. (2003), Material analysis in archaeology, *Hyperfine Interact.* **150**(1–4), 1–5.

Geçklini, A. E., N. Bozkurt, and S. Basaran (1988), Metallographic studies of archaeological metal artifacts, *Aksay Ünitesï Bilimsel Topilanti Bildinireli* Middle East Univ., Ankara, Vol. 1, pp. 229–246.

Geigl, E. M. (2002), On the circumstances surrounding the preservation and analysis of very old DNA, *Archaeometry* **4**, 337–342.

Geilmann, W. (1956), Beiträge zur Kenntnis alter Gläser, IV: Die Zersetzung der Gläser im Boden, *Glastechnik Berichte* **29**, 156–158.

Gerin, J. (1972), *Cast Metals Technology*, Addison-Wesley, New York.

Gerneay, A. M., E. R. Waite, M. J. Collins, O. E. Craig, and R. J. Sokol (2001), Survival and interpretation of archaeological proteins, in Brothwell, D. R. and A. M. Pollard (eds.), *Handbook of Archaeological Sciences*, Wiley, New York, pp. 323–330.

Gettens, R. J. (1963), *Corrosion Products of Metal Antiquities*, Smithsonian Institution, Washington, DC.

Gettens, R. J., R. L. Feller, and W. T. Chase (1972), Vermilion and cinnabar, *Stud. Conserv.* **17**, 45–69.

Gettens, R. J., H. Kuhn, and W. T. Chase (1967), Lead white, *Stud. Conserv.* **12**, 125–139.

Gettens, R. J. and G. L. Stout (1966, 1942), *Painting Materials, a Short Encyclopedia*, Dover, New York.

Getty Museums (1990), Marble: Art historical and scientific perspectives on ancient sculpture; a symposium, *J. Paul Getty Museum* (April 28–30, 1988), Malibu.

Geyh, M. A. and H. Schleicher (1990), *Absolute Age Determination: Physical and Chemical Dating Methods and Their Application*, Springer, New York.

Ghisalberti, E. L. (1979), Propolis: a review, *Bee World* **60**, 50–84.

Giakoumaki, A., K. Melessanaki, and D. Anglos (2007), Laser-induced breakdown spectroscopy (LIBS) in archaeological science–applications and prospects, *Analyt. Bioanalyt. Chem.* **387**(3), 1618–2642.

Giauque, R. D., F. Asaro, F. H. Stross, and T. R. Hester (1993), High-precision non-destructive X-ray fluorescence method applicable to establishing the provenance of obsidian artifacts, *X-Ray Spectrom.* **22**, 44–53.

Gilberg, M. (1988), History of bronze disease and its treatment, in Daniels, V. (ed.), *Early Advances in Conservation*, British Museum, London.

Gilbert, K. R. (1954), Rope making, in Singer, C. (ed.), *History of Technology*, Vol. 1, Oxford Univ. Press, Oxford, UK.

Gilbert, R. I. and J. H. Mielke (eds.) (1985), *The Analysis of Prehistoric Diets*, Academic Press, London.

Gillard, R. D., S. M. Hardman, R. G. Thomas, and D. E. Watkinson (1994), The detection of dyes by FTIR microscopy, *Stud. Conserv.* **39**(3), 187–192.

Gillard, R. D., A. M. Pollard, R. A. Sutton, and D. K. Whittaker (1990), An improved method for age at death determination from the measurement of D-aspartic acid in dental collagen, *Archaeometry* **32**, 61–70.

Gillespie, R. J., A. J. Gowlett, E. T. Hall, and R. E. M. Edges (1984), Radiocarbon dating of bone by accelerator mass spectrometry, *Archaeometry* **26**, 15–20.

Gilmour, B. and E. Worrall (1995), Paktong: The trade in Chinese nickel brass to Europe, in Hook, D. R. and D. R. M. Gaimster (eds.), *Trade and Discovery: The Scientific Study of Artefacts from Post-Medieval Europe and Beyond*, British Museum, London, pp. 279–282.

Giuliani, G., M. Chaussidon, H. J. Schubnel, D. H. Piat, C. Rollion-Bard, C. France-Lanord, D. Giard, D. de Narvaez, and B. Rondeau (2000), Oxygen isotopes and emerald trade routes since antiquity, *Science* **287**, 631–633.

Giuliani, G., C. France-Lanord, P. Coget, D. Schwarz, A. Cheilletz, Y. Branquet, D. Giard, A. Martin-Izard, P. Alexandrov, and D. H. Piat (1998), Oxygen isotope systematics of emerald: Relevance for its origin and geological significance, *Mineralium Deposita* **33**(5), 513–519.

Glascock, M. D. (1994), New World obsidian; recent investigation, in Scott, D. A. and P. Meyers (eds.), *Archaeometry of Pre-Columbian Sites and Artifacts*, Getty Conservation Institute, Los Angeles, pp. 113–134.

Glascock, M. D., G. E. Braswell, and R. H. Cobean (1998), A systematic approach to obsidian source characterization, in Shackley, M. S. (ed.), *Archaeological Obsidian Studies: Method and Theory*, Plenum, New York, pp. 15–65.

Glasser, W., R. Northey, and T. P. Schultz (eds.) (2000), *Lignin: Historical, Biological, and Materials Perspectives*, ACS Symposium Series, ACS, Washington, DC.

Glob, P. V. (1972), *The Bog People: Iron Age Man Preserved*, Yale Univ. Press, New Haven.

Glover, E. (1979), *The Gold and Silver Wire Drawers*, Phillimore, Chichester, UK.

Goedvriend, G. M. (1988), Papermaking, past and present, *Endevour* **12**, 38–43.

Goffer, Z. (1996), *Dictionary of Archaeological Materials and Archaeometry*, Elsevier, Amsterdam.

Goffer, Z. (1980), *Archaeological Chemistry*, Wiley, New York.

Gohl, C. and M. S. Vilensky (1987), *Textile Science; Fiber Properties*, Longmans, London.

Goldberg, P., V. T. Holliday, and C. R. Ferring (2000), *Earth Sciences and Archaeology*, Kluwer, New York.

Goldman, A. I. (1991), The crystalline state, in Trig, G. L. (ed.), *Encyclopaedia of Applied Physics*, VCH Verlag GmbH, Weinheim, Vol. 4, pp. 365–383.

Goldstein, S. M., L. S. Rakow, and J. K. Rakow (1982), *Cameo Glass: Masterpieces from 2000 Years of Glassmaking*, The Corning Museum of Glass, Corning, NY.

Goodburn-Brown, D. and J. Jones (1998), *Scientific Methods for Investigation of Coin Surfaces for Conservation*, Archetype, London.

Goodfriend, G. A., M. J. Collins, M. L. Fogel, S. A. Macko, and J. F. Wehmiller (eds.), *Perspectives in Amino Acid and Protein Geochemistry*, Oxford Univ. Press, New York.

Goodhew, P. J., J. Humphreys, R. Beanland, F. J. Humphreys, and R. Beanland (2000), *Electron Microscopy and Analysis*, Taylor and Francis Group.

Goodway, M. (1987), Fiber identification in practice, *J. Am. Inst. Conserv.* **26**(1).

Gordon, B. (1980), *Feltmaking*, Watson-Guptill, New York.

Gordon, R. B. (1988), Strength and structure of wrought iron, *Archaeomaterials* **2**, 109–137.

Goss, R. J. (1983), *Deer Antlers*, Academic Press.

Gottsegen, M. D. (1987), *The Painters Handbook*, Watson Guptill, New York.

Gourdin, W. H. and W. D. Kingery (1975), The beginnings of pyrotechnology – Neolithic and Egyptian lime plaster, *J. Field Archaeol.* **2**, 133–150.

Gove, H. E. (1999), *From Hiroshima to the Iceman: The Development and Applications of Accelerator Mass Spectrometry*, Institute of Physics, Bristol.

Gove, H. E. (1996), *Relic, Icon or Hoax? Carbon Dating the Turin Shroud*, Institute of Physics, Bristol.

Gowland, W. (1912), *J. Inst. Metals* **7**, 23–31.

Grace, R. (1996), Use-wear analysis; the state of the art, *Archaeometry* **38**, 209–229.

Grace, R. (1989), *Interpreting the Function of Stone Tools: The Quantification and Computerization of Microwear Analysis*, British Archaeological Reports, 474, Oxford.

Grant, J. (1954), Materials of ancient textiles and baskets, in Singer, C. (ed.), *History of Technology*, Vol. 1, Oxford Univ. Press, Oxford, UK.

Grattan, D. W. and J. C. McCawley (eds.) (1981), *Proc. Int. Council of Museums, Waterlogged Woodworking Group Conf.*, The International Council of Museums, Ottawa.

Gratuze, B., J. N. Barrandon, K. Al Isa, and M. C. Cauvin (1993), Non-destructive analysis of obsidian artefacts using nuclear techniques: Investigation of provenance of Near Eastern artefacts, *Archaeometry* **35**, 11–21.

Gray, J. (1985), The use of stable isotope data in climate reconstruction, in Wigley, T. M. L., M. J. Ingram, and G. Farmer (eds.), *Climate and History: Studies in Ancient Climates and Their Impact on Man*, Cambridge Univ. Press, Cambridge, UK.

Greathouse, G. A. and C. J. Wessel (1954), *Deterioration of Materials – Causes and Preventive Techniques*, Reinhold, New York.

Green, L. R. and V. Daniels (1995), Shades of the past, *Chem. Britain* **31**, 183–189.

Greenwood, N. N. and A. Earnshaw (1998), *Chemistry of the Elements*, Wiley, New York.

Grierson, S. (1989), *Dyeing and Dyestuffs*, Shire, Princess Risborough.

Griffiths, D. R. (1980), *The Deterioration of Ancient Glass*, B.Sc. thesis, Univ. Wales, Cardiff.

Griffiths, P. R. and J. A. de Haseth (1986), *Fourier Transform Infrared Spectrometry*, Wiley, New York.

Grimaldi, D. A. (1996a), The ancient allure of amber, *Natural History* (Feb.), 97–99.

Grimaldi, D. A. (1996b), Captured in amber, *Sci. Am.* (April), 84–91.

Grimanis, A. P., N. Kalogeropoulos, V. Kilikoglou, and M. Vassilaki-Grimani (1997), Use of NAA in a marine environment and in archeology in Greece, *J. Radioanal. Nuclear Chem.* **219**(2), 177–185.

Grimshaw, R. W. (1980), *The Chemistry and Physics of Clays*, McGraw-Hill, New York.

Grinsted, M. J. and A. T. Wilson (1979), Variations of 13C/12C ratio in cellulose of *Agathus australis* (Kauri) and climatic change in New Zealand during the last millennium, *New Zealand J. Sci.* **22**, 55–61.

Grissom, C. A. (1986), Green earth, in R. L. Feller (ed.), *Artists Pigments*, Cambridge Univ. Press, New York, pp. 101–140.

Groenman van Waateringe, J. C. and M. Robinson (1988), *Man Made Soils*, British Archaeological Reports, Oxford.

Groom, N. (1981), *Frankincense and Myrrh: A Study of the Arabian Incense Trade*, Logman, London.

Grootes, P. M., M. Stuiver, J. W. C. White, S. Johnsen, and J. Jouzel (1993), Comparison of oxygen isotope records from GISP2 and GIRP Greenland ice cores, *Nature* **366**, 552–554.

Grosjean, D., P. M. Whitmore, C. P. de Moor, and J. R. Druzik (1988), Ozone fading of organic colorants: Products and the mechanism of the reaction, *Environ. Sci. Technol.* **22**, 1357–1361.

Grun, R. (2001), Trapped charge dating, in Brothwell, D. R. and A. M. Pollard (eds.), *Handbook of Archaeological Sciences*, Wiley, New York, pp. 63–72.

Grun, R. (1997), Electron spin resonance dating, in Taylor, R. E. and M. J. Aitken, *Chronometric Dating in Archaeology*, Advances in Archaeological and Museum Science Series, Vol. 2, Plenum, New York.

Grun, R. and C. B. Stringer (1991), Electron spin resonance dating, *Archaeometry* **33**, 153–159.

Guerra, M. F. and Th. Calligaro (2003), Gold cultural heritage objects: A review of studies of provenance and manufacturing technologies, *Meas. Sci. Technol.* **14**, 1527–1537.

Gugerli, F., L. Parducci, and J. Petit Remy (2005), Ancient plant DNA: Review and prospects, *New Phytologist* **166**, 409–418.

Guillot, P. Y. (1985), Potassium-argon upper Pleistocene dating, *Terra Cognita* **5**, 234–236.

Guillot, P. Y. and Y. Cornette (1986), The Cassignoll technique for potassium-argon dating, precision and accuracy: Examples from the late Pleistocene to recent volcanics from southern Italy, *Chem. Geol. (Isotopes Geoscience Section)* **59**, 205–222.

Guineau, B. (1989), Nondestructive analysis of organic pigments and dyes using raman microprobe, microfluorimeter and absorption microspectrophotometer, *Stud. Conserv.* **34**, 38–44.

Gunstone, F. (2004), *The Chemistry of Oils and Fats: Sources, Composition, Properties and Uses*, Blackwell, Oxford, UK.

Guo Shilun, Liu Shunsheng, Sun Shengfen, Zhang Feng, Zhou Shuhua, Hao Xiuhong, Hu Ruiying, Meng Wu, Zhang Pengfa, and Liu Jinfa (1990), Age and duration of Peking Man site by fission track methods, *Abstracts, 15th Int. Conf. Particle Tracks in Solids*, Marburg.

Guo Shilun, Zhou Shuhua, Meng Wu, Zang Pengfa, Hao Xiuhong, Liu Shunsheng, Zang Feng, Hu Ruiying, and Liu Jinga (1980), Fission track dating of Peking Man, *Kexue Tongba* **25**, 770–772.

Guo Yanyi (1987), Row materials for making porcelain and the characteristics of porcelain wares in north and south China in ancient times, *Archaeometry* **29**, 3–19.

Gurke, K. (1987), *Bricks and Brickmaking*, Univ. Idaho, Moscow.

Gustafson, K. H. (1956), *The Chemistry of Tanning Processes*, Academic Press, New York.

Guy, A. G. (1960), *Elements of Physical Metallurgy*, Addison-Weseley, New York.

Hackens, T., H. McKerrell, and M. Hours (eds.) (1977), *X-Ray Fluorescence Spectroscopy Applied to Archaeology*, Vol. 1, PACT Rixensart.

Hackens, T. and G. Moucharte (1987), *Technology and Analysis of Ancient Gemstones*, Council of Europe, Strasbourg.

Hackens, T., A. V. Munaut, and C. Till (1988), *Wood and Archaeology*, Council of Europe, Strasbourg.

Hackens, T. and M. Schvoerer (eds.) (1984), *Datation – Caractérisation des Ceramiques Anciennes*, Centre National de la Recherche Scientifique, Paris.

Hall, D. O., D. D. Hall, and K. K. Rao (eds.) (1994), *Photosynthesis*, Cambridge Univ. Press, New York.

Hallas, G. and D. R. Waring (eds.) (1994), *The Chemistry and Application of Dyes*, Plenum, New York.

Hamer, F. and J. Hamer (1977), *Clays*, Pitman, Cambridge, MA.

Hamer, F. and J. Hamer (2004), *The Potter's Dictionary of Materials and Techniques*, Black, London.

Han, Y. H., T. Enomae, A. Isogai, H. Yamamoto, S. Hasegawa, J. J. Song, and S. W. Jang (2006), Traditional papermaking techniques revealed by fiber orientation in historical papers, *Stud. Conservation* **51**, 267–276.

Hancock, R. G. V. (2000), Elemental analysis, in Ciliberto, E. and G. Spoto (eds.), *Modern Analytical Methods in Art and Archaeology*, Chemical Analysis Series, Vol. 155, Wiley, New York, pp. 11–20.

Hanel, J. W., T. Kryst-Widzgowska, and J. Klinowski (1998), *A Primer of Magnetic Resonance Imaging*, Imperial College Press, London.

Harbottle, G. (1990), Neutron activation analysis in archaelogical chemistry, in Yoshihara, K. (ed.), *Chemical Applications of Nuclear Probes, Topics in Current Chemistry*, Springer, Berlin, Vol. 157, 57–91.

Harbottle, G. (1982), Chemical characterization in archaeology, in Ericson, J. E. and T. K. Earle (eds.), *Contexts for Prehistoric Exchange*, Academic Press, London, pp. 13–51.

Harbottle, G., E. T. Hall, and H. W. Catling (1969), Brookhaven National Laboratory Report 13,740.

Harbottle, G. and P. C. Weigard (1992), Turquoise in pre Columbian America, *Sci. Am.* **266**(2), 78–85.

Hare, P. E., D. W. Von Endt, and J. E. Kokis (1997), Protein and amino acid diagenesis dating, in Taylor, R. E. and M. J. Aitken (eds.), *Chronometric Dating in Archaeology*, Advances in Archaeological and Museum Science Series, Vol. 2, Plenum, New York.

Harris, D. C. (2002), *Quantitative Chemical Analysis*, Freeman, New York.

Harris, D. R. (1987), The impact on archaeology of radiocarbon dating by accelerator mass spectrometry, *Phil. Trans. Roy. Soc.* **A323**, 23–43.

Harris, E. C. (ed.) (1997), *Principles of Archaeological Stratigraphy*, Academic Press, New York.

Harris, J. E. and K. R. Weeks (1973), *X Raying the Pharaohs*, Scribner, New York.

Harris, J. E. and E. F. Wente (1980), *An X-Ray Atlas of the Royal Mummies*, Univ. Chicago Press.

Harrison, A. W. C. (1936), *The Manufacture of Lakes and Precipitated Pigments*, Hill, London.

Harrison, R. (1992), *Pollution: Causes, Effect and Control*, Royal Society of Chemistry, London.

Haslam, E. (1966), *Chemistry of Vegetable Tannins*, Academic Press, London.

Hassan, A. A. and D. J. Ortner (1977), Inclusions in bone materials as a source of error in radiocarbon dating, *Archaeometry* **19**, 131–135.

Hassett, J. J. and W. L. Banwart (1992), *Soils and their Environment*, Prentice-Hall, Englewood Cliffs, NJ.

Hauptmann, A., E. Pernicka, Th. Rehren, and U. Yalçin (eds.) (1999), *The Beginning of Metallurgy, Proc. Int. Conf.* Bochum, 1995.

Hayden, B. (ed.) (1979), *Lithic Use-Wear Analysis*, Academic Press, London.

Hecht, A. D. (1985), *Palaeoclimate Analysis and Modeling*, Wiley, New York.

Hedges, R. E. M. (2002), Bone diagenesis: An overview of processes, *Archaeometry* **44**(3), 319–328.

Hedges, R. E. M. (2001), Dating in archaeology, past, present and future, in Brothwell, D. R. and A. M. Pollard (eds.), *Handbook of Archaeological Sciences*, Wiley, New York, pp. 3–8.

Hedges, R. E. M. (2000), Radiocarbon dating, in Ciliberto, E. and G. Spoto (eds.), *Modern Analytical Methods in Art and Archaeology*, Chemical Analysis Series, Vol. 155, Wiley, New York, pp. 465–502.

Hedges, R. E. M. (1987), Radiocarbon dating by accelerator mass spectrometry: Some recent results and applications, *Phil. Trans. Roy. Soc.*

Hedges, R. E. M. (1979), Analysis of the Drake plate: Comparison with the composition of Elizabethan brass, *Archaeometry* **21**, 21–26.

Hedges, R. E. M. and A. R. Millard (1995a), Bones and groundwater: Towards the modeling of diagenetic processes, *J. Archaeol. Sci.* **22**, 155–164.

Hedges, R. E. M. and A. R. Millard (1995b), Measurements and relationships of diagenetic alteration of bone from three archaeological sites, *J. Archaeol. Sci.* **22**, 201–209.

Hedges, R. E. M. and C. J. Salter (1979), Source determination of iron currency bars through analysis of the slag inclusions, *Archaeometry* **21**, 161–175.

Hedges, R. E. M. and B. C. Sykes (1992), Biomolecular archaeology: Past, present and future, *Proc. Br. Acad.* **77**, 267–283.

Heicklen, J. (1976), *Atmospheric Chemistry*, Academic Press, New York.

Heidenreich, C. E. and V. A. Konrad (1973), Soil analysis at the Robataille site Part II: A method useful in determining the location of Longhouse patterns, *Ontario Archaeol.* **21**, 33–62.

Heidke, J. M. and E. J. Miksa (2000), Correspondence and discriminant analyses of sand and sand temper compositions, *Archaeometry* **42**, 273–299.

Heimann, R. B. (1982), Archaeothermometry, in Olin, J. S. and A. D. Franklin (eds.), *Archaeological Ceramics*, Smithsonian Institution, Washington, DC, pp. 84–96.

Heimann, R. B. and U. M. Franklin (1981), Archaeothermometry: The assessment of firing temperatures of ancient ceramics, *Int. Ins. Conserv. Can. Group* **4**(2), 23–45.

Heizer, R. F., F. Stross, T. R. Hester, A. Albee, I. Perlman, F. Asaro, and H. Bowman (1973), The Colossi of Memnon revisited, *Science* **182**, 1219–1221.

Hellborg, R., M. Faarinen, M. Kiisk, C. E. Magnusson, C. E. Persson, G. Skog, and K. Stenstrom (2003), Accelerator mass spectrometry – an overview, *Vacuum* **70**(2–3), 365–372.

Heller, J. H. (1983), *Report on the Shroud of Turin*, Houghton Mifflin, New York.

Henderson, J. (2000), *The Science and Archaeology of Materials: An Investigation of Inorganic Materials*, Routledge, London.

Henderson, J. (ed.) (1989), *Scientific Analysis in Archaeology and its Interpretation*, Oxford Univ. Committee for Archaeology Monograph 19, Oxford, UK.

Henderson, J., J. A. Evans, H. J. Sloane, M. J. Leng, and C. Doherty (2005), The use of oxygen, strontium, and lead isotopes to provenance ancient glasses in the Middle East, *J. Archaeol. Sci.* **32**, 665–673.

Hendy, C. H. and A. T. Wilson (1968), Paleoclimatic data from speleothems, *Nature* **216**, 48–53.

Hennel, J. W. and J. Klinowski (1993), *Fundamentals of Nuclear Magnetic Resonance*, Longmans, London.

Hennel, J. W. and T. Kryst-Widzgowska (1998), *A Primer of Magnetic Resonance Imaging*, Imperial College, London.

Herbert, E. W. (1984), *Red Gold in Africa: Copper in Pre-Colonial History and Culture*, Univ. Wisconsin, Madison.

Herbert Smith, G. F. and M. Phillips (1962), *Gems*, Methuen, London.

Herbig, R. (1933), *Romische Mitteilungen* **48**, 312–320.

Herodotus (1958), *The Histories*, Book II, Vol. 1 (transl. H. Carter), Heritage, New York.

Heron, C. (2001), Geochemical prospecting, in Brothwell, D. R. and A. M. Pollard (eds.), *Handbook of Archaeological Sciences*, Wiley, New York, pp. 565–574.

Heron, C. and R. P. Evershed (1993), The analysis of organic residues and the study of pottery use, *Archaeol. Meth. Theory* **5**, 247–273.

Herrmann, B. and S. Hummel (eds.) (1994), *Ancient DNA*, Springer, New York.

Herz, N. and E. G. Garrison (1998), *Geological Methods for Archaeologists*, Oxford Univ. Press, Oxford, UK.

Herz, N. and M. Waelkens (1988), *Classical Marble: Geochemistry, Technology, Trade*, NATO ASI Series E, Vol. 153, Plenum, Dordrecht.

Hess, J. and I. Perlman (1974), Mössbauer spectra of iron in ceramics and their relation to pottery colours, *Archaeometry* **16**(2), 137–152.

Hess, K., A. Hauptmann, H. Wright, and R. Whallon (1998), Evidence of fourth millennium B.C. silver production at Fatmali-Kalecik, East Anatolia, in Rehren, Th., A. Hauptmann, and J. D. Muhly (eds.), *Metallurgica Antiqua* (Der Anschnitt, Zeitschrift für Kultur im Bergbau, Bochum) **8**, 57–67.

Heyworth, M. P., J. R. Hunter, S. E. Warren, and J. N. Walsh (1988), The analysis of archaeological materials using inductively coupled plasma spectrometry, in Slater, E. A. and J. O. Tate (eds.), *Science in Archaeology*, Glasgow.

Hicks, E (1961), *Shellac*, Chemical Publishing Co., New York.

Hicks, S., U. Miller, S. Nilsson, and I. Vuorela (eds.) (1991), Airborne particles and gases and their impact on the cultural heritage and its environment, *Proc. European Workshop at Ravello, Italy*, Council of Europe, Strasbourg.

Hill, A. D., A. H. Lehman, H. Ann, and M. L. Parr (2007), Using scanning electron microscopy with energy dispersive x-ray spectroscopy to analyze archaeological materials, *J. Chem. Educ.* **84**(5), 810–813.

Hill, J. (1774), *Theophrastus' History of Stones*, London, p. 227.

Hill, N., S. Holmes, and D. Mather (eds.) (1992), *Lime and Other Alternative Cements*, Intermediate Technology Publications, London.

Hillam, J. (1987), Dendrochronology, *Current Archaeol.* **107**, 358–383.

Hillel, D. (1991), *Out of the Earth: Civilization and Life on the Soil*, Univ. California Press, Berkeley.

Hillman, G. (1989), Wild plant food economy and seasonality in Wadi Kubbanyia, in Harris, D. R. and G. C. Hillman (eds.), *Foraging and Farming*, Unwin Hyman, London.

Hillman, G. (1986). Plant foods in ancient diet: The archaeological role of palaeofaeces in general and Lindow Man's gut contents in particular, in Stead, I. M., J. B. Bourke, and D. Brothwell (eds.), *Lindow Man: The Body in the Bog*, Cornell Univ., Ithaca, NY, pp. 100–115.

Hilson, S. (1986a), *Teeth*, Cambridge Univ. Press, Cambridge, UK.

Hilson, S. (1986b), Archaeology and the study of teeth, *Endeavour* (New Series) **10**(3), 145–149.

Hinsley, J. F. (1959), *Non Destructive Testing*, McDonald and Evans, London.

Hoare, W. E. (1948), *Hot Tinning*, Tin Research Institute, Greenford.

Hofenk de Graff, J. H. (1969), *Natural Dyestuffs: Origin, Chemical Constitution, Identification*, International Council of Museums, Committee for Conservation report 5, Amsterdam.

Hofenk de Graff, J. H. and W. G. T. Roelofs (1978), The analysis of flavonoids in natural yellow dyestuffs occurring in ancient textiles, *Proc. Int. Council of Museums Committee for Conservation, 5th Trienial Mtg.*, Zagreb.

Hoffman, H., E. Torres William, and D. Ernst Randy (2002), Paleoradiology – Advanced CT in the evaluation of nine Egyptian mummies, *Radiographics* **22**(2), 377–385.

Hoffmann, R. (1990), Blue as the sea, *Am. Sci.* **78**, 308–309.

Hoffmann, U. (1962), *Angewandte Chemie* (Int. Ed.) **1**, 341–345.

Holden, M. (1991), *The Encyclopedia of Gems and Minerals*, Facts in File, New York.

Holland, R. (2007), Event review: what's cooking in archeological chemistry? Digging deep for clues about life in ancient times, *Chem. and Ind.* **5**, 32.

Holliday, V. T. and W. G. Gartner (2007), Methods of soil P analysis in archaeology, *J. Archaeol. Sci.* **34**(2), 301–333.

Holliday, V. T. (2004), *Soils in Archaeological Research*, Oxford Univ. Press, Oxford, UK.

Horai, S. (1995), Evolution and the origins of man: Clues from complete sequences of hominoid mitochondrial DNA, *Southeast Asian J. Tropical Med. Public Health* **26**(Suppl. 1), 146–154.

Hornak, J. P. (ed.) (2002), *Encyclopedia of Imaging Science and Technology*, Wiley, New York.

Höss, M. (2000), Ancient DNA: Neanderthal population genetics, *Nature* **404**, 453–454.

Howard, S. (1992), Fakes, intention, proof and impulsion to know, in Jones, M. (ed.), *Why Fakes Matter – Essays on Problems of Authenticity*, British Museum, London, pp. 51–62.

Howell, F. C., G. H. Cole, M. R. Kleindienst, B. J. Szabo, and M. P. Oakley (1972), *Nature* **237**, 51.

Howes, F. N. (1953), *Vegetable Tanning Materials*, Butterworth, London.

Hughen, K. A., M. G. L. Baillie, E. Bard, A. Bayliss, J. W. Beck, C. Bertrand, P. G. Blackwell, C. E. Buck, G. Burr, K. B. Cutler, P. E. Damon, R. L. Edwards, R. G. Fairbanks, M. Friedrich, T. P. Guilderson, B. Kromer, F. G. McCormac, S. Manning, C. Bronk Ramsey, P. J. Reimer, R. W. Reimer, S. Remmele, J. R. Southon, M. Stuiver, S. Talamo, F. W. Taylor, J. van der Plicht, and C. E. Weyhenmeyer (2004), Marine radiocarbon age calibration, *Radiocarbon* **46**(3), 1059–1086.

Hughen, K. A., J. T. Overpeck, S. J. Lehman, M. Kashgarian, J. Southon, L. C. Peterson, R. Alley, and D. M. Sigman (1998), Deglaciation changes in ocean circulation from an extended radiocarbon calibration, *Nature* **391**, 65–68.

Hughes, M. J. (1991), Tracing the source, in Bowman, S. (ed.), *Science and the Past*, British Museum, London, pp. 99–116.

Hughes, M. J. (ed.) (1981), *Scientific Studies in Ancient Ceramics*, British Museum, London.

Hughes, M. J., M. R. Cowell, and P. T. Craddock (1976), Atomic absorption techniques in archaeology, *Archaeometry* **18**, 19–37.

Hughes, M. J., M. R. Cowell, and D. R. Hook (eds.) (1991), *Neutron Activation Analysis and Plasma Emission Spectroscopy Analysis in Archaeology*, British Museum, London.

Hughes, M. K., P. M. Kelly, J. R. Pilcher, and V. C. LaMarche, Jr., (eds.) (1982), *Climate from Tree Rings*, Cambridge Univ. Press.

Hughes, R. W. (1997), *Ruby and Sapphire*, R. W. H. Publishing, Boulder, CO.

Hughes, S. (1978), *Washi, the World of Japanese Paper*, Kosansha, Tokyo.

Hummel, S. (2002), *Fingerprinting the Past: Research on Highly Degraded DNA and Its Applications*, Springer, New York.

Hummel, S. (1994), General aspects of sample preparation, in Herrmann, B. and S. Hummel (eds.), *Ancient DNA*, Springer, New York, pp. 59–68.

Hunt, L. B. (1980), The long history of lost wax casting, *Gold Bull.* **13**, 63–81.

Hunter, D. (1978), *Papermaking, the History and Technique of an Ancient Craft*, Dover, New York.

Hunter, P. (2007), Dig this – biomolecular archaeology provides new insights into past civilizations, cultures, and practices, *EMBO Reports* **8**, 215–217.

Huntley, D. J., D. I. Godfrey-Smith, and M. L. W. Thewalt (1985), Optical dating of sediments, *Nature* **313**, 105–107.

Huntley, J. P. and S. Stallibrass (eds.) (2000), *Taphonomy and Interpretation*, Oxbow, Oxford, UK.

Hurcombe, L. (1992), *Use – Wear Analysis and Obsidian: Theory, Experiment and Results*, Collis, Sheffield.

Hurley, P. M. (1959), *How Old is the Earth?*, Doubleday, Garden City, NJ.

Hurley, W. M. (1979), *Prehistoric Cordage*, Taraxacum, Washington, DC.

Hurry, J. B. (1930), *The Woad Plant and Its Dye*, Oxford Univ. Press, Oxford, UK.

Hütt, G., I. Jaek, and J. Tchonka (1988), Optical dating: K-feldspars optical response stimulation spectra, *Quatern. Sci. Rev.* **7**, 381–385.

Iannone, J. C. (1998), *The Mystery of the Shroud of Turin*, Alba House.

Inaba, M. and R. Sugisita (1988), Permanence of Washi, *Proc. Conf. Conservation of Far Eastern Art*, Kyoto.

Ingo, G. M., E. Angelini, G. Bultrini, T. de Caro, L. Pandolfi, and A. Mezzi (2002), Contribution of surface analytical techniques for the microchemical study of archaeological artefacts, *Surf. Interface Anal.* **34**(1), 328–336.

Issar, A. S. and M. Zohar (2004), *Climate Change – Environment and Civilization in the Middle East*, Springer, Heidelberg.

Issar, R. A. (1990), *Atlas of Opaque and Ore Minerals in Their Associations*, Open University, London.

Ivanovich, M. and R. S. Harmon (eds.) (1992), *Uranium Series Disequilibrium: Application of Earth, Marine and Environmental Sciences*, Oxford Univ. Press, Oxford, UK.

Jackson, C. M., C. A. Booth, and J. W. Smedley (2005), Glass by design – Raw materials, recipes and compositional data, *Archaeometry* **47**, 781–795.

Jackson, H. (1978), *Lost Wax Bronze Casting: A Photographic Essay*, Van Nostrand-Reinhold, New York.

Jackson, R. (1982), *Basic Scientific Photography*, Dept. Agriculture, Ottawa.

Jacob, F. B. (1928), *The Abrasives Handbook*, Penton, Cleveland, OH.

Jacobson, D. M. and M. P. Weitzman (1992), What is Corinthian bronze?, *Am. J. Archaeol.* **96**, 237–247.

Jaffé, H. L. C. and L. H. Van der Tweel (eds.) (1979), *Authentication in the Visual Arts: A Multidisciplinary Symposium*, Israël, Amsterdam.

Jäger, E., C. W. Ji, A. J. Hurford, L. R. Xin, J. C. Hunziker, and L. D. Ming (1985), A quaternary age-standard for potassium argon dating, *Chem. Geolo. (Isotope Geoscience Section)* **52**, 275–279.

Jakes, K. (ed.) (2002), *Archaeological Chemistry*, Vol. 6, *Materials, Methods, and Meaning*, Advances in Chemistry Series, ACS, Washington, DC.

Jamieson, G. S. (1932), *Vegetable Fats and Oils: The Chemistry, Production and Utilization of Vegetable Fats and Oils*, Cornell Univ., Cornell, NY.

Janssens, K. and R. Van Grieken (2004), *Non-destructive microanalysis of cultural heritage materials*, Elsevier, Amsterdam.

Janssens, K. and F. Adams (2000), Applications in art and archeology, in Janssens, K. H. A., F. C. V. Adams, and A. Rindby (eds.), *Microscopic X-Ray Fluorescence Analysis*, Wiley, New York, pp. 291–314.

Járó, M. (2003), Metal threads in historical textiles, *NATO Science Series, II: Mathematics, Physics and Chemistry*, Vol. 117 (*Molecular and Structural Archaeology: Cosmetic and Therapeutic Chemicals*), pp. 163–178.

Járó, M. (1990), Gold embroidery and fabrics in Europe, *Gold Bull.* **23**, 40–57.

Járó, M. and A. Tóth (1991), Scientific identification of European metal thread manufacturing techniques, *Endeavour* (New Series) **15**(4), 175–179.

Jedrzejewska, H. (1990), Ancient mortars as criterion in analysis of old architecture, in *Mortars, Cements and Grouts Used in the Conservation of Historical Buildings Symp.*, Rome, pp. 311–329.

Jeffery, G. H., J. Bassett, J. Mendham, and R. C. Denney (1989), *Vogel's Textbook of Quantitative Chemical Analysis*, Addison-Wesley, New York.

Jeffery, P. G. and D. Hutchison (1981), *Chemical Methods of Rock Analysis*, Pergamon, Oxford, UK.

Jehaes, E. (1998), *Optimization of Methods and Procedures for the Analysis of mt-DNA Sequences and Their Applications in Molecular Archaeological and Historical Finds*, Leuven Univ., Leuven.

Jembrih, D., M. Peev, M. Schreiner, P. Krejsa, and Ch. Clausen (2000), Identification and classification of iridescent glass artifacts with XRF and SEM/EDX, *Microchimica Acta* **133**, 151–157.

Jenkins, K., H. Bishop, and D. Storer (2007), Forensic anthropology and chemistry: Determining the authenticity of ancient ceramic figurines: Reel Project, *Abstracts, 39th Central Regional Meeting of the Am. Chem. Soc. (May 20–23)*, Covington, Am. Chem. Soc., Washington, DC.

Jenkins, R. (1999), *X-Ray Fluorescence Spectrometry*, Wiley, New York.

Jensen, P. and D. J. Gregory (2006), Selected physical parameters to characterize the state of preservation of waterlogged archaeological wood: A practical guide for their determination, *J. Archaeol. Sci.* **33**, 551–559.

Johansson, S. A. E., J. L. Campbell, and K. G. Malmqvist (eds.) (1995), *Particle-Induced X-Ray Emission Spectroscopy (PIXE)*, Wiley, New York.

Johnson, J. B., S. J. Haneef, B. J. Hepburn, A. J. Hutchison, G. E. Thomson, and G. C. Wood (1990), Laboratory exposure systems to simulate atmospheric degradation of stone, *Atm. Environ.* **24a**, 2582–2592.

Johnson, J. B. and G. H. Miller (1997), Archaeological applications of amino acid racemization (1997), *Archaeometry* **39**, 265–287.

Johnson, K. (1997), Chemical dating of bones based on diagenetic changes in bone apatite, *J. Archaeol. Sci.* **24**, 431–437.

Jolly, M. S., S. K. Sen, T. N. Sonwalkar, and G. K. Prasad (1979), *Non-Mulberry Silks*, Food and Agriculture Organization Service Bulletin 29.

Jones, D. A. (1996), *Principles and Prevention of Corrosion*, Prentice-Hall, Saddle River, NJ.

Jones G. T., D. G. Bailey, and C. Beck (1997), Source provenance of andesite artifacts using non-destructive XRF analysis, *J. Archaeol. Sci.* **24**, 923–943.

Jones, M. (2001), *The Molecule Hunt: Archaeology and the Hunt for Ancient DNA*, Penguin, London.

Jones, M. (ed.) (1992), *Why Fakes Matter – Essays on Problems of Authenticity*, British Museum, London.

Jones, M. (1990), *Fake?, The Art of Deception*, British Museum, London.

Jordan, B. A. (2001), Site characteristics impacting the survival of historic waterlogged wood – a review, *Int. Biodeter. Biodegrad.* **47**(1), 47–54.

Jose-Yacaman, M. and J. A. Ascencio (2000), Electron microscopy and its application to the study of archaeological materials and art preservation, in Ciliberto, E. and G. Spoto (eds.), *Modern Analytical Methods in Art and Archaeology*, Chemical Analysis Series, Vol. 155, Wiley, New York, pp. 405–443.

Joseph, M. L., P. Hudson, A. Clapp, and D. Kness (1992), *Joseph's Introductory Textile Science*, Harcourt Brace, Fort Worth, TX.

Jull, A. J. T., D. J. Donehue, and P. E. Damon (1996), Factors affecting the apparent radiocarbon age of textiles: A comment on, in D. E. Kouznetsov, A. A. Ivanov, and P. R. Veletksy (eds.), *Effects of Fires and Biofractionation of Carbon Isotopes on Results of Radiocarbon Dating of Old Textiles: The Shroud of Turin*; reprinted in *J. Archaeol. Sci.* **23**, 157–160.

Kaczmarczyk, A. and R. E. M. Hedges (1983), *Ancient Egyptian Faience*, Aris and Philips, Warminster.

Kaestle, F. and K. Ann Horsburgh (2002), Ancient DNA in anthropology: Methods, applications and ethics, *Yearbook of Physical Anthropology* (45), 92–130.

Kane, R. E. (1981), Hornbill ivory, *Gems Gemol.* **17**(2), 97.

Karp, C. (1983), Calculating atmospheric humidity, *Stud. Conserv.* **28**, 24–28.

Kars, H. and E. Burke (eds.) (2005), *Proceedings of the 33rd Intenatonal Symposium on Archaeometry*, Vrije Universiteit, Amsterdam.

Katzenberg, M. A. (2000), Stable isotope analysis: A tool for studying past diet, demography and life history, in Katzenberg, M. A. and S. R. Saunders (eds.), *Biological Anthropology of the Human Skeleton*, Wiley-Liss, New York, pp. 305–328.

Katzenberg, M. A. (1988), Stable isotope analysis of animal bone and the reconstruction of human palaeodiet, in Kennedy, B. and G. Le Moine (eds.), *Diet and Subsistence, Current Archaeological Perspectives*, Univ. Calgary, pp. 307–324.

Katzenberg, M. A. and R. G. Harrison (1997), What's in a bone? Recent advances in archaeological bone chemistry, *J. Archaeol. Res.* **5**(3), 265–293.

Kawohl, G. (1988), Zinnpest und Korrosion am Zinn, *Arbeitsblätter für Restauratoren* **23**, 246–251.

Keegan, W. F. (ed.) (1987), *Emergent Horticultural Economies of the Eastern Woodlands*, Southern Illinois Univ. at Carbondale, Center for Archaeological Investigations, Occasional Papers, Southern Illinois Univ.

Keeley, L. H. (1980), *Experimental Determination of Stone Tool Uses – Microwear Analysis*, Univ. Chicago Press, Chicago.

Keisch, B. (2003), *The Atomic Fingerprint: Nuclear Activation Analysis*, University Press of the Pacific.

Keisch, B. (1972), *Secrets of the Past: Nuclear Energy Applications in Art and Archaeology*, U.S. Atomic Energy Commission, Office of Information Services.

Keller, W. D. (1957), *The Principles of Chemical Weathering*, Lucas, Columbia.

Kelley, D. W. (1986), *Charcoal and Charcoal Burning*, Shire, Aylesbury.

Kempe, D. R. C. and A. P. Harvey (1983), *The Petrology of Archaeological Artifacts*, Clarendon, Oxford, UK.

Kent Kirk, T. and E. B. Cowling (1984), Biological decomposition of solid wood, in Rowell, R. (ed.), *The Chemistry of Solid Wood*, Advances in Chemistry Series, American Chemical Society, Washington.

Kerr, R. and N. Wood (with contributions by Ts'ai Meifen and Zhang Fukang) (2005), *Science and Civilisation in China*, Vol. 5, *Chemistry and Chemical Technology*, Cambridge University Press, New York.

Keyser-Tracqui, C. and B. Ludes (2005), Methods for the study of ancient DNA, *Meth. Mol. Bio.* **297** (Forensic DNA Typing Protocols), 253–264.

Killick, D. (2001), Science, speculation and the origins of extractive metallurgy, in Brothwell, D. R. and A. M. Pollard (eds.), *Handbook of Archaeological Sciences*, Wiley, New York, pp. 483–492.

Killick, D. and S. M. M. Young (1997), Archaeology and archaeometry: From casual dating to a meaningful relationship?, *Antiquity* **71**, 518–524.

King, R. J. (1989), Native copper, *Geology Today* (May/June), pp. 104–106.

Kingery, W. D. (ed.) (1986–2000), *Ceramics and Civilization*, Vols. 1–10, The American Ceramic Society, Cincinnati.

Kingery, W. D. (1991), Attic pottery technology, *Archaeomaterials* **5**, 47–54.

Kingery, W. D., H. K. Bowen, and D. R. Uhlmann (1976), *Introduction to Ceramics*, Wiley, New York.

Kingery, W. D. and J. D. Frierman (1974), The firing temperature at a Karanovo sherd and inferences about south-east European chalcolithic refractory technology, *Proc. Prehist. Soc.* **40**, 204–205.

Kirby, J. (ed.) (2002), Dyes in history and archaeology; papers presented at the 18th Meeting, Brussels, 1999, Archetype, London.

Kirby, J. (1988), The preparation of early lake pigments: A survey, in *Dyes on Historical and Archaeological Textiles*, 6th Mtg. (1987), National Museum of Scotland, Edinburgh.

Kirby, J. (1977), A spectrophotometric method for the identification of lake pigment dyestuffs, *Natl. Gallery Bull.* **1**, 35–45.

Kirk Othmer (1966), *Encyclopaedia of Chemical Technology*, Wiley, New York.

Kitson, F. G., B. S. Larsen, and C. N. McEwen (1996), *Gas Chromatography and Mass* Spectrometry, Academic Press, New York.

Klein, C. (2000), *Mineral Science*, Wiley, New York.

Klein, L. S. (1982), *Archaeological Typology*, British Archaeological Reports.

Kleinmann, B. (1986), History and development of early Islamic pottery glazes, in Olin, J. S. and M. K. Blackman, *Proc. 24th Int. Archaeometry Symp., 1984*, Smithsonian Institute, Washington, DC, pp. 73–84.

Klokkernes, T. (1991), The influence of air pollution on ancient monuments, buildings and museum objects, in Hicks, S., U. Miller, S. Nilsson, and I. Vuorela (eds.),

Airborne Particles and Gases and Their Impact on the Cultural Heritage and Its Environment, Proc. European Workshop at Ravello, Italy, Council of Europe, Strasbourg, pp. 121–129.

Knappett, C. (2005), Pottery, in Maschner, H. D. G. and C. Chippindale (eds.), *Handbook of Archaeological Methods*, Altamira Press, Walnut Creek, CA, Chapter 18.

Knox, R. (1987), On distinguishing meteoritic from man made nickel-iron in ancient artifacts, *Museum Appl. Sci. Center Archaeol. J.* **4**(4), 178–184.

Koch, P. L., A. K. Behrensmeyer, A. W. Stott, N. Tuross, R. P. Evershed, and M. L. Fogel (2000), The effects of weathering on the stable isotope composition of bones, *Ancient Biomol.* **3**(2), 117–134.

Kockelmann, W., L. C. Chapon, R. Engels, J. Schelten, C. Neelmeijer, H.-M. Walcha, G. Artioli, S. Shalev, E. Perelli-Cippo, M. Tardocchi, G. Gorini, and P. G. Radaelli (2006), Neutrons in cultural heritage research, *J. Neutron Res.* **14**, 37–42.

Koenig, G. and H. J. Metz (2003), The fascination of color: The story of pigments – from ancient times to the present, *Pitture e Vernici, European Coatings* **79**(20), 21–28.

Kohl, P. L., G. Harbottle, and E. V. Sayre (1979), Physical and chemical analysis of soft stone vessels from southwest Asia, *Archaeometry* **21**, 131–159.

Koller, J., U. Baumer, Y. Kaup, M. Schmid, and U. Weser (2003), Analysis of a pharaonic embalming tar, *Nature* **425**(6960), 784.

Koren, Z. (1996), Historico-chemical analysis of plant dyestuffs used in textiles from ancient Israel, in Orna, M. V. (ed.), *Archaeological Chemistry*, Vol. 5, *Organic, Inorganic and Biochemical Analysis*, Advances in Chemistry Series, American Chemical Society, Washington, DC, pp. 269–310.

Kouznetsov, D. A., A. A. Ivanov, and P. R. Veletsky (1996), Analysis of cellulose chemical modification: A potentially promising technique for characterizing cellulose archaeological textiles, *J. Archaeol. Sci.* **23**, 23–34, 109–121.

Krings, M., A. Stone, R. W. Schmitz, H. Krainitzki, M. Stoneking, and S. Paabo (1997), Neanderthal DNA sequences and the origin of modern humans, *Cell* **90**, 19–30.

Krumbein, W. C. and L. Sloss (1951), *Stratigraphy and Sedimentation*, Freeman, San Francisco.

Kucukkaya, A. G. (2004), Photogrammetry and remote sensing in archeology, *J. Quan. Spectrosc. Rad. Transf.* **88**(1–3), 83–88.

Kuhn, H. (1987a), Naturwissenschaftliche Untersuchung von Mörtelproben, in Lobbedey, K. (ed.), *Die Ausgrabungen im Dom zu Paderborn: 1987/80 und 1983, 1986*, pp. 309–317.

Kuhn, K. (1987b), in R. Mayne and R. Burgeson (eds.), *Structure and Function of Collagen Types*, Academic Press, New York.

Kuhn, H. (1986), *Conservation and Restoration of Works of Art and Antiquities*, Butterworth, London, pp. 186–192.

Kuhn, H. (1970), Verdigris, *Stud. Conserv.* **5**, 12–36.

Kuhn, H. (1960), Detection and identification of waxes, including punic wax, by infrared spectrography, *Stud. Conserv.* **5**, 71–80.

Kuleff, I. and M. Djingova (1990), Activation analysis of archaeological and art objects, in Alfassi, Z. (ed.), *Activation Analysis*, Chemical Rubber Company, Boca Raton, FL.

Kurtén, B. (1982), *Teeth: Form, Function and Evolution*, Columbia Univ., New York.

Kurz, O. (1967), *Fakes*, Dover, New York.

Kutschera, W. and W. Rom (2000), Ötzi, the prehistoric iceman, *Nuclear Instrum. Meth.* (*B*) **164–165**, 12–22.

Lajtha, K. and R. Michener (eds.) (1994), *Stable Isotopes in Ecology and Environmental Science*, Blackwell, Oxford, UK.

Lamb, H. H. (1969), *Nature* **234**, 1209.

Lamb, H. H. (1977), *Climate History and the Future*, Princeton Univ., Princeton, NJ.

Lambert, J. B., C. E. Shawl, and J. A. Stearns (2000), Nuclear magnetic resonance in archeology, *Chem. Soc. Rev.* **29**(3), 175–182.

Lambert, J. B. (1997), *Traces of the Past: Unraveling the Secrets of Archaeology through Chemistry*, Addison-Wesley, Reading, MA.

Lambert, J. B. (ed.) (1984), *Archaeological Chemistry*, Advances in Chemistry Series, Vol. 3, ACS, Washington, DC.

Lambert, J. B. and J. S. Fry (1982), Amber, *Science* **217**, 55–57.

Lambert, J. B., J. S. Frye, and A. Jurkiewicz (1992), The provenance and coal rank of jet by carbon-13 nuclear magnetic resonance spectroscopy, *Archaeometry* **34**, 121–128.

Lambert, J. B., E. Graham, M. T. Smith, and J. S. Frye (1994), Amber and jet from Tipu, Belize, *Ancient Mesoamerica* **5**, 55–60.

Lambert, J. B. and C. Grupe (eds.) (1993), *Prehistoric Human Bone: Archaeology at the Molecular Level*, Springer, Berlin.

Lambert, J. B. and C. D. McLaughlin (1978), Analysis of early Egyptian glass by atomic absorption and X-ray photoelectron spectroscopy, in Carter, G. F. (ed.), *Archaeological Chemistry*, Advances in Chemistry Series, Vol. 2, ACS, New York.

Lambert, J. B., S. C. Johnson, and G. O. Poinar, Jr. (1996), Nuclear magnetic resonance sediment characterization of cretaceous amber, *Archaeometry* **38**, 325–335.

Lambourne, R. (ed.) (1987), *Paint and Surface Coatings*, Halstead, New York.

La Niece, S. (1995), Depletion gilding from third millennium B.C. Ur, *Iraq* **57**, 41–47.

La Niece, S. (1983), Niello: An historical and technical survey, *Antiquities J.* **63**, 279–298.

Lancaster, J. G. (1980), *Metallurgy of Welding*, Allan & Unwin, Boston.

Lanford, W. A. (1986), Glass hydration: A method of dating glass objects, *Science* **166**, 975–976.

Lang, J. and A. Middleton (eds.) (1997), *Radiography of Cultural Material*, Butterworth-Heinemann, London.

Langenheim, J. H. (1969), Amber, a botanical enquiry, *Science* **163**(3), 1157–1169.

Larsen, C. S. (1997), *Bioarchaeology: Interpreting Behavior from the Human Skeleton*, Cambridge Univ. Press, Cambridge, UK.

Latham, A. G. (2001), Uranium series dating, in Brothwell, D. R. and A. M. Pollard (eds.), *Handbook of Archaeological Sciences*, Wiley, New York, pp. 63–72.

Laubengayer, A. W. (1931), *J. Am. Ceramic Soc.* **14**, 833.

Laufer, B. (1930), The early history of felt, *Am. Anthropol.* **32**, 1–18.

Laufer, B. (1912), *Jade*, Field Museum of Natural History, Chicago.

Laurie, A. P. (1978), *Greek and Roman Methods of Painting*, Longwood, Boston.

Lawrence, J. C. (1982), *Trace Analysis*, Academic Press, New York.

Layer, P. W., C. M. Hall, and D. York (1987), The derivation of argon-40/argon-39 age spectra of single grains of hornblende and biotite by laser step heating, *Geophys. Research Lett.* **14**, 757–760.

Lazzarini, L., G. Moschini, and B. M. Stievano (1980), A contribution to the identification of Italian, Greek and Anatolian marbles through a petrological study and the evaluation of the Ca/Sr ratio, *Archaeometry* **22**, 173–183.

Lea, F. M. (1962), *The Chemistry of Cement and Concrete*, Arnold, London.

Leach, B. and J. Tait (2000), Papyrus, in Nicholson, P. T. and I. Shaw (eds.), *Ancient Egyptian Materials and Technology*, Cambridge Univ. Press, Cambridge, UK.

Lechtman, H. (1991), The production of copper-arsenic alloys in the central Andes: Highland ores and coastal smelters?, *J. Field Archaeol.* **18**(1), 43–76.

Lechtman, H. (1984), Pre-Columbian surface metallurgy, *Sci. Am.* **250**(June), 38–45.

Lechtman, H. (1979), A pre-Columbian technique for electrochemical replacement plating of gold and silver on objects of copper, *J. Metals* **31**, 154–160.

Lechtman, H. and S. Klein (1999), The production of copper-arsenic alloys (arsenic bronze) by co-smelting: Modern experiment, ancient practice, *J. Archaeol. Sci.* **26**, 497–526.

Lee, R. R., D. A. Leich, T. A. Trombrello, J. E. Ericson, and I. Friedman (1974), Obsidian hydration profile measurements using a nuclear reaction technique, *Nature* **250**, 44–47.

Lee-Thorp, J. and M. Sponheimer (2006), Contributions of biogeochemistry to understanding hominid dietary ecology, *Am. J. Phys. Anthro.* **43**(Supplement), 131–148.

Lee-Whitmann, L. and M. Skelton (1984), Where did all the silver go?, *Textile Museum J.* **22**, 33–52.

Lefferts, K. C., L. J. Majewski, E. V. Sayre, P. Meyers, R. M. Organ, C. S. Smith, R. H. Brill, I. L. Barnes, T. J. Murphy, and F. R. Matson (1981), Technical examination of the classical bronze horse from the Metropolitan Museum of Art, *J. Am. Inst. Conserv.* **21**(1), 1–42.

LeGeros, R. Z., O. R. Trautz, J. P. LeGeros, E. Klein, and W. P. Shirra (1967), Apatite crystallites: Effects of carbonate on morphology, *Science* **155**, 1409–1411.

Levey, M. (ed.) (1967), *Symp. Archaeological Chemistry*, Univ. Pennsylvania, Philadelphia.

Levey, M. (1959), *Chemistry and Chemical Technology in Ancient Mesopotamia*, Amsterdam.

Levey, M. (1958), Alum in ancient Mesopotamia, *Isis* **49**, 166–169.

Lewin, S. Z. (1973), A new approach to establishing the authenticity of patinas on copper-base Artifacs, in *Application of Science in the Examination of Works of Art, Proc. Seminar*, June 15–19, 1970, Museum of Fine Arts, Boston, pp. 62–66.

Lewin, S. Z. and S. M. Alexander (1967, 1968), The composition and the nature of natural patinas, *A.A.T.A.* **6**, 201–366; **7**, 151–189.

Lewis, N. (1990), Papyrus and ancient writing: The first hundred years of papyrology, *Archaeology* **36**(4), 31–37.

Lewis, P. A. (1988), *Pigments Handbook*, Wiley, New York.

Lewkowitsch, J. (1922), *Chemical Technology and Analysis of Oils, Fats, and Waxes*, Macmillan, London.

Libby, W. F. (1952), *Radiocarbon Dating*, Univ. Chicago Press, Chicago.

Lifshin, E. (ed.) (1999), *X-ray Characterization of Materials*, Wiley, New York.

Lillie, R. D. (1979), The red dyes used by ancient dyers: Their probable identity, *J. Soc. Dyers Colourists* 57–61.

Limbrey, S. (1975), *Soil Science and Archaeology*, Academic Press, London.

Lindahl, A. and O. Stilborg (1995), *Aim of Laboratory Analyses of Ceramics in Archaeology*, Almqvist and Wiksell, Stockholm.

Linderholm, J. and E. Lundberg (1994), Chemical characterization of various archaeological soil samples using main and trace elements determined by inductively coupled plasma atomic emission spectrometry, *J. Archaeol. Sci.* **21**, 303–314.

Lippard, S. J. and J. M. Berg (1994), *Principles of Bioinorganic Chemistry*, University Science Books, Herndon, VA.

Lipson, E. (1953), *A Short History of Wool*, Heineman, London.

Liritzis, I. (2006), SIMS-SS: A new obsidian hydration dating method: Analysis and theoretical principles, *Archaeometry* **48**, 533–547.

Long, D. A. (1977), *Raman Spectroscopy*, McGraw-Hill, New York.

Longworth, G. (1984), Study of ceramics and archeological materials, in Long, G. L., *Mössbauer Spectroscopy Applied to Inorganic Chemistry*, Plenum, New York, pp. 511–526.

Longworth, G. and S. E. Warren (1979), The application of Mössbauer spectroscopy to the characterization of western Mediterranean obsidian, *J. Archaeol. Sci.* **6**, 1–15.

Loveland, R. (1970), *Photomicrography – a Comprehensive Treatise*, Wiley, New York.

Lowe, J. J. (ed.) (1997), *Radiocarbon Dating: Recent Applications and Future Potential*, Wiley, New York.

Lowenstam, H. A. and S. Weiner (1989), *On Biomineralization*, Oxford Univ. Press, Oxford, UK.

Loy, T. H. (1993), The artifact as site: An example of the biomolecular analysis of organic residues on prehistoric tools, *World Archaeol.* **25**, 44–63.

Loy, T. H. (1983), Prehistoric blood residues: Detection on tool surfaces and identification of species of origin, *Science* **220**, 1269–1271.

Lu, P. J., N. Yao, J. F. So, G. E. Harlow, J. F. Lu, G. F. Wang, and P. M. Chaikin (2005), The earliest use of corundum and diamond in prehistoric China, *Archaeometry* **47**, 1–12.

Lucas, A. (1962), *Ancient Egyptian Materials and Industries*, Arnold, London.

Lucas, A. (1932), The use of natron in mummification, *J. Egypt. Archaeol.* **18**, 125–140.

Luckenback, A. H., C. G. Holland, and R. O. Allen (1975), Soapstone artifacts: Tracing prehistoric trade patterns in Virginia, *Science* **187**, 57–58.

Ludi, A. and R. Giovanoli (1967), Egyptian Blue, *Naturwissenschaften* **54**, 88–91.

Luedtke, B. E. (1992), *An Archaeologists Guide to Chert and Flint*, Institute of Archaeology, Univ. California, Los Angeles.

Lutz, J. and E. Pernicka (1996), Energy dispersive X-ray fluorescence analysis of ancient copper alloys: Empirical values for precision and accuracy, *Archaeometry* **38**, 313–323.

Lynch, G. (1993), *Brickwork: History, Technology and Practice*, Donhead, London.

MacAlister, R. A. S. (1912), *The Excavations of Gezer*, Macmillan, London.

MacGregor, E. A. and C. T. Greenwood (1980), *Polymers in Nature*, Wiley, New York.

MacHugh, D. E., C. J. Edwards, J. F. Bailey, D. R. Bancroft, and D. G. Bradley (2000), The extraction and analysis of ancient DNA from bone and teeth: A survey of current methodologies, *Ancient Biomol.* **3**, 81–103.

MacKerrel, H. (1972), *Educ. Chem.* **9**, 54.

Madder, S. S. (1998), *Biology*, McGraw-Hill, Boston.

Maddin, R. (1988), *The Beginning of the Use of Metals and Alloys*, Massachusets Instiute of Technology (MIT), Cambridge.

Maeder, F. and M. Halbeisen (2001), Muschelseide – Auf der Suche nach einem vergessenen Material, *Waffen. und Kostumkunde* **43**(1), 33–41.

Malinovsky, R., A. Slatkine, and M. Ben Yair (1961), *Proc. Int. Symp. Reliability of Concrete*, RILEM, pp. 531–533.

Malley, J. (1987), The use of modern analytical techniques in the identification of ancient gems, in Hackens, T. and G. Mucharte (eds.), *Technology and Analysis of Ancient Gemstones*, Council of Europe, Strasbourg, pp. 41–53.

Malmqvist, K. G. (1995), Applications of PIXE in art and archaeology, in Johansson, S. A. E., J. L. Campbell, and K. G. Malmqvist (eds.), *Particle-Induced X-Ray Emission Spectroscopy*, Wiley, New York.

Manchester, K. (1984), Tuberculosis and leprosy in antiquity: An interpretation, *Med. Hist.* **28**, 162–173.

Mando, P. A. (2005), Nuclear physics and archaeometry, *Nuclear Physics A* **751**, 393c–408c.

Maneta, C. (1997), Porcelain: Technology, deterioration and conservation, *Glass Ceramics Conserv.* **3**, 5–8.

Maniatis, Y. (2004), Scientific techniques and methodologies for the provenance of white marbles, *Proc. Int. School of Physics "Enrico Fermi"* **154** (*Physics Methods in Archaeometry*), pp. 179–202.

Manley, W. F., G. H. Miller, and J. Czywczynski (2000), Kinetics of aspartic acid racemization, in Goodfriend, G. A., M. J. Collins, M. L. Fogel, S. A. Macko, and J. F. Wehmiller (eds.), *Perspectives in Amino Acid and Protein Geochemistry*, Oxford Univ. Press, New York, pp. 202–218.

Mann, S. and D. Walsh (1996), Feigning nature's sculptures, *Chem. Britain* **32**(11), 31–34.

Mann, W. B., R. L. Ayres, and S. B. Garfinkel (eds.) (1980), *Radioactivity and its Measurement*, Pergamon, Oxford.

Marlar, R. A., B. L. Leonard, B. R. Billman, P. L. Lambert, and J. E. Marlar (2000), Biochemical evidence of cannibalism at a prehistoric pueblo site in southwestern Colorado, *Nature* **407**, 74–78.

Marotta, I. and F. Rollo (2002), Molecular paleontology, *Cell. Molec. Life Sci.* **59**(1), 97–111.

Marschner, R. F. and H. T. Wright (1978), Asphalts from Middle Eastern archaeological sites, in Carter, G. F. (ed.), *Archaeological Chemistry*, Advances in Chemistry Series, Vol. 2, ACS, Washington, DC.

Martin, R. E. (1999), *Taphonomy, a Process Approach*, Cambridge Univ. Press, Cambridge, UK.

Martinetz, D., K. Lhos, and J. Janzen (1988), *Weihrauch und Myrrhe: Botanik, Chemie*, Wissenschaftliche Verlag, Stutgart.

Maschner, H. D. G. and C. Chippindale (eds.) (2005), *Handbook of Archaeological Methods*, Altamira Press, Walnut Creek, CA.

Mason, R. B. and M. S. Tite (1997), The beginning of tin-opacification of pottery glazes, *Archaeometry* **39**, 41–58.

Mason, R. B. and M. S. Tite (1994), The beginnings of islamic stone paste technology, *Archaeometry* **36**, 77–91.

Masschelein-Kleiner, L. (1985), *Ancient Binding Media, Varnishes and Adhesives*, International Center for the Study of the Preservation and Restoration of Cultural Property, Rome.

Masschelein-Kleiner, L. and J. B. Heylen (1968), *Stud. Conserv.* **13**, 87–91.

Masterton, W. L. and E. J. Slowinski (1986), *Qualitative Analysis*, Saunders, Philadelphia.

Mathews, K. J. (1997), The establishment of a data base of neutron activation analysis of white marble, *Archaeometry* **39**, 321–332.

Matson, F. R. (1981), Archaeological ceramics and the physical sciences: Problem definition and results, *J. Field Archaeol.* **8**, 447–456.

Matsunaga, M. and I. Nakai (2000), Study of the origin of the silver luster on gray iron age pottery, *Anatolian Archaeol. Stud.* **9**, 207–211.

Mattera, J. (2000), *The Art of Encaustic Painting*, Watson Guptill, New York.

Mayell, H. (2003), "Jesus Box" is a fake, Israeli experts rule, *National Geographic News*, June 18.

Mayer, R. (1981), *The Artist's Handbook to Materials and Techniques*, Faber and Faber, London.

Mays, S. (1998), *The Archaeology of Human Bones*, Routledge, London.

McDougall, I. and T. M. Harrison (1999), *Geochronology and Thermochronology by the Ar-40/A-39 Method*, Oxford Univ. Press, Oxford, UK.

McGovern, P. E. (2003), *Ancient Wine: The Search for the Origins of Viniculture*, Princeton Univ. Press, Princeton, NJ.

McGovern, P. E. and R. H. Michel (1991), Royal purple dye: Its identification by complementary physicochemical techniques, *Inorg. Chem.* **3**(1), 69–76.

McGovern, P. E., T. L. Sever, J. Wilson Myers, E. E. Myers, B. Bevan, N. F. Miller, S. Bottema, H. Hongo, R. H. Meadow, P. Kuniholm, S. G. E. Bowman, M. N. Leese, R. E. M. Hedges, F. R. Matson, I. C. Freestone, S. J. Vaughan, J. Henderson, P. B. Vandiver, C. S. Tumosa, C. W. Beck, P. Smith, A. M. Child, A. M. Pollard, I. Thuesen, and C. Sease (1995), Science in archaeology: A review, *Am. J. Archaeol.* **99**(1), 79–142.

McGregor, A. (1985), *Bone, Amber, Ivory and Horn*, Croom Helm, London.

McIlraith, S. J. (1990), *Diagenesis*, Canada Geoscience Reprint Series, Toronto.

McKeever, S. W. (1988), *Thermoluminescence of Solids*, Cambridge Univ. Press, Cambridge, UK.

McLaren, A. (1994), *A History of Contraception*, Blackwell, Oxford.

McLaren, K. (1986), *The Colour Science of Dyes and Pigments*, Hilger, Bristol.

McLaughlin, W. L., A. Scharmann, and D. Regulla (eds.) (1990), *Proc. 2nd Int. Sympo. e.s.r. Dosimetry and Applications*.

McLeod, I. D. (1981), Bronze disease: An electrochemical explanation, *Bull. Inst. Conserva. Cultural Mater.* **7**, 16–26.

McPherson, M. J., S. G. Møller, R. Beynon, and C. Howe (2000), *PCR – the Basics: From Background to Bench*, Springer, New York.

Meiggs, R. (1982), *Trees and Timber in the Ancient Mediterranean World*, Oxford Univ. Press, Oxford, UK.

Melessanaki, K., S. Kotoulas, A. Petrakis, A. Hatziapostolou, D. Anglos, S. Ferrence, and P. P. Betancourt (2002), LIBS (laser induced breakdown spectroscopy): A new tool in archaeometry?, *Trends Opt. Photon.* **81**, 57–59.

Mell, C. C. (1932), A brief historical account of weld, *Textile Colorist* **35**, 33–51.

Merkel, J. F. (1990), Experimental reconstruction of Bronze Age copper smelting based on archaeological evidence from Timna, in Rothenberg, B. (ed.), *The Ancient Metallurgy of Copper*, International Association of Meteorology and Atmospheric Sciences, London, pp. 78–122.

Merrit, J. (1994), Fiber identification, *Textile Conserv. Newsl.* **27**, 9–12.

Metcalfe, E. and F. E. Prichard (1987), *Atomic Absorption and Emission Spectroscopy*, Wiley, New York.

Michael, H. N. and E. K. Ralph (eds.) (1971), *Dating Techniques for the Archaeologist*, MIT, Cambridge, MA.

Michel, H. V. and F. Asaro (1979), Chemical study of the plate of brass, *Archaeometry* **21**, 3–19.

Michel, R. H. and P. E. McGovern (1990), The chemical processing of royal purple dye: Ancient descriptions as elucidated by modern science, *ArchaeoMaterials* **4**, 97–104.

Michels, J. W. and S. T. Tsong (1980), Obsidian dating, a coming of age, in Schiffer, M. B. (ed.), *Advances in Archaeological Method and Theory*, Vol. 3, Academic Press, New York, pp. 405–444.

Middendorf, B., J. J. Hughes, K. Callebaut, G. Baronio, and I. Papayianni (2005), Investigative methods for the characterization of historic mortars, *Mater. Struct.* **38**, Part 1: Mineralogical Characterization, 761–769; Part 2: Chemical Characterization, 771–780.

Middlemost, E. A. K. (1985), *Magmas and Magmatic Rocks*, Longman, London.

Middleton, J. (1845), On fluorine in bones, its source and its application to the determination of the geological age of fossil bones, *Quart. J. Geol. Soc.* **1**, 214–216.

Mie Jianjun (1995), The history, metallurgy and spread of paktong, *Bull. Metals Museum* **24**, 43–55.

Miles, G. and S. Pollard (1985), *Lead and Tin: Studies in Conservation and Technology*, UK Institute of Conservation, London.

Millard, A. R. (2001), The deterioration of bone, in Brothwell, D. R. and A. M. Pollard (eds.), *Handbook of Archaeological Sciences*, Wiley, New York, pp. 637–647.

Millard, A. R. (1993), *Diagenesis of Archaeological Bone: The Case of Uranium Uptake*, Oxford Univ. Press, Oxford, UK.

Miller, D. S. and G. A. Wagner (1981), Fission track ages applied to obsidian artifacts from South America, *Nucl. Tracks* **5**, 147–155.

Miller, J. M. (1988), *Chromatography: Concepts and Contrasts*, Wiley, New York.

Millet A. and H. W. Catling (1967), *Archaeometry* **10**, 70–75.

Mills, J. S. (1972), Identification of organic materials in museum objects, in *Conservation in the Tropics, Proc. Asia-Pacific Seminar on Conservation of Cultural Property*, International Center for the Study of the Preservation and Restoration of Cul and Central Conservation Laboratory, New Delhi, pp. 159–170.

Mills, J. S. and R. White (1996), *The Organic Chemistry of Museum Objects*, Butterworth-Heinemann, London.

Mills, J. S., R. White, and L. J. Gough (1985), The chemical composition of Baltic amber, *Chem. Geol.* **47**, 15–39.

Mischara, J. and P. Myers (1974), Ancient Egyptian silver: A review, in Bishay, A. (ed.), *Recent Advances in Science and Technology of Materials*, Vol. 3, Plenum, New York, pp. 29–45.

Mitchell, C. A. (1924), *Inks: Their Composition and Manufacture*, Griffin, London.

Mitra, S. (ed.) (1992), *Applied Mössbauer Spectroscopy – Theory and Practice for Geochemists and Archaeologists*, Pergamon, Oxford, UK, pp. 341–363.

Mix, P. E. (1987), *Non-Destructive Testing*, Wiley, New York.

Modugno, F., E. Ribechini, and M. P. Colombini (2006), Aromatic resin characterization by gas chromatography-MSS spectrometry, *J. Chromatog. A* **1134**, 298–304.

Moens, L., A. Von Bohlen, and P. Vandenabeele (2000), X ray fluorescence, in Ciliberto, E. and G. Spoto (eds.), *Modern Analytical Methods in Art and Archaeology*, Chemical Analysis Series, Vol. 155, Wiley, New York.

Mognonov, D. M., T. N. Dorzhieva, L. V. Lbova, I. V. Zvontsov, A. I. Buraev, and V. V. Khakhinov (2002), Dating of osteological material from archaeological sources using thermal analysis, *Russian J. Appl. Chem.* (transl. of *Zhurnal Prikladnoi Khimii*), **75**(3), 504–506.

Mohen, J. P. (1996), *L'Art et la Science, L'Esprit des Chefs-D'Oeuvre Decouvertes*, Gallimard-Reunion des Musees Nationaux, Paris.

Moldenke, H. N. and A. L. Moldenke (1952), *Plants of the Bible*, Ronald, New York.

Molera, J., M. Vendrel Saz, and J. Perez (2001), Chemical and textural characterization of tin glazes in Islamic ceramics from eastern Spain, *J. Archaeol. Sci.* **28**, 331–340.

Mommsen, H. (2001), Provenance determination of pottery by trace element analysis: Problems, solutions and applications, *J. Radioanal. Nucl. Chem.* **247**(3), 657–662.

Mommsen, H., A. Brining, H. Dittmann, A. Hein, A. Rosenberg, and G. Sarrazin (1997a), Early Roman cameo glass: X ray fluorescence analysis induced by synchrotron radiation (S.I.X.R.F.), *Glass Sci. Technol.* (*Glasstechnische Berichte*) (July), 211–219.

Mommsen, H., H. Dittmann, A. Hein, and A. Rosenberg (1997b), X-ray fluorescence analysis induced by synchrotron radiation (S.Y.X.R.F.) and first archaeometric applications, in Dirksen, D. and G. von Bally (eds.), *Optical Technologies in the Humanities*, Springer, Berlin, pp. 119–121.

Montes, L. T., M. R. Garcia, J. Gazzola, and S. Gomez (2005), Analysis of stucco floors from the citadel of the archaeological zone of Teotihuacan, Mexico, *Materials Research Society Symposium Proceedings (Materials Issues in Art and Archaeology VII)* **852**, 353–359.

Montgomery, J., J. A. Evans, and T. Neighbour (2003), Strontium isotope evidence for population movement within the Hebridean Norse community of NW Scotland, *J. Geol. Soc.* **160**(5), 649–653.

Moore, D. T. and W. A. Oddy (1985), Touchstones: Some aspects of their nomenclature, petrography and provenance, *J. Archaeol. Sci.* **12**(1), 59–80.

Morel, R. S. (1939), *The Scientific Aspects of Artists and Decorators Materials*, Oxford Univ. Press, Oxford, UK.

Morell, V. (1994), Mummy settles TB antiquity debate, *Science* **263**(5154), 1686–1687.

Morgan, E. D., C. Cornford, D. R. J. Pollock, and P. Isaacson (1973), The transformation of fatty materials buried in the soil, *Sci. Archaeol.* **10**, 9–10.

Morgan, S. W. K. (1985), *Zinc, Its Alloys and Compounds*, Wiley, New York.

Moropoulou, A., A. Bikulas, and S. Anagnostopoulou (2005), Composite materials in ancient structures, *Cement Concrete Compos.* **27**, 295–300.

Morris, E. T. (1984), *The Story of Perfume*, Scribner, New York.

Mottana, A., R. Crespi, and G. Liborio (1978), *Simon and Schuster's Guide to Rocks and Minerals*, Simon and Schuster, New York.

Mottran, H. R., S. N. Dudd, G. J. Lawrence, A. W. Stott, and R. P. Evershed (1999), New chromatographic, mass spectrometric and stable isotope approaches to the classification of degraded animal fats preserved in archaeological pottery, *J. Chromatogr. A* **833**, 209–221.

Mountain, J. L., A. A. Lin, A. M. Bowcock, and L. Cavalli-Sforza (1993), Evolution of modern humans: Evidence from nuclear DNA polymorphisms, in Aitken, M. J., Stinger, C. B. and P. A. Mellars (eds.), *The Origin of Modern Humans and the Impact of Chronometric Dating*, Princeton University Press, Princeton, NJ.

Mueller, J. W. (ed.) (1975), *Sampling in Archaeology*, Univ. Arizona, Tucson.

Mueller, M., B. Murphy, M. Burghammer, C. Riekel, M. Roberts, M. Papiz, D. Clarke, J. Gunneweg, and E. Pantos (2004), Identification of ancient textile fibers from Khirbet Qumran caves using synchrotron radiation microbeam diffraction, *Spectrochimica Acta* **59B**(10–11), 1669–1674.

Mühlethaler, B. (1973), *Conservation of Waterlogged Wood and Wet Leather*, Paris.

Muhly, J. D. (1973), *Copper and Tin in the Bronze Age*, Archon, Hamden.

Muller, H. (1987), *Jet*, Butterworth, London.

Mullis, K. (1990), The unusual origin of the polymerase chain reaction, *Sci. Am.* (April), 56–65.

Murray, H. H. (1986), Clays, in *Ullman's Encyclopaedia of Industrial Chemistry*, Vol. 7, VCH Verlag GmbH, Weinheim, pp. 109–136.

Murray, M. L. and M. J. Schoeninger (1988), Diet, status and complex social structure in Iron Age central Europe: Some contributions of bone chemistry, in Gibson, D. B. and M. N. Geselowitz (eds.), *Tribe and Polity in Late Prehistoric Europe*, Plenum Press, New York, pp. 155–176.

Murray-Wallace, C. V. (1993), A review of the application of the amino acid racemisation reaction to archaeological dating, *The Artefact* **16**, 19–26.

Muscarella, O. W. (2000), *The Lie Became Great: The Forgery of Ancient Middle Estern Cultures*, Styx, Groningen.

Needham, J. (1980), *Science and Civilization in China*, Vol. 5, Cambridge Univ. Press, Cambridge, UK.

Needham, J. (1958), *Science and Civilization in China*, Vol. 1, Cambridge Univ. Press, Cambridge, UK.

Needles, H. L. and S. H. Zeronian (eds.) (1986), *Historical Textile and Paper Materials: Conservation and Characterization*, ACS, Washington, DC.

Neff, H. (2000), Neutron activation analysis for provenance determination in archaeology, in Ciliberto, E. and G. Spoto (eds.), *Modern Methods in Art and Archaeology*, Chemical Analysis Series, Vol. 155, Wiley, New York, pp. 81–134.

Neff, H. (ed.) (1992), *Chemical Characterization of Ceramic Pastes in Archaeology*, Monographs in World Archaeology, Vol. 7, Prehistory Press, Madison, WI.

Nelson, B., M. A. Goni, J. I. Edges, and R. A. Blanchette (1995), Soft rot fungal degradation of lignin in 2.700 years old archaeological woods, *Holzforschung* **41**(1), 1–10.

Neri, A. (1662, 2000 reprint), *L'Arte Vetraria* (*The Art of Glass*, transl. C. Merrett), Society of Glass Technology, London.

Neuberger, A. (1969), *The Technical Arts and Sciences of the Ancient*, Methuen, London.

Newall, R. S. (1959), *Stonehenge*, Her Majesty's Stationery Office, London.

Newbury, D. (1986), *Advances in Scanning Electron Microscopy and X-Ray Micro Analysis*, Plenum, New York.

Newman, A. (ed.) (1987), *Chemistry of Clays and Clay Minerals*, Longmans, London.

Newton, R. G. (1985), The durability of glass: A review, *Glass Technol.* **26**, 21–38.

Newton, R. G. (1980), Recent views on ancient glass, *Glass Technol.* **21**, 173–183.

Newton, R. G. and S. Davison (1997), *The Conservation of Glass*, Butterworth-Heinemann, London.

Nicholson, E. D. (1981), The ancient craft of gold beating, *Gold Bull.* **14**, 161–166.

Nicholson, P. T. (1993), *Egyptian Faience and Glass*, Shire Publications, Princess Risborough.

Nicholson, P. T. and E. Peltenburg (2000), Egyptian faience, in Nicholson, P. T. and I. Shaw (eds.), *Ancient Egyptian Materials and Technology*, Cambridge Univ. Press, Cambridge, UK.

Nicholson, P. T. and I. Shaw (eds.) (2000), *Ancient Egyptian Materials and Technology*, Cambridge Univ. Press, Cambridge, UK.

Nielsen-Marsh, C. (1997), *Studies in Archaeological Bone Diagenesis*, Oxford Univ. Press, Oxford, UK.

Nigam, J. K. and P. N. Prasad (eds.) (1992), *Frontiers of Polymer Research*, Kluwer, New York.

Nir-El, Y. and M. Broshi (1996), The red ink of the Dead Sea scrolls, *Archaeometry* **38**, 97–102.

Nishimura, S. (1971), Fission track dating of archaeological materials from Japan, *Nature* **230**, 242–243.

Nissenbaum, A. (1992), Molecular archaeology: Organic geochemistry of Egyptian mummies, *J. Archaeol. Sci.* **19**, 1–6.

Nixon, T. (ed.) (2004), Preserving archaeological remains *in situ, Proceed. 2nd Conference,* 12–14 September 2001, English Heritage, London.

Noble, J. V. (1987, 1965), *The Technique of Attic Vase Painting,* Watton Guptill, New York.

Noble, J. V. (1969), The technique of Egyptian faience, *Am. J. Archaeol.* **73**, 435–439.

Noble, J. V. (1960), The technique of attic vase painting, *Am. J. Archeol.* **64**, 307–318.

Noll, W. (1982), Techniken der Dekoration antiker Keramic, *Berichte Deutsche Keramische Geselltschaft* **59**, 3–17.

Noll, W., R. Holm, and H. Born (1975), Painting on ancient ceramics, *Angewandte Chemie* (Int. Ed.) **14**(9), 602–613.

Noll, W., R. Holm, and H. Born (1973), *Berichte Deutsche Keramische Geselltschaft* **50**, 328.

Noller, J. S., J. M. Sowers, and W. R. Lettis (eds.) (2002), *Quaternary Geochronology Methods and Applications,* American Geophysical Union, Washington, DC.

Nord, A. G. and K. Tronner (2000), A note on the analysis of gilded metal embroidery threads, *Stud. Conserv.* **45**(4), 274–279.

North, N. A. and I. D. Macleod (1987), Corrosion of metals, in Pearson, C. (ed.), *Conservation of Marine Archaeological Objects,* Butterworth, London, pp. 68–98.

Northover, P. (1989), Non ferrous metallurgy, in Henderson, J. (ed.), *Scientific Analysis in Archaeology,* Oxford Univ. Committee for Archaeology, Monograph 19, Oxford, UK, pp. 213–236.

Norton, N. A. Conservation of metals, in Pearson, C. (eds.), *Conservation of Marine Archaeological Objects,* Butterworth, London, pp. 207–252.

Notis, M. (1988), The Japanese alloy shakudo: Its history and patination, in Maddin, R. (ed.), *The Beginning of the Use of Metals and Alloys,* Massachusets Instiute of Technology (MIT), Cambridge, MA, pp. 315–327.

Nriagu, J. O. (1983), *Lead and Lead Poisoning in Antiquity,* Wiley, New York.

Oakley, K. P. (1963), Analytical methods of dating bones, in Brothwell, D., E. Higgs, and G. Clark (eds.), *Science in Archaeology,* Thames and Hudson, London, pp. 24–34.

Oakley, K. P. and C. R. Hoskins (1950), New evidence on the antiquity of Piltdown Man, *Nature* **165**, 379–382.

Oakley, K. P. and M. F. A. Montagu (1949), A consideration of the Galley Hill skeleton, *Bull. Br. Museum* (*Natural Hist.*) *Geology* **1**, 25–48.

Oberlies, F. (1968), *Naturwissenschaften* **55**, 277–289.

O'Brien, M. J. and R. L. Lyman (1999), *Seriation, Stratigraphy, and Index Fossils: The Backbone of Archaeological Dating,* Kluwer Academic/Plenum, New York.

O'Connell T. C., R. E. M. Hedges, and G. J. van Klinken (1997), An improved method for measuring racemization of amino acids from archaeological bone collagen, *Ancient Biomolec.* **1**, 215–220.

O'Connor, T. (2005), *Biosphere to Lithosphere, New Studies in Vertebrate Taphonomy,* Oxbow, Oxford.

O'Connor, T. (2000), *The Archaeology of Animal Bones*, Stroud-Sutton, Cambridge, UK.

O'Connor, T. P., K. Starling, K. Watkinson, and D. Watkinson (eds.) (1987), *Archaeological Bone, Antler and Ivory*, UK Institute for Conservation, London.

Oddy, W. A. (2004), The gilding of metals since 3000 BC – "All that glisters is not gold," *Proc. Int. School of Physics "Enrico Fermi,"* Vol. 154, *Physics Methods in Archaeometry*, pp. 251–256.

Oddy, W. A. (1991), Gilding: An outline of the technological history of the plating of gold on to silver or copper in the Old World, *Endeavour* **15**, 29–33.

Oddy, A. (1983) Assaying in antiquity, *Gold Bull.* **16**(2), 52–59.

Oddy, W. A. (1980), *Methods of Chemical and Metallurgical Investigation of Ancient Coinage*, Royal Numismatic Society, London.

Oddy, W. A., T. G. Padley, and N. D. Meeks (1979), Some unusual techniques of gilding in antiquity, *ArkaeoPhysica* **10**, 108–113.

O'Donoghue, M. (1970), *Gemstones*, Chapman & Hall, London.

O'Flaherty, F., W. T. Roddy, and R. M. Lollar (1965), *The Chemistry and Technology of Leather*, Reinhold, New York.

Ogden, J. M. (1976), Platinum group metals inclusions in ancient gold artifacts, *J. Hist. Metallurg. Soc.* **11**(2), 53–72.

O'Grady, C. (2005), The occurrence of rock varnish on stone and ceramic artifacts, *Rev. Conservation* **6**, 31–38.

Oguchi, H. (1983), Japanese shakudo, its history, properties and production from gold containing alloys, *Gold Bull.* **16**, 125–132.

O'Keefe, S. F. (2000), An overview of oils and fats, in Kiple, K. F. and K. C. Ornelas, *Cambridge World Hist. Food* **1**, 375–386.

Oleson, J. P. (1988), The technology of Roman harbours, *Int. J. Nautical Archaeol.* **17**, 117–129.

Olin, J. S. and A. D. Franklin (eds.) (1982), *Archaeological Ceramics*, Smithsonian Institution, Washington, DC.

Olin, J. S. and E. V. Sayre (1974), Neutron activation analysis of some ancient glass, in Beck, C. W. (ed.), *Archaeological Chemistry*, Advances in Chemistry Series, Vol. 1, ACS, Washington, DC.

Ollier, C. (1969), *Weathering*, Oliver and Boyd, Edinburgh.

Olson, G. W. (1981), *Soils and the Environment: A Guide to Soil Surveys and Their Applications*, Chapman & Hall, New York.

Oppenheimer, L., D. P. Barag, R. H. Brill, and A. V. Saldern (1970), *Glass and Glassmaking in Ancient Mesopotamia*, Corning Museum of Glass, Corning, NY.

Orna, M. V. (ed.) (1996), Archaeological chemistry, in *Organic, Inorganic and Biochemical Analysis*, Advances in Chemistry Series, Vol. 5, ACS, Washington, DC.

Orna, M. V. (1976), The molecular basis of form and color, *J. Chem. Educ.* **53**, 638.

Ortner, D. J. and W. G. Putschar (1981), Identification of pathological conditions in human skeleton remains, *Smithsonian Contributions Anthropol.* **28**, 142–176.

Orton, C. (2000), *Sampling in Archaeology*, Cambridge Univ. Press, Cambridge, UK.

Orton, C., P. Tyers, and A. Vince (1993), *Pottery in Archaeology*, Cambridge Univ. Press, Cambridge, UK.

O'Toole, M. (1995), *Water*, Cambridge Univ. Press, Cambridge, UK.

Ottaway, J. H. and M. R. Mathews (1988), Trace element analysis of soil samples from a stratified archaeological site, *Environ. Geochem. Health* **10**, 105–112.

Overheim, R. D. and D. L. Wagner (1982), *Light and Color*, Wiley, 1982.

Pääbo, S. (1999), Human evolution, *Trends Genet.* **15**(12), M13–M16.

Pääbo, S. (1993), Ancient DNA, *Sci. Am.* (Nov.), 18–28.

Pääbo, S. (1989), Ancient DNA: Extraction, characterization, molecular cloning and enzymatic amplification, *Proc. Natl. Academy of Sciences of the USA* **86**, 939–943.

Pääbo, S., H. Poinar, D. Serre, V. Jaenicke-Despres, J. Hebler, N. Rohland, M. Kuch, J. Krause, L. Vigilant, and M. Hofreiter (2004), Genetic analyses from ancient DNA, *Annu. Rev. Genet.* **38**, 645–679.

Pacifici, A. M. (1986), Il marmo: Materia, techniche, conservazione, restauro; Una bibliografia (a marble bibliography), in *OPD Restauro: Restauro del Marmo: Opere e Problemi*, Firenze, pp. 205–240.

Padfield, T. and S. Landi (1966), The fastness to light of the natural dyes, *Stud. Conserv.* **11**, 181–196.

Panagiotakopulu, E. (1999), An examination of biological materials from coprolites from XVIII Dynasty Amarna, Egypt, *J. Archaeol. Sci.* **26**(5), 547–551.

Parker, A. and A. M. Pollard (2006), Archaeological geochemistry, *App. Geochem.* **21**, 1625.

Parker, A. and B. W. Selwood (1983), *Sediment Diagenesis*, Kluwer, New York.

Parkinson, R. B. (1995), *Papyrus*, British Museum Press, London.

Parry, E. J. (1935), *Shellac*, Pitman, London.

Parry, E. J. (1918), *Gums and Resins*, Pitman, London.

Patton, W. W. and T. P. Miller (1970), Obsidian in Alaska, *Science* **169**, 760–761.

Paynter, S. and M. S. Tite (2001), The evolution of glazing technologies in the Ancient Near East and Egypt, in Shortland, A. J. (ed.), *The Social Context of Technological Change – Egypt and the Near East 1650–1550 BC*, Oxbow, Oxford, UK.

Pearsall, D. M. and D. R. Piperno (1993), *Current Research in Phytolith Analysis: Applications in Archaeology and Paleoecology, MASCA*, Univ. Pennsylvania, Philadelphia.

Pearson, C. (ed.) (1987), *Conservation of Marine Archaeological Objects*, Butterworth, London.

Pearson, G. W. and M. Stuiver (1993), High-precision bidecadal calibration of the radiocarbon time scale 500–2500 BC, *Radiocarbon* **35**, 25–33.

Pedley, J. G. (1997), *Greek Art and Archaeology*, Prentice-Hall, New York.

Peigler, R. S. (1993), Wild silks of the world, *Am. Entomol.* (Fall), 51–161.

Peltenberg, E. J. (1987), Early faience: Recent studies, origins and relations with glass, in Bimson, M. and I. C. Freestone (eds.), *Early Vitreous Materials*, British Museum, London.

Peltz, C. and M. Bichler (2000), Abrasive supply for ancient Egypt revealed, *Proc. 16th European TRIGA Conf.*, Pitesti, Romany, Sept. 25–28. 2000.

Penhallurick, R. D. (1986), *Tin in Antiquity*, The Institute of Metals, London.

Percy, J. (1861), Metallurgy: *The Art of Extracting Metals from their Ores and Adapting Them to Various Purposes of Manufacture*, Murray, London; reprint in two parts (ca. 1985): Vol. 1, *Fuel: Fire-Clays; Copper; Zinc; Brass, etc.*; Vol. 2, *Iron; Steel*, De Archaeologische Pers Nederland, Eindhoven.

Perdue, R. E. and C. J. Kraebel (1961), The rice paper plant, *Econ. Botany* **15**(2), 165–179.

Perez y Jorba, M. and J. P. Dallas (1984), Composition et altération des Grisailles anciennes: 3 examples de grisaille du XIIIe siècle étudiés par rayons X et microsonde électronique, *CVMA Newsl.*, **37/38**, 8–12.

Perlman, I. (1984), Modern neutron activation analysis and ancient history, in Lambert, J. B. (ed.), *Archaeological Chemistry*, Advances in Chemistry Series, Vol. 3, ACS, Washington, DC, pp. 117–132.

Perlman, I. and J. Yellin (1980), The provenience of obsidian from Neolithic sites in Israel, *Israel Explor. J.* **30**, 83–88.

Pernicka, E. (2004), Archaeometallurgy: Examples of the application of scientific methods to the provenance of archaeological metal objects, *Proc. Int. School of Physics "Enrico Fermi,"* Vol. 154, *Physics Methods in Archaeometry*, pp. 309–329.

Pernicka, E. (2000), Isotope archaeology, in Rammlmair, D. (ed.), *Applied Mineralogy in Research, Economy, Technology, Ecology and Culture, Proc. 6th Int. Congress on Applied Mineralogy*, Goettingen, Balkema, Rotterdam, pp. 1025–1028.

Pernicka, E. (1996), Analytical techniques, in Meyers, M. (ed.), *The Oxford Encyclopedia of Archaeology in the Near East*, Vol. 1, Oxford Univ. Press, Oxford, UK, pp. 118–122.

Persson, K. B. (1997), Soil phosphate analysis: A new technique for measurement in the field using a test strip, *Archaeometry* **39**, 441–443.

Peter, J. (2006), Ancient soldering techniques, *Przeglad Spawalnictwa* **78**, 19–22.

Peters, K. E., C. C. Walters, and J. M. Moldowan (2005), *The Biomarker Guide*, Vol. 1, *Biomarkers and Isotopes in the Environment and Human History*, 2nd ed., Cambridge Univ. Press, Cambridge, UK.

Pettijohn, F. J. (1949), *Sedimentary Rocks*, Harper, New York.

Pfister, R. (1935), Tissues coptes du Musee du Louvre, *Seminarium Kondakovianum* **7**, 40–41.

Photos, E., R. E. Jones, and Th. Papadopoulus (1994), The black decoration on a Mycenean dagger, *Archaeometry* **36**, 267–275.

Piaskowski, J. (1988), The earliest iron in the world, *PACT* **21**, 37–46.

Pike, A. W. G. and P. B. Pettit (2005), Dating techniques, in Maschner, H. D. G. and C. Chippindale (eds.), *Handbook of Archaeological Methods*, Altamira Press, Walnut Creek, CA, Chapter 10.

Piperno, D. R. (1988), *Phytolith Analysis: An Archaeological and Geological Perspective*, Academic Press, San Diego.

Pires, J. and A. J. Cruz (2007), Techniques of thermal analysis applied to the study of cultural heritage, *J. Thermal Anal. and Calorim.* **87**(2), 411–415.

Plenderleith, H. J. (1966), *The Conservation of Antiquities and Works of Art*, Oxford Univ. Press, London, pp. 116–123.

Plisson, H. and M. Mauger (2001), Chemical and mechanical alteration of microwear polishes: An experimental approach, *Chimia* **55**, 931–937.

Plotnick, R. E. (1993), Taphonomy – perfecting the fossil record, *Geotimes* **38**(11), 14.

Plumb, R. C. (1989), Antique window panes and the flow of supercooled liquids, *J. Chem. Educ.* **66**(12), 994–996.

Pluta, M. (1989), *Advanced Light Microscopy*, Elsevier, Amsterdam.

Poinar, G. O. and R. Milki (2001), *Lebanese Amber: The Oldest Insect Ecosystem in Fossilized Resin*, Oregon State University, Corvallis, OR.

Poinar, H. N. (2002), The genetic secret some fossils hold, *Acc. Chem. Res.* **35**, 676–684.

Poinar, H. N., M. Hofreiter, W. G. Spaulding, P. S. Martin, B. A. Stankiewicz, H. Bland, R. P. Evershed, G. Possnert, and S. Pääbo, (1998), Molecular coproscopy: Dung and diet of the extinct ground sloth *Nothrotheriops shasrensis*, *Science* **281**, 402–406.

Poinar, H. N., M. Höss, J. L. Bada, and S. Pääbo (1996), Amino acid racemization and the preservation of ancient DNA, *Science* **272**, 864–866.

Pollard, A. M. (2005), What is this thing we call archaeology?, *British Archaeology*, July/August, 55.

Pollard, A. M., C. M. Batt, B. Stern, and S. M. M. Young (2007), *Analytical Chemistry in Archaeology*, Cambridge University Press, Cambridge, UK.

Pollard, A. M. (1979), *X-Rays Fluorescence and Surface Studies of Glass with Application to the Durability of Medieval Window Glass*, Ph.D. thesis, Univ. York.

Pollard, A. M. and C. Heron (1996), *Archaeological Chemistry*, Royal Chemical Society (RCS), London.

Pollman, H. O. (1999), *Bibliographie zum Obsidian-Artefakt und Provenienz*, Deutscher Bergbau Museum, Bochum.

Polosmak, N. (1994), A mummy unearthed from the pastures of heaven, *Natl. Geogr.* (Oct.), 80–103.

Pool, C. A. (2000), Why a Kiln?, *Archaeometry* **42**(1), 61–76.

Porat, N. and Sh. Ilani (1998), A Roman period palette: Composition of pigments from King Herod's palaces in Jerico and Massada, Israel, *Israel J. Earth Sci.* **47**, 75–85.

Poupeau, G., L. Bellot-Gurlet, O. Dorighel, T. Calligaro, J. C. Dran, and J. Salomon (1996), Obsidian circulation in Prehispanic times in Colombia and Ecuador: A coupled PIXE/fission track dating approach, *C. R. Acad. Sci.*, Series IIa, *Sciences de la Terre et des Planetes* **323**(5), 443–450.

Powledge, T. M. and M. Rose (1996), The great DNA hunt: Colonizing the Americas, *Archaeology* **49**(6), 58–68.

Prandtstetten, R. (1980), Historiche Mauerwerke und Sein Restaurirung, *Restaurator Renblätter* **4**, 63–85.

Price, T. D. (ed.) (1989), *The Chemistry of Prehistoric Human Bone*, Cambridge Univ. Press, Cambridge, UK.

Price, T. D., J. H. Burton, and R. A. Bentley (2002), The characterization of biologically available strontium isotopes ratios for the study of prehistoric migrations, *Archaeometry* **44**(1), 117–135.

Procacci, U. and L. Guarnieri (1975), *Come Nasce un Affresco* (*How a Fresco is Born*), Bonechi, Florence.

Pullman, B. (2001), *The Atom in the History of Human Thought*, Oxford Univ. Press, Oxford, UK.

Putnam, J. L. (1960), *Isotopes*, Penguin, London.

Putnam, R. J. (1983), *Carrion and Dung: The Decomposition of Animal Wastes*, Arnold, London.

Raffi, A., M. Spigelman, J. Stanford, E. Lemma, H. Doneghue, and J. Zias (1994), *Mycobacerium leprae* DNA from ancient bone detected by PCR, *Lancet* **343**, 1360–1361.

Ragab, H. (1994), Resurrection of papyrus as a national heritage and the harvest of 35 years experience with papyrus, *Bull. Center Papyrol. Stud.* (Cairo) **10**, 151–167.

Ragab, H. (1980), *Le Papyrus*, thesis, Cairo Univ., Cairo.

Raman, C. V. and V. S. Rajagoplan (1940), *Proc. Indian Acad. Sci.* **11a**, 469.

Ramoutsakis, I. A., Y. A. Ramoutsakis, S. Haniotakis, and A. M. Tsatsakis (2000), Remedies used in Hellenic history, *Veterinary Human Toxicol.* **42**(4), 238–241.

Ramsey, C. B. (2005), *OxCal Radiocarbon Calibration Software*, http://www.rlaha.ox.ac.uk/orau/oxcal.htm.

Rapp, G. Jr. (1985), The provenance of artificial raw materials, in Rapp, G. Jr. and J. A. Gifford (eds.), *Archaeological Geology*, Yale Univ., New Haven.

Rapp, G. Jr. and J. A. Gifford (eds.) (1985), *Archaeological Geology*, Yale Univ., New Haven.

Rapp, G. Jr. and C. L. Hill (1998), *Geoarchaeology: The Earth Science Approach to Archaeological Interpretation*, Yale Univ., New Haven.

Rathburn, T. A. (2000), Chemical approaches to dietary representation, in Kiple, K. F. and K. C. Ornelas (eds.), *The Cambridge World History of Food*, Vol. 1, Cambridge, UK, pp. 58–62.

Raymond, L. A. (1995), *Petrology: The Study of Igneous, Sedimentary and Metamorphic Rocks*, W. C. Brown, Dubuque.

Raymond, R. (1986), *Out of the Fiery Furnace*, Pennsylvania State Univ., Philadelphia.

Reader, J. (1981), *Missing Links: The Hunt for Earliest Man*, Little, Brown, New York.

Reed, R. (1975), *The Nature and Making of Parchment*, Elmete, Leeds.

Reed, R. (1972), *Ancient Skin, Parchments and Leather*, Seminar, London.

Reed, S. J. B. (1997), *Electron Microprobe Analysis,* Cambridge Univ., Cambridge, UK.

Rees-Jones, J. and M. S. Tite (1997), Optical dating results for British archaeological sediments, *Archaeometry* **39**, 177–187.

Reeves, R. D. and G. C. Armitage (1973), *New Zealand J. Sci.* **16**, 521.

Regert, M. and C. Rolando (2002), Identification of archeological adhesives using direct inlet electron ionization mass spectrometry, *Anal. Chem.* **74**(5), 965–975.

Regert, M., N. Garnier, O. Decavallas, C. Cren-Olivé, and Ch. Rolando (2003), Structural characterization of lipid constituents from natural substances preserved in archaeological environments, *Meas. Sci. Technol.* **14**, 1620–1630.

Rehder, J. E. (1987), Natural drought furnaces, *Archaeomaterials* **3**, 27–37.

Rehder, J. E. (1986), Primitive furnaces and the development of metallurgy, *J. Hist. Metallurg. Soc.* **20**(2), 87–92.

Reibold, M., A. A. Levin, D. C. Meyer, P. Paufler, and W. Kochmann (2006), Microstructure of a Damascene sabre after annealing, *Int. J. Mat. Res.* **97**, 1172–1182.

Reiche, I., L. Favre-Quattropani, C. Vignaud, H. Bocherens, L. Charlet, and M. Menu (2003a), A multi-analytical study of bone diagenesis, *Meas. Sci. Technol.* **14**(9), 1608–1619.

Reiche, I., M. Radtke, and C. Brouder (2003b), X-ray analysis in art – glass and ivory (in German), *Physik in Unserer Zeit* **34**(2), 80–86.

Reinhard, J. (1996), Peruvian ice maiden, *Natl. Geogr.* (June).

Reinhard, K. J. and V. M. Bryant (Jr.) (1992), Coprolite analysis: A biological perspective on archaeology, in Schiffer, M. B. (ed.), *Archaeological Method and Theory,* Vol. 4, Univ. Arizona Press, Tucson, pp. 245–288.

Renfrew, C. (2001), From molecular genetics to archaeogenetics, *Proc. Natl. Acad. Sci. USA* **98**(9), 4830–4832.

Renfrew, C. (1967), Cycladic metallurgy and the Aegean early Bronze Age, *Am. J. Archaeol.* **71**, 1–20.

Renfrew, C. and P. Bahn (1996), *Archaeology: Theories, Methods and Practice,* Thames and Hudson, London.

Renfrew, C. and J. E. Dixon (1976), Obsidian in eastern Asia – a review, in Sieveking, G., I. H. Longworth, and K. E. Wilson (eds.), *Problems in Economic and Social Archaeology,* Duckworht, London, pp. 137–150.

Renne P., W. D. Sharp, A. L. Deino, G. Orsi, and L. Civetta (1997), Argon-40 – Argon-39 dating into the historical realm – calibration against Pliny the Younger, *Science* **277**, 1279–1280.

Retallack, G. J. (2000), *A Color Guide to Palaeosols,* Wiley, New York.

Retallack, G. J. (1990), *Soils of the Past,* Unwin Hyman, Boston.

Reyes-Valerio, C. (1993), *De Bonampak al Templo Mayor: El Azul Maya en Mesoamerica,* Siglo Veintiuno, Madrid.

Rhodes, G. (2000), *Crystallography Made Crystal Clear*, Academic Press, New York.

Rice, P. (1980), *Amber, the Golden Gem of the Ages*, Van Nostrand-Reinhold, New York.

Rice, P. M. (ed.) (1997), *The Prehistory and History of Ceramic Kilns*, The American Ceramic Society.

Rice, P. M. (1987), *Pottery Analysis: A Sourcebook*, Univ. Chicago Press, Chicago.

Rice, P. M. (ed.) (1982), *Pots and Potters: Current Approaches to Ceramic Archaeology*, Pennsylvania State Univ., Philadelphia.

Richards, J. F. (ed.) (1983), *Precious Metals in the Later Medieval and Early Modern Worlds*, Carolina Academic, Durham.

Riddle, J. M. (1992), *Contraception and Abortion from the Ancient World to the Renaissance*, Harvard Univ. Press, Cambridge, MA.

Rieth, A. (1970), *Archaeological Fakes*, Praegen, New York.

Rink, W. J. (ed.) (2002), *Proc. 10th Int. Specialist Conf. Luminescence and Electron Spin Resonance Dating Abstracts*, School of Geography and Geology, McMaster Univ., Hamilton.

Rink, W. J. (2000), *Beyond C-14 Dating: A User's Guide to Long-Range Dating Methods in Archaeology*, School of Geography, McMaster Univ., Hamilton.

Roberts, C. and K. Manchester (2005), *The Archaeology of Disease*, 2nd Ed., Cornell University, Ithaca, New York.

Roberts, J. C. (1998), *The Chemistry of Paper*, Royal Society of Chemistry, Cambridge.

Roberts, S. J., C. I. Smith, A. Millard, and M. J. Collins (2002), The taphonomy of cooked bone: Characterizing boiling and its physico-chemical effects, *Archaeometry* **44**(3), 485–494.

Robins, D., A. Alstin, and D. Fletton (1987), The examination of organic components in historical non-metallic seals with C-13 Fourier transform nuclear magnetic resonance spectroscopy, in Grimstad, K. (ed.), *8th Triennial Mtg., Int. Council of Museums Committee for Conservation*, Sidney, pp. 82–87.

Robinson, S. (1969), *A History of Dyed Textiles*, MIT, Cambridge, MA.

Robinson, W. S. (1951), A method of chronologically ordering archaeological deposits, *Am. Antiquity* **16**, 293–301.

Rocchi, G. (1985), *Instituzione di Restauro dei Benni Architettonici*, Hoepli, Milano.

Rollo, F., M. Ubaldi, I. Marota, S. Luciani, and L. Ermini (2002), DNA diagenesis: Effect of environment and time on human bone, *Ancient Biomolec.* **4**(1), 1–7.

Rolls, D. and W. J. Bland (1998), *Weathering*, Arnold, London.

Romer, A. (1982), *The Restless Atom*, Dover, New York.

Römich, H. (1999), Historic glass and its interaction with the environment, in Tennent, N. H. (ed.), *The Conservation of Glass and Ceramics: Research, Practice and Training*, James and James, London.

Rooksby, H. P. (1962), Opacifiers in opal glass through the ages, *GEC J.* **29**, 20–26.

Rostoker, W. and B. Bronson (1990), *Pre-Industrial Iron: Its Technology and Ethnology*, ArchaeoMaterials Monograph 1, Philadelphia.

Roth, E. and B. Poty (1989), *Nuclear Methods of Dating*, Kluwer, Boston.

Rottländer, R. C. A. (1989), A new method for the investigation of the content of archaeological vessels, *Chemie für Labor und Betrieb* **40**, 237–238.

Rottländer, R. C. A. (1985), Detection and identification of archaeological fats, *Fette, Seifen* **87**, 314–317.

Rottländer, R. C. A. (1983), *Einführung in die Natuwissenschaftichen Methoden in der Archäologie*, Univ. Tübingen, Tübingen.

Rouessac, F. and A. Rouessac (2000), *Chemical Analysis: Modern Instrumentation Methods and Techniques*, Wiley, New York.

Rowell, D. L. (1994), *Soil Science: Methods and Application*, Longman, London.

Rowell, R. M. (1990), The chemistry of solid wood, in Rowell, R. M. and R. J. Barbour (eds.), *Archaeological Woods: Properties, Chemistry and Preservation*, Advances in Chemistry Series, ACS, Washington, DC, pp. 455–487.

Rowell, R. M. and R. J. Barbour (eds.) (1990), *Archaeological Woods: Properties, Chemistry and Preservation*, Advances in Chemistry Series, American Chemical Society, Washington, DC.

Roy, A. (ed.) (1993), *Artists Pigments: A Handbook to Their History and Characterization*, Vol. 2, National Gallery of Art, Washington, DC.

Ruek, P. (1991), *Pergament. Geschichte. Struktur, Restaurierung, Herstellung;* Historische Hilfswischenschaften, Jan Thorbecke Verlag, Sigmaringen.

Rutgers, L. V., K. van der Borg, A. F. M. de Jong, and I. Poole (2005), Radiocarbon dating: Jewish inspiration of Christian catacombs?, *Nature* **436**(7049), 339.

Rutter, N. W. and B. Blackwell (1995), Amino acid racemization dating, in Rutter, N. W. and N. R. Catto (eds.), *Dating Methods for Quaternary Deposits*, Vol. 2, Geological Association of Canada, Geotext, St. John's, Newfoundland, pp. 125–164.

Ryan, F. (1994), *Tuberculosis: The Greatest Story Never Told*, Swift Publishers, Worcestershire.

Ryder, M. L. (1964), Parchment – its history, manufacture and composition, *J. Soc. Archivists* **3**, 391–399.

Ryder, M. L. and T. Gabra-Sanders (1985), The application of microscopy to textile history, *Textile History*, **16**(2), 123–140.

Sabroff A. M., F. W. Boulger, and H. L. Henning (1968), *Forging: Material and Practices*, Reinhold, New York.

Saiz-Jimenez, C. (ed.) (2003), *Molecular Biology and Cultural Heritage, Proc. Int. Congress on Molecular Biology and Cultural Heritage in Sevilla, Spain*, March 4–7, Balkema, Rotterdam.

Salem, L. (1987), *Marvels of the Molecule*, VCH Verlag GmbH, Weinheim.

Saltzman, M. (1992), Identifying dyes in textiles, *Am. Sci.* **80**, 474–481.

Saltzman, M., A. M. Keay, and J. Christiansen (1963), The identification of colorants in ancient textiles, *Dyestuffs* **44**, 241–251.

Sanchez del Rio, M., M. Picquart, E. Haro-Poniatowski, E. van Elslande, and V. H. Uc (2006), On the Raman spectrum of Maya blue, *J. Raman Spect.* **37**, 1046–1053.

Sandberg, G. (1997), *The Red Dyes: Cochineal, Madder and Murex Purple; A World Tour of Textile Techniques*, Lark Books, Asheville.

Sandu, I., F. Diaconescu, I. G. Sandu, A. Alexandru, A. Sandu, and V. Andrei (2006), The authentication of old bronze coins and the structure of the archaeological patina, *Analele Universitatii "Dunarea de Jos" din Galati, Fascicula IX: Metalurgie si Stiinta Materialelor,* **24**(1), 38–48.

Sanford, M. K. (ed.) (1993), *Investigations of Ancient Human Tissue – Chemical Analyses in Anthropology*, Gordon & Breach, Langhorne.

Sarp Tuncoku, S., E. N. Caner-Saltik, and H. Boke (1998), Pozzolanic properties of some medieval masonry mortars, *31st Int. Symp. Archaeometry Abstracts*, Budapest, pp. 139–142.

Sax, M. and A. P. Middleton (1992), A system of nomenclature for quartz and its application to the material of cylinder seals, *Archaeometry* **34**, 11–20.

Saxby, G. (2001), *The Science of Imaging: An Introduction*, Institute of Physics, Bristol.

Sayre, E. V. (2000), Determination of provenance, in Williamson, R. A. and P. R. Nickens (eds.), *Science and Technology in Historic Preservation, Advances in Archaeological and Museum Science*, Vol. 4, Kluwer Academic, New York.

Sayre, E. V. and R. V. Smith (1974), Analytical studies of ancient Egyptian glass, in Bishay, A. (ed.), *Recent Advances in Science and Technology of Materials*, Vol. 3, Plenum, New York.

Schaefer, G. (1941), Madder, *Ciba Rev.* 407–417.

Schmitt, H. (1959), *The Art of the Faker, 3000 Years of Deception*, Little, Brown, Boston.

Schmitz, R. L. (1996), *Introduction to Water Pollution*, Gulf, Houston.

Schoeninger, M. J. (1989), Reconstructing prehistoric human diet, in Price, S. P. (ed.), *The Chemistry of Ancient Bone*, Cambridge Univ. Press, Cambridge, UK, pp. 38–67.

Schramm, H. P. (1985), Ultramicro analysis, an effective procedure for substantive tests of works of art, *Wiener Berichte Über Naturwissenschaften in der Kunst* **2**, 14–39.

Schreiner, M., B. Fruhmann, D. Jembrih-Simburger, and R. Linke (2004), X-rays in art and archaeology: An overview, *Powder Diffraction* **19**(1), 3–11.

Schreiner, M., I. Prohaska, J. Rendl, and C. Weigel (1998), Leaching studies of potash-lime-silica glass with medieval glass composition using AAS and IRRS, in Tennent, N. (ed.), *Conservation of Glass and Ceramics*, James and James, London.

Schreiner, M., G. Woisetschläger, I. Schmitz, and M. Wadsak (1999), Characterization of surface layers formed under natural conditions on medieval stained glass and ancient copper alloys using SEM, SIMS and AFM, *J. Anal. Atomic Spectrom.* **14**(3), 395–403.

Schuler, F. (1963), Ancient glassmaking techniques: The evolution of techniques, *Adv. Glass Technol.* **2**, 379–383.

Schuler, F. (1962), Ancient glassmaking techniques: The Egyptian cane vessel process, *Archaeology* **15**, 32–36.

Schuler, F. (1959), Ancient glassmaking techniques: The blowing process, *Archaeology* **12**, 47–53, 116–120.

Schurr, M. R. (1989), Fluoride dating of prehistoric bones by ion selective electrode, *J. Archaeol Sci.*, **16**, 265–270.

Schuster, P. F., M. M. Reddy, and S. I. Sherwood (1991), A quantitative field study of the role of acid rain and sulfur dioxide in marble dissolution, *La Conservation des Monuments dans le Bassin Méditerranéen, Proc. 2nd Int. Symp.*, Genève.

Schwarcz, H. P. (2002), Chronometric dating in archaeology; a review, *Acc. Chem. Research* **35**(8), 637–643.

Schwarcz, H. P. (1997), Uranium series dating, in Taylor, R. E. and M. J. Aitken (eds.), *Chronometric Dating in Archaeology*, Plenum, New York, pp. 159–182.

Schwarcz, H. P. and B. Blackwell (1991), Archaeological applications, in Ivanovich, M. and R. S. Harmon (eds.), *Uranium Series Disequilibrium Applications to Environmental Problems*, 2nd ed., Oxford Univ. Press, Oxford, UK, pp. 513–552.

Schwarcz, H. P. and M. J. Schoeninger (1991), Stable isotope analysis in human nutritional ecology, *Yearbook Phys. Anthropol.* **34**, 283–321.

Schweingruber, F. H. (1988), *Tree Rings, Basics and Applications of Dendrochronology*, Reidel, Dordrecht.

Schweppe, H. (1989), Identification of red madder and insect dyes by thin layer chromatography, in *Historic Textile and Paper Materials II: Conservation and Characterization*, ACS, pp. 111–219.

Schweppe, H. and R. Runge (1986), Carmine: Cochineal carmine and kermes, in Feller, R. E. (ed.), *Artists Pigments*, Cambridge Univ. Press, New York, pp. 225–283.

Science Library (1942), *Mercury: Occurrence, Concentration and Extraction*, Science Library Bibliographies, London.

Scollar, I., A. Tabbagh, A. Hesse, and I. Herzog (1990), *Archaeological Prospecting and Remote Sensing*, Cambridge Univ. Press, Cambridge, UK.

Scorzelli, R. B., S. Petrick, A. M. Rossi, G. Poupeau, and G. Bigazzi (2001), Obsidian archeological artefacts provenance studies in the western Mediterranean basin: An approach by Mossbauer spectroscopy and electron paramagnetic resonance, *C. R. Acad. Sci., Series IIa, Sciences de la Terre et des Planetes*, **332**(12), 769–776.

Scott, D. A. (2002), *Copper and Bronze in Art: Corrosion, Colorants and Conservation*, Getty Trust, Malibu.

Scott, D. A. (1992), *Spanish Colonial and Indigenous Platinum, Proc. International Institute of Conservation Symp.*, Madrid, pp. 148–153.

Scott, D. A. (1991), *Metallography and Microstructure in Ancient and Historic Metals*, Getty Trust, Malibu.

Scott, D. A. (1990), Bronze disease: A review of some chemical problems and the role of relative humidity, *J. Am. Inst. Conserv.* **29**(2), 193–206.

Scott, D. A. (1986), Gold and silver alloy coatings over copper, *Archaeometry* **28**, 33–50.

Scott, D. A. (1983), Depletion gilding and surface treatment of gold alloys, *J. Hist. Metallurg. Soc.* **17**, 99–115.

Scott, D. A. and W. Bray (1992), Pre-Hispanic platinum alloys: Their composition and use in Ecuador and Colombia, in Scott, D. A. and P. Meyers (eds.), *Archaeometry of Pre-Columbian Sites and Artifacts*, Getty Conservation Institute, Los Angeles.

Scott, R. P. W. (1997), *Tandem Techniques*, Wiley, New York.

Scully, J. C. (1990), *The Fundamentals of Corrosion*, Pergamon, Oxford.

Sedgwick, R. D. and D. M. Hindenlang (1988), Mass-spectrometry and chromatography-mass-spectrometry, in Sibila, J. (ed.), *A Guide to Materials Characterization and Chemical Analysis*, VCH Verlag GmbH, Weinheim, pp. 45–53.

Seefelder, M. (1994), *Indigo – Kultur, Wissenschaft und Technik*, Ecomed Verlag, Landsberg.

Seefelder, M. (1982), *Indigo*, BASF, Ludwigshafen.

Sellers, W. I. and A. T. Chamberlain (1998), Ultrasonic cave mapping, *J. Archaeol. Sci.* **25**, 867–873.

Selley, R. C. (1982), *An Introduction to Sedimentology*, Academic Press, New York.

Semenov, S. A. (1964), *Prehistoric Technology*, Adams and Dart, Bath.

Senn, M., W. Devos, W. Fasnacht, Th. Geiger, F. Michel, A. Ritter, and G. Fortunato (2001), An odyssey through time at EMPA, *Chimia* **55**, 931–937.

Serpico, M. and R. White (2000a), Resins, amber and bitumen, in Nicholson, P. T. and I. Shaw (eds.), *Ancient Egyptian Materials and Technology*, Cambridge Univ. Press, Cambridge, UK, pp. 430–474.

Serpico, M. and R. White (2000b), Oil, fat and wax, in Nicholson, P. T. and I. Shaw, *Ancient Egyptian Materials and Technology*, Cambridge Univ. Press, Cambridge, UK, pp. 390–429.

Setford, S. J. (1994), *A Basic Introduction to Separation Science*, Rapra Technologies.

Settle, F. A. (ed.) (1997), *Handbook of Instrumental Techniques for Chemical Analysis*, Prentice-Hall, New York.

Sever, T. and J. Wiseman (1985), *Remote Sensing and Archaeology*, National Aeronautics and Space Administration, Gulfport, MS.

Shackleton, N. J. (1973), Oxygen isotope analysis as a means of determining season of occupation of prehistoric hidden sights, *Archaeometry* **15**, 133–141.

Shackleton, N. J. (1970), Stable isotope study of the palaeoenvironment of the neolithic site of Nea Nikomedeia, Greece, *Nature* **227**, 943–944.

Shackleton, N. J. and N. D. Opdyke (1973), Oxygen isotope and palaeomagnetic stratigraphy of equatorial Pacific core V28–238, *Quatern. Research* **3**, 39–55.

Shakley, M. S. (2005), *What is XRF (X-Ray Fluorescence Spectrometry)?*, http://www.swxrflab.net/xrfinstrument.htm.

Shackley, M. S. (ed.) (1998), *Archaeological Obsidian Studies – Methods and Theory*, Plenum, New York.

Shackley, M. L. (ed.) (1977), *Archaeological Obsidian Studies*, Plenum, New York.

Shackley, M. L. (1975), *Archaeological Sediments*, Wiley, New York.

Shadmon, A. (1996), *Stone, an Introduction*, Intermediate Technology Publications, London.

Shafer, T. (1976), *Ceramics Monthly* **24**(7), 44–48.

Shalev, S. (1999), Recasting the Nahal Mishmar hoard, in Hauptmann, A., E. Pernicka, Th. Rehren, and U. Yalçin (eds.) (1999), *The Beginning of Metallurgy, Proc. Int. Conf.*, Bochum, 1995.

Shams, G. P. (1987), *Some Minor Textiles in Antiquity*, Aströms, Göteborg.

Shaver, C. L., R. C. Cass, and J. R. Druzik (1983), Ozone and the deterioration of works of art, *Environ. Sci. Technol.* **17**, 748–752.

Shedrinsky, A. and N. S. Baer (2007), The application of analytical pyrolysis to the study of cultural materials, in Wampler, Th. P. (ed.), *Applied Pyrolysis Handbook* 2nd Ed., CRC Press, Boca Raton, FL, pp. 105–131.

Shelby, J. E. (1997), *Introduction to Glass Science and Technology*, Royal Society of Chemistry, Cambridge.

Shepard, A. O. (1954), *Ceramics for the Archaeologist*, Carnegie Institute Publication 609, Washington, DC.

Shepard, A. O. and H. E. D. Pollock (1971), *Maya Blue*, Carnegie Institution, Washington, DC.

Shepherd, R. (1993), *Ancient Mining*, Chapman & Hall, London.

Shepherd, W. (1972), *Flint: Its Origin, Properties and Uses*, Faber and Faber, London.

Sherwood, S. I. (1995), Clearing the air: The role of environmental chemistry in the decay of cultural objects, *Proc. Int. Symp. Conservation and Restoration of Cultural Property, Tokyo 1990*, Tokyo National Research Institute of Cultural Property, pp. 41–50.

Shimura, M. (1988), The fabrication of gold leaf in Japan, in Maddin, R. (ed.), *The Beginning of the Use of Metals and Alloys*, MIT, Cambridge, MA.

Shipman, R. (1981), *Life History of a Fossil: An Introduction to Taphonomy and Palaeoecology*, Harvard Univ. Press, Cambridge, MA.

Shirono, S. and Y. Hayakawa (2006), Identification of painting materials used for mural paintings by image analysis and XRF, *Adv. in X-Ray Anal.* **49**, 213–217.

Shortland, A. (2006). Application of lead isotope analysis to a wide range of late bronze age Egyptian materials, *Archaeometry* **48**, 657–669.

Shortland, A., N. Rogers, and K. Eremin (2007), Trace element discriminants between Egyptian and Mesopotamian late Bronze Age glasses, *J. Archaeol. Sci.* **34**, 823–829.

Shortland A., L. Schachner, I. Freestone, and M. Tite (2006), Natron as a flux in the early vitreous materials industry: Sources, beginnings and reasons for decline, *J. Archaeol. Sci.* **33**, 521–530.

Shreeve, J. (1995), *The Neanderthal Enigma: Solving the Mystery of Modern Human Origins*, Avon, New York.

Shull, P. J. (ed.) (2002), *Nondestructive Evaluation: Theory, Techniques and Applications*, Marcel Dekker, New York.

Shultz, S. L. and V. Shultz (1975), *Nuclear Technology in Archaeology*, TID 3920, Oak Ridge, TN.

Shumann, Th. (1943), *Forschung und Fortschrift* **19**, 356–359.

Sieveking, G. and M. B. Bart (1986), *The Scientific Study of Flint and Chert*, Cambridge Univ. Press, Cambridge, UK.

Sigleo, A. M. (1975), Turquoise mine and artifact correlation for Snaketown site, Arizona, *Science* **189**, 459–460.

Sillen, A. and M. Kavanagh (1982), Strontium and palaeodietary research, a review, *Yearbook Phys. Anthropol.* **25**, 67–90.

Simmonds, R. J. (1992), *Chemistry of Biomolecules – an Introduction*, Royal Society of Chemistry, London.

Singer, C. (ed.), *History of Technology*, Oxford Univ. Press, Oxford, UK.

Skinner, A. F. and M. N. Rudolph (1996), Dating flint artifacts with electron spin resonance, in Orna, M. V. (ed.), *Archaeological Chemistry*, Vol. 5, *Organic, Inorganic and Biochemical Analysis,* Advances in Chemistry Series, ACS, Washington, DC, pp. 37–46.

Skinner, B. J. and S. C. Porter (1992), *The Dynamic Earth – an Introduction to Physical Geology*, Wiley, New York.

Slater, R. V. (2006), The metallurgy of archaeological wire: A tool for the modern metallurgist, *Wire J. Int.* **39**, 58–61.

Smart, P. L. (1991), Uranium series dating, in Smart, P. L. and P. D. Frances (eds.), *Quaternary Dating Methods – a Users Guide*, Technical Guides Vol. 4, Quaternary Research Association, Cambridge, UK.

Smart, P. L. and P. D. Frances (eds.) (1992), *Quaternary Dating Methods – a Users Guide,* Technical Guides Vol. 4, Quaternary Research Association, Cambridge, UK.

Smith, A. L. (1979), *Applied Infrared Spectroscopy*, Wiley, New York.

Smith, A. W. (1978), Stable carbon and oxygen isotope ratios of malachite from the patina of ancient bronze objects, *Archaeometry* **20**, 123–133.

Smith, B. C. (1995), *Fundamentals of Fourier Transform Infrared Spectroscopy*, Chemical Rubber Company, Boca Raton, FL.

Smith, B. N. and S. Epstein (1970), Two categories of C-13/C-12 ratios for higher plants, *Plant Physiol.* **47**, 380–384.

Smith, C. S. (1983), Damascus steel, *Science* **216**, 242–244.

Smith, C. S. (1981), *A Search for Structure*, MIT, Cambridge, MA.

Smith, C. S. (1968), The early history of casting, molds, and the science of solidification, *Metal Transformations: Informal Proc. 2nd Bull Int. Conf. on Materials in Pittsburgh*, 1966, New York, pp. 3–51.

Smith, C. S. (1965), *A History of Metallurgy*, Univ. Chicago Press, Chicago.

Smith, F. and R. Montgomery (1959), *The Chemistry of Gums and Mucilages,* Reinhold, New York.

Smith, G. (1995), *Beer, A History of Suds and Civilization*, Avon, New York.

Smith, R. E. (1967), *Forging and Welding*, McKnight and McKnight, Bloomington.

Smith, P. I. S. (1923), *Glue and Gelatine*, Pitman, London.

Sneddon, J. (1998), Laser induced breakdown spectrometry, *Chem. Educ.* **3**(6), S1430–S4171(98). 06260-8.

Sobolik, K. D. (2000), Dietary reconstruction as seen in coprolites, in Kiple, K. F. and K. C. Ornelas (eds.), *The Cambridge World History of Food*, Cambridge, UK, pp. 44–50.

Spanier E. (ed.) (1987), *The Royal Purple and The Biblical Blue: Argaman and Tekhelet*, Keter, Jerusalem.

Spencer, F. (1990), *Piltdown: A Scientific Forgery*, Oxford Univ. Press, London.

Spigelman, M. and E. Lemma (1993), The use of the polymerase chain reaction to detect *Mycobacterium tuberculosis* in ancient skeletons, *Int. J. Osteoarchaeol.* **3**, 137–143.

Spindler, K. (1994), *The Man in the Ice: The Discovery of a 5000 Year Old Body Reveals the Secrets of the Stone Age*, Doubleday, Canada.

Spindler, K., H. Wilfing, E. Rastbichler-Zissernig, D. zur Nedden, and H. Nothdurfter (eds.) (1996), *Human Mummies: A Global Survey of Their Status and the Techniques of Conservation*, Springer, New York.

Sposito, G. (1989), *The Chemistry of Soils*, Oxford Univ. Press, Oxford, UK.

Spoto, G. (2002), Studying archaeological remains – a great challenge for modern chemistry, *Chimica e l'Industria* **84**(10), e2/1–e2/5.

Spoto, G. and E. Ciliberto (2000), X-ray photoelectron spectroscopy and Auger electron spectroscopy in art and archaeology, in Ciliberto, E. and G. Spoto (eds.), *Modern Analytical Methods in Art and Archaeology*, Chemical Analysis Series, Vol. 155, Wiley, New York, pp. 363–404.

Stambolov, T. (1969), *Manufacture, Deterioration and Preservation of Leather: A Literature Survey*, ICOM Plenary Meeting, Central Research Laboratory for Objects of Art and Science, Amsterdam.

Stambolov, T., R. D. Bleck, and N. Eichelmann (1988, 1987), Korrosion und Konservierung von Kunst-und Kulturgut aus Metall, *Restaurirung und Museumtechnik* **9**, 1–52; **8**, 1–91.

Stanford, M. K. (ed.) (1993), *Elemental and Isotopic Analysis: Understanding Diet and Disease in Past Populations*, Gordon & Breach, Langhorn.

Starr, C. (1997), *Biology: Concepts and Applications*, Wadsworth Publishing, Belmont.

Stead, I. M., J. B. Bourke, and D. Brothwell (eds.) (1986), *Lindow Man: The Body in the Bog*, Cornell Univ., Ithaca.

Stearn C. W. and R. L. Carroll (189), *Paleontology: The Record of Life*, Wiley, New York.

Steckoll, S. H., Z. Goffer, H. Nathan, and N. Hass (1971), Red stained bones from Qumran, *Nature* **231**, 469–470.

Stein, J. K. and W. R. Farrand (2001), *Sediments in Archaeological Context*, Univ. Utah, Salt Lake City.

Stern, A. (ed.) (1976), *Air Pollution*, Academic Press, New York.

Stern, W. B. (2001), Archaeometry – analyzing the cultural heritage, *Chimia* **55**(11), 915–922.

Stevenson, C. M., D. Wheeler, S. W. Novak, R. J. Speakman, and M. D. Glascock (2007), A new dating method for high-calcium archaeological glasses based upon surface-water diffusion: Preliminary calibrations and procedures, *Archaeometry* **49**, 153–177.

Stevenson, C. M., M. Goitesnian, and M. Macko (2000), Redefining the working assumptions of obsidian hydration dating, *J. California and Great Basin Anthropology* **22**, 223–226.

Stevenson, C. M., J. J. Mazer, and B. E. Scheetz (1992), Laboratory obsidian hydration rates: Theory, method, and application, in Shackley, S. (ed.), *Method and Theory in Archaeological Volcanic Glass Studies*, Plenum, New York, pp. 181–204.

Stoeppler, M. (ed.) (1997), *Sampling and Sample Preparation*, Springer, New York.

Storey, J. (1992), *The Thames and Hudson Manual of Dyes and Fabrics*, Thames and Hudson, London.

Stos-Gale, Z. A. (1992), Isotope archaeology: Reading the past in metals, minerals and bone, *Endeavour* (New Series) **16**(2), 85–90.

Stos-Gale, Z. A. (1989), Lead isotope studies, in Henderson, J. (ed.), *Scientific Analysis in Archaeology*, Oxford Univ. Committee for Archaeology, Monograph 19, Oxford, UK, pp. 274–301.

Stott, A. W., R. Berstan, R. P. Evershed, C. Bronk-Ramsey, M. Humm, and R. E. M. Hedges (2003), Direct dating of archaeological pottery by compound specific C-14 analysis of preserved lipids, *Anal. Chem.* **75**(19), 5037–5045.

Streicher, B. (1991), *Überlegungen und Vrsuche zur Ergänzung historicher Putze mit Hilfe Kalkgebundener Massen, dargestellt an Beispiel der Wandmalereien von St. Kastor in Koblenz*, Diplomarbeit H, Köln (Cologne, Germany).

Stringer, Ch. (1993), *In Search of the Neanderthals: Solving the Puzzle of Human Origins*, Thames and Hudson, New York.

Stross, F. H. and W. J. Eisenloed (1965), *A Report on a Group of Limestone Carvings*, San Carlos.

Struik, L. C. E. (1978), *Physical Ageing in Amorphous Polymers and Other Materials*, Elsevier, New York.

Stuart, B. H., B. George, and P. McIntyre (1996), *Modern Infrared Spectroscopy*, Wiley, 1996.

Stuiver, M., P. J. Reimer, and R. Reimer (2005), *CALIB Radiocarbon Calibration*, http://radiocarbon.pa.qub.ac.uk/calib/.

Suess, H. E. (1965), *J. Geophys. Research* **70**, 5937.

Sukhov, D. A., O. Y. Derkacheva, S. A. Kazansky, and D. M. Kheyfetz (1995), The new prospects of FT-IR spectroscopy for non destructive testing of artistic and cultural objects, *Proc. 4th Int. Conf. Non Destructive Testing of Works of Art, Berlin, Oct. 1994*, Deutsche Geselltschaft fur Zertörungsfrei Prüfung, Berlin, pp. 795–798.

Sunta, C. M. and M. David (1981), *PACT* **6**, 460–467.

Sussman, C. (1985), Microwear analysis, *World Archaeol.* **17**(1), 101–111.

Sykes, R. L. (1991), The principles of tanning, in Calnan, C. B. and B. Haines (eds.), *Leather: Its Composition and Changes with Time*, The Leather Conservation Center, Northampton.

Szabo, B. J., H. E. Malde, and C. Irwin-Williams (1969), *Earth Planetary Sci. Lett.* **10**, 253–256.

Tacamán, M. J., L. Rendón, J. Arenas, and M. C. Serra Puche (1996), Maya-Blue: An ancient nanostructured material, *Science* **273**, 223–225.

Taft, W. S. and J. W. May (2000), *The Science of Paintings*, Springer, New York.

Tague, E. L. (1926), *Casein*, Van Nostrand, New York.

Tait, H. (ed.) (1991), *Five Thousand Years of Glass*, British Museum, London.

Takahata, N. and Y. Satta (1997), Evolution of the primate lineage leading to modern humans: Phylogenetics and demographic inferences from DNA sequences, *Proc. Natl. Acad. Sci. USA* **94**, 4811–4815.

Taylor, G. C., M. Crossey, J. Saldanha, and T. Waldron (1996), DNA from *Mycobacterium tuberculosis* identified in medieval human skeletal remains using polymerase chain reaction, *J. Archaeol. Sci.* **23**, 789–798.

Taylor, G. W. (1990), Ancient textile dyes, *Chem. Britain* **26**, 1155–1158.

Taylor, J. T and A. C. Bull (1986), *Ceramic Glaze Technology*, Pergamon, Oxford.

Taylor, R. E. (1997), Radiocarbon dating, in Taylor, R. E. and M. J. Aitken (eds.), *Chronometric Dating in Archaeology*, Advances in Archaeological and Museum Science Series, Vol. 2, Plenum, New York.

Taylor, R. E. (1987), *Radiocarbon Dating: An Archaeological Perspective*, Academic Press, Orlando, FL.

Taylor, R. E. (1976), *Advances in Obsidian Glass Studies*, Noyes, Park Ridge.

Taylor, R. E. and M. J. Aitken (1997), *Chronometric Dating in Archaeology*, Advances in Archaeological and Museum Science Series, Vol. 2, Plenum, New York.

Taylor, R. E., A. Long, and R. Kra (eds.) (1994), *Radiocarbon After Four Decades: An Interdisciplinary Perspective*, Springer, New York.

Taylor, R. J. and V. J. Robinson (1996), Provenance studies of Roman African red slip ware, *Archaeometry* **38**, 231–255.

Tchapla, A., P. Mejanelle, J. Bleton, and S. Goursaud (2004), Characterization of embalming materials of a mummy of the Ptolemaic era – comparison with balms from mummies of different eras, *J. Separ. Sci.* **27**(3), 217–234.

Teixeira, S. R., J. B. Dixon, G. N. White, L. A. Newsom, and A. Lee (2002), Charcoal in soils: A preliminary view, *Soil Science Society of America Book Series*, Vol. 7, pp. 819–830.

Tennent, N. (ed.) (1998), *Conservation of Glass and Ceramics*, James and James, London.

The Textile Institute (1974), *Identification of Textile Materials*, Manchester, UK.

Thomas, K. D. (1993), Molecular biology and archaeology: A prospectus for interdisciplinary research, *World Archaeol.* **25**, 1–17.

Thompson, R. (1991), *The Chemistry of Wood Preservation*, Royal Society of Chemistry, Cambridge.

Thorpe, J. F. and M. A. Whiteley (1955), *Thorpe's Dictionary of Applied Chemistry*, Vol. 5, Longmans Green, London, p. 536.

Thorpe, R. S., O. William-Thorpe, D. J. Jenkins, and J. S. Watson (1991), The geological sources and transport of bluestones of Stonhenge, Wiltshire, U.K., *Proc. Prehist. Soc.* **57**(2), 103–157.

Tite, M. S. (2004), Glazed pottery, *Proc. Int. School of Physics "Enrico Fermi,"* Vol. 154, *Physics Methods in Archaeometry*, pp. 377–384.

Tite, M. S. (2001), Materials study in archaeology, in Brothwell, D. R. and A. M. Pollard (eds.), *Handbook of Archaeological Sciences*, Wiley, New York, pp. 443–448.

Tite, M. S. (1987), Characterization of early vitreous materials, *Archaeometry* **29**, 21–34.

Tite, M. S. (1986), Egyptian Blue, faience and related materials, in Jones, R. E. and H. W. Catling (eds.), *Science in Archaeology*, British School of Athens, London.

Tite, M. S. (1972), *Methods of Physical Examination in Archaeology*, Seminar Press, London.

Tite, M. S. and M. Bimson (1991), A technological study of English porcelains, *Archaeometry* **33**, 3–27.

Tite, M. S. and M. Bimson (1986), Faience: An investigation of the microstructures associated with the different methods of glazing, *Archaeometry* **28**, 69–78.

Tite, M. S., M. Bimson, and M. R. Cowell (1984), Technological examination of Egyptian Blue, in Lambert, J. C. (ed.), *Archaeological Chemistry*, Advances in Chemistry Series, Vol. 3, ACS, Washington, DC, pp. 215–242.

Tite, M. S., I. Freestone, R. Mason, J. Molera, M. Vendrell-Saz, and N. Wood (1998), Lead glazes in antiquity – methods of production and reasons for use, *Archaeometry* **40**, 241–260.

Tite, M. S. and Y. Maniatis (1975), Examination of ancient pottery using the scanning electron microscope, *Nature* **257**, 122–123.

Tite, M. S., A. Shortland, and S. Paynter (2002), The beginning of vitreous materials in the Near East and Egypt, *Acc. Chem. Research* **35**, 585–593.

Tokarski, C., C. Cren-Olive, C. Rolando, and E. Martin (2003), Protein studies in cultural heritage, in Saiz-Jimenez, C. (ed.), *Molecular Biology and Cultural Heritage, Proc. Int. Congress on Molecular Biology and Cultural Heritage in Sevilla, Spain*, March 4–7, Balkema, Rotterdam, pp. 119–130.

Toland, J. Ch. (2000), *Radar Imaging*, Bridgeway.

Torrence, R. (1986), *Production and Exchange of Prehistoric Stone Tools: Prehistoric Obsidian in the Aegean*, Cambridge Univ. Press, Cambridge, UK.

Troja, S. O., and R. G. Roberts (2000), Luminescence dating in Gliberto, E. and G. Spoto (eds.), *Modern Analytical Methods in Art and Archaeology*, Chemical Analysis Series, Vol. 155, Wiley New York, p. 585–640.

Trueman, C. N. and D. M. Martill (2002), The long-term survival of bone: The role of bioerosion, *Archaeometry* **44**(3), 371–382.

Truncer, J., M. D. Glascock, and H. Neff (1998), Steatite source characterization in eastern North America, *Archaeometry* **40**, 23–44.

Tsoumis, G. (1991), *Science and Technology of Wood*, Chapman & Hall, London.

Turner, R. C. and R. G. Scaife (1995), *Bog Bodies: New Discoveries and New Perspectives*, British Museum, London.

Turner, W. E. S. (1959), Studies in ancient glasses and glass-making processes (VI): The composition and physical characteristics of the glass of the Portland vase, *J. Soc. Glass Technol.* **43**, T262–T284.

Turner, W. E. S. and H. P. Rooksby (1961), Further historical studies based on X-ray diffraction methods on the reagents employed in making opal and opaque glass, *Jahrbuch des Romisch-Germanischen Zentralmuseums* (Mainz) **8**, 1–6.

Turner, W. E. S. and H. P. Rooksby (1959), A study of opalizing agents of ancient glass throughout 3400 years, *Glasstechnische Berichte* **32**(*K*), 17–28.

Tykot, R. H. (2002a), Geochemical analysis of obsidian and the reconstruction of trade mechanisms in the early Neolithic period of the western Mediterranean, *Archaeological Chemistry*, ACS Symposium Series, Vol. 831, pp. 169–184.

Tykot, R. H. (2002b), *Contribution of Stable Isotope Analysis to Understanding Dietary Variation among the Maya*, in Jakes, K. (ed.), *Archaeological Chemistry*, Vol. 6, *Materials, Methods and Meaning*, Advances in Chemistry Series, ACS, Washington, DC, pp. 214–230.

Tykot, R. H. (1996), Obsidian procurement and distribution in the central and western Mediterranean, *J. Mediterranean Archaeol.* **9**(1), 39–82.

Tykot, R. H., N. J. van der Merwe, and H. Hammond (1996), Stable isotope analysis of bone collagen, bone apatite and tooth enamel in the reconstruction of human diet, in Orna, M. V. (ed.), *Archaeological Chemistry*, Advances in Chemistry Series, Vol. 5, ACS, Washington, DC, pp. 355–365.

Tykot, R. H. and S. M. M. Young (1996), Archaeological applications for inductively coupled plasma-mass spectroscopy, in Orna, M. V. (ed.), *Archaeological Chemistry*, Advances in Chemistry Series, Vol. 5, ACS, Washington, DC, pp. 116–130.

Tylecote, R. F. (1987), *The Early History of Metallurgy in Europe*, Longman, London.

Tylecote, R. F. (1986), *The Prehistory of Metallurgy in the British Isles*, Institute of Metals, London.

Tylecote, R. F. (1976), *A History of Metallurgy*, The Metals Society, London.

Tylecote, R. F. (1962), *Metallurgy in Archaeology*, Edward Arnold, London.

Tylecote, R. F. and R. F. McKarrel (1978), *The Working of Copper-Arsenic Alloys in the Early Bronze Age*, Arnold, London.

Tyson, J. (1988), *Analysis: What Analytical Chemists Do*, Royal Society of Chemistry, London.

Uemura, R., T. Kameda, K. Kimura, D. Kitamura, and K. Yamasaki (1954), Studies on the Mitsuda-e, a kind of oil-painting, *Scientific Papers on Japanese Antiquities and Art Crafts*, Vol. 9, pp. 15–21.

Unger, A., A. P. Schniewind, and W. Unger (2001), *Conservation of Wood Artifacts*, Springer, New York.

Untracht, O. (1985), *Jewelry: Concepts and Techniques*, Hale, London.

Urey, H. C. (1948), Oxygen isotopes in nature and in the laboratory, *Science* **108**, 489–496.

Valladas, H. (2003), Direct radiocarbon dating of prehistoric cave paintings by accelerator mass spectrometry, *Meas. Sci. Technol.* **14**, 1487–1492.

van Bergen, P. F., R. P. Evershed, T. M. Peakman, E. C. Leigh-Firbank, M. C. Horton, and P. A. Rowley-Conway (1998), Chemical characterization of frankincense and pine resins from Qasr Ibraim, *Abstracts, 31st Int. Symp. Archaeometry*, Budapest, p. 142.

van den Haute, P. and F. De Corte (eds.) (1998), *Advances in Fission-Track Geochronology*, Solid Earth Sciences Library, Vol. 10, a selection of papers presented at the 1996 International Workshop on Fission-Track Dating, Gent.

Van der Leeuw, S. E. and A. C. Pritchard (1984), *The Many Dimensions of Pottery; Ceramics in Archaeology and Anthropology*, Univ. Amsterdam, Amsterdam.

van der Merwe, N. J. (2005), CO_2, grasses, and human evolution, *South Africa's Ecological Studies* **177** (History of Atmospheric CO_2 and Its Effects on Plants, Animals, and Ecosystems), 293–328.

van der Merwe, N. J. (1992), Light stable isotopes and the reconstruction of prehistoric diets, in Pollard, A. M. (ed.), *New Developments in Archaeological Science*, Oxford Univ. Press, Oxford, UK, pp. 247–264.

van der Merwe, N. J (1982), Carbon isotopes, photosynthesis and archaeolgy, *Am. Sci.* **70**, 596–606.

van der Merwe, N. J. and D. H. Avery (1982b), Pathways to steel, *Am. Sci.* **70**, 146–155.

van der Merwe, N. J. (1969), *The Carbon-14 Dating of Iron*, Univ. Chicago Press, Chicago.

van der Merwe, N. J., J. A. Lee-Thorp, J. F. Thackeray, A. Hall-Martin, F. J. Kruger, H. Coetzee, R. H. V. Bell, and M. Lindeque (1990), Source-area determination of elephant ivory by isotope analysis, *Nature* **346**, 744–746.

van der Sanden, W. (1996), *Through Nature to Eternity: The Bog Bodies of Northwestern Europe*, Batavian Lion International, Amsterdam.

Vandiver, P. B., J. L. Mass, and A. Murray (eds.) (2005), *Materials Issues in Art and Archaeology VII* (Symposium held 30 November–3 December 2004, Boston, Massachusetts), in *Materials Research Symposium Proceedings*, Vol. 852, Materials Research Society, Warrendale, Pennsylvania.

Vandiver, P. (1982), Technological change in Egyptian faience, in Olin, J. S. and A. D. Franklin (eds.), *Archaeological Ceramics*, Smithsonian Institution, Washington, DC, pp. 167–179.

Varron, A. (1938), The origin and uses of silk, *Ciba Reviews* **11**.

Vaughan, P. (1985), *Use-Wear Analysis of Flaked Stone Tools*, Univ. Arizona Press, Tucson.

Velde, B. and I. C. Druc (1998), *Archaeological Ceramic Materials*, Springer, Heidelberg.

Vendrell, M., J. Molera, and M. S. Tite (2000), Optical properties of tin-opacified glazes, *Archaeometry* **42**, 325–340.

Verhecken, A. (2005), A concise history of dye analysis, *Dyes in History and Archaeology* **20**, 1–22.

Verhecken, A. and J. Wouters (1988–1989), The coccid insect dyes: Historical, geographical and technical data, *Bull. Inst. Royal Patrimoine Artistique* **22**, 207–239.

Verhoeven, J. D., A. H. Pendray, and W. E. Dauksch (1998), The key role of impurities in ancient Damascus steel blades, *J. Metals* **50**(9), 58–64.

Verhoeven, J. D., A. H. Pendray, and E. D. Gibson, (1996), Wootz Damascus steel blades, *Mater. Charact.* **37**, 9–22.

Verhoeven, J. D. and D. T. Peterson (1992), What is Damascus steel?, *Mater. Charact.* **29**, 341–355.

Vernadskii, V. I. (1998), *The Biosphere*, Copernicus Books, New York.

Von Fischer, W. (1948), *Paint and Varnish Technology*, Reinhold, New York.

von Hagen, W. (1977), *The Aztec and Maya Paper Makers*, Hacker, New York.

von Lipmann, E. O. (1937), Das Natron in Alten Aegipten, *Naturwissenschaften* **25**, 592–598.

von Rosen, L. (1990), *Lapis Lazuli in Archaeological Contexts*, Paul Astroms Forlag.

Vorst, B. (1986), Parchment making, ancient and modern, *Fine Print* **12**(4), 209–221.

Voss, R. F. (1976), Damascus Stahl, *Museumkunde* **19**, 225–233, 277–284.

Vreeland (Jr.), J. M. (1999), The revival of colored cotton, *Sci. Am.* (April), 113–118.

Vuissoz, A., M. Worobey, N. Odegaard, M. Bunce, C. A. Machado, N. Lynnerup, E. E. Peacock, and M. T. P. Gilbert (2007), The Survival of PCR-Amplifiable DNA in Cow Leather, *J. Archaeol. Sci.* **33**, 823–829.

Wadsworth, J. (2002), Ancient and modern steels and laminated composites containing steels, *Mater. Research Soc. Bull.* **27**(12), 980–987.

Wadsworth, J. and O. D. Sherby (1983), Damascus steel-making, *Science* **216**, 328–330.

Wadsworth, J. and O. D. Sherby (1980), On the Bulat – Damascus steels revisited, *Progress Mater. Sci.* **25**, 35–68.

Waelkens, M., N. Herz, and L. Moens (eds.) (1992), *Ancient Stones: Quarrying, Trade and Provenance; Interdisciplinary Studies on Stones and Stone Technology in Europe and the Near East from the Prehistoric to the Early Christian Period*, Acta Archaeologica Lovanesiana Monograph 4, Leuven Univ., Leuven.

Wagner, G. A. (2000), Isotope analysis, dating and provenance methods, in Ciliberto, E. and G. Spoto (eds.), *Modern Analytical Methods in Art and Archaeology*, Chemical Analysis Series, Vol. 155, Wiley, New York, pp. 445–464.

Wagner, G. A. (1999), *Age Determination of Young Rocks and Artifacts, Physical and Chemical Clocks in Quaternary Geology and Archaeology*, Springer, New York.

Wagner, G. A. (1983), *Thermoluminescence Dating*, Handbooks for Archaeologists, Vol. 1, European Science Foundation, Strasbourg.

Wagner, G. A., M. Aitken, and L. Mejdahl (eds.) (1983), *Thermoluminescence Dating*, European Science Foundation.

Wagner, G. A. and P. van den Haute (1992), *Fission Track Dating*, Kluwer, Boston.

Wagner, U., F. E. Wagner, and J. Riederer (1986), in Olin, J. S. and M. J. Blackman (eds.), *Proc. 24th Int. Archaeometry Symp.*, Washington, DC, pp. 129–132.

Wainwright, I. N. M., J. M. Taylor, and R. D. Harley (1986), Lead antimoniate yellow, in Feller, R. (ed.), *Artists Pigments*, Cambridge Univ. Press, New York, pp. 219–223.

Wales, S., J. Evans, and A. R. Leeds (1992), The value of using chemical analytical techniques on coprolites, in White, R. and H. Page (eds.), *Organic Residues in Archaeology: Their Identification and Analaysis*, UK Institute of Conservation, London, pp. 33–38.

Walker, M. (2005), *Quaternary Dating Methods: An Introduction*, Wiley, New York.

Walker, R. (1982), Corrosion and preservation of bronze artifacts, *J. Chem. Educ.* **59**, 943.

Walker, S. and M. Bierbrier (1997), *Ancient Faces*, British Museum, London.

Walker, S. and K. Matthews (1988), Recent work in stable isotope analysis of white marble at the British Museum, in Fant, J. C. (ed.), *Ancient Marble Quarrying and Trade*, B.A.R., International Series, Vol. 453, Oxford, UK, pp. 117–125.

Wallert, A. and R. Boytner (1996), Dyes from the Tumilaca and Chiribaya cultures, South Coast of Peru, *J. Archaeol. Sci.* **23**, 853–861.

Walter, P. H. (2003), Inventing the science of make-up, *NATO Science Series, II: Mathematics, Physics and Chemistry*, Vol. 117, *Molecular and Structural Archaeology: Cosmetic and Therapeutic Chemicals*, pp. 1–9.

Walter, R. (1997), Potassium-argon/argon-dating, in Taylor, R. E. and M. J. Aitken (eds.), *Chronometric Dating in Archaeology*, Advances in Archaeological and Museum Science Series, Vol. 2, Plenum, New York, pp. 97–126.

Walton, P. and G. W. Taylor (1991), The characterization of dyes in textiles from archaeological excavations, *Chromatogr. Anal.* **17,** 6–7.

Warashina, T., T. Higashimura, and Y. Maeda (1981), Determination of the firing temperature of ancient pottery by means esr spectrometry, *British Museum Occasional Papers*, **19**, 117–123.

Ward, F. (1994), *Emeralds*, Gem Book, Bethesda, MD.

Warren, J. (1999), *Conservation of Brick*, Butterworth-Heinemann, London.

Warth, A. H. (1956), *The Chemistry and Technology of Waxes*, Reinhold, New York.

Watanabe, N. and M. Suzuki (1969), Fission track dating of archaeological glass materials from Japan, *Nature* **222**, 1057–1058.

Watson, J. T. (1985), *Introduction to Mass Spectrometry*, Raven, New York.

Watt, M. and W. Sellar (1996), *Frankincense & Myrrh: Through the Ages, and a Complete Guide to Their Use in Herbalism and Aromatherapy Today*, C. W. Daniel Co., Saffron Walden.

Watts, S., A. M. Pollard, and G. A. Wolff (1997), Kimmeridge jet – a potential new source for British jet, *Archaeometry* **39**, 125–143.

Wayman, M. L. (2000, 2001), Archaeometallurgical contributions to a better understanding of the past, *Mater. Charact.* **45**(4/5), 259–267.

Wayman, M. L. (1989), Native copper: Humanity's introduction to metallurgy?, in Wayman, M. L. (ed.), *All That Glisters: Readings from Metallurgical History*, Canadian Institute of Mining and Metallurgy, Montreal.

Wayne, R. K., J. A. Leonard, and A. Cooper (1999), Full of sound and fury: The recent history of ancient DNA, *Annu. Rev. Ecol. Syst.* **30**, 457–477.

Weaver, C. E. and L. D. Pollard (1975), *The Chemistry of Clay Minerals*, Elsevier, Amsterdam.

Weaver, J. W. (ed.) (1984), *Analytical Methods for a Textile Laboratory*, American Association of Textile Chemists and Colorists, Triangle Park, NC.

Web, A. D. (1984), The science of making wine, *Sci. Am.* **72**, 360–367.

Weber-Partenheim, W. (1971), *Ciba Geigy Rev.* 21–45.

Webster, R. (1955), The emerald, *J. Gemmol.* **5**, 185–190.

Weil, P. D. (1977), *A Review of the History and Practice of Patination*, NBS Special Publication 479, Gaithesburg, pp. 77–92.

Weiner, J. S. (1981), *The Piltdown Forgery*, Dover, New York.

Weiner, S., W. Traub, H. Elster, and M. J. DeNiro (1989), The molecular structure of bone and its relation to diagenesis, *Appl. Geochem.* **4**, 231–232.

Weiner, S., Z. Kustanovich, E. Gil-Av, and W. Traub (1980), Dead Sea scroll parchments: Unfolding of the collagen molecules and racemization of aspartic acid, *Nature* **287**, 820–823.

Weishaupt, D. and V. Köchli (2003), *How Does MRI Work?*, Springer, New York.

Weiss, L. E. (1954), *Am. J. Sci.* **252**, 641–645.

Weisseman, S. U. and W. S. Williams (eds.) (1994), *Ancient Technologies and Archaeological Materials*, Gordon & Breach, Langhorne.

Wells, E. C. (2004), A brief history of archaeological soil chemistry, *Newsletter Int. Union of Soil Sci. Soil Sci. Soc. America* **11**, 2–4.

Wells, E. C., R. E. Terry, J. J. Parnell, P. J. Hardin, M. W. Jackson, and S. D. Houston (2000), Chemical analyses of ancient anthrosols in residencial areas in Piedras Negras, Guatemala, *J. Archaeol. Sci.* **27**(5), 449–462.

Wen, R., C. S. Wang, Z. W. Mao, Y. Y. Huang, and A. M. Pollard (2007), The chemical composition of blue pigment on Chinese blue-and-white porcelain of the Yuan and Ming Dynasties (AD 1271–1644), *Archaeometry* **49**, 101–115.

Wendell, J. E. (1995), Cotton, in Smart, J. and N. W. Simmonds (eds.), *The Evolution of Crop Plants*, Wiley, New York.

Wenite , K. (2001), *The Materials and Techniques of Painting*, Prentice Hall, New York.

Werner, A. E., M. Bimson, and N. D. Meeks (1975), The use of replica techniques and the scanning electron microscope in the study of ancient glass, *J. Glass Stud.* **17**, 158–160.

Wertime, T. A. (1961), *The Coming of the Age of Steel*, Brill, Leiden.

Wertime, T. A. and J. D. Muhly (eds.) (1980), *The Coming of the Age of Iron*, Yale Univ., New Haven.

Wertime, T. A. and S. F. Wertime (eds.) (1982), *Early Pyrotechnology*, Smithsonian Institute, Washington, DC.

Weser, U. and Y. Kaup (2002), Borate, an effective mummification agent in Pharaonic Egypt, *Zeitschrift fur Naturforschung B* **57**(7), 819–822.

West, E. G. (1982), *Copper and Its Alloys*, Ellis Horwood, Chichester.

West-Fitzhugh, E. (ed.) (1997), *Artists Pigments: A Handbook of Their History and Characterization*, Vol. 3, National Gallery of Art, Washington, DC.

Westgate, J. A., A. Sandhu, and P. Shane (1997), Fission track dating, in Taylor, R. E. and M. J. Aitken (eds.), *Chronometric Dating in Archaeology*, Advances in Archaeological and Museum Science Series, Vol. 2, Plenum, New York.

Weyl, W. A. (1999), *Coloured Glasses*, Society of Glass Technology, Sheffield.

Whewell, C. S. (1972), Felting and fulling, *Ciba-Geigy Rev.* 12–20.

Whitbread, I. K. (1995), We are what we study: Problems in communication and collaboration between ceramologists and archaeological scientists, in Lindahl A. and O. Stilborg (eds.), *The Aim of Laboratory Analyses of Ceramic in Archaeology, Workshop Proc.*, KVHAA; reprinted in *Konferenser* **34**, 91–100.

White, C. D. (ed.) (1999), *Reconstructing Ancient Maya Diets*, Univ. Utah.

White C. D. and H. P. Schwarcz (1989), Ancient Maya diet at Lamanai, Belize: As inferred from isotopic and chemical analysis of human bone, *J. Archaeol. Sci.* **16**, 451–474.

White, E. M. and L. A. Hannus (1983), Chemical weathering of bone in archaeological soils, *Am. Antiquity* **48**(2), 316–322.

White, R. (1978), The application of gas-chromatography to the identification of waxes, *Stud. Conserv.* **23**, 57–68.

Whiting, M. C. (1978), The identification of dyes in old oriental textiles, in *Proc. Int. Council of Museums Committee for Conservation, 5th Trienial Mtg*, Zagreb.

Whitmore, P. M. and G. R. Cass (1989), The fading of artists colorants, *Stud. Conserv.* **34**, 85–97.

Wiedemann, H. G. (1995), Paper investigations in Maya and Aztec cultures, in Vandiver, P. B., J. R. Druzik, J. L. Galvan Madrid, I. C. Freestone, and G. S. Wheeler (eds.), *Material Issues in Art and Archaeology*, paper 4, *Symp. Proc.* Materials Research Society, Pittsburg, Vol. 352, pp. 711–722.

Wiedemann, H. G. (1982), The Bust of Nefertiti, *Anal. Chem.* **54**, 619A–620A, 622A–624A, 626A, 628A.

Wigley, T. M. L., M. J. Ingram, and G. Farmer (eds.) (1985), *Climate and History: Studies in Ancient Climates and Their Impact on Man*, Cambridge Univ. Press, Cambridge, UK.

Wilkins, P. C. and R. G. Wilkins (1997), *Inorganic Chemistry in Biology*, Oxford Univ. Press, Oxford, UK.

Willerslev, E. and A. Cooper (2005), Ancient DNA *Proc. Royal Soc.*, (B), **272**, 3–16.

Williams-Thorpe, O., S. E. Warren, and J. G. Nandis (1997), Characterization of obsidian sources and artefacts from central and eastern Europe, using instrumental neutron activation analysis, in Korek, J. (ed.), *Proc. Int. Conf. Lithic Raw Material Characterization, Budapest and Sümeg, 1996*, Budapest.

Wills, G. (1968), *Ivory*, Arco, London.

Wilson, A. T. and C. H. Hendy (1971), Past wind strength from isotope studies, *Nature* **234**, 344–345.

Wilson, K. (1979), *A History of Textiles*, Westviews, Boulder, CO.

Wilson, L. and A. M. Pollard (2004), Making sense of diagenesis – The development of integrated laboratory and geochemical modeling strategies in studies of diagenetic change, in Nixon, T. (ed.), *Preserving Archaeological Remains in Situ; Proceedings of the 2nd Conference*, 12–14 September 2001, English Heritage, London, pp. 40–41.

Wilson, L. and A. M. Pollard (2002), Here today, gone tomorrow? Integrated experimentation and geochemical modeling in studies of archaeological diagenetic change, *Acc. Chem. Research* **35**(8), 644–651.

Winkler, E. M. (1973), *Stone: Properties, Durability in Man's Environment*, Springer, Vienna.

Winter, A. (1959), Die Technik des griechischen Töpfers in ihren Grundlagen, *Technische Beiträge zur Archäologie* **1**.

Wisseman S. and W. Williams (1994), *Ancient Technologies and Archaeological Materials*, Gordon & Breach, New York.

Wolters, J. (1983), *Die Granulation*, Callway, Munich.

Wolters, J. (1981), The ancient craft of granulation, *Gold Bull.* **14**, 119–123.

Wood, J. T. (1912), *The Puering, Bating and Drenching of Skins*, Spoon, London.

Woodhead-Galloway, J. (1980), *Collagen: The Anatomy of a Protein*, Arnold, London.

Woods, W. I. (1977), The soil phosphate analysis of anthropic soils, *Am. Antiquity* **42**, 248–252.

Woolley, C. L. (1962), *As I Seem to Remember*, Allen & Unwin, London.

Wright, G. R. H. (1998), *Ancient Building Technology*, Brill, Leiden.

Wright, M. and B. Wheals (1987), Pyrolysis-mass spectrometry of natural gums, resins and waxes and its use for detecting such materials in ancient Egyptian cases, *J. Appl. Pyrol.* **11**, 195–211.

Wright, V. P. (ed.) (1986), *Paleosols, Their Recognition and Interpretation*, Princeton Univ. Press, Princeton, NJ.

Wyckoff, R. W. G. (1980), Collagen in fossil bones, in Hare, P. E., T. C. Hoering, and K. King, Jr. (eds.), *Biogeochemistry of Amino Acids*, Wiley, New York.

Wyckoff, R. W. G. (1972), *The Biochemistry of Animal Fossils*, Scientechnica, Bristol.

Yahuda, A. S. (1944), The story of a forgery and the Mesha inscription, *Jewish Quart. Rev.* **35**, 139–163.

Yap, C. T. and Y. Hua (1994), A study of Chinese porcelain raw materials for Ding, Xing, Gongxian and Dehua Wares, *Archaeometry* **36**, 63–76.

Yap, C. T. and Y. Hua (1992), Geographical classification of Chinese white porcelain, in Li Jiazhi and Chen Xianqui (eds.), *Proc. 1992 Int. Symp. Science and Technology of Ancient Ceramics*, Shanghai.

Yasuda, H., I. Iuchi, and K. Mukai (1976), *J. Archaeol. Soc. Japan* **61**, 277–281.

Yegingil, Z. and T. Lunel (1990), Provenance studies of obsidian artifacts determined by using fission track ages and trace element analysis, *Nucl. Tracks Rad. Meas.* **17**(3), 433.

Yelon, A., A. Saucier, J. P. Lerocque, P. E. L. Smith, and P. Vandiver (1992), Thermal analysis of early Neolithic pottery from Tepe Ganj Dareh, *Material Research Society Symp. Proc.*, Vol. 267, pp. 591–607.

Young, R. J. and P. Lovell (1991), *Introduction to Polymers*, Chapman & Hall, New York.

Young, S. M. M., P. Budd, R. Hagerty, and A. M. Pollard (1997), Inductively coupled plasma mass-spectrometry for the analysis of ancient metals, *Archaeometry* **39**(2), 379–392.

Young, S. M. M. and M. Pollard (2000), Atomic spectroscopy and spectrometry, in Ciliberto, E. and G. Spoto (eds.), *Modern Analytical Methods in Art and Archaeology*, Chemical Analysis Series, Vol. 155, Wiley, New York, pp. 21–54.

Zachariasen, W. H. (1932), The atomic arrangement of glass, *J. Am. Chem. Soc.* **54**, 3841–3851.

Zerdoun, B. Y. M. (1983), Les encres noires au moyen age (jusqu'à 1600), Centre National de la Recherche Scientifique, Paris.

Zeronian, S. H. and T. P. Nevell (eds.) (1985), *Cellulose Chemistry and Its Applications*, Ellis Harwood, Chichester.

Zhadkevich, A. M. (2004) History of origination, technological features and technical capabilities of the first methods of soldering and brazing, *Welding J.* **11**, 39–44.

Zhou, W. (1996), Chinese traditional zinc smelting technology and the history of zinc production in China, *Bull. Metals Museum* **25**, 36–47.

Zimmerman, D. W., M. P. Yuhas, and P. Meyers (1974), Thermoluminescence authenticity measurements on core material from the bronze horse of the New York Metropolitan Museum of Art, *Archaeometry* **16**, 19–30.

Zollinger, H. (1991), *Colour Chemistry*, VCH, Verlag GmbH, Weinheim.

Zouridakis, N., J. F. Saliege, A. Pearson, and S. E. Philipakis (1987), Radiocarbon dating of mortars from ancient Greek palaces, *Archaeometry* **29**, 60–68.

Zucchiatti, A. (2004), X-Ray spectrometry in archaeometry, *X-Ray Spectrom.* 533–552.

Zussmann, J. (1972), Asbestos: Nature and history, *Ciba-Geigy Rev.* 3–8.

INDEX

Archaeological Chemistry, Second Edition By Zvi Goffer

CHEMICAL ANALYSIS

A SERIES OF MONOGRAPHS ON ANALYTICAL CHEMISTRY
AND ITS APPLICATIONS

Series Editor
J. D. WINEFORDNER

VOLUME 170

CHEMICAL ANALYSIS

A SERIES OF MONOGRAPHS ON ANALYTICAL CHEMISTRY
AND ITS APPLICATIONS

Series Editor
J. D. WINEFORDNER

Vol. 1 **The Analytical Chemistry of Industrial Poisons, Hazards, and Solvents**. *Second Edition.* By the late Morris B. Jacobs
Vol. 2 **Chromatographic Adsorption Analysis**. By Harold H. Strain (*out of print*)
Vol. 3 **Photometric Determination of Traces of Metals**. *Fourth Edition*
Part I: General Aspects. By E. B. Sandell and Hiroshi Onishi
Part IIA: Individual Metals, Aluminum to Lithium. By Hiroshi Onishi
Part IIB: Individual Metals, Magnesium to Zirconium. By Hiroshi Onishi
Vol. 4 **Organic Reagents Used in Gravimetric and Volumetric Analysis**. By John F. Flagg (*out of print*)
Vol. 5 **Aquametry: A Treatise on Methods for the Determination of Water**. *Second Edition* (*in three parts*). By John Mitchell, Jr. and Donald Milton Smith
Vol. 6 **Analysis of Insecticides and Acaricides**. By Francis A. Gunther and Roger C. Blinn (*out of print*)
Vol. 7 **Chemical Analysis of Industrial Solvents**. By the late Morris B. Jacobs and Leopold Schetlan
Vol. 8 **Colorimetric Determination of Nonmetals**. *Second Edition.* Edited by the late David F. Boltz and James A. Howell
Vol. 9 **Analytical Chemistry of Titanium Metals and Compounds**. By Maurice Codell
Vol. 10 **The Chemical Analysis of Air Pollutants**. By the late Morris B. Jacobs
Vol. 11 **X-Ray Spectrochemical Analysis**. *Second Edition.* By L. S. Birks
Vol. 12 **Systematic Analysis of Surface-Active Agents**. *Second Edition.* By Milton J. Rosen and Henry A. Goldsmith
Vol. 13 **Alternating Current Polarography and Tensammetry**. By B. Breyer and H.H.Bauer
Vol. 14 **Flame Photometry**. By R. Herrmann and J. Alkemade
Vol. 15 **The Titration of Organic Compounds** (*in two parts*). By M. R. F. Ashworth
Vol. 16 **Complexation in Analytical Chemistry: A Guide for the Critical Selection of Analytical Methods Based on Complexation Reactions**. By the late Anders Ringbom
Vol. 17 **Electron Probe Microanalysis**. *Second Edition.* By L. S. Birks
Vol. 18 **Organic Complexing Reagents: Structure, Behavior, and Application to Inorganic Analysis**. By D. D. Perrin
Vol. 19 **Thermal Analysis**. *Third Edition.* By Wesley Wm.Wendlandt
Vol. 20 **Amperometric Titrations**. By John T. Stock
Vol. 21 **Reflctance Spectroscopy**. By Wesley Wm.Wendlandt and Harry G. Hecht
Vol. 22 **The Analytical Toxicology of Industrial Inorganic Poisons**. By the late Morris B. Jacobs
Vol. 23 **The Formation and Properties of Precipitates**. By Alan G.Walton
Vol. 24 **Kinetics in Analytical Chemistry**. By Harry B. Mark, Jr. and Garry A. Rechnitz
Vol. 25 **Atomic Absorption Spectroscopy**. *Second Edition.* By Morris Slavin
Vol. 26 **Characterization of Organometallic Compounds** (*in two parts*). Edited by Minoru Tsutsui
Vol. 27 **Rock and Mineral Analysis**. *Second Edition.* By Wesley M. Johnson and John A. Maxwell
Vol. 28 **The Analytical Chemistry of Nitrogen and Its Compounds** (*in two parts*). Edited by C. A. Streuli and Philip R.Averell
Vol. 29 **The Analytical Chemistry of Sulfur and Its Compounds** (*in three parts*). By J. H. Karchmer
Vol. 30 **Ultramicro Elemental Analysis**. By Güther Toölg
Vol. 31 **Photometric Organic Analysis** (*in two parts*). By Eugene Sawicki
Vol. 32 **Determination of Organic Compounds: Methods and Procedures**. By Frederick T. Weiss
Vol. 33 **Masking and Demasking of Chemical Reactions**. By D. D. Perrin

Vol. 34. **Neutron Activation Analysis**. By D. De Soete, R. Gijbels, and J. Hoste
Vol. 35 **Laser Raman Spectroscopy**. By Marvin C. Tobin
Vol. 36 **Emission Spectrochemical Analysis**. By Morris Slavin
Vol. 37 **Analytical Chemistry of Phosphorus Compounds**. Edited by M. Halmann
Vol. 38 **Luminescence Spectrometry in Analytical Chemistry**. By J. D.Winefordner, S. G. Schulman, and T. C. O'Haver
Vol. 39. **Activation Analysis with Neutron Generators**. By Sam S. Nargolwalla and Edwin P. Przybylowicz
Vol. 40 **Determination of Gaseous Elements in Metals**. Edited by Lynn L. Lewis, Laben M. Melnick, and Ben D. Holt
Vol. 41 **Analysis of Silicones**. Edited by A. Lee Smith
Vol. 42 **Foundations of Ultracentrifugal Analysis**. By H. Fujita
Vol. 43 **Chemical Infrared Fourier Transform Spectroscopy**. By Peter R. Griffiths
Vol. 44 **Microscale Manipulations in Chemistry**. By T. S. Ma and V. Horak
Vol. 45 **Thermometric Titrations**. By J. Barthel
Vol. 46 **Trace Analysis: Spectroscopic Methods for Elements**. Edited by J. D.Winefordner
Vol. 47 **Contamination Control in Trace Element Analysis**. By Morris Zief and James W. Mitchell
Vol. 48 **Analytical Applications of NMR**. By D. E. Leyden and R. H. Cox
Vol. 49 **Measurement of Dissolved Oxygen**. By Michael L. Hitchman
Vol. 50 **Analytical Laser Spectroscopy**. Edited by Nicolo Omenetto
Vol. 51 **Trace Element Analysis of Geological Materials**. By Roger D. Reeves and Robert R. Brooks
Vol. 52 **Chemical Analysis by Microwave Rotational Spectroscopy**. By Ravi Varma and Lawrence W. Hrubesh
Vol. 53 **Information Theory as Applied to Chemical Analysis**. By Karl Eckschlager and Vladimir Stepanek
Vol. 54 **Applied Infrared Spectroscopy: Fundamentals, Techniques, and Analytical Problemsolving**. By A. Lee Smith
Vol. 55 **Archaeological Chemistry**. By Zvi Goffer
Vol. 56 **Immobilized Enzymes in Analytical and Clinical Chemistry**. By P. W. Carr and L. D. Bowers
Vol. 57 **Photoacoustics and Photoacoustic Spectroscopy**. By Allan Rosencwaig
Vol. 58 **Analysis of Pesticide Residues**. Edited by H. Anson Moye
Vol. 59 **Affity Chromatography**. By William H. Scouten
Vol. 60 **Quality Control in Analytical Chemistry**. *Second Edition.* By G. Kateman and L. Buydens
Vol. 61 **Direct Characterization of Fineparticles**. By Brian H. Kaye
Vol. 62 **Flow Injection Analysis**. By J. Ruzicka and E. H. Hansen
Vol. 63 **Applied Electron Spectroscopy for Chemical Analysis**. Edited by Hassan Windawi and Floyd Ho
Vol. 64 **Analytical Aspects of Environmental Chemistry**. Edited by David F. S. Natusch and Philip K. Hopke
Vol. 65 **The Interpretation of Analytical Chemical Data by the Use of Cluster Analysis**. By D. Luc Massart and Leonard Kaufman
Vol. 66 **Solid Phase Biochemistry: Analytical and Synthetic Aspects**. Edited by William H. Scouten
Vol. 67 **An Introduction to Photoelectron Spectroscopy**. By Pradip K. Ghosh
Vol. 68 **Room Temperature Phosphorimetry for Chemical Analysis**. By Tuan Vo-Dinh
Vol. 69 **Potentiometry and Potentiometric Titrations**. By E. P. Serjeant
Vol. 70 **Design and Application of Process Analyzer Systems**. By Paul E. Mix
Vol. 71 **Analysis of Organic and Biological Surfaces**. Edited by Patrick Echlin
Vol. 72 **Small Bore Liquid Chromatography Columns: Their Properties and Uses**. Edited by Raymond P.W. Scott
Vol. 73 **Modern Methods of Particle Size Analysis**. Edited by Howard G. Barth
Vol. 74 **Auger Electron Spectroscopy**. By Michael Thompson, M. D. Baker, Alec Christie, and J. F. Tyson
Vol. 75 **Spot Test Analysis: Clinical, Environmental, Forensic and Geochemical Applications**. By Ervin Jungreis
Vol. 76 **Receptor Modeling in Environmental Chemistry**. By Philip K. Hopke

Vol. 77 **Molecular Luminescence Spectroscopy: Methods and Applications** (*in three parts*). Edited by Stephen G. Schulman
Vol. 78 **Inorganic Chromatographic Analysis**. Edited by John C. MacDonald
Vol. 79 **Analytical Solution Calorimetry**. Edited by J. K. Grime
Vol. 80 **Selected Methods of Trace Metal Analysis: Biological and Environmental Samples**. By Jon C.VanLoon
Vol. 81 **The Analysis of Extraterrestrial Materials**. By Isidore Adler
Vol. 82 **Chemometrics**. By Muhammad A. Sharaf, Deborah L. Illman, and Bruce R. Kowalski
Vol. 83 **Fourier Transform Infrared Spectrometry**. By Peter R. Griffiths and James A. de Haseth
Vol. 84 **Trace Analysis: Spectroscopic Methods for Molecules**. Edited by Gary Christian and James B. Callis
Vol. 85 **Ultratrace Analysis of Pharmaceuticals and Other Compounds of Interest**. Edited by S. Ahuja
Vol. 86 **Secondary Ion Mass Spectrometry: Basic Concepts, Instrumental Aspects, Applications and Trends**. By A. Benninghoven, F. G. Rüenauer, and H.W.Werner
Vol. 87 **Analytical Applications of Lasers**. Edited by Edward H. Piepmeier
Vol. 88 **Applied Geochemical Analysis**. By C. O. Ingamells and F. F. Pitard
Vol. 89 **Detectors for Liquid Chromatography**. Edited by Edward S.Yeung
Vol. 90 **Inductively Coupled Plasma Emission Spectroscopy: Part 1: Methodology, Instrumentation, and Performance; Part II: Applications and Fundamentals**. Edited by J. M. Boumans
Vol. 91 **Applications of New Mass Spectrometry Techniques in Pesticide Chemistry**. Edited by Joseph Rosen
Vol. 92 **X-Ray Absorption: Principles,Applications,Techniques of EXAFS, SEXAFS, and XANES**. Edited by D. C. Konnigsberger
Vol. 93 **Quantitative Structure-Chromatographic Retention Relationships**. By Roman Kaliszan
Vol. 94 **Laser Remote Chemical Analysis**. Edited by Raymond M. Measures
Vol. 95 **Inorganic Mass Spectrometry**. Edited by F.Adams,R.Gijbels, and R.Van Grieken
Vol. 96 **Kinetic Aspects of Analytical Chemistry**. By Horacio A. Mottola
Vol. 97 **Two-Dimensional NMR Spectroscopy**. By Jan Schraml and Jon M. Bellama
Vol. 98 **High Performance Liquid Chromatography**. Edited by Phyllis R. Brown and Richard A. Hartwick
Vol. 99 **X-Ray Fluorescence Spectrometry**. By Ron Jenkins
Vol. 100 **Analytical Aspects of Drug Testing**. Edited by Dale G. Deustch
Vol. 101 **Chemical Analysis of Polycyclic Aromatic Compounds**. Edited by Tuan Vo-Dinh
Vol. 102 **Quadrupole Storage Mass Spectrometry**. By Raymond E. March and Richard J. Hughes (*out of print: see Vol. 165)*
Vol. 103 **Determination of Molecular Weight**. Edited by Anthony R. Cooper
Vol. 104 **Selectivity and Detectability Optimization in HPLC**. By Satinder Ahuja
Vol. 105 **Laser Microanalysis**. By Lieselotte Moenke-Blankenburg
Vol. 106 **Clinical Chemistry**. Edited by E. Howard Taylor
Vol. 107 **Multielement Detection Systems for Spectrochemical Analysis**. By Kenneth W. Busch and Marianna A. Busch
Vol. 108 **Planar Chromatography in the Life Sciences**. Edited by Joseph C. Touchstone
Vol. 109 **Fluorometric Analysis in Biomedical Chemistry: Trends and Techniques Including HPLC Applications**. By Norio Ichinose, George Schwedt, Frank Michael Schnepel, and Kyoko Adochi
Vol. 110 **An Introduction to Laboratory Automation**. By Victor Cerdá and Guillermo Ramis
Vol. 111 **Gas Chromatography: Biochemical, Biomedical, and Clinical Applications**. Edited by Ray E. Clement
Vol. 112 **The Analytical Chemistry of Silicones**. Edited by A. Lee Smith
Vol. 113 **Modern Methods of Polymer Characterization**. Edited by Howard G. Barth and Jimmy W. Mays
Vol. 114 **Analytical Raman Spectroscopy**. Edited by Jeanette Graselli and Bernard J. Bulkin
Vol. 115 **Trace and Ultratrace Analysis by HPLC**. By Satinder Ahuja
Vol. 116 **Radiochemistry and Nuclear Methods of Analysis**. By William D. Ehmann and Diane E.Vance

Vol. 117 **Applications of Fluorescence in Immunoassays**. By Ilkka Hemmila
Vol. 118 **Principles and Practice of Spectroscopic Calibration**. By Howard Mark
Vol. 119 **Activation Spectrometry in Chemical Analysis**. By S. J. Parry
Vol. 120 **Remote Sensing by Fourier Transform Spectrometry**. By Reinhard Beer
Vol. 121 **Detectors for Capillary Chromatography**. Edited by Herbert H. Hill and Dennis McMinn
Vol. 122 **Photochemical Vapor Deposition**. By J. G. Eden
Vol. 123 **Statistical Methods in Analytical Chemistry**. By Peter C. Meier and Richard Züd
Vol. 124 **Laser Ionization Mass Analysis**. Edited by Akos Vertes, Renaat Gijbels, and Fred Adams
Vol. 125 **Physics and Chemistry of Solid State Sensor Devices**. By Andreas Mandelis and Constantinos Christofides
Vol. 126 **Electroanalytical Stripping Methods**. By Khjena Z. Brainina and E. Neyman
Vol. 127 **Air Monitoring by Spectroscopic Techniques**. Edited by Markus W. Sigrist
Vol. 128 **Information Theory in Analytical Chemistry**. By Karel Eckschlager and Klaus Danzer
Vol. 129 **Flame Chemiluminescence Analysis by Molecular Emission Cavity Detection**. Edited by David Stiles, Anthony Calokerinos, and Alan Townshend
Vol. 130 **Hydride Generation Atomic Absorption Spectrometry**. Edited by Jiri Dedina and Dimiter L. Tsalev
Vol. 131 **Selective Detectors: Environmental, Industrial, and Biomedical Applications**. Edited by Robert E. Sievers
Vol. 132 **High-Speed Countercurrent Chromatography**. Edited by Yoichiro Ito and Walter D. Conway
Vol. 133 **Particle-Induced X-Ray Emission Spectrometry**. By Sven A. E. Johansson, John L. Campbell, and Klas G. Malmqvist
Vol. 134 **Photothermal Spectroscopy Methods for Chemical Analysis**. By Stephen E. Bialkowski
Vol. 135 **Element Speciation in Bioinorganic Chemistry**. Edited by Sergio Caroli
Vol. 136 **Laser-Enhanced Ionization Spectrometry**. Edited by John C. Travis and Gregory C. Turk
Vol. 137 **Fluorescence Imaging Spectroscopy and Microscopy**. Edited by Xue Feng Wang and Brian Herman
Vol. 138 **Introduction to X-Ray Powder Diffractometry**. By Ron Jenkins and Robert L. Snyder
Vol. 139 **Modern Techniques in Electroanalysis**. Edited by Petr Vanýek
Vol. 140 **Total-Reflction X-Ray Fluorescence Analysis**. By Reinhold Klockenkamper
Vol. 141 **Spot Test Analysis: Clinical, Environmental, Forensic, and Geochemical Applications**. *Second Edition.* By Ervin Jungreis
Vol. 142 **The Impact of Stereochemistry on Drug Development and Use**. Edited by Hassan Y. Aboul-Enein and Irving W.Wainer
Vol. 143 **Macrocyclic Compounds in Analytical Chemistry**. Edited by Yury A. Zolotov
Vol. 144 **Surface-Launched Acoustic Wave Sensors: Chemical Sensing and Thin-Film Characterization**. By Michael Thompson and David Stone
Vol. 145 **Modern Isotope Ratio Mass Spectrometry**. Edited by T. J. Platzner
Vol. 146 **High Performance Capillary Electrophoresis: Theory, Techniques, and Applications**. Edited by Morteza G. Khaledi
Vol. 147 **Solid Phase Extraction: Principles and Practice**. By E. M. Thurman
Vol. 148 **Commercial Biosensors: Applications to Clinical, Bioprocess and Environmental Samples**. Edited by Graham Ramsay
Vol. 149 **A Practical Guide to Graphite Furnace Atomic Absorption Spectrometry**. By David J. Butcher and Joseph Sneddon
Vol. 150 **Principles of Chemical and Biological Sensors**. Edited by Dermot Diamond
Vol. 151 **Pesticide Residue in Foods: Methods, Technologies, and Regulations**. By W. George Fong, H. Anson Moye, James N. Seiber, and John P. Toth
Vol. 152 **X-Ray Fluorescence Spectrometry**. *Second Edition.* By Ron Jenkins
Vol. 153 **Statistical Methods in Analytical Chemistry**. *Second Edition.* By Peter C. Meier and Richard E. Züd
Vol. 154 **Modern Analytical Methodologies in Fat- and Water-Soluble Vitamins**. Edited by Won O. Song, Gary R. Beecher, and Ronald R. Eitenmiller
Vol. 155 **Modern Analytical Methods in Art and Archaeology**. Edited by Enrico Ciliberto and Guiseppe Spoto

Vol. 156 **Shpol'skii Spectroscopy and Other Site Selection Methods: Applications in Environmental Analysis, Bioanalytical Chemistry and Chemical Physics**. Edited by C. Gooijer, F. Ariese and J.W. Hofstraat

Vol. 157 **Raman Spectroscopy for Chemical Analysis**. By Richard L. McCreery

Vol. 158 **Large (C> = 24) Polycyclic Aromatic Hydrocarbons: Chemistry and Analysis**. By John C. Fetzer

Vol. 159 **Handbook of Petroleum Analysis**. By James G. Speight

Vol. 160 **Handbook of Petroleum Product Analysis**. By James G. Speight

Vol. 161 **Photoacoustic Infrared Spectroscopy**. By Kirk H. Michaelian

Vol. 162 **Sample Preparation Techniques in Analytical Chemistry**. Edited by Somenath Mitra

Vol. 163 **Analysis and Purification Methods in Combination Chemistry**. Edited by Bing Yan

Vol. 164 **Chemometrics: From Basics to Wavelet Transform**. By Foo-tim Chau, Yi-Zeng Liang, Junbin Gao, and Xue-guang Shao

Vol. 165 **Quadrupole Ion Trap Mass Spectrometry**. *Second Edition.* By Raymond E. March and John F. J. Todd

Vol. 166 **Handbook of Coal Analysis**. By James G. Speight

Vol. 167 **Introduction to Soil Chemistry: Analysis and Instrumentation**. By Alfred R. Conklin, Jr.

Vol. 168 **Environmental Analysis and Technology for the Refining Industry**. By James G. Speight

Vol. 169 **Identification of Microorganisms by Mass Spectrometry**. Edited by Charles L. Wilkins and Jackson O. Lay, Jr.

Vol. 170 **Archaeological Chemistry**. *Second Edition.* By Zvi Goffer

Vol. 171 **Fourier Transform Infrared Spectrometry**. *Second Edition.* By Peter R. Griffiths and James A. de Haseth